ENVIRONMENTAL HEALTH – PHYSICAL, CHEMICAL AND BIOLOGICAL FACTORS

SEWAGE SLUDGE MANAGEMENT

FROM THE PAST TO OUR CENTURY

ENVIRONMENTAL HEALTH – PHYSICAL, CHEMICAL AND BIOLOGICAL FACTORS

Additional books in this series can be found on Nova's website under the Series tab.

Additional E-books in this series can be found on Nova's website under the E-books tab.

ENVIRONMENTAL REMDIATION TECHNOLOGIES, REGULATIONS AND SAFETY

Additional books in this series can be found on Nova's website under the Series tab.

Additional E-books in this series can be found on Nova's website under the E-books tab.

SEWAGE SLUDGE MANAGEMENT

FROM THE PAST TO OUR CENTURY

ANTONIS A. ZORPAS
AND
VASSILIS J. INGLEZAKIS
EDITORS

Nova Science Publishers, Inc.
New York

Library of Congress Cataloging-in-Publication Data

Sewage sludge management : from the past to our century / editors, Antonis A. Zorpas, Vassilis J. Inglezakis.
p. cm.
Includes bibliographical references and index.
ISBN 978-1-61324-393-0 (hardcover)
1. Sewage--Purification--History. 2. Sewage sludge--History. I. Zorpas, Antonis A. II. Inglezakis, Vassilis J.
TD745.S45 2011
631.8'69--dc22
2011012570

Published by Nova Science Publishers, Inc. † New York

CONTENTS

PREFACE

It is more than clear that during the last decades the interest in environmental issues is constantly increasing and environmental issues have gradually been broadened including economic and social responsibilities. Among several issues, the treatment of wastewater and consequently, the handling of sewage sludge generated, are of the most significant challenges in environmental management. Sludge is a semi-solid material composed of by-products collected at different stages of the wastewater treatment process. Only in European Union, more than 50 000 wastewater treatment plants are operating producing more than 8 million tons of dry solids per year. Naturally, sewage sludge became a problem only when cities grown and consequently, waste quantities grown. The large part of the population in the developed countries is connected via sewerage systems to wastewater treatment plants, which generally consists of a mixture of domestic, storm water and industrial wastewater. A glance in the past shows that since the half of the century the sludge was discharged into rivers or the sea without any treatment. Gradually, the dumping sites moved from inland or coastal waters to open sea. Only the last 30 years and only in the developed world there are steps towards a sustainable management of sludge.

The properties of sludge depend on the contents of the wastewater treated and also on the operational characteristics of the wastewater and sludge treatment facilities. In any case, sludge contains both compounds of agricultural value and pollutants. The quantity of sludge produced in a wastewater treatment plant is approximately 1% of the quantity of treated wastewater while sludge management costs 40 to 50% of the total wastewater treatment costs. Several treatment processes exist, such as thickening, dewatering, stabilisation and disinfection, and thermal drying. Treated sludge can be recycled or disposed of using three main routes: recycling to agriculture (land spreading), incineration or land filling.

The aim of the present book is to review current scientific knowledge on the sewage sludge topic and assess the environmental and economic impact of the main disposal and recycling routes for sewage sludge.

It is an honour to be invited to provide this preface.

Yangsheng Liu,
Professor, Ph.D., Peking University

A WORD FROM THE EDITORS

"The Most exciting phrase to hear in science, the one that heralds the most discoveries, is not 'Eureka' but "That's funny …""

Issac Asimov

Oscar Wilde state that "Anybody can sympathize with the sufferings of a friend but it requires a very fine nature to sympathize with a friend's success". What we learn form life is that practices make perfects and that the real principles are behind a long lasting friendship.

My dad (Andreas) always told me that "Antonis, there are two basic principles in life which are (1st) Second thought is always better than the first one and (2nd) Learn how to apply the first principle".

When I met Vasilis for first time during 1995 the year that we both started our PhD at the National Technical University I didn't imagine that 15 years later we would give to the international scientific community more than 30 common scientific research papers, several common contributions to other scientists' edited books, research programs and finally our edited books. This is the second edited book that Vasilis and I are offering to the scientific community.

This book started back in time, during our PhD while Vassilis was working in the Chemical Reaction Engineering lab and I (Antonis) in the General Chemistry lab both experimentalists, both on natural zeolites but on different subjects: waster/wastewater and sewage sludge treatment, respectively. However, it took more than 15 years for us to mature as scientists and to come up with the ambition and the determination to write a book on Sewage Sludge Management.

It is an honor for us to have in our book an impressive number of authors coming from different countries, namely Cyprus, Greece, Saudi Arabia, USA, Spain, Romania, Germany, and Belgium. Thus, we consider that the geographical, cultural, scientific and professional diversity of the book guarantees the expression of divergent views on the subject.

The participation of Professor Yangsheng Liu from the University of Beijing in China, in our book as signatory of the Prefaces is an event that we will never forget and will always be an ultimate honor for us.

From this position we have the need to thank by heart all those who have supported us, as well as all those who have discouraged us throughout our carrier. We have learned that the most important values of life are the ones which are without price; thereby after 15 years we offer you the output of a long-lasting friendship. And thus why friendship is unnecessary like philosophy like art… it has no survival value; rather is one of those things that give value to survival.

We hope the readers to appreciate our efforts and to enjoy reading.

I (Antonis) and Vasilis, would like to say THANKS to all of you and especially to Mrs Stella Zorpa (Managing Director of the SMART LEARNING Institute in Cyprus) and Mrs Irene Voukalli Director of the Quality Control Department in the Institute of Environmental Technology and Sustainable Development for their personal help. Also, many thanks to our Publishers NOVA SCIENCE for trusting us.

Antonis A. Zorpas
Vassilis J. Inglezakis

ABOUT THE EDITORS AND AUTHORS

Authors CVs are presented in alphabetical base

Dr. Alonso Esteban
Dr. Aparicio Irene
Dr. Appels Lise
Dr. Athanasiou Costas
Dr. Costa N. Costas
Dr. Dewil Raf
Dr. Finkbeiner Matthias
Dr. Gómez Lucas Ignacio
Mr. Güney Kenan
Dr. Karagiannidis Avraam
Mrs. Lehmann Annekatrin
Dr. Leverenz Harold
Mrs. Martín Julia
Dr. Martín Soriano-Disla José
Dr. Muhammad H. Al-Malack
Mr. Muhammad Muhitur Rahman
Dr. Navarro Pedreño Jose
Dr. Nicolaides Demetris
Mrs Papadimitriou Chrysi
Dr. Papadopoulos Ioannis
Dr. Papastergiadis Efthimios
Dr. Poulopoulos G.Stavros
Dr. Samaras Petros
Dr. Santos L. Juan
Dr. Savvas Petros
Dr. Tchobanoglous George
Dr. Traverso Marzia
Mr. Venetis Christos
Mrs. Voukalli Irene
Dr. Zouboulis Anastasios

Dr. Antonis A. Zorpas is Chemical Engineering, holding MSc degree in the field of Sludge Treatment (1995) and a PhD degree (2000) in the field of Environmental Engineering / Management from the National Technical University of Athens in the Department of Chemical Engineer. He has more than 15 years experience in the field of Environmental Engineering/Management, treatment of Solid and Liquid Waste using several methods like composting, chemical and electrochemical oxidation, reed beds as well as has significant impact in the field of sustainable development the last 8 years. He carries out more than 200 Environmental Impact and Risk Assessment Analysis Studies as well as he implement more than 100 Environmental Management Systems in several industrial activities in. He is presented with significant co-operations with Several Universities and Research Centres in the EU as well as with the European Commission

He has a rich academic record having more than 110 publications in academic journals (45) and international conferences (70) and published books (2). He participates in more than 55 research programs. He is also the Vice president of Cyprus Environmental Engineers in Cyprus (2007-2013). He is the managing Director of the Institute of Environmental Technology and Sustainable Development (NGO). He is also scientific cooperators with the EU, the Cyprus University, National Technical University of Athens, Aristotle University of Thessaloniki and other Universities in Europe. He has taught several scientific lessons in the scientific area of Solid and Liquid Waste management, Sustainable Development, Environmental Management Systems etc.

Generally I have researched or continue to research in the following scientific fields:

Mechanical simulation of several technologies in the Laboratory and in real scale

Development of Innovation Technologies

In the design of several technologies for the treatment of several waster streams (Liquid and Solid)

Development of methodologies for the determination of Environmental Impact Assessment

Development of methodologies for the determination of Risk Assessment Analysis

Management of Liquid and Solid waste.

Sustainable Development and Coastal Environment

Environmental Resource Management

Waste to Energy

Dr. Inglezakis J. Vassilis is Chemical Engineer holding PhD in the field of environmental technologies. His post-doc research focused on wastewater treatment and waste management fields. He has published 5 books, 34 papers in academic Journals and 49 announcements in conferences while he is reviewer in 28 academic Journals. Apart from research, he has 15 years of experience in international environmental projects working in several counties and several employers. Since 2002 he is acts as external collaborator of the Unit of Environmental Science and Technology within the National Technical University of Athens, Greece and since 2007 he is director of the consulting company SC European Focus Consulting srl, Romania.

Dr. Yangsheng Liu is a full professor at College of Environmental Science and Engineering at Peking University (China). He received his BSc degree (1990), MSc degree (1996) and Ph.D. degree (1999) in chemical engineering, and accomplished two-year postdoctoral research on solid waste pollution control and management (1999~2001). Currently, Dr. Liu is the executive vice director of Beijing Municipal Key Laboratory for Solid Waste Utilization and Management, the executive vice director of Shenzhen Municipal Key Laboratory for Sustainable Development, and the vice director of Institute of Solid Waste Utilization and Ecological Restoration at Peking University Shenzhen Graduate School. Dr. Liu is a famous environmental scientist and environmental engineering educator. His research and writings cover a wide range of environmental problems including solid waste incineration and landfilling, organic wastes composting and anaerobic digestion for methane production, industrial wastes reutilization, and biomass energy development. He has owned 20 invention patents, and published more than 80 papers at national and international journals, and has ever been invited to review manuscripts for international famous journals including Environmental Science & Technology, Chemosphere, Energy, Journal of Hazardous Materials, Waste Management, Atmospheric Environment, Journal of Environmental management, Industrial & Engineering Chemistry Research, etc. Dr. Liu is the member of editorial board for two journals "Urban Management Science and Technology" and "Environmental Protection of Oil & Gas Fields", and also the member of China Society of Environmental Sciences, China Society of Environmental Protection and Industry, and China Association of Urban Environmental Sanitation.

Dr. Alonso Esteban is an Analytical Chemist holding PhD in the field of the environmental analytical chemistry. His research is focused on the development of methodologies for the determination of majority and emergent pollutants in environmental matrices such as sewage sludge, sediments, wastewater and surface water; and their application to evaluate the occurrence, fate and risk assessment of these pollutants in the environment. He has directed 21 research projects on environmental sciences and technologies and published more than 60 scientific papers in International Journals with high impact factors. He worked at the Center for New Water Technologies, a non-profit association of private firms and government agencies focused on the field of novel water technologies, before joining the University of Seville. Since 2004, he is Professor of Analytical Chemistry and Head of the Research Group ANQUIMED at the Department of Analytical Chemistry of the University of Seville (Spain).

Dr. Aparicio Irene is an Analytical Chemist holding PhD in the field of the development and validation of methodologies for the analysis of pharmaceutical compounds. Her research is now focused on the development of methodologies for the determination of majority and emergent pollutants in environmental matrices such as sewage sludge, sediments, wastewater and surface water. The methodologies developed are then applied to evaluate the occurrence and fate of the pollutants in the environment and to obtain information about their potential toxicological effects to the ecosystems. She has participated in 11 research projects and collaborated in 4 book chapters. She has published 23 peer reviewed papers in International Journals with high impact factors and reported her scientific results as 58 congress communications. In 2002, she obtained an assistant professor position in the Department of Analytical Chemistry of the University of Seville (Spain). In 2010, she raised to the rank of Analytical Chemistry Professor. She gives lessons about Analytical Chemistry especially about instrumental techniques of analysis and has participated in courses about chromatographic analysis.

Dr. Appels Lise holds a M.Sc. in Environmental Engineering from the University of Antwerp (2006) and a Ph.D. in Chemical Engineering of the Katholieke Universiteit Leuven (2010). She is currently doing research as a postdoctoral research fellow at the Chemical Engineering department of the Katholieke Universiteit Leuven. Her work mainly focuses on anaerobic digestion processes and more specifically on the enhancement of anaerobic digestion by pre-treatment methods and the modeling of these processes. She has (co-

)authored 10 publications in international peer-reviewed journals and has presented more than 20 contributions on international conferences. She is member of the "Process Technology for Sustainable Energy" section of the European Federation of Chemical Engineers.

Dr. Athanasiou Costas obtained his Ph.D. in 2003, from the Department of Chemical Engineering of Aristotle University of Thessaloniki. Since June 2010, he is appointing an academic position as a Lecturer, in the Department of Environmental Engineering at the Democritus University of Thrace (Laboratory of Renewable Energy Sources), and he is a collaborating researcher of the Chemical Process Engineering Research Institute at the Centre for Research & Technology Hellas (Laboratory of Process Systems Design and Implementation). He is the co-author of more 45 papers in international scientific journals and international conference proceedings, in the fields of biomass-to-energy science and engineering.

Dr. Costa N. Costas is an Associate Professor at the Department of Environmental Management of the School of Geotechnical Sciences and Environmental Management. He has obtained his BSc and MSc in Chemistry and his PhD in Environmental Catalysis from the University of Cyprus. He has obtained several National and International honors and awards including an Honor from the *Denmark Royal Academy of Science*s for the development of a Novel Sustainable de-pollution technology. Dr. Costa has more than 35 publications in High Impact, International, Refereed Journals, and over 45 publications in International Conferences Proceedings. He is currently the Coordinator of seven (7) Research Projects, while he also participated in more than 20 Research Projects. He owns six (6) patents for the Development of Novel Technologies for the Reduction of Nitrogen Oxides (NO_X).

Dr. Dewil Raf studied Chemical Engineering at the Katholieke Universiteit Leuven (2003) and holds a Ph.D. in Bioscience Engineering of the University of Antwerp (2006). He is currently working as an Assistant Professor at the Chemical Engineering Department of the Katholieke Universiteit Leuven. His research mainly focuses on conversion methods for the valorization of biomass and organic wastes. Key topics are anaerobic digestion and thermo-chemical conversion by pyrolysis. He had (co-)authored over 30 publications in peer-reviewed international journals and about 40 presentations at international conferences. He is member of the editorial board of "Environmental Progress and Sustainable Energy" and two other international journals, and member of the "Process Technology for Sustainable Energy" section of the European Federation of Chemical Engineers.

Dr. Finkbeiner Matthias is currently Chair of Sustainable Engineering and Vice-Director of the Department of Environmental Technology at Technical University Berlin. He is also Advisory Professor at Aalto University in Finland, Chairman of the ISO-Committee TC207/SC5 for Life Cycle Assessment, member of the International Life Cycle Board (ILCB) of the UNEP/SETAC Life Cycle Initiative and the Steering Committee of the Greenhouse Gas Protocol Product/Supply Chain Initiative of the World Business Council for Sustainable Development. Earlier in his career, he was Manager for Life Cycle Engineering at the Design-for-Environment Department for Mercedes-Benz Cars at Daimler AG in Stuttgart and Vice-Director Environmental Management at PE International.

Dr. Gómez Lucas, Ignacio is Graduate in Chemistry (Alicante University, 1981) , holding PhD degree in the field of organic residues as amendment of soils (1986). His post-doc research focused on waste management fields as amendment of soils and it environ and nutritional effects on the soil-plant system. He has 20 years experience in the field of use of organic waste in agricultural and degraded soils, characterization of the properties of soils and organic wastes, and evaluation of salinity and bioavailability of heavy metal in soils. He has published 2 books and more than 100 papers academic Journals. He has participated in numerous nationals and international projects in the soil science and environmental areas. Since 1993 he is Professor of the Alicante University and since 2002 is the Head of the Agrochemical and Environmental Department, of Miguel Hernández University (Elche, Spain).

Mr. Güney Kenan is Environmental Engineer, holding M.Sc. degree in the field of "Water Resources Engineering and Management" (2007) from University of Stuttgart. Mr. Güney is a researcher in the "Institute for Sanitary

Engineering, Water Quality and Solid Waste Management" in Germany. He is the manager of an ongoing bilateral project "Industrial Wastewater Reuse in Textile Industry by Application of Appropriate Membrane Treatment Technology and Investigating Pre-treatment Methods, Fouling Phenomenon, and Cleaning of Fouled Membranes" between University of Stuttgart and Istanbul Technical University.

Dr. Karagiannidis Avraam is Associate Professor at the Department of Mechanical Engineering of Aristotle University of Thessaloniki. He works mainly on: (a) treatment and integrated management of solid wastes, (b) location of noxious and obnoxious facilities, allocation of flows and impact assessment, (c) energy conservation and utilization of waste heat, and (d) multi-criteria decision support on energetic and environmental applications. He is the head of the Solid Waste Management group at the Laboratory of Heat Transfer and Environmental Engineering since 1994 and has participated in over 50 national and international research projects since 1991. He is the author and co-author of over 130 scientific publications and member of the Hellenic Technical Chamber and the Hellenic Solid Waste Management Association. He has served as reviewer for international journals and as scientific committee member for international conferences.

Mrs Lehmann Annekatrin is an environmental engineer. After finishing her studies in 2007 with a diploma thesis on Sewage Sludge Treatment for Nutrient Recovery, she worked as a research assistant in waste management department of Technical University of Berlin. During that time she published papers in academic journals and announcements in conferences. Since 2009 she is working on her Phd study on life cycle based sustainability assessment on water and waste water treatment technologies for developing regions at the Institute of Technology Assessment and Systems Analysis (ITAS), Karlsruhe Institute of Technology (KIT) and at the department of Sustainable Engineering (SEE), Technical University of Berlin.

Dr. Leverenz Harold is a post-doctoral researcher in the Department of Civil and Environmental Engineering at the University of California at Davis (UC Davis). He studies decentralized and satellite wastewater reuse systems, natural

	treatment systems for nutrient removal from wastewater, modeling of wastewater process based greenhouse gas emissions, and source control systems as a wastewater management strategy for nutrient and energy recovery. He received a B.S. in Biosystems Engineering from Michigan State University and an M.S. and Ph.D. in Environmental Engineering from UC Davis. Harold is a registered civil engineer in California.
	Mrs. Martín Julia is Chemical sciences PhD student in the area of Analytical Chemistry (2007). Her research is now focused on the development of methodologies for the determination of majority and emergent pollutants in environmental matrices. The methodologies developed are then applied to evaluate the occurrence and fate of the pollutants in the environment and to obtain information about their potential toxicological effects to the ecosystems. She has participated in 2 research project. She has collaborated in 1 book chapters, 4 papers in International Journals with high impact factors and has participated in more than 15 announcements in national and international conferences.
	*Dr. Martín Soriano-Disla José* studied a Bachelor of Environmental Sciences (1999-2004) obtaining a PhD with European mention (2010) at the University Miguel Hernández of Elche, Department of Agrochemistry and Environment. He has 6 years experience in the fields of the organic wastes management, heavy metal mobility, bioavailability and toxicity, near-infrared spectroscopy, organic matter quality and soil degradation. He has enjoyed three international stays at the INRA (Paris, France), CNRS (Montpellier, France), ESR and Lincoln University (Porirua and Christchurch, New Zealand). He has published 3 book chapters, 7 articles in peer-reviewed international (ISI) Journals, 5 articles in peer-reviewed non-ISI Journals, 16 conference abstracts, 1 refereed conference paper and 4 plenary and session keynotes. He has received 11 awards/fellows, he has been involved in 7 research projects and he has supervised 12 student projects. In addition, he has lecturing experience at the University Miguel Hernández of Elche and he is reviewer of major international journals (6).
	Dr. Muhammad H. Al-Malack received his Ph.D. degree from the University of Newcastle upon Tyne (England) in 1993 in the field of Environmental Engineering. Presently,

Dr. Al-Malack is working as a Professor in the Civil Engineering Department of King Fahd University of Petroleum & Minerals. He received his B.Sc. degree from King Fahd University of Petroleum & Minerals (KFUPM) in the field of Applied Civil Engineering. His M. Sc. degree was in the field of Water Resources and Environmental Engineering and was obtained from KFUPM. Between 1987 and 2006, Dr. Al-Malack joined the Research Institute where he was exposed to different research projects. During that period, he managed and participated in several projects related to liquid and solid waste management and environmental impact assessment. Dr. Al-Malack has published more than thirty-five technical papers in refereed journals and sixteen papers in national and international conferences. Beside reviewing technical papers and reports for national and international conferences and institutions, Dr. Al-Malack has conducted refereeing activities for national and international journals such as the Arabian Journal for Science and Engineering, Water Research, Journal of Membrane Science, Journal of Environmental Management, Journal of Hazardous Materials and Journal of Water, Air, and Soil Pollution.

Mr. Muhammad Muhitur Rahman completed his MSc in Civil Engineering with the specialization of Environmental Engineering from King Fahd University of Petroleum & Minerals (KFUPM) in 2005. After completing the MSc, he joined the Department of Civil Engineering, KFUPM in 2006 as Research Engineer. Mr. Rahman has extensive experience in development of wastewater feasibility and strategic studies, conceptual and detail design and associated reporting of municipal and industrial wastewater project. Mr. Rahman is the technical author of several journal and conference papers and co-author of a book chapter on use of membrane technology in the treatment of wastewater.

Dr. Navarro Pedreño Jose is professor of Soil and Environmental Science at the University Miguel Hernández of Elche. He was fellowship of the Spanish Government (FPI) and has PhD degree in Environmental Sciences at the University of Alicante (1992). He was associate and assistant professor of the Autonomous University of Madrid (1993-95) and Polytechnic University of Valencia (1995-96), coordinator of Environmental Sciences career (2001-2003) and Vice chancellor of the University Miguel Hernandez of

Elche (2003-05). He has been collaborating with EU and USA universities. Nowadays, he is coordinator of a Master Program (Analysis and Management of Mediterranean Ecosystems). He has participated in more than 50 research projects, published more than 150 articles in academic journals, conferences, published books and congresses communications. He is member of the International Union of Soil Sciences, European Societe for Soil Conservation and DesertNet International.

Dr. Nicolaides Demetris holds Diploma in Civil Engineering from the National Technical University of Athens (NTUA), as well as MSc in Structural Engineering from the Cardiff University, UK. From the same school he received his PhD in Civil Engineering in 2004, in the research area of fracture and fatigue of Ultra High Performance Fibre Reinforced Cementitious Composites (UHPFRCCs). He is currently a Lecturer of Civil Engineering at Frederick University in Cyprus. Dr. Nicolaides's research focuses on 1) the development of cementless concrete mixtures, 2) the use of waste materials as aggregates for the production of recycled concrete, 3) the effectiveness of supplementary cementitious materials in mitigating thermal cracking of concrete and 4) the impact/blast resistance of UHPFRCCs. He has published research work in the field UHPFRCCs, including both material development and experimentation, modeling of the constitutive behaviour using non-linear fracture mechanics theories, investigation of the autogenous shrinkage of this class of materials and finally retrofitting of concrete structures with the use of UHPFRCCs.

Mrs Papadimitriou Chrysi obtained an MSc in Biology of Water Resource Management at Napier University and a BSc (Hons) in Biological Sciences at Anglia Polytechnic University. She has been an adjunct professor at the Department of Pollution control technologies at Technological Educational Institute of West Macedonia in Greece since 2004. Since 2003 she has been working as a research assistant in the Department of Chemical Engineering at Aristotle University of Thessaloniki, Greece in the fields of wastewater and sludge treatment technologies and ecotoxicology.

Dr. Papadopoulos Ioannis is a Professor at the Department of Environmental Management of the Cyprus University of Technology. His expertise focus in the field of Soil Science and Soil Chemistry. He received his BSc in Horticulture and his MSc in Plant physiology from the University of Belgrade. He received his PhD in Soil Science (soil fertility) from Davis University, California (USA) and a post-graduate title in water quality from the University of Newcastle (UK). Dr. Papadopoulos worked as an Agricultural Research officer and as the Director of the Agricultural Research Institute (ARI) of Cyprus in the fields of soil science and use (reuse) of treated wastewaters in agriculture. He has more than 100 publications in International Refereed Scientific Journals, while he has participated in more than 100 International Conferences. Dr. Papadopoulso participated in a very large number of National and International Reserach projects, while he is acting as an appointed consultant of EU, FAO, IFAD, IAEA, UNEP, WHO, GTZ, CEFIGRE and CIHEAM. His research interests fall in the fields of soil fertility, fertilization of arid and irrigated cultures with contemporary methods as well as fertilization-irrigation (Water Fertilization) with the use of modern methods and systems of irrigation and use (reuse) of treated wastewaters for irrigation.

Dr. Papastergiadis Efthimios is a graduate of the Department of Chemistry of the Aristotle University of Thessaloniki. He has completed his Ph.D. studies in the same Department in 1996 in the field of Chemical Technology. He is currently employed as a Lecturer in the Department of Food Technology of the Alexander Technological Educational Institute of Thessaloniki. He is responsible of the laboratory of material analysis using advanced characterization techniques (XRD, SEM, EDS). He has experience in related research projects and his current research interest is focused on the development and characterization of novel materials for food packaging. Dr. Papastergiadis has published his research work (more than 20 scientific papers) in referred journals and proceedings of international conferences.

	Dr. Poulopoulos G. Stavros has studied Chemical Engineering and holds M.Sc. and Ph.D. from the National Technical University of Athens in the field of environmental protection. He has conducted research for 15 years on the treatment of gaseous emissions and toxic wastewaters. His work as an author includes 23 articles in international scientific journals, 24 announcements in scientific conferences and one book on adsorption, ion exchange and catalysis. Besides his academic activities, he has been working as environmental specialist at the Hellenic Ministry of Infrastructure, Transport and Networks since 2007.
	*Dr. Samaras Petros,* Chemical Engineer, is associate professor of water technology, at the Department of Food Technology, of the Technological Educational Institute of Thessaloniki, with expertise on the application of physical, chemical and biological methods for water and wastewater treatment. He has participated in more than 40 national and international research and demonstration projects, funded by E.U. and from local sources (ministries, industries, companies etc.). He has over than 50 publications in journals; more than 100 presentations in international conferences; 6 chapters in books; 1 EU patent holder.
	*Dr. Santos L. Juan* is an Analytical Chemist holding PhD in the field of the environmental analytical chemistry. His research is focused on the development of methodologies for the determination of majority and emergent pollutants in environmental matrices such as sewage sludge, sediments, wastewater and surface water; and their application to evaluate the occurrence, fate and risk assessment of these pollutants in the environment. He has participated in 11 research project and collaborated in 4 book chapters. He has published 17 papers in international journals; most of them in peer reviewed international journals with high impact factors, and reported his scientific results as 40 announcements in national and international conferences. In 2008, he obtained an assistant professor position in the Department of Analytical Chemistry of the University of Seville (Spain).
	Dr. Savva G. Petros is a Teaching and Research Associate at the Department of Environmental Management of the School of Geotechnical Sciences and Environmental Management. He has obtained his BSc and MSc in Chemistry and his PhD in Environmental Chemistry from the University of Cyprus. He has obtained several National and International honors

and awards including an Honor from the *Denmark Royal Academy of Sciences* for the development of a Novel Sustainable de-pollution technology. Dr. Savva has more than 10 publications in High Impact, International, Refereed Journals, and over 25 publications in International Conferences Proceedings. He has participated in more than 15 Research Projects. He owns three (3) patents for the Development of Novel De-Pollution Technologies.

Dr. Tchobanoglous George is a Professor Emeritus in the Department of Civil and Environmental Engineering at the University of California at Davis. His research interests include wastewater treatment and reuse, UV disinfection, aquatic systems, decentralized wastewater management, and solid waste management. He has authored or coauthored over 450 publications, including 22 text and reference books. Among his many honors, he received the Clarke Prize in 2003. In 2004, he was inducted into the National Academy of Engineering. In 2005, he received an Honorary Doctor of Engineering degree from the Colorado School of Mines. In 2007, he received the Frederick George Pohland Medal. In 2010, he was made an honorary member of the Greek Technical Chamber. He received a B.S. in Civil Engineering from the University of the Pacific, an M.S. in Sanitary Engineering from the University of California, Berkeley, and a Ph.D. in Environmental Engineering from Stanford University.

Dr. Traverso Marzia. Ph.D. of `Environmental Applied Physics´ of University of Palermo, and environmental engineer, working in the field of Life Cycle Assessment on building sector and strategic environmental assessment to the transportation urban systems. She was a responsible of EMS and Ecolabel school of Sicily and lecturer in an Ecolabel course for the building products. She wrote several papers for national and international scientific magazines and congresses. Presently she is researcher as post doctoral scientific assistant at the Chair of Sustainable Engineering of Technical University of Berlin. Her main research topic is the Social Life Cycle Assessment and Life Cycle Sustainability Assessment.

Dipl.-Ing Venetis Christos is Waste Management Engineer. He studied and worked as a student assistant in the RWTH Aachen Technical University. He has three years experience in the field of Waste Treatment Technologies such as anaerobic digestion, composting, mechanical and thermal treatment and 5 publications in international conferences. Since 2008 he is associate of EPTA Ltd. Environmental Engineers - Consultants where he has participated in numerous studies and projects in the field of waste management.

Mrs Voukkali Irene has Food Science in the Aristotle University of Thessaloniki and the department of Agriculturist. She is a Lead Auditor of ISO 9001:2008 (Quality Management Systems) and ISO 14001:2004 (Environmental Management Systems). Since 2007 she has collaborated with the Institute of Environmental Technology and Sustainable Development and she the Director of the Quality Control Department of the Institute. She has implemented more than 50 Environmental Management Systems (ISO 14001/EMAS), Environmental Impact Assessments, Quality Management Systems, and Risk Assessment Analysis in several activities in Cyprus. She organized several seminal for SMEs most of theme focus on Quality Management, Environmental Management and Health Safety. Until now she has 4 scientific Chapters in 3 several books. She has published 2 scientific papers and she has 6 presentations in International Conferences.

Dr. Zouboulis Anastasios is Professor of Chemical & Environmental Technology, Department of Chemistry, Aristotle University of Thessaloniki is performing research in the field of water and wastewater treatment technologies, wastewater management, reclamation and reuse. He is author/co-author of more than 150 papers, published in scientific journals and of more than 100 papers published in the proceedings of national & International conferences. He has supervised 8 Ph.D. and 12 M.Sc. Theses and more than 50 Diploma theses. He has participated in more than 50 national and international research and demonstration projects, funded by E.U. and from local sources (ministries, industries, companies etc.), while he was Scientific Responsible in more than 30 of them, related mainly with the fields of wastewater treatment processes, environmental biotechnology, as well as the treatment of industrial solid toxic wastes by the application of appropriate stabilization methods. His international reputation was recognised by his election as (foreign) member in the Russian Academy of

	Sciences (since 2003). He has received several prizes and scholarships from national and international organizations. He is also an active consultant for several local industries in the field of environmental protection, a national expert for the horizontal activities of FP6 and FP7 (EU), and he is acting often as a reviewer/evaluator for scientific publications, as well as for several EU or national research projects. He is also member of the Editorial Board of scientific publications (Separation Science & Technology, Water Research, Journal of Hazardous Materials etc.), as well as a member of the Organizing and Scientific Committees of several national and international conferences (over 20).

In: Sewage Sludge Management
Editors: A. A. Zorpas and V. J. Inglezakis

ISBN: 978-1-61324-393-0
© 2012 Nova Science Publishers, Inc.

Chapter 1

ABOUT SEWAGE SLUDGE

Antonis A. Zorpas[1] and Vassilis J. Inglezakis

Institute of Environmental Technology and Sustainable Development,
Paralimni, Cyprus

The large part of the population in EU is connected via sewerage systems to treatment plants, which treat the municipal wastewater. The municipal wastewater generally consists of a mixture of domestic, industrial wastewater and storm water. More than 50,000 wastewater treatment plants are operating in the European Union, producing more than 8.1 millions tons of dry solids per year. The main disposal routes for sewage sludge arising from the wastewater treatment plants are application to land, composting, disposal in a landfill and incineration as well as sea disposal. Sewage sludge has been used in agriculture as fertilizer as sustainable practice utilizing, recycling the nutrients back to land and to improve soil quality.

Sludge is a natural byproduct of anaerobic biological digestion. This dead microbial material, which accumulates over many years in the bottom of anaerobic lagoons and is rich in nutrients and organic matter, periodically must be removed from the lagoon. Typically, it is land-applied to crops. Sludge (also known as biosolids or residuals) is a thick, black, viscous substance that is rich in organic material and nutrients. It is comprised of the dead and

[1] E-mail: antoniszorpas@envitech.org. Website: www.envitech.org.

degraded microbial cells that anaerobically digested the manure influent and of any other materials (excess feed, debris, rocks, etc.) that were placed in the manure collection system and have settled to the bottom of the lagoon.

Sludge is a generic term for solids separated from suspension in a liquid. This 'soupy' material usually contains significant quantities of 'interstitial' water (between the solid particles). Commonly sludge refers to the residual, semi-solid material left from industrial wastewater or sewage treatment processes. It can also refer to the settled suspension obtained from conventional drinking water treatment, and numerous other industrial processes.

When fresh sewage or wastewater is added to a settling tank, approximately 50% of the suspended solid matter will settle out in an hour and a half. This collection of solids is known as raw sludge or primary solids and is said to be "fresh" before anaerobic processes become active. The sludge will become putrescent in a short time once anaerobic bacteria take over, and must be removed from the sedimentation tank before this happens.

This is accomplished in one of two ways. In an Imhoff tank, anaerobic bacteria, resulting in liquefaction and reduced volume of the sludge, pass fresh sludge through a slot to the lower story or digestion chamber where it is decomposed. After digesting for an extended period, the result is called "digested" sludge and may be disposed of by drying and then landfilling. More commonly with domestic sewage, the fresh sludge is continuously extracted from the tank mechanically and passed to separate sludge digestion tanks that operate at higher temperatures than the lower story of the Imhoff tank and, as a result, digest much more rapidly and efficiently.

Excess solids from biological processes such as activated sludge may still be referred to as sludge, but the term biosolids, is more commonly used to refer to the material, particularly after further processing such as aerobic composting. Industrial wastewater solids are also referred to as sludge, whether generated from biological or physical-chemical processes. Surface water plants also generate sludge made up of solids removed from the raw water.

Sludge is formed during wastewater treatment. Wastewater is a combination of the liquid- or water-carried wastes removed from residential, institutional, commercial and industrial establishments, together with ground water, surface water and storm water, as may be present. Wastewater may contain certain undesirable components, including organic, inorganic and toxic substances, as well as pathogenic or disease-causing micro-organisms. Table 1 shows a typical composition of untreated municipal wastewater.

The main groups of the organic solids in wastewater are proteins, carbohydrates, fats and oils. Protein contains about 16% nitrogen, and together with urea are the major sources of nitrogen in wastewater. In its untreated form, wastewater can not be disposed of for several

Table 1. Domestic characteristics of sewage in mg/l

	Weak	Medium	Strong
Total suspended solids	100	200	350
Volatile suspended solids	75	135	210
BOD5	100	200	400
COD	175	300	600
Ammonia-N	5	10	20
Organic-N	8	20	40
PO4-P	7	10	20

reasons. First, the biological decomposition of the organic materials in wastewater consumes oxygen and thus reduces the quantity available in the receiving waters for the aquatic life.

There are many types of sludge; sewage sludge, biological sludge, drilling mud, sludge at the bottom of rivers, manure, industrial sludge, sludge drying bed, etc. Some types of sludge are a result of natural processes, while other types of sludge are produced in industrial processes such as by filtering of liquids. Most types of sludge contain lots of water, and are more or less foul smelling. Many of these types are also characterised as special waste--i.e. Waste that is subjected to restrictions regarding collection and destruction by the authorities. The reason for a number of sludge types having to be processed as special waste is that the content, completely or partially, is comprised of components damaging to the environment and/or health. These components can be hydrocarbons as in "bio-sludge", or bacteria and parasite eggs as in sewage sludge. Sludge can be a wonderful source of nutrients for the soil. Using this material as a fertilizer can benefit the environment by turning wastes into valuable resources. These sludges would otherwise have to be disposed of by landfilling, lagooning, incineration, or ocean dumping. On the other hand, heavy metals sometimes found in sewage sludge may present environmental problems. Several practices, similar to those used with other organic fertilizers, will maximize the benefits of using sewage sludge while minimizing the risks.

There are regulations governing the quality of the effluent discharged from Sewage Treatment Works (STW). The regulations set limits for the concentrations of BOD5, COD, total suspended solids, phosphorus, nitrogen, mercury and phenol, among others, in the discharge from STW. In the European Union (EU), the quality of effluent is currently regulated through the Urban Wastewater Treatment Directive (91/271/EEC) of May 1991. The directive also requires that all communities with above 2000 population equivalents install adequate collection, treatment and disposal systems for the wastewater generated. Table 2 gives some values of the national standards of some countries for illustration.

The world is currently witnessing a rapid increase in sludge production, and this is expected to continue up to the early part of the next century. In both developed and NICs, the increase in sludge production is driven by the increase in the percentage of households connected to central treatment plants, the increasingly tightening of pollution limits on the effluent discharges, as well as the availability of technologies capable of achieving higher efficiency of wastewater treatment. The effect of the increase in environmental awareness in general can also not be ignored. On the other hand, due to ever increasing transportation and disposal costs, every effort is being made to reduce the quantity of sludge for disposal by

Table 2. Comparison of national standards for effluents in mg/l

	Hong Kong	Japan	Malaysia	Singapore	Taiwan	France	Netherlands	Portugal	Cyprus	EU
BOD$_5$	20–40	160	20–50	100	100	30	20	40	20-100	25
COD	80–85					90		150	50-100	125
SS	30–40					30	30	60	1-10	35
N	20–50					10	10–15	15	2	10–15
Hg	0.001	0.05	0.005–0.05	0.001–0.05	0.005				----	
Phenols	0.1	5	0.001–1	5	5					

increasingly lowering the water content of the sludge. The net quantity of sludge available for disposal therefore depends on the financial, environmental and technological limitations of a given city or country.

The progressive implementation of the Urban Waste Water Treatment Directive 91/271/EEC in all Member States is increasing the quantities of sewage sludge requiring disposal. From an annual production of some 5.5 million tonnes of dry matter in 1992, the Community is heading towards nearly 9 million tonnes by the end of 2005. This increase is mainly due to the practical implementation of the Directive as well as the slow but constant rise in the number of households connected to sewers and the increase in the level of treatment (up to tertiary treatment with removal of nutrients in some Member States). The Directive sets the following targets for secondary treatment of waste waters coming from agglomerations:

at the latest by 31 December 2000 for agglomerations of more than 15,000 p.e. (Population equivalent);

at the latest by 31 December 2005 for agglomerations between 10,000 and 15,000 p.e.;

at the latest by 31 December 2005 for agglomerations of between 2,000 and 10,000 p.e. Discharging to fresh waters and estuaries.

There are more stringent provisions for agglomerations discharging into sensitive areas such as fresh waters or estuaries.

The Sewage Sludge Directive 86/278/EEC seeks to encourage the use of sewage sludge in agriculture and to regulate its use in such a way as to prevent harmful effects on soil, vegetation, animals and man. To this end, it prohibits the use of untreated sludge on agricultural land unless it is injected or incorporated into the soil. Treated sludge is defined as having undergone "biological, chemical or heat treatment, long-term storage or any other appropriate process so as significantly to reduce its fermentability and the health hazards resulting from its use". To provide protection against potential health risks from residual pathogens, sludge must not be applied to soil in which fruit and vegetable crops are growing or grown, or less than ten months before fruit and vegetable crops are to be harvested. Grazing animals must not be allowed access to grassland or forage land less than three weeks after the application of sludge. The Directive also requires that sludge should be used in such a way that account is taken of the nutrient requirements of plants and that the quality of the soil and of the surface and groundwater is not impaired. The Directive specifies rules for the sampling and analysis of sludges and soils. It sets out requirements for the keeping of detailed records of the quantities of sludge produced, the quantities used in agriculture, the composition and properties of the sludge, the type of treatment and the sites where the sludge is used. Limit values for concentrations of heavy metals in sewage sludge intended for agricultural use and in sludge-treated soils are in Annexes I A, I B and I C of the Directive. Although at Community level the reuse of sludge accounts for about 40% of the overall sludge production, landfilling as well as incineration in some Member States are the most widely used disposal outlets despite their environmental drawbacks.

Directive 86/278/ EEC was adopted over 20 years ago with a view to encourage the application of sewage sludge in agriculture and to regulate its use, so as to present harmful effects on soil, vegetation, animals and humans. The European Commission is currently assessing whether the current Directive should be reviewed – and if so, the extent of this review. For example, Directive 86/278/EEC sets limit values for seven heavy metals. Since its adoption, several Member States have enacted and implemented stricter limit values for heavy metals and set requirements for other contaminants. For its assessment, the European Commission has launched a study to gather existing information on the environmental, economic, and social as well as health impacts of present practices of sewage sludge use on land.

Generally there are several recommendations for the use of sewage sludge in soils as follows:

- ✓ Farmers should not accept sludge or apply it to land without a complete waste analysis.
- ✓ The soil pH on sludge application sites should be maintained above 6.5 to lower plant uptake of heavy metals.
- ✓ Sludge should be applied by directly injecting it in the soil or incorporating it into the soil within 24 hours to reduce the potential for odor and loss of nutrients in runoff.
- ✓ For surface application, use only sludge that has been properly digested and stabilized at the treatment plant.
- ✓ Do not apply sludge long before the crop will utilize it.
- ✓ Keep accurate records of sludge applications. Sewage sludge containing heavy metals should be monitored to prevent contamination of soil or groundwater.
- ✓ Excessive rates can contribute to leaching, erosion, and runoff losses; therefore, do not apply more than is needed.

This book will also assess the risks and opportunities that can be foreseen in coming years. The book will identify possible options for European policy and estimate their costs and benefits.

Sludge management has become a key issue in urban and industrial wastewater treatment for two main reasons: (1) large amounts of sludge are generated as a waste or by-product of wastewater treatment processes; and (2) solid waste management and disposal are among the most complex problems of wastewater treatment facilities. In general, sludges and biosolids resulting from wastewater treatment operations are in liquid form, typically containing 0.5–15% total solids (TS). Most of them are organic compounds, with a broad range of volatile solids contents (50–80%), commonly from 75% to 80% of TS. Sludge production and characteristics are highly dependent on the wastewater composition and the treatment used. The main sludge treatment operations are aimed at increasing the concentration of total solids in order to reduce the sludge volume (i.e. sludge thickening and dewatering) or decreasing the concentration of volatile solids and stabilising the biodegradable fraction of organic matter (i.e. sludge stabilisation via anaerobic digestion or composting). Decreasing sludge volume by means of dewatering technologies reduces the costs of sludge handling, transportation and final disposal. Furthermore, sludge dewatering is always required prior to treatments such as composting, incineration or landfilling. Dewatering may be carried out by using conventional mechanical processes such as centrifugation and filtration, or by using other processes such as water evaporation, evapotranspiration (ET) and percolation. Sludge treatment wetlands are extensive treatments that achieve sludge dewatering and mineralisation by means of the latter processes. In general, sludge treatment systems involve high costs, ranging from 20% to 60%

of the total operating cost of wastewater treatment plants (WWTPs). This is particularly critical in the case of WWTPs of small rural communities, which, in practice, may then transport raw sludge to larger WWTPs instead of implementing their own sludge treatment line. The use of sludge treatment wetlands may provide an opportunity to treat the sludge within the WWTPs of this type of communities. Sludge treatment wetlands, also known as sludge drying reed beds, are rather new sludge treatment systems based on treatment wetlands (TW). TW are being used in many regions of the world for wastewater treatment, and are made up of shallow ponds, beds or trenches filled with a gravel layer and planted with emergent rooted wetland vegetation such as Phragmites australis (common reed). Sludge treatment wetlands have been used in Europe for sludge dewatering and stabilisation since the late 1980s. The largest experience comes from Denmark, where there are over 140 fullscale systems currently in operation. Other systems implemented in northern Europe are located in Poland, Belgium and the United Kingdom. In the Mediterranean region, full-scale systems are operating in Italy, France and Spain. Several pilot plant trials have been carried out in Palestine and Cameroon. Also in China, several studies reporter on sludge treatment wetlands at pilot scale with good results.

The characteristics of sludge to be measured are strongly related to its ultimate fate. For example, if the sludge is to be thickened by gravity, its settling and compaction characteristics are important. On the other hand, if the sludge is to be digested anaerobic ally, the concentrations of volatile solids and heavy metals are of importance. The first characteristic, solids concentration, is perhaps the most important variable in defining the volume of sludge to be handled, and determining whether the sludge behaves as a liquid or a solid. The rheological characteristics of sludge are very important because they are one of only few truly basic parameters describing the physical nature of sludge. Sludge varies from a Newtonian fluid, where shear is proportional to the velocity gradient, to a plastic fluid, where a threshold shear must be reached before the sludge starts to move. Most wastewater sludges are pseudoplastic. The chemical characteristics of sludge are of great importance for several reasons. Characteristics of sludge that affect its suitability for beneficial use include organic content (usually measured as volatile solids), nutrients, pathogens, metals, and toxic organics. With respect to microbiological characteristics of municipal sludges, wastewaters generally contain four major types of pathogens: bacteria, protozoa, viruses, and helminths. The concentrations of these pathogenic organisms in wastewater depend on the health condition of the community. In times of epidemics, these concentrations would be high. Since pathogens come from relatively large volumes of wastewater, when ending up in the sludge, they, in general, become concentrated and very infectious.

This edit book tries to collect and presents all the relevant information on sludge management from the past to our century. It concerns a bibliography trip around the world as the most significant Professors and Researches give their experiences and knowledge on the subject.

In: Sewage Sludge Management
Editors: A. A. Zorpas and V. J. Inglezakis

ISBN: 978-1-61324-393-0
© 2012 Nova Science Publishers, Inc.

Chapter 2

MUNICIPAL SLUDGE: GENERATION AND CHARACTERISTICS

Muhammad H. Al-Malack[*,1] *and Muhammad Muhitur Rahman*[2]

[1] Professor, Civil Engineering Department King
Fahd University of Petroleum & Minerals,
Dhahran, Saudi Arabia
[2] Research Engineer, Civil Engineering Department King
Fahd University of Petroleum & Minerals,
Dhahran, Saudi Arabia

ABSTRACT

The rapid increase in sludge production creates the need for adoption of economically and environmentally acceptable sludge management schemes. This Chapter presents a review and investigation results concerning sludge generation, characteristics and effect of natural drying on sludge characteristics. The data presented in this chapter are mainly taken from an extensive research, conducted in order to determine the physical, chemical and microbiological characteristics of municipal sludges produced at three major cities, namely, Qateef, Dammam and Khobar in the Eastern Province of Saudi Arabia between April 2000 and February 2001. Sludge samples were collected, from sand drying beds, twice a season for one year and were analyzed for certain potential parameters. The chapter describes the effect of initial sludge depth in drying beds on the fate of pathogen, which is a unique approach for formulating mathematical representation to describe the pathogen die-off with respect to time and sludge initial depth.

---

[*] E-mail: mhmalack@kfupm.edu.sa.

1. INTRODUCTION

Wastewater treatment facilities have been, or are being, expanded to provide higher degree of wastewater treatment and to cope with the increasing quantities of wastewater being delivered to the treatment facilities. This, of course, translates into increased volumes of produced wastewater sludges, which must be managed properly. If sludge is not characterized by the concerned authority and is disposed in landfills, then it may pose great hazards to the surrounding environment. Furthermore, farmers may utilize municipal sludge as a fertilizer or soil conditioner after an uncontrolled drying period and without proper characterization or treatment. This practice may result in crops that are polluted with pathogens, toxic organics, and heavy metals. Consequently, serious health hazards are expected if consumers continue to bio-accumulate toxicants through this practice. Accordingly, the issue of municipal sludge should be fully investigated and addressed in order to help decision makers become familiarized with the characteristics of those sludges, the possibilities of treatment, and the consequences of the aforementioned fertilizing practices on public health.

2. SLUDGE GENERATION

Solids generation in biological wastewater treatment plants is difficult to estimate, where all suspended solids in the influent do not appear in the sludge. Some are biologically metabolized to soluble or gaseous end-products in the biological treatment process or sludge digestion process. Some soluble wastewater components will be transformed into biological solids that can be reduced during digestion. Solids generation in a biological treatment process is a function of the type and the operation of the process.

The sources of sludge in a wastewater treatment plant vary according to the type of plant and its method of operation [1]. The principal sources of sludge at municipal wastewater treatment plants are the primary sedimentation basin and the secondary clarifiers. Additional sludge may also come from chemical precipitation, nitrification-denitrification facilities, screening, grinder, and filtration devices. Table 1 shows the principal sources of solids and sludge and the types generated [1].

Table 1. Sources of Solids and Sludge from a Conventional Wastewater Treatment Plant [Metcalf & Eddy, 2003]

Unit Operation or Process	Types of Solids or Sludge	Remarks
Screening	Coarse solids	Coarse solids are removed by mechanical and hand-cleaned bar screens. In small plants, screenings are often comminuted for removal in subsequent treatment units
Grit removal	Grit and scum	Scum removal facilities are often omitted in grit removal facilities.

Unit Operation or Process	Types of Solids or Sludge	Remarks
Pre-aeration	Grit and scum	In some plants, scum removal facilities are not provided in pre-aeration tanks. If the pre-aeration tanks are not preceded by grit removal facilities, grit deposition may occur in pre-aeration tanks.
Primary sedimentation	Primary sludge and scum	Quantities of sludge and scum depend upon the nature of the collection system and whether industrial wastes are discharged to the system.
Biological treatment	Suspended solids	Suspended solids are produced by the biological conversion of BOD. Some form of thickening may be required to concentrate the waste sludge stream from biological treatment.
Secondary sedimentation	Secondary sludge and scum	Provision for scum removal from secondary settling tanks is a requirement of the U.S. Environmental Protection Agency.
Sludge-processing facilities	Sludge, compost, and ashes	The characteristics of the end products depend on the characteristics of the sludge being treated and the operations and processes used. Regulations for the disposal of residuals are becoming increasingly stringent.

The per capita production of suspended solids could be assumed as 91 g/capita/day. If there is a moderate amount of industrial wastes, this may rise to 100 g/capita/day, and if it is a combined sewage with considerable industrial waste, it may be 113 g/capita/day. If ground garbage is added to the sewage, then there will be an additional 32 to 50 g/capita/day. Without sludge digestion, the per capita sludge production rates are approximately 80 and 115 g/d for primary and secondary treatment, respectively. Adding iron or aluminum salts for phosphorus removal will increase sludge production by about 40% for primary treatment or 25% for secondary treatment. For primary treatment, the quantities of sludge produced may be 0.25 to 0.35 percent by volume of wastewater treated. When treatment is upgraded to activated sludge, the quantities increase to 1.5 to 2.0 percent of the volume of wastewater treated. Use of chemicals for phosphorous removal can add another 1.0 percent. The observed yields are 35 to 52 percent higher for the anoxic reactors than they were for the aerobic ones. There is effect of temperature on wastewater treatment under aerobic and anoxic conditions. Additional uptake of phosphorus resulted in an increase of the inorganic sludge mass, but the organic sludge mass did not change.

In Europe, dry sludge production range from 30 to 124 g/capita/day. In Metropolitan area of Seattle, the amount of solids generate per household is 125 g/day (about 35 g/capita/day). In the United Kingdom, raw primary sludge produces 52 g/capita/day, co-

Table 2. Sludge Production and Methods of Disposal in 1990 [Kiely, 1997]

Country	Population (total) (million)	Population served (%)	Sludge produced (TDS×1000/yr)	Disposal Method (%)			
				Agriculture	Landfill	Incineration	Other
Austria	7.8	48	320	13	56	31	0
Belgium	9.9	33	75	31	56	9	4
Denmark	5.1	100	130	37	33	28	2
France	56	64	700	50	50	0	0
West	62	90	2500	25	63	12	0
Germany	10	-	15	3	97	0	0
Greece	3.	44	24	28	18	0	54
Ireland	57	30	800	34	55	11	0
Italy	0.4	92	15	81	18	0	1
Luxembou	15	90	282	44	53	3	0
rg	10.3	47	(871)[a]	80	13	0	7
Netherland	39	47	200	10	15	10	30
s	6.4	80	280	60	30	20	0
Portugal	57	84	215	51	16	5	28
Spain	123	42	1075	24	41	22	13
Switzerlan	249	-	2440	16	43	21	21
d			800-				
UK			1600[b]				
Japan							
USA							

[a] The production of sludge in the Netherlands is expected to go from 282000TDS/yr in 1990 to 871000 TDS/yr in 2000.

[b] A range of 800 million to 1.6 billion wet tons in the United States.

Table 3. Typical Data for the Physical Characteristics and Quantities of Sludge Produced from Various Wastewater Treatment Operations and Processes [Metcalf & Eddy, 2003]

Treatment operation or Process	Specific gravity of sludge solids	Specific gravity of sludge	Dry Solids, lb/10^3 gal	
			Range	Typical
Primary sedimentation	1.4	1.02	0.9-1.4	1.25
Activated sludge (waste sludge)	1.25	1.005	0.6-0.8	0.7
Trickling Filtration (waste	1.45	1.025	0.5-0.8	0.6

sludge)	1.30	1.015	0.7-1.0	0.8[a]
Extended aeration (waste sludge)	1.30	1.01	0.7-1.0	0.8[a]
Aerated lagoons (waste sludge)	1.20	1.005	0.1-0.2	0.15
Filtration				
Algae removal	1.20	1.005	0.1-0.2	0.15
Chemical addition to primary sedimentation tanks for phosphorus removal Low lime (350-500 mg/l) High lime (800-1600 mg/l)				
	1.9	1.04	2.0-3.3	2.5[b]
Suspended-growth nitrification	2.2	1.05	5.0-11.0	6.6[b]
Suspended-growth denitrification	-	-	-	-
Roughing filter	1.20	1.005	0.1-0.25	0.15
	1.28	1.02	-	-[d]

[a] Assuming no primary treatment.
[b] Sludge in addition to that normally removed by primary sedimentation.
[c] Negligible.
[d] Included in sludge production from biological secondary treatment processes.

settled activated produces 74 g/capita/day and co-settled activated tertiary sludge produces 76 g/capita/day. Research has showed that no excess wastewater sludge is produced in a full-scale wastewater treatment plant if the return sludge is ozonated. Soluble solids concentration of 2 to 15 mg/l in the effluent was higher than those in the treatment processes without ozonation. The annual sludge production in 1990 for a number of countries is shown in Table 2. It should be noted that while the population of each country is given, the population served by wastewater facilities is typically about one half of the total.

Table 3 shows data on the quantities of sludge produced from various processes and operations. It is worth mentioning that the quantity of sludge produced varies widely. To reliably predict sludge production, more detailed data needed to be collected on wastewater characteristics, the treatment process, and the operating parameters. Corresponding data on the sludge solids concentration to be expected from various processes are given in Table 4. [2-12]

Table 4. Sludge Concentrations from Various Treatment Operations and Processes [Metcalf & Eddy, 2003].

Operation or process application	Sludge solids concentration, % dry solids	
	Range	Typical
Primary settling tank		
Primary sludge	4.0-10.0	5.0
Primary sludge to a cyclone	0.5-3.0	1.5
Primary sludge and trickling-filter humus	3.0-8.0	4.0
Primary sludge with iron addition for phosphorus removal	4.0-10.0	5.0
Primary sludge with low lime addition for phosphorus removal	2.0-8.0	4.0
Primary sludge with high lime addition for phosphorus removal	4.0-16.0	10.0
Scum	3.0-10.0	5.0
Secondary settling tank		
Waste activated sludge		
With primary settling	0.5-1.5	0.8
Without primary settling	0.8-2.5	1.3
High purity oxygen activated sludge		
With primary settling	1.3-3.0	2.0
Without primary settling	1.4-4.0	2.5
Trickling-filter humus sludge	1.0-3.0	1.5
Rotating biological contractor waste sludge	1.0-3.0	1.5
Gravity thickener		
Primary sludge only	5.0-10.0	8.0
Primary and waste activated sludge	2.0-8.0	4.0
Primary sludge and trickling-filter humus	4.0-9.0	5.0
Dissolved-air flotation thickener		
Waste activated sludge only		
With chemical addition	4.0-6.0	5.0
Without chemical addition	3.0-5.0	4.0
Centrifuge thickener		
Waste activated sludge only	4.0-8.0	5.0
Gravity belt thickener		
Waste activated sludge only with chemical addition	3.0-6.0	5.0
Anaerobic digester		
Primary sludge only	5.0-10.0	7.0
Primary and waste activated sludge	2.5-7.0	3.5
Primary sludge and trickling-filter humus	3.0-8.0	4.0

Operation or process application	Sludge solids concentration, % dry solids	
	Range	Typical
Aerobic digester		
Primary sludge only	2.5-7.0	3.5
Primary and waste activated sludge	1.5-4.0	2.5
Waste activated sludge only	0.8-2.5	1.3

3. SLUDGE CHARACTERISTICS

The characteristics of sludge to be measured are strongly related to its ultimate fate. For example, if the sludge is to be thickened by gravity, it's settling and compaction characteristics are important. On the other hand, if the sludge is to be digested anaerobically, the concentrations of volatile solids and heavy metals are of importance.

3.1 Physical Characteristics

The first characteristic, solids concentration, is perhaps the most important variable in defining the volume of sludge to be handled, and determining whether the sludge behaves as a liquid or a solid. The specific gravity of inorganic solids is about 2-2.5 and that of the organic fraction is 1.2-1.3. Table 5 shows solids concentrations in various municipal sludges [13].

The rheological characteristics of sludge are very important because they are one of only few truly basic parameters describing the physical nature of sludge. Sludge varies from a Newtonian fluid, where shear is proportional to the velocity gradient, to a plastic fluid, where a threshold shear must be reached before the sludge starts to move. Most wastewater sludges are pseudoplastic. The physical characteristics of sludge from different treatment processes are given in Table 6. [11].

Table 5. Solids Concentration in Various Sludge Types [Droste, 1997]

Sludge type	Solids concentration %
Primary sludge	5-8%
Waste activated sludge	0.5-2.0%
Fixed film waste sludge	3-10%
Primary and waste activated sludge	2.5-4%
Primary and fixed film sludge	3-5%
Aerobically digested sludge (thickened)	1-2%
Anaerobically digested sludge (thickened)	6-12%

Table 6. Physical Characteristics of Sludge [Kiely, 1997]

Parameter	Primary sludge	Secondary sludge	Dewatered sludge
Dry solids	2-6%	0.5-2%	15-35%
Volatile solids	60-80%	50-70%	30-60%
Sludge specific gravity	1.02	1.05	1.1
Solids specific gravity	1.4	1.25	1.2-1.4
Shear strength (kN/m2)	<5	<2	<20
Energy content (MJ/kg VS)	12-22	12-20	25-30
Particle size (90%)	<200 μm	<100 μm	<100 μm

Table 7. Characteristics of dried sludge collected from three cities of the Kingdom

Parameters	Mean Values		
	Qateef	Dammam	Khobar
Moisture Content %	26	25	53
Total solids %	74	74	47.0
Total Volatile Solids %	45	44.8	29.3
pH	6.9	6.9	6.9

Table 7 shows mean values of the physical characteristics of dried sludge samples collected from the cities of Qateef, Dammam, and Khobar during a one-year study. Total dried solids were found to be 74.4, 51.8, and 47.0 per cent for dried sludge samples collected from Qateef, Dammam, and Khobar, respectively. It is clear that samples collected from Qateef wastewater treatment plant contained the highest percentage of dried solids among the other cities. This could be attributed to the longer drying period (15 days) adopted at Qateef when compared to Dammam and Khobar (7 days). Another reason for the higher percentage could be the better physical condition of the drying beds at Qateef plant compared to those at Dammam and Khobar plants.

Volatile Solids as a percentage of total solids ranged between about 30 to 45 percent, with highest and lowest values belonged to Qateef and Khobar, respectively. Due to drying, especially in the case of Saudi Arabia's severe weather conditions, portions of the volatile solids may have been lost.

3.2 Chemical Characteristics

The chemical characteristics of sludge are of great importance for several reasons. Characteristics of sludge that affect its suitability for beneficial use include organic content (usually measured as volatile solids), nutrients, pathogens, metals, and toxic organics. Table 8 shows the mean concentrations of several elements in different sludge types, while typical nutrient values of sludge as compared to commercial fertilizers are reported in Table 9. Moreover, Table 10 gives a range of typical chemical compositions of different sludges [1, 11]

Table 8. Concentrations of Several Elements in Different Sludge Types [Kiely, 1997]

Parameter	Mean Concentration (% of dry solids)		
	Anaerobic	Aerobic	All
K	0.52	0.46	0.4
Na	0.7	1.1	0.57
Ca	5.8	3.3	4.9
Mg	0.58	0.52	0.54
Ba	0.08	0.02	0.06
Fe	1.6	1.1	1.3
Al	1.7	0.7	1.2

Table 9. Comparison of Nutrient Levels in Commercial Fertilizers and Wastewater Sludge [Metcalf & Eddy, 2003]

	Nutrients, %		
	Nitrogen	Phosphorous	Potassium
Fertilizers for typical agricultural use	5	10	10
Typical values for stabilized sludge	3.3	2.3	0.3

Table 10. Typical Chemical Composition of Sludges [Kiely, 1997]

Parameter	Primary sludge	Anaerobically digested sludge	Aerobically digested sludge
pH	5-8	6.5-7.5	
Alkalinity (mg/l $CaCO_3$)	500-1500	2500-3500	
Nitrogen (N% of TS)	1.5-4	1.6-6	0.5-7.6
Phosphorous (P_2O_5% of TS)	0.8-2.8	1.5-4	1.1-5.5
Fats, grease (% of TS)	6-30	5-20	
Protein (% of TS)	20-30	15-20	
Organic acids (mg/l as Hac)	6800-10000	2700-6800	

Trace elements in sludge are those inorganic chemical elements that, in very small quantities, can be essential or detrimental to plants and animals.

The term 'heavy metals' has been used to denote some of the trace elements present in sludge. Several investigators described analytical methods for measuring trace metals and organic compounds [14-19]. Table 11 shows a typical composition of heavy metal in sludge.

Table 11. Concentrations of Heavy Metals in Sludge [Metcalf & Eddy, 2003]

Metal	Dry sludge, mg/kg	
	Range	Median
Arsenic	1.1-230	10
Cadmium	1-3410	10
Chromium	10-99000	500
Cobalt	11.3-2490	30
Copper	84-17000	800
Iron	1000-154000	17000
Lead	13-26000	500
Manganese	32-9870	260
Mercury	0.6-56	6
Molybdenum	0.1-214	4
Nickel	2-5300	80
Selenium	1.7-17.2	5
Tin	2.6-329	14
Zinc	101-49000	1700

Table 12 summarizes some chemical properties of dried sludge for three cities in Saudi Arabia. Oil and grease content was found to be between 0.64 and 1.27 per cent, with Dammam and Qateef containing the highest and lowest values, respectively. The reason for Dammam sludge containing the highest oil and grease content could be attributed to the industrial activities in the city, where the industrial wastewater produced at Dammam first industrial city is being discharged to the municipal sewer line. Total nitrogen (TKN) was found to range between 3.38 and 3.83 percent of total solids.

Table 12. Characteristics of dried sludge collected from three cities of the Kingdom

Parameters	Mean Values		
	Qateef	Dammam	Khobar
Oil and Grease %	0.6	0.6	1.1
Total alkanity (mg/l)	289	289	258
Ammonia (mg/l)	0.3	0.3	0.3
Nitrite (mg/l)	2.9	2.9	2.7
Nitrate (mg/l)	39	38.8	39.3
Organic Nitrogen %	3.2	3.2	2.9
TKN %	3.5	3.5	3.4
Arsenic (mg/l)	10	9.9	11.1
Boron (mg/l)	50	50	42.8
Cadmium (mg/l)	35	34.5	28.6
Chromium (mg/l)	72	71.6	72.9
Copper (mg/l)	145	145	194
Iron (mg/l)	982	982	2321
Mercury (mg/l)	2	2.387	8.4
Manganese (mg/l)	45	45	35.8

Parameters	Mean Values		
	Qateef	Dammam	Khobar
Nickle (mg/l)	36	36	46.8
Lead (mg/l)	39	39	131
Zinc (mg/l)	329	329	662

With respect to heavy metal concentrations in dried sludge samples collected from the three cities, Table 12 shows that arsenic concentration ranges between 9.86 and 12.28 ppm (mg/kg) of dried weight. Mercury content in dried sludge samples collected from the three cities was found to range between 2.39 ppm (in Qateef) and 8.43 ppm (in Khobar). However, Khobar sludge samples contain an average mercury concentration (8.43 ppm) that is higher than the median value (6 ppm).

3.3   Microbiological Characteristics

With respect to microbiological characteristics of municipal sludges, wastewaters generally contain four major types of pathogens: bacteria, protozoa, viruses, and helminths. The concentrations of these pathogenic organisms in wastewater depend on the health condition of the community. In times of epidemics, these concentrations would be high. Since pathogens come from relatively large volumes of wastewater, when ending up in the sludge, they, in general, become concentrated and very infectious. It is found through a parasitological analysis of helminths on 89 sludge samples, three sediments, and seven composts, that the average concentration of helminths was 130 eggs per 100 grams of dry matter. Sludge from all types of treatment (mesophilic anaerobic and aerobic digestion, composting, and liming) contained 10 or more viable eggs per 100 grams of dry matter. Antibiotic resistance of *Escherichia coli* in sludge and wastewater is affected by location but not the digestion process. The *E.coli* strains in digested municipal sludge from El Paso, Texas have more antibiotic resistance than those from any other site. Table 13 shows the levels of indicator pathogenic organisms in municipal sludges, while Table 14 shows some disease-producing protozoa and helminths in sludges. Table 15 shows survival times of various pathogens in soil and on plant surfaces. [10, 20, 21, 22, 23]

Table 13. Levels of Indicator and Pathogenic Organisms in Different Sludges (no. per gram of dry weight) [Kiely, 1997]

Sludge (untreated)	Total coliform	Faecal coliform	Faecal Streptococci	Salmonella species	Pseudomonas aeruginosa	Enteric viruses
Primary	10^6-10^8	10^6-10^7	10^6	4×10^2	3×10^3	0.002-
Secondary	10^7-10^8	10^7-10^9	10^6	9×10^2	1×10^4	0.004
Mixed	10^7-10^9	10^5-10^6	10^6	5×10^2	10^3-10^5	0.015-0.026 --

Table 14. Pathogenic Organisms in Wastewater and Sludge [U.S.EPA, 1992]

Organism	Disease/symptoms
Protozoa	
Cryptosporidium	Gastroenteritis
Entamoeba	Acute enteritis
histolytica	Giardiasis
Giardia lamblia	Diarrhea and dysentery
Balantidium coli	Toxoplasmosis
Toxoplasma gondii	
Helminths	Digestive and nutritional disturbances; abdominal pain, vomiting
Ascaris	May produce symptoms such as coughing, chest pain, and fever
lumbricoides	Abdominal pain, diarrhea, anemia, weight loss
Ascaris suum	Fever, abdominal discomfort, muscle aches, neurological symptoms
Trichuris trichiura	Nervousness, insomnia, anorexia, abdominal pain, digestive
Toxocara canis	disturbances
Taenia saginata	Nervousness, insomnia, anorexia, abdominal pain, digestive
Taenia solium	disturbances
Necatur americanus	Hookworm disease
Hymenolepis nana	Taeniasis

**Table 15. Survival Times of Various Pathogens in Soil
and on Plant Surfaces [U.S.EPA, 1992]**

Pathogen	SOIL		PLANTS	
	Absolute Maximum	Common Maximum	Absolute Maximum	Common Maximum
Bacteria	1 year	2 months	6 months	1 month
Viruses	6 months	3 months	2 months	1 month
Protozoan cysts	10 days	2 days	5 days	2 days
Helminth ova	7 years	2 years	5 months	1 month

4. EFFECT OF NATURAL DRYING ON SLUDGE CHARACTERISTICS

Sludge dewatering is a physical unit process used to remove as much water as possible from sludge to produce a highly concentrated cake. Dewatering differs from thickening, as the sludge should behave as a solid after it has been dewatered. Metcalf and Eddy (2003) reported that dewatering is performed for one or more of the following reasons:

1. The costs for trucking sludge to the ultimate disposal site become substantially lower when sludge volume is reduced by dewatering.
2. Dewatered sludge is generally easier to handle than thickened or liquid sludge.
3. Dewatering is required normally prior to incineration of the sludge to increase the energy content by removal of excess moisture.

4. Dewatering is required before composting to reduce the requirements for supplemental bulking agents or amendments.

5. In some cases, dewatering may be required to render the sludge odorless or nonputrescible.

6. Dewatering is required prior to landfilling to reduce leachate production at the landfill site.

A number of sludge dewatering techniques are currently in use. The selection of sludge-dewatering system depends on:

1. Characteristics of the sludge to be dewatered,
2. Available space, and
3. Moisture content requirements of the sludge cake for ultimate disposal.

When land is available and the sludge quantity is small, natural dewatering systems are most attractive. These include drying beds and drying lagoons. The mechanical dewatering systems are generally selected where land is not available. [24]

Drying Beds

Sludge drying beds are the oldest method of sludge dewatering and are still used extensively in small-to-medium sized plants to dewater sludge. They are relatively inexpensive and provide dry sludge cake. In the recent years, much advancement has been made to the conventional drying beds, and new systems are used on medium- and large-sized plants. Theses variations of the drying beds are

1. Conventional sand
2. Paved
3. Wire-wedge
4. Vacuum assisted

Typical sand beds consist of a layer of coarse sand 15-25 cm in depth and supported on a gravel bed (0.3-2.5 cm) that incorporates selected tiles or perforated pipe under-drain. Sludge is placed on the bed in 20-30 cm layers and allowed to dry. Sludge cake removal is manual by shoveling into wheelbarrows or trucks or a scraper or front-end loader. The under-drained liquid is returned to the plant. The drying period is 10-15 days, and the moisture content of the cake is 60-70 percent. Sludge loading rate is 100-300 kg dry solids per m^2 per year for uncovered beds. In Alexandria Metropolitan Area, solar energy is used for sludge drying. Solar air heats drying beds which saves about 55 percent of the bed area as compared to conventional drying beds, and the drying time is reduced from 18 to 8 days. Moreover, the solar air heated beds are less expensive than conventional open sand beds. The dewatering of aerobically digested sludge using drying beds study shows that larger portions of the moisture could be removed by drainage from thin sludge layers. When the initial sludge layer depth is 350 mm, only one third of the moisture can be removed. Moreover, the evaporation of thin or thick sludge layers is independent of sludge surface cracking, until sludge total solids reaches a value of 30% or greater. Above critical moisture content between 600 and 1100%, the rate of evaporation from sludge equals the rate of evaporation from free water surface. Below the critical moisture content, the rate of evaporation decreases rapidly. Temperature and moisture have significant main and interactive effects on the inactivation rate of *Escherichia coli* in biosolids. Now-a-days high-speed sludge drying beds are becoming popular. The main principle of high-speed sludge drying bed is the destruction of cake layer by aeration and/or mechanical stirring to maintain high dewatering rate. Study reported that high-speed sludge drying beds could treat higher sludge loading rate than the conventional drying beds. An assessment of sand drying bed shows that it has become less popular as a dewatering system because it is subjected to uncontrollable conditions such as temperature, rainfall and sludge drainage rate. There is a distinctive effect of solar drying process for a tilled and untilled biosolids in Nitrogen losses from sludge. Nitrogen losses vary depending on the solar drying season and tillage. Although not directly measured, the majority of nitrogen loss occurs through ammonia volatilization [24-32].

4.1 Effect of Drying on Physical and Chemical Characteristics

There is distinct effect of natural drying on physico-chemical properties of sludge. The discussion of the effects of solar drying in this section are summarized with the help of data collected from wastewater treatment plants in three cities of Saudi Arabia, namely Qatif, Dammam and Khobar. Different physical characteristics are presented individually to discuss the effect of drying [1].

Water Content

The initial water contents of the sludge were about 97-98, 95-99, and 96-99 percent for Qateef, Dammam, and Khobar, respectively. At the end of the drying period, the water content of sludge reached values of about 13-20, 29-42, and 45-48 percent. Based on the results, sludge water contents were reduced by about 78-87, 55-70, and 50-54 percent. This continuous decrease in sludge water content is attributed to the two main mechanisms, namely, infiltration and evaporation. Moreover, after four days of drying, a sharp decrease in sludge water content was noticed to take place in the wastewater treatment plants of the three cities. The trend of sludge water content with drying time is attributed to the types of water associated with sludge flocs, namely, unbound and bound waters. Unbound water, which comprises the largest water content, can be separated mechanically. Dewatering devices primarily remove free unbound water, some interstitial water can be removed as well, but it is likely that the major fraction of bound water is vicinal water that cannot be removed mechanically. Bound water, which contributes to the smallest water content, has the sternest physical-chemical bonding to the particles and can only be removed thermally. Based on that, the results obtained after four days of drying could be attributed to a combined effect of infiltration and evaporation.

Total Solids Content

The initial values of sludge total solids were found to be 2.2-3.1, 1.5-4.7, and 0.85-3.8 percent for sludge samples collected from Qateef, Dammam, and Khobar wastewater treatment plants, respectively. At the end of drying periods, total solids concentrations were found to reach values of about 80-87, 58-71, and 45-52 percent. The trend of sludge total solids is attributed to the same reasons given on water content. The values of solid content in the dewatered sludge samples collected from the three cities is much greater than those reported in literature, which could be attributed to the weather conditions of the Kingdom of Saudi Arabia.

Volatile Solids Content

The concentration of the total volatile solids in sludge samples in Qateef increased from 1.4 to 7.4 and from 1.9 to 39.8 percent, respectively, during the first four days of drying. A relatively sharp increase in the concentration of total volatile solids was noticed between the fourth and the eighth days, which could be attributed to the same reasons given on water content and total solids. Similar trends of increase in the concentration of volatile solids were also noticed in samples collected from Dammam and Khobar wastewater treatment plants. Total volatile solids were found to range between 51 and 65 percent of the total solids in sludge samples collected from the three cities. Moreover, the total volatile solids content, as a percent of total solids, was found to decrease over the drying period. In Qateef, the total volatile solids decreased from 62 to 53 percent, during the first session, while in Dammam and Khobar, it decreased from 64 to 54 percent and from 66 to 51 percent, respectively. This could be attributed to weathering effects and biodegradation activities.

Sludge pH and Total Alkalinity

In Qateef and Khobar, the sludge pH was seen to decrease during the first days of drying, after which, the pH was increasing. On the other hand, the pH of sludge samples collected from Dammam was behaving with an opposite trend. In general, the variations in sludge pH, for the three cities, were not significant and that the pH was shifting towards neutrality. The reasons for pH fluctuations could be attributed to several reasons such as evaporation and other weathering effects. Such effects will affect the ionic strength of the collected sludge samples, which, in turn, will affect the pH values. Total alkalinity of sludge samples collected from Qateef, Dammam and Khobar was decreasing with respect to drying period. This can be attributed to infiltration and evaporation of water.

Oil and Grease Content

Concentrations of oil and grease in sludge samples collected from Qateef, Dammam, and Khobar were found to decrease with respect to drying time, which is attributed to infiltration and weathering effects. Moreover, the rate of decrease in oil and grease concentration is high at the start of the drying period and tends to slow down over the course of drying. This is because of the fact that, at the start of the drying period, the water infiltration rate will be relatively high and will decrease with respect to time. In sludge samples collected from Qateef during the first session, the oil and grease concentration was seen to decrease from 4.2 to 0.21 percent in the first eight days of drying, and by the end of the drying period, it reached 0.08 percent. In Dammam, which was found to contain the highest concentration among all, oil and grease decreased from 8.66 to 1.41 during the first four days of drying and by the end of the drying period it reached a value of 0.68 percent. Similarly, Khobar sludge was found to initially contain 6.37 percent oil and grease, which decreased to 1.06 and 0.92 after four and seven days of drying, respectively. The presence of high concentration of oil and grease in sludge samples collected from Dammam could be attributed to the industrial activities, in the first industrial city, which discharge their wastewater into the municipal sewer line.

Ammonia-Nitrogen Content

The results show that initially the raw sludge has a very high concentration of ammonia-nitrogen ranging from about 1.5% at Dammam to 3.9% at Khobar, based on dry weight. This high concentration may be attributed to the release of ammonia-nitrogen from the death and decay of microorganisms during the thickening process before the sludge is brought to the drying beds. It is clear that as the sludge dries off, the concentration of ammonia-nitrogen decreases rapidly. However, for the city of Qateef, the concentration was reduced by about 47% after two days of drying and continued decreasing until the end of the drying period (15 days) where the final reduction was about 70% of the initial value. The reduction of ammonia-nitrogen concentration is mainly due to water infiltration into the sand layer underneath the sludge layer. Another reason for ammonia-nitrogen decrease could be attributed to oxidation of NH4-N to NO3 due to exposure of sludge to air in the drying bed.

Nitrite Content

Nitrite concentration in sludge samples collected from the three cities of Qateef, Dammam, and Khobar is relatively low ranging from 18.6 ppm at Khobar to 15.3 ppm dry weight at Dammam. This trend is expected, since nitrite is an intermediate compound for the nitrification process during the secondary biological treatment in which it is continually converted to nitrate. The results show that as the sludge dries off, the concentration of nitrite reduces in the first four days of drying and then stabilizes until the end of the drying period. The nitrite reduction in the first few days of drying is mainly due to water infiltration into the sand layer underneath the sludge layer. Volatilization losses of nitrite could be another reason for its continuous decrease.

Nitrate Content

Nitrate concentration in sludge samples ranged from about 151 ppm at Khobar to 132 ppm of dry weight at Dammam. The main source of nitrate in the sludge is expected to be the nitrification process in the secondary biological treatment and in the thickening tank prior to the drying beds. Therefore, nitrate concentrations are expected to be much higher than those of nitrite. The results show that as the sludge dries off, the concentration of nitrate reduces in the same way as with nitrite.

Organic Nitrogen and TKN Content

The relationship between drying period and both of organic nitrogen and TKN in sludge samples demonstrate a consistent trend when compared with other nitrogen compounds. The concentrations of these parameters were highest at the city of Dammam followed by Khobar, while Qateef samples contained the lowest concentrations among all. As sludge dries, the concentrations of TKN and organic nitrogen were found to decrease rapidly in the first two

Table 16. Seasonal Effect on Total Solids in Dried Sludge

City	Parameters	Session	Season			
			Spring	Summer	Fall	Winter
Qateef	Total Solids %	1	80.42	85.7	89.92	41.7
		2	87.4	88.2	--	47.4
Dammam	Total Solids %	1	71.2	55.3	49.1	27.6
		2	57.7	59.4	--	42.2
Khobar	Total Solids %	1	54.8	50.9	51.1	34.6
		2	51.6	55.2	--	31

days of the drying period. This is attributed to the high rate of water infiltration in the first days of the drying period. By the end of the drying period, TKN reached values of 0.965%, 1.25%, and 0.658% for sludge samples collected from Qateef, Dammam, and Khobar respectively.

Heavy Metals Contents

The concentration of heavy metals, namely arsenic, boron, cadmium, chromium, copper, iron, lead, mercury, manganese, nickel, and zinc, in sludge samples collected from the cities demonstrate that the concentrations of heavy metal in dewatered sludge collected from the three cities are within or even less than those reported in literature. Concentrations of heavy metals of sludge in drying beds of the three cities were decreasing with respect to drying period, which could be attributed to water infiltration.

The seasonal effect on physical characteristics is prominent from collected data which is represented in Table 16. With respect to total solids in dried sludge, and beside the effect of the physical conditions of the drying beds, the table clearly demonstrates the seasonal effect on the contents of total solids. As an example, Qateef sludge was found to contain the lowest percentage of total solids during the winter season, while the highest percentage was obtained during summer and fall seasons.

4.2 Effect of Drying on Microbiological Characteristics

With respect to the survival of pathogens in drying beds it is reported that drying beds retain 80 per cent of solids and 100 per cent of helminth eggs. The treatment of sludge in drying beds is efficient in eliminating pathogenic micro-organisms such as *fecal coliforms*, *protozoan cysts* and helminth eggs. Survival of nonembryonated eggs of *A. suum* depends on different climatic-geographical conditions [33-35].

Regarding the microbiological quality of dried sludge samples collected from the three cities, Table 8 shows that *fecal coliform* density ranged between 4.97×10^5 and 2.11×10^6 MPN per gram of dried sludge, which is less than that reported by Kiely, 1997, for secondary untreated sludge (10^7 to 10^8). Similarly, *salmonella* species were found to range between 22 MPN per gram of dried sludge (in Qateef samples) to 127 MPN per gram of dried sludge (in Khobar samples). According to the U.S.EPA regulations, class A sludge should have either *fecal coliform* density under 1000 MPN per gram of dry solids or *salmonella* species density

Table 17 General characteristics of dried sludge collected from three cities of the Kingdom

Parameters	Mean Values		
	Qateef	Dammam	Khobar
Fecal Coliform (MPN / g)	4.87E+5	2.11E+06	4.97E+05
Streptococcus (Colony / g)	3.87E+03	2.99E+03	1.37E+03
Shigella (Colony / g)	23	157	177
Salmonella (MPN / g)	22	107	127
Clostridium (Colony / g)	300	3900	3900
Ascaris (Parasite / g)	0	0	0
Trichuris (Parasite / g)	0	0	0
Hymenolepis nana (Parasite / g)	0	0	0
Entaamoeba histolytica (Parasite / g)	0	0	0

under 3 MPN per 4 grams of dried solids, while class B sludge should have *fecal coliform* density under 2×10^6 MPN per gram of dry solids (EPA, 1993). From the results on *salmonella*, it is clear that dried sludge from the three cities can not be considered as class A sludge and, therefore, can not be applied to lawn or home garden. Results on parasites in dried sludge collected from the three cities showed that all parasites under investigation were less than 1 per gram of dried solids [36].

Regarding the seasonal effect on total coliform in dried sludge samples collected from the three cities, Table 18 clearly shows that as the ambient temperature decreases the density of total coliform increases. In all sludge samples, the winter samples were found to contain the highest densities of total coliform. As have been shown before, the increase in ambient temperatures will result in increasing the rate of evaporation, which in turn will increase the content of total solids in sludge samples. Moreover, the increase in ambient temperatures will also result in increasing the rate of disinfection or inactivation of microorganisms in sludge samples.

Table 18. Seasonal Effect on Total Solids and Total Coliform in Dried Sludge

City	Parameters	Session	Season			
			Spring	Summer	Fall	Winter
Qateef	*Total Coliform* / gram	1	2.40×10^6	1.7×10^5	1.1×10^5	5.4×10^7
		2	4×10^5	1.1×10^7	--	5.2×10^7
Dammam	*Total Coliform* / gram	1	1.30×10^7	2×10^5	$1.1 * 10^5$	7.4×10^7
		2	1.1×10^6	1.1×10^5	--	7.1×10^7
Khobar	*Total Coliform* / gram	1	3.00×106	6×105	$3 * 105$	5×107
		2	1.1×106	4×104	--	4.8×107

4.3 Effect of Sludge Initial Depth on the Characteristics of Dried Sludge

Effect of sludge initial depth on the fate of pathogen in drying bed during the drying period is a vital issue. In this section, attempt is taken to investigate relationship between the decreases of microorganism with time at different sludge thickness.

For this purpose, different pathogens were considered namely, total *coliform, streptococci, shigella, salmonella,* and *clostridium* and helminthic and protozoan pathogens, namely, *Ascaris lumbricoides, Enterobious vermicularis, Ancylostoma doudenale, Trichuris trichura, Hymenolepis nana*, and *Entamoeba histolytica*

A typical effect of initial sludge depth on the survival of total *coliform* in sludge samples is presented in Figure 1. Generally, the figure shows that total *coliform* count was decreasing, in sludge samples, with respect to drying period. Moreover, the figure demonstrates a trend where total *coliform* was found to survive longer in higher sludge depths. The reason for longer survival of *coliform* bacteria in thicker sludge could be due to penetration of solar heat and radiation, which will be reduced in thicker sludge depths. Regarding the effect of initial sludge depth on the survival of *streptococci*, Figure 2 shows that as the initial sludge depth was increased, the *streptococci* would survive longer. By the end of the 30-day drying period, the *streptococci* count reached values of 25, 37, 49, 59, 71, and 90 organisms per gram dry weight, for initial depths of 10, 15, 20, 25, 30, and 35 cm, respectively.

The effect of initial sludge depth on the survival of *shigella* in sludge samples is presented in Figure 3. Generally, the figure shows that *shigella* count was decreasing, in sludge samples, with respect to drying period. Moreover, the results demonstrate a trend where *shigella* was found to survive longer in higher sludge depths. The reason for longer survival of *shigella* bacteria in thicker sludge could be due to the same reasons given above.

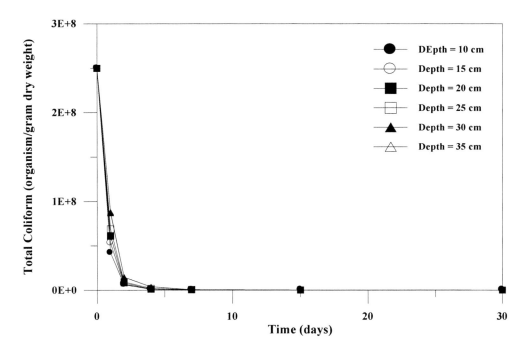

Figure 1. Coliform Die-off at Different Initial Sludge Depths.

The results show that sludge samples contain *shigella* densities ranging between 50 and 200 organisms per gram dry weight, after 30 days of drying period. Regarding the effect of initial sludge depth on the survival of *salmonella*, Figure 4 shows that as the initial sludge depth was increased, the *salmonella* would survive longer. At the end of the 30-day drying period, the *salmonella* count reached values of 15, 25, 35, 35, 50, and 60 organisms per gram dry weight, for initial depths of 10, 15, 20, 25, 30, and 35 cm, respectively.

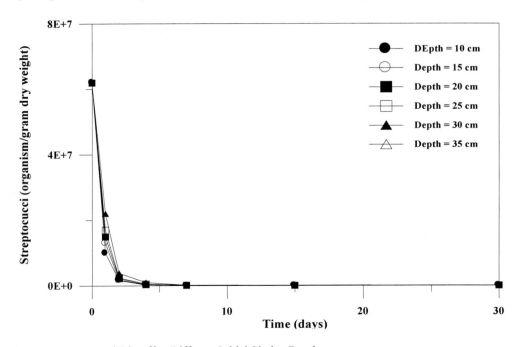

Figure 2. Streptococci Die-off at Different Initial Sludge Depths.

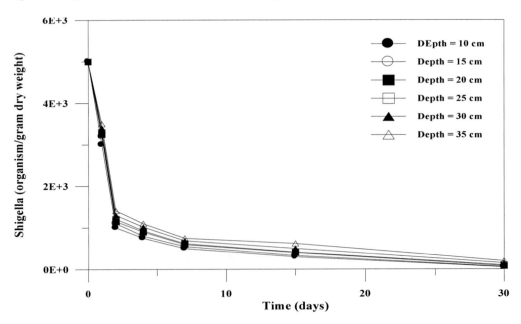

Figure 3. Shigella Die-off at Different Initial Sludge Depths.

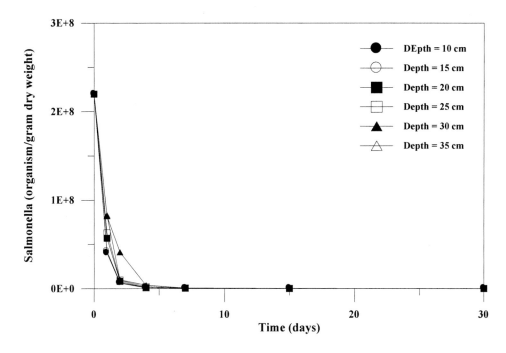

Figure 4. Salmonella Die-off at Different Initial Sludge Depths.

The effect of initial sludge depth on the survival of *clostridium* in sludge samples is presented in Figure 5. Generally, the figure shows that *clostridium* count was decreasing, in sludge samples, with respect to drying period. Moreover, the results demonstrate a trend where *clostridium* was found to survive longer in higher sludge depths. The reason for longer survival of *clostridium* bacteria in thicker sludge could be due to penetration of solar heat and radiation, which will be reduced in thicker sludges. The results show that sludge samples contain *clostridium* densities ranging between 7 and 45 organisms per gram dry weight, after 30 days of drying period.

Protozoan and helminthic pathogens are of great concern due to their probable effects on the public health. The following discussion will focus on the effect on helminthic and protozoan pathogens such as *Ascaris lumbricoides, Enterobious vermicularis, Ancylostoma doudenale, Trichuris trichura, Hymenolepis nana*, and *Entamoeba histolytica*. Regarding the effect of initial sludge depth on the survival of *Ascaris lumbricoides*, Figure 6 shows that those parasites were decreasing with respect to drying period in sludge samples. The figure also shows that sludge samples collected from higher initial depths were containing higher number of *Ascaris lumbricoides*. After 2 days of drying, the number of *Ascaris lumbricoides* were 3, 4, 6, 7, 7, and 10 in sludge samples collected from drying beds with initial sludge depth of 10, 15, 20, 25, 30, and 35 cm, respectively. This can be attributed to the same reasons given on total *coliform*. The figure also demonstrates that the maximum rate of decrease in *Ascaris lumbricoides* was taking place during the first two days of drying. This could be attributed to the fact that water infiltration is maximum at the start of the drying time, which will result in washing out those parasites from sludge solids. Moreover, the figure shows that all sludge samples were free from *Ascaris lumbricoides*, after 30 days of drying.

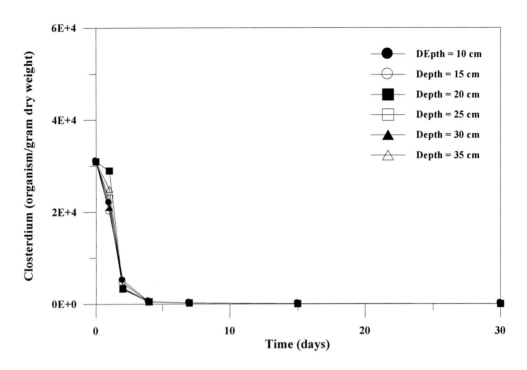

Figure 5. Closterdium Die-off at Different Initial Sludge Depths.

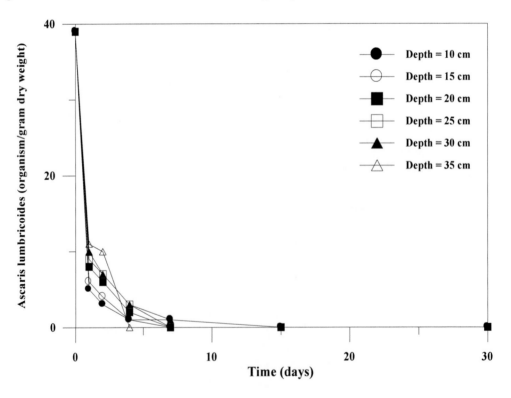

Figure 6. Ascaris Die-off at Different Initial Sludge Depths.

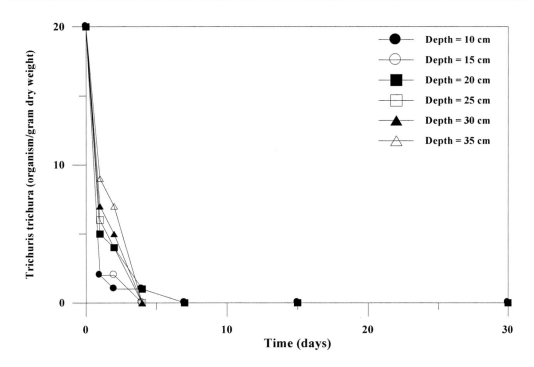

Figure 7. Trichuris Die-off at Different Initial Sludge Depths.

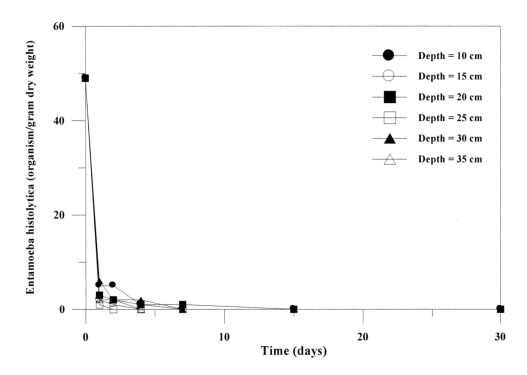

Figure 8. Histoclytica Die-off at Different Initial Sludge Depths.

Figure 7 shows the effect of initial sludge depth on the content of *Trichuris trichura* in sludge samples. The figure clearly demonstrates that as initial sludge depth was increased, the count of surviving *Trichuris trichura* in sludge samples increases, during the first days of drying.

Sludge samples collected after two days of drying were found to contain *Trichuris trichura* counts of 1, 2, 4, 4, 5, and 7 when collected from drying beds with initial sludge depth of 10, 15, 20, 25, 30, and 35 cm, respectively.   This is attributed to the same reasons given on *Ascaris lumbricoides.* The figure clearly demonstrates that all sludge samples were free from *Trichuris trichura*, after 30 days of drying.

The effect of initial sludge depth on the contents of the protozoan parasite *Entamoeba histolytica* in sludge samples is shown in Figure 8. As in the case of *Ascaris lumbricoides* and *Trichuris trichura*, the count of surviving *Entamoeba histolytica* was found to be affected by the initial sludge depth. After two days of drying, the *Entamoeba histolytica* counts were 1, 2, 2, 3, 5, and 8 for samples collected from drying beds with initial sludge depth of 10, 15, 20, 25, 30, and 35 cm, respectively. After 30 days of drying, the *Entamoeba histolytica* parasites were not detected in all sludge samples.

After the through investigation of the results presented above, it is understood that there is a definite relationship between the sludge initial depth and fate of microorganism in drying bed. This relationship can be mathematically represented as below:

$$\ln(y) = A - B \times \ln(time)$$

Where   y = microorganism / gram of dry sludge
A and B = constants depend on initial sludge depth

Generally, the values of constants A and B were found to increase with the increase in the initial sludge depth in the drying beds. In the case of the total coliform, the constants A and B were found to increase in the initial stage of increasing the sludge depth, but after that the values were almost constant. In order to come up with one approximated values of constants A and B, average values were found to be 8.05 and 1.79, respectively. Based on these results, the general mathematical representation of the process is the following form:

$$\ln(y) = 8.05 - 1.79 \times \ln(time)$$

However, it is better to represent the value of A and B on the basis of individual species rather than overall average values of all species.

CONCLUSION

It is clear that sludge is an unavoidable part of wastewater treatment. Measures have to be implemented in order to deal with sludge complications. Nowadays, the area of sludge disposal is a growing concern due to the substantial increase in sludge production. This increase in sludge production is due to the increasing awareness concerning environmental pollution and the consequent increasingly stringent effluent discharge standards. Hence, the use of secondary and, in some locations, tertiary treatment of wastewater was found to be a necessary practice in order to comply with those standards. As the quantity of sludge increases, so does the number of regulations attempting to control this material, as well as other potentially toxic or hazardous wastes. Sludge could pose health hazards if not handled properly, due to its harmful characteristics. On the other hand, sludge can be usefully utilized as a fertilizer and soil conditioner, if properly treated.

REFERENCES

[1]   Al-Malack, M. H., Bukhari, A. A. and Abuzaid, N. S. (2007), Physico-Chemical Characteristics of Municipal Sludge Produced at Three Major Cities of the Eastern Province of Saudi Arabia, J. King Saud Univ. Vol 0, Eng Sci (2), pp (A.H. 1427/1428)

[2]   Metcalf and Eddy. 2003. Wastewater Engineering: Treatment, Disposal, and Reuse, 4th Editionn. New York: McGraw-Hill.

[3]   Steel, E.W. 1960. Water Supply and Sewerage, 4th Edition, Tokyo, Japan: McGraw-Hill.

[4]   Schmidtke, N.W. 1981. "Sludge Generation, Handling and Disposal at Phosphorus Control Facilities in Ontario, in Characterization, Treatment and Use of Sewage Sludge", in Proceedings of Second European Symposium, Vienna: Commission of the European Communities, pp. 190–225.

[5]   Koch, C.M.; Lee, J.S.; Bratby, J.R. and Barber, D.B. 1997. "A Critical Evaluation of Procedure for Estimating Biosolids Production", in Proceedings of the Conference on Water Residuals Biosolids Management: Approaching the Year 2000, Philadelphia: Water Environment Federation, 1997.

[6]   Lishman, L.A.; Legge, R.L. and Farquhar, G.J. 2000. "Temperature Effects on Wastewater Treatment under Aerobic and Anoxic Conditions", Water Research, 34(8), pp. 2263–2276.

[7]   Jardin, N. and Popel, J. 1997. "Waste Activated Sludge Production of the Enhanced Biological Phosphorus Removal Process", Water Environ. Res., 69, pp. 375.

[8]   Davis, M.L. and Cornwell, D.A. 1998. Environmental Engineering, 3rd edn, Boston, Mass: McGraw-Hil.

[9]   Vesilind, P.A.; Hartman, G.C. and Skene, E.T. 1986. Sludge Management and Disposal for the Practicing Engineer, Chelsea, MI, U.S.A., Lewis Publishers.

[10]   Machno, P.S. 1997. "Biosolids Quantity and Quality: A Comparison of Primary and Secondary Treatment", in Proceedings of Sludge 2000, Cambridge, 1992.

[11]   Kiely, G. 1997. Environmental Engineering. London: McGraw-Hill.

[12]   Sakai, Y.; Fukase, T.; Yasui, H. and Shibata, M. 1997. "Activated Sludge Process without Excess Sludge Production", Wat. Sci. Technol., 36, p.163.

[13]   Droste, R.L. 1997. Theory and Practice of Water and Wastewater Treatment, 1st edn, New York: John Wiley & Sons, Inc.

[14] Xiang, L.; Chan, L.C. and Wong, J.W.C. 2000. "Removal of Heavy Metals from Anaerobically Digested Sewage Sludge by Isolated Indigenous Iron-Oxidizing Bacteria", Chemosphere, 41(1), pp. 283–287.

[15] Zorpas, A.A.; Constantinides, T.; Vlyssides, A.G. Haralambous, I. and Loizidou, M. 2000. "Heavy Metal Uptake by Natural Zeolite and Metals Partitioning in Sewage Sludge Compost", Bioresource Technology, 72(2), pp. 113–119.

[16] Alonso, J.I.G.; Camblor, M.G.; Bayon, M.M. and Marchantegayon, J.M.S.A. 1997. "Different Quantification Approaches for the Analysis of Biological and Environmental Samples Using Inductively Coupled Plasma Mass Spectrometry", J. Mass Spectrom., 32, p. 556.

[17] Maaskant, J.F.N.; Boekholt, A.H.; Jenks, P.J. and Rucinski, R.D. 1998. "An International Inter-laboratory Study for the Production of a Sewage Sludge Certified Reference Material for Routine Use in Inorganic Quality Control", Fresenius J. Anal. Chem., 36, p. 406.

[18] Laschka, D. and Nachtwey, M. 1992. "Platinum in Municipal Sewage Treatment Plants", Chemosphere, 34(1997), p. 1803.

[19] Steiger, M. T. 1992. "Fate of Chlorinated Organic Compounds During Thermal Conversion of Sewage Sludge", Proceedings of Water Quality International 92, Washington, D.C., U.S.A, pp. 2261–2264.

[20] Al-Malack, M. H., Bukhari, A. A., Abuzaid, N. S. and Essa M.H (2002). "Characterization, Utilization and Disposal of Municipal Sludge: The State-of-the-art", The Arabian Journal for Science and Engineering, Vol 27, Number 1B.

[21] Gaspard, P.; Wiart, J. and Schwartzbrod, J. 1997. "Parasitological Contamination of Urban Sludge Used for Agricultural Purposes", Waste Manage. Res., 14, p. 429.

[22] Pillai, S.D.; Widmer, K.W.; Maciorowski, K.G. and Ricke, S.C. 1997. "Antibiotic Resistance Profiles of Escherichia coli Isolated from Rural and Urban Environments", J. Environ. Health, Part A., 32, p. 1665.

[23] United States Environmental Protection Agency. 1992. Control of Pathogens and Vector Attraction in Sewage Sludge, Washington, DC: U.S. EPA.

[24] Al-Malack, M. H. 2010. "Effect of Sludge Initial Depth on the Fate of Pathogens in Sand Drying Beds in the Eastern Province of Saudi Arabia", Int. J. Environ. Res., 4(4): 825-836.

[25] Al-Muzaini, S. 2003. "Performance of Sand Drying beds for sludge dewatering", Arabian Journal for Science and Engineering 28 (2B), 161-169.

[26] Hossam, A. and Saad, S. (1990), Solar Energy for Sludge Drying in Alexandria Metropolitan Area, Wat. Sci. Technol. 22 (12), 193-204

[27] Marklund, S. (1990). Dewatering of Sludge by Natural Methods, Wat. Sci. Technol. 22 (3-4), 239-246.

[28] Nishimura, O., Gotoh, K. and Sato, A. (1994). Gravity Dewatering Mechanism – Application to High Speed Sludge Drying Beds, Proceedings of the Japan Society of Civil Engineers, no. 497 (2-2), 119-126

[29] Marklund, S. (1993). Dewatering of Drying Beds – Combined Biological-Chemical Sludge Behaviour.Wat. Sci. Technol., 28 (10), 65-72.

[30] Mehrdadi, N., Joshi, S. G., Nasrabadi, T. and Hoveidi, H. (2007). Aplication of solar energy for drying of sludge from pharmaceutical industrial waste water and probable reuse, Int. J. Environ, Res., 1 (1), 42-48

[31] O'Shaughnessy, S. A, Kim, M. Y. and Choi, C. Y. (2008), Mathematical model to predict pathogen die-off in biosolids, Journal of Residuals Science & Technology, 5 (2), 87-93

[32] Yamaoka, M. and Hata, K. (2003), Improvements in drying beds for non-concentrated sludge. Advances in Environmental Research, 7 (3), 721-725.

[33] Cofie, O. O., Agbottah, S., Strauss, M., Esseku, H., Montangero, A., Awuah, E. and Kone, D. (2006). Solidliquid separation of faecl sludge using drying beds in Ghana: Implications for nutrient recycling in urban agriculture. Water Research, 40 (1), 75-82.

[34] Fars, S., Oufdou, K., Nejmedddine, A., Hassani, L., Melloul, A., Bousselhaj, K., Amahmid, O., Bouhoum, K., Lakmichi, H. and Mezrioui, N., (2005). Antibiotic resistance and survival of fecal coliforms in activated sludge system in a semiarid region (Beni Mellal, Morocco).World J. Microb. Biot., 21 (4), 493-500.

[35] Plachy, P. and Juris, P. (1995). Survival of the Model Helminth Ascaris-Suum Eggs in the Sludge Drying Beds of Sewage-Treatment Plants, Veterinarni Medicina, 40 (1), 23-27.

[36] Al-Malack, M. H., Bukhari, A. A., Abuzaid, N. S. and Essa M.H (2002), Fate of Pathogens in Sludge Sand Drying Beds at Qateef, Khobar and Dammam: A Case Study, Int. J. Environ. Res.1 (1): 19-27, Winter 2007.

In: Sewage Sludge Management
Editors: A. A. Zorpas and V. J. Inglezakis

ISBN: 978-1-61324-393-0
© 2012 Nova Science Publishers, Inc.

Chapter 3

SEWAGE SLUDGE PROPERTIES

Vassilis J. Inglezakis[*]

SC European Focus Consulting srl, Bacau, Romania

ABSTRACT

Wastewater treatment necessarily produces two end products: effluent and sewage sludge. In the present Chapter the several types of sewage sludges and the relevant psysicochemical properties are presented. It is important to mention that is generally difficult to identify which type of sludge was studied in the related literature and the relevant data should be used as general guide rather than exact figures and facts.

1. INTRODUCTION

1.1 General

In general, water quality is influenced by both (direct) point source, as for example industrial discharge and diffuse pollution, caused by a variety of activities that have no specific point of discharge, as for example agriculture, atmospheric deposition, rural dwellings etc. Inputs of pollutants to the urban wastewater system occur from three generic sources: domestic wastewater (point source), light industrial/commercial wastewater (point source) and urban runoff (diffuse source). While large point sources are easily quantifiable as they are connected to urban wastewater system the contribution from small point sources, such as households and small businesses, is much more difficult to identify and quantify. The pollutants of interest can be divided into two main groups [1]:

---

[*] E-mail: www.efcon-group.ro, inglezakis@efcon-group.ro.

· potentially toxic elements (PTEs) including cadmium (Cd), chromium (Cr III and CrVI), copper (Cu), mercury (Hg), nickel (Ni), lead (Pb) and zinc (Zn)
· organic pollutants including PAHs, PCBs, DEHP, LAS, NPE, dioxins (PCDD) and furans (PCDF).

Potentially toxic element contamination of urban wastewater and sewage sludge is usually attributed to discharges from major commercial activities. Faeces contribute 60 – 70 % of the load of Cd, Zn, Cu and Ni in domestic wastewater and more than 20% of the input of these elements in mixed wastewater from domestic and industrial premises. Faecal matter typically contains 250 mg Zn/kg, 70 mg Cu/kg, 5 mg Ni/kg, 2 mg Cd/kg1 and 10 mg Pb/kg (DS). The other principal sources of metals in domestic wastewater are body care products, pharmaceuticals, cleaning products and liquid wastes. Furthermore, plumbing is the main source of Cu in hard water areas while high Pb inputs have been reported in districts with extensive networks of Pb pipe work [1].

In general, urban runoff is not a major contributor of potentially toxic elements but inputs of the main persistent organic pollutants, as for example PAHs and PCBs, are principally from atmospheric deposition onto paved surfaces and runoff as well as combustion from traffic and commercial sources. These compounds are hydrophobic and thus they are easily removed during urban wastewater treatment but they are transferred to the sludge generated. Detergent residues, surfactants, plasticising agents and polyacrylamide compounds are of the most abundant organic contaminants present in urban waste water and/or sewage sludge while a number of other organic compounds are potentially resistant to wastewater and sewage sludge treatment, as for example the brominated diphenyl ethers (PBDEs) and chlorinated paraffins. Finally, the degree of removal and biodegradation of pharmaceutical compounds during waste water treatment varies considerably, although many common analgesic drugs rapidly biodegrade [1].

1.2 Wastewater and Sludge Treatment

Each person discharging human waste to a wastewater treatment system produces approximately 21 kilograms of sewage sludge each year [14]. Total solids in wastewater comprise dissolved solids, suspended solids (settleable and colloidal) and floatable solids (scum). The sources and types of solids generated in a treatment plant with primary, biological, and chemical treatment facilities are illustrated in Figure 1 [2]. In general, wastewater sludge can be classified as primary, secondary (biological) and chemical (tertiary). Raw sludge is called the sludge which is untreated biologically or chemically for volatile solids or pathogen reduction while when treated, the resulting biosolids can be classified by the treatment, such as aerobically or anaerobically digested, composted etc. The treated sludge can come from a single type or a mixture of several sludges [2].

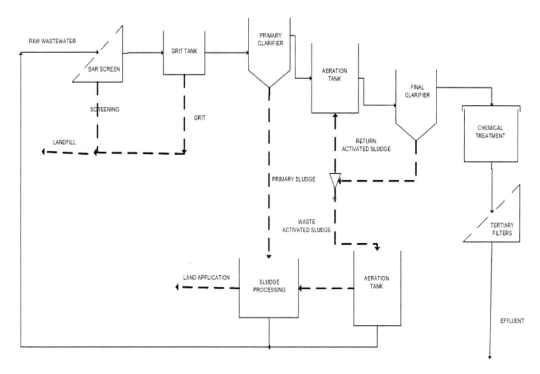

Figure 1. The sources and types of solids generated in a treatment plants [2]. Dotted lines are for solids and smooth lines for water flows.

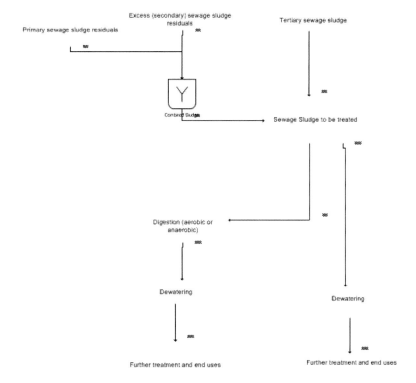

Figure 2. Sewage sludge treatment alternatives (general scheme) [14].

Most wastewater treatment plants use the physical process of primary settling to remove the settleable solids from raw wastewater, generating the *primary sludge*. In a typical plant with primary settling and a conventional activated sludge secondary treatment process, the dry weight of the primary sludge solids is about 50% of that for the total sludge solids. The total solids concentration in raw primary sludge can vary between 2 and 7% [2]. *Secondary sludge*, also known as *biological sludge*, is produced by biological treatment processes such as activated sludge and membrane bioreactors and generally contains solids concentrations of 0.4-4%, in dry solids weight. Finally, *chemical sludge* is generated when chemicals are used in wastewater treatment, especially in industrial wastewater treatment, to precipitate and remove several substances.

Table 1. Sludge treatment methods [12]

Unit operation	Sub-types	Function
Preliminary operations	Grinding	Size reduction
	Degritting	Grit removal
	Blending	Blending
Thickening	Gravity	Volume reduction
	Flotation	Volume reduction
	Centrifugation	Volume reduction
Stabilization	Chlorine oxidation	Stabilization
	Lime stabilization	Stabilization
	Heat treatment	Stabilization
	Anaerobic digestion	Stabilization and mass reduction
	Aerobic digestion	Stabilization and mass reduction
Conditioning	Chemical	Conditioning
	Elutriation	Leaching
	Heat treatment	Conditioning
Disinfection	-	Disinfection
Dewatering	Vacuum filter	Volume reduction
	Fliter press	Volume reduction
	Horizontal belt filter	Volume reduction
	Centrifuge	Volume reduction
	Drying bed	Volume reduction
	Lagoon	Volume reduction
Drying	Flash dryer	Weight and volume reduction
	Spray dryer	Weight and volume reduction
	Rotary dryer	Weight and volume reduction
	Multiple-hearth dryer	Weight and volume reduction
	Oil emersion dehydration	Weight and volume reduction
Composting	Sludge-only	Product recovery and volume reduction
	Co-composting with solid wastes	Product recovery and volume reduction

Unit operation	Sub-types	Function
Thermal reduction	Multiple-hearth incineration	Volume reduction, resource recovery
	Fluidized bed incineration	Volume reduction
	Flash combustion	Volume reduction
	Co-incineration with solid wastes	Volume reduction
	Co-pyrolysis with solid wastes	Volume reduction, resource recovery
	Wet air oxidation	Volume reduction

PHYSICOCHEMICAL PROPERTIES OF SLUDGE

2.1 General

Fresh primary sludge is a gray or light brown suspension with solids of different sizes and composition and due to the high organic content it decays quickly and becomes septic, which can be identified by its change to a dark gray or black color and an objectionable sour odor (Table 2). The quantity of raw primary sludge can be approximately 0.4 to 0.5% by volume of the plant influent flow, or approximately 1.1 m^3 per 1000 people. Activated sludge contains mostly bacterial cells that are viscous and difficult to dewater. The sludge is light gray or dark brown in color. Suspended solids concentration is 0.4 to 1.5%. Raw primary sludge particle size distribution is: greater than 7 mm (5 to 20%), 1 to 7 mm (9 to 33%), and smaller than 1 mm (50 to 88%), of which about 45% is less than 0.2 mm. In activated sludge, the approximate distribution is: 90% below 0.2 mm, 8% between 0.2 and 1 mm, 1.6% between 1 and 3 mm, and 0.4% over 3 mm. The density of primary sludge is 1.0 to 1.03 g/cm^3, and the density of activated sludge is about 1.0 g/cm^3. The density of dry sludge solids is 1.2 to 1.4 g/cm^3 [2].

Table 2. Sludge qualitative characteristics [12]

Solids or sludge	Description
Screenings	Include all types of organic and inorganic materials large enough to be removed on bar racks.
Grit	Usually made up of the heavier inorganic solids that settle with relatively high velocities and may also contain significant amounts of fats and grease.
Scum	Consists of floatable materials skimmed from the surface of primary and secondary settling tanks.
Primary sludge	Sludge from primary sedimentation tanks is usually gray and slimy and in most cases has an extremely offensive odor.
Chemical-precipitation sludge	Sludge from chemical precipitation tanks is usually dark in color, though its surface maybe red if it contains much iron. While is somewhat slimy, the hydrate of iron or aluminum in it makes it gelatinous.

Table 2. (Continued)

Solids or sludge	Description
Activated sludge	In general has a brown flocculant appearance and if in good condition has an inoffensive characteristic odor.
Trickling-filter sludge	Trickling-filter humus is brownish, flocculant and relatively inoffensive when fresh. It is readily digested.
Digested sludge (aerobic)	Is brown to dark brown and has a flocculant appearance. The odour is not offensive and is often characterized as musty.
Digested sludge (anaerobic)	Is dark brown to black and contains an exceptionally large quantity of gas.
Septage	Sludge from septic tanks is black and is generally offensive in odour because of the hydrogen sulfide and other gases it gives off.

Table 3. Wastewater sludge characteristics in Europe and Russia

Property	Raw primary sludge	Digested primary sludge	Unthickened activated sludge	Digested mixture of primary and thickened activated sludge
Total dry solids (TS, %)	4.5	6	0.5	3
Volatile solids (% of TS)	70	50	75	60
Grease and fats (%)	18	11	6	5
Protein (% of TS)	25	18	37	22
Ammonia nitrogen (% of TS)	3	2	5	3.5
Phosphoric acid (% of TS)	1.4	2	4	3
Potash (% of TS)	0.5	0.4	0.4	0.4
pH	6	7	7	7

Table 4. Wastewater sludge characteristics in North America

Property	Untreated sludge	primary	Digested sludge	primary	Activated sludge	
	Range	Typical	Range	Typical	Range	Typical
Total dry solids (TS, %)	2-8	5	6-12	10	0.4-1.5	1
Volatile solids (% of TS)	60-80	65	30-60	40	60-80	75
Grease and fats (ether-soluble, % of TS)	6-30	-	5-20	-	-	-
Protein (% of TS)	20-30	25	15-20	18	32-41	-
Nitrogen (N, % of TS)	1.5-6	2.5	1.6-6	4	2.4-5	-
Phosphorus (P_2O_5, % of TS)	0.8-3	1.6	1.5-4	2.5	2.8-11	-
Potash (K_2O, % of TS)	0-1	0.4	0-3	1	0.5-0.7	-
Cellulose (% of TS)	8-15	10	8-15	10	-	-
Iron (not a sulfide)	2-4	2.5	3-8	4	-	-

Property	Untreated sludge		Digested sludge		Activated sludge	
	Range	Typical	Range	Typical	Range	Typical
Silica (SiO_2, % of TS)	15-20	-	10-20	-	-	-
Thermal content (MJ/kg)	14-23	16.5	6-14	9	-	-
pH	5-8	6	6.5-7.5	7	6.5-8.5	7

Table 5. Physicochemical characteristics of sewage sludge [7]. Data expressed on dry matter basis, except dry matter and humidity that are expressed on wet weight

Parameter	Unit	Anaerobic sludge		Aerobic sludge	
		Mean	STD	Mean	STD
Dry matter	%	20.3	8.2	22.1	12.4
Humidity	%	79.7	8.2	77.2	12.7
Ash	%	40.2	12	45.2	8.4
Organic matter	%	59.9	12	55.1	8.1
Organic C	%	30.4	7.6	26.6	3.9
Total N	%	4.1	1.6	3.2	1.1
Total P	%	0.9	0.5	2.1	1.4
Total K	%	0.4	0.2	0.4	0.1
pH	-	7.4	0.4	7.1	0.7

Sludge characteristics vary widely from one treatment plant to another. Tables 3 and 4 list the characteristics of typical sludge samples from European and North America wastewater treatment plants [2, 12].

When the sludge is treated, the resulting biosolids can be classified by the treatment, such as aerobically digested (mesophilic and thermophilic), anaerobically digested (mesophilic and thermophilic), alkaline stabilized, composted, and thermally dried. In Table 5, the typical physicochemical characteristics of anaerobic and aerobic sludge are presented.

In 2009 the United States Environmental Protection Agency recently published the results of a survey contacted in 2006-2007. In total 84 sewage sludge samples were collected from 74 randomly selected publicly owned treatment works in 35 states and 145 analytes were measured [13]. It is important to note that the survey was focused on targeted analytes. The results show that 4 anions, namely nitrite/nitrate, fluoride and water extractable phosphorus, 27 metals, 3 pharmaceuticals, namely cyprofloxacin, diphenhydramine, and triclocarban, 3 steroids namely campesterol, cholestanol, and coprostanol and 11 flame retardants were essentially found in every sample.

2.2 Metals

Review studies have shown that the main urban wastewater pollution sources of potentially toxic elements are from industrial point sources and thus, metal concentrations in sewage sludge mainly depend on the type and amount of industrial waste discharged into system. Because metals are generally insoluble they usually present at higher levels in sewage sludge than in wastewater and dewatering of sewage sludge has a minimal impact on reducing metal concentrations [6]. Domestic sources of potentially toxic elements in wastewater are frequently not quantified due to the difficulty in isolating them. Domestic sources include the potentially toxic elements discharged from the household to wastewater collecting systems and, in addition, corrosion from materials used in distribution and plumbing networks, tap water and detergents [1]. The main domestic sources of potentially toxic elements in wastewater are (in order of importance):

- **cadmium:** faeces > bath water > laundry > tap water > kitchen
- **chromium:** laundry > kitchen > faeces > bath water > tap water
- **copper:** faeces > plumbing >tap water > laundry > kitchen
- **lead:** plumbing > bath water > tap water > laundry > faeces > kitchen
- **nickel:** faeces > bath water > laundry > tap water > kitchen
- **zinc:** faeces > plumbing > tap water > laundry > kitchen

Finally, runoff to wastewater collecting systems and waterways has been intensely studied due to its potentially high loading of potentially toxic elements. The main sources of pollution in urban precipitation runoff can be summarized as follows [1]:

- Road and vehicle related pollution
- Degradation of roofing materials
- Construction
- Litter, vegetation and associated human activities
- Erosion of soil

Some mean indicative values collected the years 1975-1977 in urban areas are 0.21 mg/L for Pb, 0.317-4.1 mg/L for Fe, 0.03 mg/L for Cu, 0.11 mg/L for Mn and 0.271 mg/L for Zn. A study carried out around the region of Nantes in France in 1999, analyzed road runoff from a major highway for a year showing that lead and zinc are the main pollutants present in runoff waters. The mean concentrations found are 0.058 mg/L for Pb, 0.045 mg/L for Cu, 1 µg/L for Cd and 0.356 mg/L for Zn.

It should be noted that the when reviewing the sludge data in the related literature it is generally difficult to identify which type of sludge was studied (for example if it was primary or secondary sludge or if it was domestic or municipal sludge or activated or digested sludge etc.) a fact that is of importance especially for organic compounds as they decompose during several treatment processes. In Table 6, the treated sewage sludge is in a separate column to avoid confusion.

Potentially toxic elements are also removed with the solids during the primary and secondary sedimentation stages of conventional wastewater treatment. In Table 7, the removal of metals and transfer to sewage sludge during conventional urban wastewater treatment is presented.

**Table 6. Average values of metals in sewage sludge (mg/kg DW) [1, 6, 9, 10].
Note that the first two columns are for wastewater and the rest for sewage sludge**

Metal	Domestic wastewater (mg/L)	Commercial wastewater (mg/L)	Typical median[2]	Mean EU[2] and US[3] (in parenthesis)		Mean US[4]	Treated sludge[5] EU and US (in parenthesis)
Zn	0.1-1	0.03-133	1700	1222 (1740)		1202	817-1000 (1285-1490)
Fe			17000	-		-	
Cu	0.2	0.04-26	800	337 (850)		741	365-380 (616-639)
Ni	0.04	≤ 7.3	80	37 (82)		42.7	33-44 (71-90.6)
Cd	< 0.03	0.003-1.3	10	2.8 (16)		6.9	2.2-4 (25-38.1)
Pb	0.1	≤ 13	500	124 (500)		134	97 (170-204)
Cr	0.03	≤ 20	500	141 (890)		119	74-145 (178-589)
Se			5	-		5.2	- (6-6.14)
Hg			6	2.2 (5)		5.2	2-2.7 (2.3-3.24)
Co			30	-		-	
Mo			4	-		9.2	
As		3.4 (μg/L)	10	-		9.9	- (4.9-11)
Sn			14	-		-	
Mn			260	-	(260)	-	

Source: [10].

Data are reported for 13 countries: Austria, Denmark, Finland, France, Germany, Greece (Athens), Ireland, Luxembourg, Norway, Poland, Sweden, The Netherlands and UK. Source: [9].

Total elemental composition of over 200 sewage sludge samples from eight US states. Source: [9].

Source: [6].

Source: [1].

**Table 7. Removal of metals and transfer to sewage sludge
during conventional urban wastewater treatment**

Metal	Primary (%)	Secondary (%)	Primary and secondary (%)
Zn	50	56	70-78
Cu	52	57	75-79
Ni	24	26	40-44
Cd	40	40	64-75

Table 7. (Continued)

Metal	Primary (%)	Secondary (%)	Primary and secondary (%)
Pb	56	60	70-80
Cr	40	64	75-78
Hg	55	55	70-80
Se	N/A[1]	N/A	70
As	N/A	N/A	70
Mo	N/A	N/A	70

[1]N/A: data not available.

2.3 Organics

The general effect of wastewater treatment processes is to concentrate the organic pollutants in the sewage sludge and the extent of this removal depends on the properties of the organic species [1]. The overall result of this process is wastewater relatively free of contaminants and sewage sludge rich in contaminants, organic and inorganic. There are a large number of organic pollutants from a wide range of sources which may enter the urban wastewater system. A study carried out in France in 1995, showed the sources of the main organic micropollutants in sludge from wastewater treatment plants were mainly domestic and commercially related (Table 8). Sludge's of different origin may differ substantially in quality, for example PAHs in industrial and non-industrial sewage sludge [11].

Table 8. Principal sources of organic micropollutants in urban wastewater treatment works [1]. Likely is marked by (++) and less likely by (+)

Pollutant	Origin	Domestic usage	Storm runoff	Commercial effluent
Aliphatic hydrocarbons	Fuel	++	++	++
Monocyclic Aromatic hydrocarbons	Solvents, phenols	+	+	++
PAHs	By-products of petrol transformation and insecticides			
Halogens	Solvents, plastics, chlorination	++	+	++
Chlorophenols and chlorobenzenes	Solvents, pesticides	+	+	++
Pesticides	-	+	+	++
Chlorinated PAHs	PCB, hydraulic fluids	+	+	++
Phthalate esters	Plastifier	+	+	++
Detergents	-	++	+	++
Nitrosamines	Industrial by-products (rubber)	-	+	++

Linear alkylbenzenesulphonate (LAS)

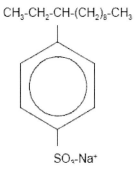

Polychlorinated Dibenzo-p-dioxins and Dibenzofurans (PCDD/PCDF)

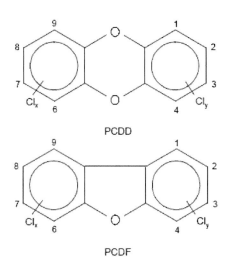

Polychlorinated Biphenyls (PCBs)

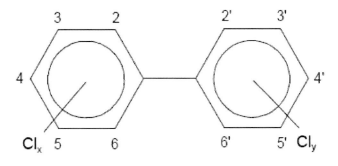

Di-(2-ethyhexyl)phthalate (DEHP)

Figure 3. (Continued).

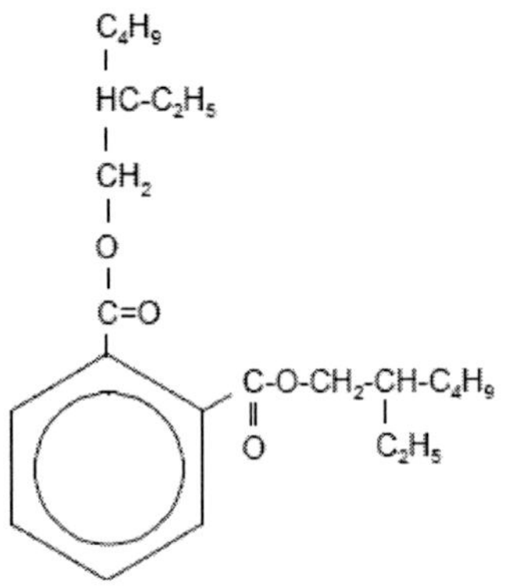

$$C_4H_9$$
$$|$$
$$HC\text{-}C_2H_5$$
$$|$$
$$CH_2$$
$$|$$
$$O$$
$$|$$
$$C{=}O$$

C-O-CH₂-CH-C₄H₉

Structure formula of selected PAHs

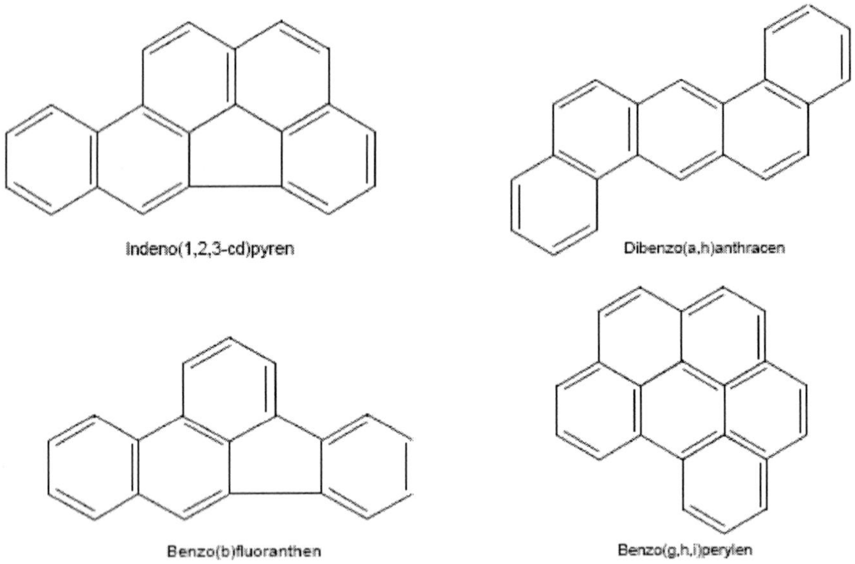

Indeno(1,2,3-cd)pyren

Dibenzo(a,h)anthracen

Benzo(b)fluoranthen

Benzo(g,h,i)perylen

Figure 3. Some basic types of organic pollutants found in sewage sludge.

A recent literature review showed that potentially 541 XOCs xenobiotic organic compounds (XOCs) could be found in sewage sludge, of which only 192 compounds have been quantified, approximately 36%, a fact that indicates the large number of dangerous chemicals not analyzed yet [11]. XOCs compounds are grouped into 14 groups: aliphatic hydrocarbons, dioxins and furans, endocrine disruptors, flame-retardants, organotins, PAHs, PCBs, pesticides, pharmaceuticals, phthalates and plasticizers, as well as miscellaneous compounds. It is important to mention that that 99 XOCs could be classified as being hazardous with regard to the solid phase and 23 were found to be priority pollutants in the subsequent hazard assessment. Since hydrophilic compounds primarily not adhere to sludge, not all compounds found in the water phase will be present in the solid phase and this is the reason why the 364 compounds in Table 9 may be an overestimation. Furthermore, the number of compounds present in both sewage sludge and potential sources to sewage sludge (15) indicates that there is a potentially vast number of XOCs not analysed yet.

Table 9. Present and potentially present compounds in sewage sludge [11]

Compound	Observed	Potentially present[1]	Overlap between the two categories[3]
Aliphatic hydrocarbons	28	16	4
Antioxidants	-	13	-
Dioxins and furans	17[2]	-	-
Endocrine disruptors	11	18	1
Flame retardants	10	4	-
Fragrances and flavours	-	49	-
Organotins	7	-	-
PAH	56	-	-
PCB	12	-	-
Perticides	13	13	1
Pharmaceuticals	20	128	5
Phthalates and plasticizers	9	7	2
UV filters	-	2	-
Miscellaneous	9	114	2
Total number of compounds	**192**	**364**	**15**

Due to the presence of untreated and treated wastewater, grey wastewater, wastewater discharge affected surface waters or household consumption of pharmaceuticals, household chemicals and personal care products.

Contains individual compounds evaluated as sum parameters.

The number of compounds present in both sewage sludge and potential sources to sewage sludge.

In EU level, the following compounds are considered of primary importance as limits are to be set in the revision of Sewage Sludge Directive [5]:

- AOX (sum of halogenated organic compounds)
- linear alkylbenzene sulphonates (LAS)
- di(2-ethylhexyl)phthalate (DEHP)
- NPE (nonylphenole and nonylphenole ethoxylates with 1 or 2 ethoxy groups)
- polynuclear aromatic hydrocarbons (PAHs)
- polychlorinated biphenyls (PCBs)
- polychlorinated dibenzo-p-dioxins and -furans (PCDD/Fs)

In a literature review including 900 papers published in the period 1977-2001, residue data about the level of organic pollutants in German sewage sludges were collected [5]. The results show that 332 organic compounds with known or suspected toxic effects have been detected in sewage sludges, 42 of them regularly, most of them within the level range of g/kg to mg/kg dry matter. In Table 10 data from several studies are presented. As it is underlined above for the case of metals, it is generally difficult to identify which type of sludge that was studied in the given references a fact that is of importance especially for organic compounds as they decompose during several treatment processes. In Table 10, the type of sewage sludge is indicated in parenthesis, if the information is available.

Table 10. Organic pollutants in sludge [1, 8, 5]

Pollutant	Concentration (mg/kg DW)	Year of survey	Country
AOX	75-890	1995	Denmark
	196-206	1994-1996	Germany
NPE	25-2298	1989	Norway
	44-7214	1989-1991	Sweden
	26-1100 (treated)	1990	Sweden
	13-27 (treated)	1995-1998	Sweden
	0.3-537	1993-1995	Denmark
	60-120 (treated)	1988-1989	Germany
	3.8-96.3 (treated)	1996	Germany
	24 (treated)	1994-1995	Austria
LAS	11-16100 (various)	1995	Denmark
	< 1-424	1996-1997	Norway
	2199-17955(treated)	1994-1995	Austria
	1600-11800 (anaerobically digested)	2000	Germany
	182-432 (aerobic)	2000	Germany
	11500-14000 (anaerobically digested)	2000	Italy
	12100-17800 (anaerobically digested)	2000	Spain
	400-700 (untreated)	2000	Spain
	2900-11900 (anaerobically digested)	2000	Switzerland
	9300-18800 (anaerobically digested)	2000	UK
DEHP	27-1115	1989	Norway
	<1-140	2000	Norway
	25-661	1989-1991	Sweden
	3.9-170	1995	Denmark
	20-80 (treated)	1991-1996	Germany
PAHs (sum)	<0.01-8.5	1995	Denmark
	0.7-30	2000	Norway
	1.2-2.2 (treated)	1995-1996	Sweden
PAHs (sum)	1.2-2.2 (treated)	1995-1996	Sweden
	0.25-16.28	2000	Germany
	2.6-15.3 (treated)	1996	Germany
	6-83.8 (treated)	1994	UK
PCBs (sum)	0.017-0.10	2000	Norway
	0.02-0.27 (treated)	1994-1995	Austria

Pollutant	Concentration (mg/kg DW)	Year of survey	Country
PCBs (sum) (Continued)	0.03-0.4 (treated)	1994	France
	0.0006-0.232	1993	Sweden
	0.080-7	1989-1991	Sweden
	0.1 (treated)	1995-1996	Sweden
	< 0.2	2000	Germany
	0.01-0.04 (treated)	1991-1996	Germany
	0.01-22 (various)	1984	UK
Aldrin	0.01-0.2 (various)	1984	UK
Dieldrin	0.01-53 (various)		
Endrin	0.01-0.7 (various)		
PCDD/F (ng/kg dm)	5.7-115	1989-1991	Sweden
	10.3-34.2	1993-1994	Denmark
	0.7-1207	1999	Germany
	8-38	1999	Austria
	64	1999	Spain
	9-192	1999	UK

Mesophilic anaerobic digestion is the principal sludge stabilisation process adopted in most European countries, where approximately 50% of sludge production is treated by this method. Volatile compounds are generally lost to the atmosphere or transferred to the supernatant during digestion, whereas PAHs and phthalate acid esters are conserved [1]. Many organic contaminants are biodegraded under anaerobic conditions while in general, biodegradation during anaerobic digestion destroy 15 to 35% of the organics present. The biodegradation of relatively persistent organic compounds such as PAHs up to to 13-50% has been reported for composted sludge [1].

2.4 Pathogens

Pathogens, in the main, are parasites (microorganisms) in the wide sense of the word and in the general case they are disease-causing. Humans may be exposed to pathogens in biosolids from ingestion of contaminated food, water, or soil, dermal contact and inhalation of bioaerosols [14].

There are five main types of pathogens observed in sludge: bacteria, viruses, fungi and yeast, parasitic worms, and protozoa. Humans and animals are sensitive to some of these organisms, which may cause numerous pathologies ranging from simple digestion troubles to lethal infections [3]. The nature and concentrations of pathogens in sewage depend on the health and the size of the population in the catchment and is very likely that pathogens will be in the sewage at all times [4]. The routine examination of biosolids for the presence of human pathogens is often difficult, and therefore, indicator microorganisms whose presence would suggest that human pathogens might also be present are used, such as fecal coliforms used in US [14].

Table 11. Typical concentrations of micro-organisms (wet weight) in untreated sewage sludge and anaerobically digested liquid sludges (in parenthesis) [4, 6]

Main type	Sub-type	Concentration (g^{-1})
Bacteria	Escherichia coli	10^6-10^7 (300-60·10^4)
	Salmonella	80-1000 (0.03-0.62)
Viruses	Entero-	2.5-10^4 (1-10)
Protozoa	Giardia	100-1000
Helminths	Ascaris	2-1000 (0-10)
	Toxacara	10-100
	Taenia	5

Several types of E. coli are pathogenic to human. Enterohaemorrhagic E. coli of the serotype 0157:H7 has been of the greatest concern in the United States [14]. Bacteria are capable of independent existence, mammalian ones have optimum growth temperatures around the body temperature, which is 35-40°C and most are inactivated at temperatures in excess of 70°C over a relatively short period of time [4,14]. Viruses are unable to multiply outside of the living cells of their host, but they can survive in adverse conditions. More than 140 enteric viruses can be transmitted by biosolids. The caliciviruses, adenoviruses, hepatitis A and E viruses, astroviruses, and rotaviruses are of particular concern [4, 14]. Worms and pathogenic protozoa are extremely resistant to the stresses of the ambient environment but they cannot reproduce outside a suitable host and this means that their concentrations in sludges are relatively low. Cryptosporidium and Giardia are the protozoan parasites most often associated with biosolids. They are parasites of the small intestine that cause diarrhea [4, 14]. Finally, for some pathogens the dangerous properties comes from their ability to produce stable exotoxins e.g. *Clostridium botulinum* [4].

REFERENCES

[1] Pollutants in urban waste water and sewage sludge, European Commission, DG Environment, Luxembourg: Office for Official Publications of the European Communities, 2001

[2] Izrail S. Turovskiy, P. K. Mathai, Wastewater sludge processing, John Wiley & Sons, Inc., USA, 2006

[3] Disposal and Recycling Routes for Sewage Sludge, Synthesis report, DG Environment B/2, European Commission, 2002

[4] Evaluation of sludge treatments for pathogen reduction, Luxembourg: Office for Official Publications of the European Communities, 2001

[5] H. Langenkamp, P. Part, W. Erhardt, A. Prüeß, Organic contaminants in sewage sludge for agricultural use, European Commission, Joint Research Centre, Institute for Environment and Sustainability, Soil and Waste Unit, 2001

[6] Process Design Manual, Land Application of Sewage Sludge and Domestic Septage, United States Environmental Protection Agency, 1995

[7] Workshop on problems around sludge, Jointly organised by: the Directorate-General for the Environment, the Joint Research Centre of the European Commission, Edited by

Heinrich Langenkamp, Luca Marmo, Proceedings, 18-19 November 1999 - Stresa (NO) Italy, 1999

[8] Howard R. Rogers, Sources, behaviour and fate of organic contaminants during sewage treatment and in sewage sludges, The Science of the Total Environment, 185, 3-26, 1996

[9] M.A. Stylianou, V.J. Inglezakis, K.G. Moustakas, M.D. Loizidou, Improvement of the quality of sewage sludge compost by adding natural clinoptilolite, Desalination, 224, 240–249 (2008)

[10] D. Fytili, A. Zabaniotou, Utilization of sewage sludge in EU application of old and new methods—A review, Renewable and Sustainable Energy Reviews, 12, 116–140 (2008)

[11] Eva Eriksson, Nina Christensen, Jens Ejbye Schmidt, Anna Ledin, Potential priority pollutants in sewage sludge, Desalination 226, 371–388 (2008)

[12] Metcalf and Eddy, Inc, revised by George Tchobanoglous, Wastewater Engineering: Treatment, Disposal and Reuse, Tata McGraw-Hill Publishing Company Limited, Second Edition, New Delhi (1979)

[13] Targeted National Sewage Sludge Survey Overview Report, EPA-822-R-08-014, U.S. Environmental Protection Agency, 2009

[14] Biosolids Applied To Land: Advancing Standards And Practices, National Research Council, National Academy Press, Washington, DC USA, July 2002

In: Sewage Sludge Management
Editors: A. A. Zorpas and V. J. Inglezakis

ISBN: 978-1-61324-393-0
© 2012 Nova Science Publishers, Inc.

Chapter 4

WASTEWATER TREATMENT RESIDUALS: SOURCES, CHARACTERISTICS AND QUANTITIES

George Tchobanoglous[1], Avraam Karagiannidis[2] and Harold Leveren[3]

[1]Professor Emeritus Department of Civil and Environmental Engineering,
University of California at Davis, Davis, CA, US
[2]Associate Professor Department of Mechanical Engineering,
Aristotle University of Thessaloniki, Thessaloniki, Greece
[3]Research Associate Department of Environmental Engineering,
Technical University of Crete, Crete, Greece

ABSTRACT

Wastewater residuals are produced as a result of treatment. To design treatment and disposal facilities properly, the sources, characteristics, and quantities of the waste residuals to be handled must be known. Therefore, the purpose of this chapter is to present background data and information on treatment plant residuals that will serve as a basis for the material to be presented in the subsequent sections.[*]

SOURCES OF WASTEWATER RESIDUALS

The principal residuals resulting from the treatment of wastewater are: screenings (coarse solids), grit, scum, sludge, and biosolids. The sources of wastewater residuals are identified in Table 1 and illustrated on Figure 1. Processes used for filtering, thickening, digesting, conditioning, and dewatering of solids produced from primary and secondary settling tanks also constitute sources. The term *sludge* is used to describe the liquid, semi-liquid, and semi-solid organic material removed from wastewater, usually by settling, before it has undergone.

---

[*] Adapted from Tchobanoglous et al., 2003.

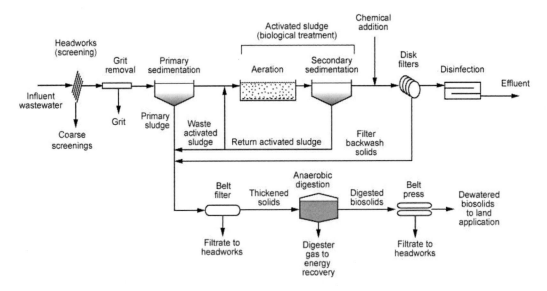

Figure 1. Typical flow diagram for the biological treatment of wastewater with residual waste steams identified (see also Table 1).

any treatment to stabilize it for beneficial use. The term *sludge* is also generally used in conjunction with a process descriptor, such as primary sludge, waste activated sludge, and secondary sludge. The term *biosolids* is used to describe the organic material after it has been stabilized so that it can be used beneficially in a variety of applications. In cases where it is uncertain whether beneficial use criteria have been met, the term solids or sludge is used

CHARACTERISTICS OF RESIDUALS

To treat and dispose of the residuals produced from wastewater treatment plants in the most effective manner, it is important to know the physical and chemical characteristics of the residuals that will be processed.

Physical Characteristics

The physical characteristics of the residuals identified in Table 1 are described in Table 2. The characteristics will vary depending on the origin of the solids, the amount of aging that has taken place, and the type of processing to which they have been subjected.

Chemical Composition of Residuals

Typical data on the chemical composition of untreated primary sludge, digested biosolids, and untreated activated sludge are reported in Table 3. Many of the chemical constituents, including nutrients, are important in considering the ultimate disposal of the processed solids and the liquid removed during processing. The measurement of pH, alkalinity, and organic acid content is important in process control of anaerobic digestion. The content of heavy metals, pesticides, and hydrocarbons has to be determined when incineration and land application methods are contemplated. The thermal content of solids is important where a thermal reduction process such as incineration is considered.

Trace Constituents

Trace elements are those inorganic chemical elements that, in very small quantities, can be essential or detrimental to plants and animals. The term "heavy metals" is used to denote several of the trace elements present in sludge and biosolids. Concentrations of heavy metals may vary widely, as indicated in Table 4 For the application of biosolids to land, concentrations of heavy metals may limit the application rate and the useful life of the application site.

Fertilized Value of Biosolids

Solids characteristics that affect their suitability for application to land and for beneficial use include organic content (usually measured as volatile solids), nutrients, pathogens, metals, and toxic organics. The fertilizer value of the sludge and solids, which should be evaluated where they are to be used as a soil conditioner, is based primarily on the content of nitrogen, phosphorus, and potassium (potash). Typical nutrient values of wastewater biosolids as compared to commercial fertilizers are reported in Table 5. In most land application systems, biosolids provide sufficient nutrients for good plant growth. in some applications, the phosphorus and potassium content may be low and require augmentation.

ENERGY CONTENT OF SLUDGE AND BIOSOLIDS

The energy content of wastewater, comprised of the heat and the organic constituents found in wastewater, is an important consideration as new technologies and concepts are evaluated. Typically, the total energy content in wastewater is two to four times the energy needed to operate a WWTP. The following discussion is focused on the energy content of sludge and biosolids.

Chemical Energy in Primary Sludge and Biosolids

The chemical energy content of the organic constituents in primary sludge and biosolids can be determined by: (1) using a full scale boiler as a calorimeter, (2) using a laboratory bomb calorimete), and (3) by calculation, if the elemental composition is known. Because of the difficulty in instrumenting a full-scale boiler, most of the experimental data on the energy content of the organic constituents of wastewater, sludge and biosolids are based on the results of bomb calorimeter tests. Typical data, based on bomb calorimeter tests, on the energy content found in primary sludge, digested primary sludge, and untreated activated sludge are reported in Table 3. As shown, the energy content of sludge will vary significantly; depending on whether volatile solids (VS) or total solids (TS) are considered. Total solids will typically contain a significant concentration of silica and other inert material. As reported in Table 3, the VS are about 65 perent of the total solids in primary sludge.

Estimation of Energy Content

The energy content of primary sludge and biosolids can be estimated from an elemental analysis of the constituents in organic compounds using the following expression, which is a modified form of the DuLong formula developed by Channiwala (1992).

$$\text{HHV (MJ/kg)} = 34.91\ C + 117.83\ H - 10.34\ O - 1.51\ N + 10.05\ S - 2.11A \qquad (1)$$

Where HHV is the high heating value and C is the weight fraction of carbon; H of hydrogen; O of oxygen; N of nitrogen S of sulfur, and A of ash as derived from an ultimate analysis or from the chemical formula, if known.

When the HHV is used, it is assumed that the water component is in the liquid state at the end of combustion. Another estimate of the heating value of a combustible material is the lower LHV (lower heating value) in which it is assumed that the latent heat of vaporization is not recovered. In general, the LHV is about 6 to 8 percent lower than the corresponding HHV. For stationary combustion units with exhaust heat recovery, use of the HHV is the most appropriate. Where exhaust heat is not recovered, use of the LHV is most appropriate. Also, in most European literature LHVs are reported, whereas HHVs are reported in the American literature. The application of Eq. 1 is illustrated in Example 1.

EXAMPLE 1   Estimate the chemical energy content of dry biosolids comprised of bacterial cell biomass on the basis of MJ/kg dry biosolids. Assume the chemical composition of cell biomass is $C_5H_7NO_2$, the ash content is 5 percent, and the silica content is 20 percent.

Analysis

1. Determine the energy content of the biosolids using Eq. 1
   a. Determine the weight fractions of the elements and ash comprising the biosolids.

   Computation table, Example 1, Step 1

Component	Coefficient	mw	Molecular mass	Weight fraction
Carbon	5	12	60	0.398[a]
Hydrogen	7	1	7	0.047
Oxygen	2	16	32	0.212

Nitrogen	1	14	14	0.093
Sulfur	0	0	0	0
Ash + silica	0	0		0.250
			113	1.000

[a](60/113) x 0.75 = 0.52

b. The energy content of the biosolids using Eq. 1 is:

HHV (MJ/kg) = 34.91 (0.398) + 117.83 (0.047) - 10.34 (0.212)
- 1.51 (0.093) - 2.11 (0.250)
HHV (MJ/kg) = 13.89 + 5.54 – 2.19 – 0.14 – 0.53
= 16.85 MJ/kg dry biosolids

Comment

In practice, because the HHV is difficult to recover, the LHV is often used in place of the HHV. If the difference between the HHV and LHV is assumed to be eight percent, the corresponding LHV of biosolids is 15.5 kg biosolids, which is consistent with the values reported in Table 3. The energy content of wet sludge or biosloids is computed the same way, but the moisture content is taken into account.

Recovery of Energy

The key question is what is the best way to extract and utilize the chemical energy in wastewater, a renewable source of recoverable energy. The most common method in use today is by means of anaerobic digestion and the generation of electrical energy using the digester gas. However, depending on the technologies involved for wastewater management, thermal processes such as combustion and pyrolysis, as discussed in this book, appear to offer significant advantages when the complete management of wastewater sludge and biosolids is evaluated, especially in large metropolitan areas.

QUANTITIES OF SCREENINGS AND GRIT

The quantity of screenings removed from wastewater will vary with the type of rack or screen used, the type of collection system, and the geographic location. For estimating purposes, the values given in Table 6 can be used as astarting point. Similarly, the quantities of grit will vary greatly from one location to another, depending on the type of wastewater collection system, the characteristics of the drainage area, the condition of the sewers, the frequency of street sanding to counteract icing conditions, the types of industrial wastes, the number of household food waste grinders, and the proximity of and sandy beaches. Typical values of grit will vary from 0.002 – 0.200 m3/103 m3, with a typical value of 0,015 m3/103 m3.

QUANTITIES OF SLUDGE AND BIOSOLIDS

Data on the quantities of solids produced from various processes and operations are presented in Table 7. Corresponding data on the solids concentrations to be expected from various processes are given in Table 8. Although the data in Table 7 are useful as presented, it should be noted that the quantity of solids produced will vary widely.

Quantity Variations

The quantity of solids entering and produced at the wastewater treatment plant daily may be expected to fluctuate over a wide range. To ensure capacity capable of handling these variations, the designer of solids processing and disposal facilities should consider (1) the average and maximum rates of solids production, and (2) the potential storage capacity of the treatment units within the plant. The variation in daily quantity that may be expected in large cities is shown in Figure. 2. The curve is characteristic of large cities having a number of large collection lines laid on flat slopes; even greater variations may be expected at small plants.

A limited quantity of solids may be stored temporarily in the sedimentation and aeration tanks. This storage provides capacity for equalizing short-term peak loads. In solids processing systems where digestion tanks are used, their large storage capacity provides a substantial dampening effect on peak solids loads. In solids treatment systems where digestion is used, the design is usually based on maximum monthly loadings. Where digestion is not used, the solids treatment process should be capable of handling the solids production

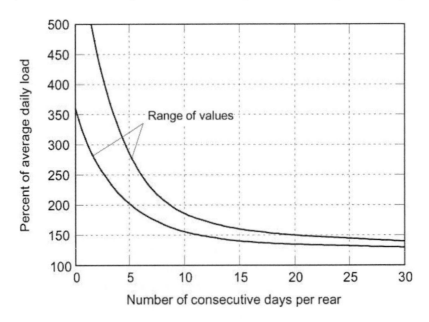

Figure 2. Peak sludge loading as a percentage of the average daily load.

of the maximum week. Certain components of the solids operations system, such as sludge pumping and thickening, may need to be sized to handle the maximum day conditions.

Volume-Weight Relationships

The volume of sludge depends mainly on its water content and only slightly on the character of the solid matter. A 10 percent sludge, for example, contains 90 percent water by weight. If the solid matter is composed of fixed (mineral) solids and volatile (organic) solids, the specific gravity of all of the solid matter can be computed using Eq. 2.

$$\frac{W_s}{S_s \rho_w} = \frac{W_f}{S_f \rho_w} + \frac{W_v}{S_v \rho_w} \tag{2}$$

where W_s = weight of solids
S_s = specific gravity of solids
r_w = density of water
W_f = weight of fixed solids (mineral matter)
S_f = specific gravity of fixed solids
W_v = weight of volatile solids
S_v = specific gravity of volatile solids

Therefore, if one-third of the solid matter in a sludge containing 90 percent water is composed of fixed mineral solids with a specific gravity of 2.5, and two-thirds is composed of volatile solids with a specific gravity of 1.0, then the specific gravity of all solids S_s would be equal to 1.25, as follows:

$$\frac{1}{S_s} = \frac{0.33}{2.5} + \frac{0.67}{1.0} = 0.802$$

$$S_s = \frac{1}{0.802} = 1.25$$

If the specific gravity of the water is taken to be 1.0, the specific gravity of the sludge, S_{sl}, is 1.02, as follows:

$$\frac{1}{S_{sl}} = \frac{0.1}{1.25} + \frac{0.9}{1.0} = 0.98$$

$$S_{sl} = \frac{1}{0.98} = 1.02$$

The volume of a sludge may be computed with the following expression:

$$V = \frac{M_s}{\rho_w S_{sl} P_s}$$
(3)

where V = volume, m3
M_s = mass of dry solids, kg
r_w = specific weight of water, 103 kg/m3
S_{sl} = specific gravity of the sludge
P_s = percent solids expressed as a decimal

For approximate calculations for a given solids content, it is simple to remember that the volume varies inversely with the percent of solid matter contained in the sludge as given by

$$\frac{V_1}{V_2} = \frac{P_2}{P_1} \quad \text{(approximate)}$$
(4)

where V_1, V_2 = sludge volumes

P_1, P_2 = percent of solid matter

The application of these volume and weight relationships is illustrated in Example 2.

EXAMPLE 2   Determine the liquid volume before and after digestion and the percent reduction for 1000 kg (dry basis) of primary sludge with the following characteristics:

Item	Primary	Digested
Solids, %	5	10
Volatile matter, %	65	60 (destroyed)
Specific gravity of fixed solids	2.55	2.5
Specific gravity of volatile solids	~ 1.0	~ 1.0

Analysis

1. Compute the average specific gravity of all the solids in the primary sludge using Eq. 2.

$$\frac{1}{S_s} = \frac{0.35}{2.55} + \frac{0.65}{1.0} = 0.787$$

$$S_s = \frac{1}{0.787} = 1.27$$

2. Compute the specific gravity of the primary sludge.

$$\frac{1}{S_{sl}} = \frac{0.05}{1.27} + \frac{0.95}{1.0} = 0.99$$

$$S_{sl} = \frac{1}{0.99} = 1.01$$

3. Compute the volume of the primary sludge using Eq. 3.

$$V = \frac{1,000\,kg}{(1,000\,kg/m^3)(1.01)(0.05)} = 19.8\,m^3$$

4. Compute the percentage of volatile matter after digestion.

$$Volatile\ matter,\% = \frac{total\ volatile\ solids\ after\ digestion}{total\ solids\ after\ digestion} \times 100$$

$$Volatile\ matter,\% = \frac{0.4(0.65 \times 100)}{35 + 0.4(65)} \times 100 = 42.6$$

5. Compute the average specific gravity of all the solids in the digested sludge using Eq. 2.

$$\frac{1}{S_s} = \frac{0.574}{2.5} + \frac{0.426}{1.0} = 0.656$$

$$S_s = \frac{1}{0.656} = 1.52$$

6. Compute the specific gravity of the digested sludge.

$$\frac{1}{S_{sl}} = \frac{0.1}{1.52} + \frac{0.9}{1.0} = 0.97$$

$$S_{sl} = \frac{1}{0.97} = 1.03$$

7. Compute the volume of digested sludge using Eq. 3.

$$V = \frac{350 + 0.4(650)}{(1000\,kg/m^3)(1.03)(0.10)} = 5.9\,m^3$$

8. Determine the percentage reduction in the sludge volume after digestion.

$$\%\,reduction = \frac{(19.8 - 5.9)\,m^3}{19.8\,m^3} \times 100 = 70.2$$

NEED FOR CURRENT DATA

The information presented in this chapter is meant to serve as an introduction to the sources, characteristics and quantities of residuals that can be expected from the treatment of wastewater. However, because the characteristics of untreated wastewater are so variable, depending on local ordnances, source control programs, and the nature of the collection system, it is imperative that current data be collected. If a new treatment plant is being proposed where one has not existed, residuals data should be gathered from nearby treatment plants. Also careful attention must be paid to proposed future plans. For example, to optimize the recovery of energy from wastewater, the residents could be encouraged to grind food waste and discharge it to the collection system. Such a practice would change the composition of the primary sludge significantly.

REFERENCES

Channiwala, S.A. (1992) Ph.D Thesis, Indian Institute of Technology, Department of Mechanical Engineering, Bombay, India.

Shizas, I., and D.M. Bagley (2004) "Experimental Determination of Energy Content of Unknown Organics in Municipal Wastewater Streams," J. Energy Eng., 130, 2, 45-53.

Tchobanoglous, G., and F.L. Burton (1991) Wastewater Engineering. Treatment, Disposal, Reuse, 3rd ed., McGraw-Hill Inc., New York.

Tchobanoglous, G., F.L. Burton, and H.D. Stensel (2003) Wastewater Engineering: Treatment and Reuse, 4th ed., Metcalf and Eddy, Inc., McGraw-Hill Book Company, New York.

U.S. EPA (1979) Process Design Manual Sludge Treatment and Disposal, EPA 625/1 - 79-011, US. Environmental Protection Agency, Office of Research and Development, Washington, DC.

U.S. EPA (1985) Seminar Publication Composting of Municipal Wastewater Sludges, EPA/625/4-85/014, U.S. Environmental Protection Agency

Zanoni, A.E., and D.L. Mueller (1982) "Calorific Value of Wastewater Plant Sludges," J. Environ. Eng. Div. (Am. Soc. Civ. Eng.), 108, 187-195.

In: Sewage Sludge Management
Editors: A. A. Zorpas and V. J. Inglezakis

ISBN: 978-1-61324-393-0
© 2012 Nova Science Publishers, Inc.

Chapter 5

SEWAGE SLUDGE AND GLOBAL CHANGE

Jose Navarro Pedreño, José Martin Soriano Disla and Ignacio Gómez Lucas

Departamento de Agroquímica y Medio Ambiente,
Universidad Miguel Hernández de Elche, Elche, Alicante, Spain

ABSTRACT

One of the options that have generated broad interest to combat global change is C sequestration in agricultural soils through the increment of soils organic carbon (SOC). One of the management practices that enhance the SOC pool thus contributing to C sequestration by soils is the agricultural application of sewage sludge. The application of sewage sludge is a current practice for disposing of an unwanted waste, contributing to C sequestration and returning valuable nutrients and organic matter to reclaim degraded soils. However, a very important controversial exists about the effects (positive or negative) of sewage sludge applications on the SOC pool and, in general, in terms of global warming potential. This chapter is focussed on the study of the contribution of land application of biosolids (sewage sludge) on soil carbon stock that positively combats climate change meanwhile C storage.

SEWAGE SLUDGE APPLICATIONS AS A SOURCE OF SOC AND POTENTIAL RISKS

Sewage sludge is a biosolid usually rich in organic matter (OM) and nutrients produced in the wastewater treatment process. Thus, the application of sewage sludge is a current practice for disposing of an unwanted waste, contributing to C sequestration and returning valuable nutrients and OM to reclaim degraded soils (Roldán *et al.*, 1996; Bernal *et al*,. 1998; Speir *et al*,. 2003; Singh and Agrawal, 2008; Ramlal *et al.*, 2009). This process of utilization rather than disposal these requires small amounts of energy and offers to farmers a source of free or inexpensive fertilizer (Tian *et al.*, 2009). The use of sewage sludges for reclaiming degraded soils must be considered as one of the most important strategies in developed countries which can enhance the soil organic carbon (SOC) pool thus contributing to C sequestration (Tian *et al.*, 2009).

The wastewater treatment in developed countries has produced an increment of solid wastes. The production of sewage sludge has increased dramatically in Europe due to the progressive implementation of the European Directive 91/271/EEC. The increment in the production of sewage sludge and the potential benefits for the soil-plant system has derived in great amounts applied to soils. Indeed, the implementation of European Union regulations such the European Directive 99/31/EEC, which restricts the disposal of organic-rich materials in landfills, encourages the member states to use the sewage sludge as soil amendments. However, the application of sewage sludge is of great environmental concern due to the potential transference of heavy metals, nutrients, microorganisms and organic pollutants that can result in soil and water contamination, phytotoxicity, transfer to the food supply, and reduction of soil biological activity (Giller *et al.*, 1998; Horswell *et al.*, 2003; Vogeler *et al.*, 2006; Speir *et al.*, 2007). Sewage sludge amended soils generally contain elevated levels of OM and heavy metals (Cd, Cu, Ni, Pb and Zn) compared to control soils (Ashworth and Alloway, 2008). Another important factor to be considered is the nutrient content in the sewage sludge, especially N. The presence of this element is a limiting factor associated to mineralization-immobilization processes produced by soil microorganism (directly dependent of N sources transformed into inorganic forms) and the possible derived effect of groundwater pollution by soluble N-compounds (Navarro-Pedreño *et al.*, 1996; Countinho *et al.*, 1997; Luczkiewicz, 2006).

It is well recognized that sewage sludge contains high levels of labile OM. The dissolved OM has been reported to be very mobile through the soil system (Dunivant *et al.*, 1992) and to form stable, soluble complexes with heavy metals (Antoniadis and Alloway, 2002; Ashworth and Alloway, 2004). In addition, the rapid decomposition of an OM not sufficiently stabilized may increment the production of phytotoxic substances, produce unpleasant odours, decrease the O_2 concentration and create anaerobic environments at the root system (Iglesias-Jiménez and Pérez-García, 1989; Hue and Liu, 1995). The enhancement of reducing conditions can also increase the heavy metal solubility (Cottenie, 1981). Another potential negative effect derived form applications of wastes with high levels of labile OM is the mineralization of the native OM (Bernal *et al.*, 1998). This can be related to the rapid mineralization of the organic C after sewage sludge application, a phenomenon known as priming effect (Dalenberg and Jager, 1989).

Sewage sludge characteristics are the main limiting factor for its application in agriculture and natural soils. Although special attention is given to the heavy metal content of the sludge and soil pH, other characteristics should be considered (Wang *et al.*, 2008). For instance, the OM content and type in soils and sewage sludge is an important factor controlling the mineralization processes and C sequestration in soils. Some of the limiting factors have been commented in other chapters. Despite these limitations, the application of sewage sludge in agricultural soils is just regulated by the total heavy metal content in both sewage sludge and soils, depending on soil pH (European Directive 86/278/EEC). Thus, it is necessary to define the limiting factors of sewage sludge applications, focusing the attention in the biosolid characteristics at first. In this way, and prior to their application, the following characteristics, in this order, should be considered:

1. The stability of the biosolid, if it is fresh or composted.
2. Pollutant content, paying special attention to heavy metals (i.e. Cd, Cr, Cu, Hg, Mn, Ni, Zn, Pb).
3. Organic matter content.
4. Nitrogen content.
5. Salinity, due to the presence of soluble salts in some sewage sludge than can affect soil-plant system.

In the case of soils, pH is the main limiting factor (mainly for its contribution to the solubility of heavy metals) but some others should be considered: OM content, type of minerals, water content and also the relations with other environmental compartments due to the possibility of contamination, especially water pollution.

Thus, and although the application of sewage sludge in soils can derive in many negative impacts, some of these impacts could be substantially reduced if some factors are considered: application rate, crop type, sewage sludge and soil characteristics, environmental conditions and management practices.

What it seems clear is that the application of sewage sludge can result in many benefits for the soil; however, also negative impacts can be derived. A very important controversial exists about the effects (positive or negative) of sewage sludge applications on the SOC pool and, in general, in terms of global warming potential (Jones *et al.*, 2006). Our concern in this chapter is focused on the contribution of sewage sludge applications to the soil C stock and on the mechanisms involved in this sequestration.

FACTORS AFFECTING THE CONTRIBUTION OF SEWAGE SLUDGE APPLICATIONS TO C STOCK

If a maximal benefit from a sewage sludge application wants to be achieved in terms of C sequestration, there is a need to known the factors that can influence the permanence of this C in the soil. This permanence is influenced by the balance between stabilization and decomposition processes. Many of the mechanisms involving stabilization and decomposition act at the same time, which makes difficult to study the dynamic of the organic C in soil. The factors influencing the C sequestration after application of sewage sludge are:

Sewage Sludge Characteristics

The OM characteristics of the sewage sludge, given by the source materials, condition their biodegradability. Anyway, sewage sludge after conventional treatments present high levels of labile OM. Prior to the application, the high levels of biodegradability of these wastes should be reduced and one of the best ways is by composting. This process is a way of managing and reducing organic wastes, obtaining a stable and sanitized product form biological oxidative transformation of the organic matter (de Bertoldi *et al.*, 1983). There is a general agreement that composted materials are preferable in SOC conservation terms (Bernal et al., 1998; Sánchez-Monedero et al., 2004). The application of raw organic materials is expected to induce greater C increases in the short-term, in contrast to composted materials which are expected to have more influence in the long-term C storage (Robin et al., 2008).

Application Conditions and Management Practices

The application conditions are very important in terms of C conservation. Normally, surface applications promote the OM mineralization in contrast to applications where the sewage sludge is mixed with the soil. The water content of the biosolid need to be considered since the mineralization of the OM is highly dependent on this parameter. Another important point is the way in which the sewage sludge is presented (as pellets, powder, as a cake…). The application rate also conditions the mineralization, presenting normally a positive relationship (Giusquiani et al., 1995).

The subsequent management of the soils applied with sewage sludge is also important in terms of C sequestration. It has been proved that intensive cropping systems have significantly reduced the levels of SOM (CEC, 2006).

Environmental Conditions

Environmental conditions such precipitation, humidity and temperature influence decisively the mineralization of the OM. It is expected that the higher the temperature and the closer to the optimal humidity conditions for microorganisms the higher the mineralization.

Soil Properties

Apart from the environmental factors, soil C sequestration has been related to soil properties such as texture and mineralogy (Lal, 2004a). Higher mineralization rates are expected in soils with light textures (i.e. sandy soils) compared with those observed in soils with heavier textures. This can be related to the higher oxygen availability of the sandy soils and to the lower levels of silt and clay observed in these soils. Indeed, and as it was stated in the chapter "Global change and soil carbon", SOM can be protected by intimate association with silt and clay particles and physically stabilized through aggregate formation (John et al., 2005).

The OM contents of soil can also influence the C sequestration after sewage sludge applications since it has been reported that soils having high levels of SOC, where a high microbial density and competition between different populations is expected, experience larger priming effects that those observed in poor soils (Hart et al., 1986). Soil pH can also influence the OM mineralization since it is one of the most influential factors affecting the microbial communities (Rousk et al., 2009).

STUDIES OF ORGANIC WASTE APPLICATIONS
FOR ENHANCING SOIL ORGANIC CARBON

Despite the existence of a general agreement about the necessity of increasing the SOC pool in agricultural soils and the potential of sewage sludge for achieving this objective, the identification and quantification of the contribution of these applications to the SOC pool still remain unanswered (Freibauer *et al.*, 2004). This biosolid is available for soil disposal and should be use to increase SOC. Tian *et al.* (2009) studied long-term applications from 1971 of digested sludges applied first as liquid biosolids (1971-1980), after that as dewatered solids (1980-1995) and finally as air-dried solids. Following this application, soils samples were collected at a depth of 0-15 cm each spring, before the annual application of biosolids. High SOC improvement and soil C sequestration in biosolid-amended fields of a large-scale and long-term land reclamation project was obtained. Authors concluded that biosolids are an effective material in increasing SOM in low-productivity soils and could turn the Midwest corn-soybean systems soils for C-neutral to C-sink.

Preliminary studies in Europe determined that increasing the amount of OM in soils, C sequestration may be enhanced. The potential of European agricultural soils for C sequestration through the application of organic wastes has been found to be higher for south-west and south-east Europe (i.e. Spain and Turkey), where low soil C contents occur (Vleeshouwers and Verhagen, 2002).

In European soils, amendments with sewage sludge at 1 and 2.59 t ha^{-1} y^{-1} increased SOC contents by 1.96 and 2.14 % y^{-1}, respectively (Smith *et al.*, 1997). The measures for increasing soil C stocks in agricultural soils and potential yearly soil C sequestration rates indicate a potential rate of 0.3 t C ha^{-1}year^{-1} for the use of sewages sludge (Freibauer *et al.*, 2004).

The potential of agricultural soils for contributing to the short-term C sequestration following the application of sewage sludge was proved in the study of Soriano-Disla *et al.* (2010) using 60 Spanish agricultural soils. This was especially evident for soils having low initial SOC contents. However, caution is required for sewage sludge applications in soils with high initial levels of SOC since absence of variation, or even a decrease of SOC was detected following the application of sewage sludge, indicating degradation of the native SOC.

Albaladejo *et al.* (2008) developed a study whose main objective was to evaluate the long-term effect of a single addition of two different doses of urban solid refuse (P1: 13 kg m^{-2} and P2: 26 kg m^{-2}) on the potential C sequestration. The land reclamation strategy of this study was based on the hypothesis that organic additions improve soil physical and biological properties and soil fertility, leading to higher availability of water and nutrients for plants. This favours natural recovery by spontaneous vegetation, which increases biomass production and, consequently, the deposition of plant litter-fall in the soil. In this way, it can be expected that the initial effect of external OM may be sustained by new inputs from the enhancement of the re-established vegetation. Ultimate reclamation will only occur when an active biological system is established in the form of vegetative cover and soil micro and macro fauna (Logan, 1992). The addition of the single organic amendment to a degraded semiarid soil was effective in improving the potential for C sequestration, representing a total C sequestration of 3.4 Mg ha^{-1} in P1 and 9.5 Mg ha^{-1} in P2.

It is important to bear in mind that the use of sewage sludge in agricultural soils is not the single alternative for these materials as soil amendments. The application of sewage sludge for soil rehabilitation and other opportunities for improving soil quality and C sequestration should be considered. The sink capacity of SOM for atmospheric CO_2 can be greatly enhanced when degraded soils and ecosystems are restored, marginal agricultural soils are converted to a restorative land use or replanted to perennial vegetation, and recommended management practices are adopted on agricultural soils (Lal, 2004b). Sewage sludge can be the source of OM for these land uses, restoring the OM levels of the soils.

FUTURE TRENDS IN THE USE OF SEWAGE SLUDGE FOR ENHANCING SOIL ORGANIC CARBON

Research related to sewage sludge application to soils has been centred mainly in two lines: benefits associated to biological aspects (nutrient disposal, biological activity, pollution…) and physical properties (stability of soils, aggregation…), as the references of this book revealed. However, more research is needed to determine the role of sewage sludge and its contribution for enhancing soil C sequestration.

The C sequestration in soil via sludge application needs of more experimental data in long-term experiments, as Rothamsted's long term experiments (Rothamsted Research, 2006). The dynamics of OM added with sewage sludge is crucial to determine if a given soil where the biosolid is applied has the possibility of enhancing soil C and reducing the levels of CO_2 in the atmosphere, forming protected soil C and recalcitrant soil OM.

One interesting future line of research in Soil Science is the development of tools for predicting the effect of exogenous OM applications, such sewage sludge, on soil organic C sequestration. The proportion of exogenous OM remaining in soils over long-term following organic waste applications can be evaluated by the application of a recently developed indicator (Lashermes *et al.*, 2009). The inputs for the determination of this index were different biochemical fractions of the OM and C mineralization.

There is a broad consensus that future efforts should stress the performance of full-scale land application sites rather than laboratory-generated information (O'Connor *et al.*, 2005). Field studies involve high cost and environmental risk of failure for reasons beyond the control of researcher but offer the ultimate scenario for addressing problems. Moreover, the obtained results suggest faster degradation of exogenous OM under field conditions compared with laboratory (Chabbi and Rumpel, 2009).

CONCLUSION

The application of sewage sludge is one of the best strategies to increase the input of OM. According to the objectives of combating climate change, the final proposal of the use of sewage sludge is to increase soil C in non-hydrolizable forms, in stable compounds forming part of the humic substances of the soil and the recalcitrant soil OM for carbon sequestration.

Several studies have demonstrated the viability of using sewage sludge as amendment that can improve the soil properties, increase the OM and the nutrient content and favour plant growth. All of these studies agree to present sewage sludge application as an attractive option for the disposal of this waste, favouring the C sequestration. However, the quantification of the contribution of sewage sludge applications to the C sequestration remains still unknown and long-term land studies are needed.

If a maximal benefit wants to be achieved from a sewage sludge application in terms of C sequestration, there is a need to consider the factors that can influence the permanence of this C in the soil: sewage sludge treatment, application conditions and management practices, environmental conditions and soil properties.

Despite the potential benefits, the application of sewage sludge raises environmental problems related to the high levels of labile OM and the presence of microorganisms, organic contaminants and heavy metals in these wastes. Since the regulation about sewage sludge application just includes the heavy metal content in soils and sludge, more characteristics need to be considered for environmental-safety applications.

REFERENCES

Albaladejo, J.; López, J.; Boix-Fayos, C.; Barbera, G.G.; Martínez-Mena, M. (2008). Long-term effect of a single application of organic refuse on carbon sequestration and soil physical properties. Journal of Environmental Quality 37 : 2093–2099.

Antoniadis, V.; Alloway, B.J. (2002). Leaching of cadmium, nickel, and zinc down the profile of sewage sludge-treated soil. Communication in Soil Science and Plant Analysis 33: 273–286.

Ashworth D.J.; Alloway, B.J. (2004). Soil mobility of sewage sludge-derived dissolved organic matter, copper, nickel and zinc. Environmental Pollution 127: 137–144.

Ashworth, D.J.; Alloway, B.J. (2008). Influence of dissolved organic matter on the solubility of heavy metals in sewage-sludge-amended soils. Communications in Soil Science and Plant Analysis 39: 538–550.

Bernal, M.P.; Sánchez-Monedero, M.A.; Paredes, C.; Roig, A. (1998). Carbon mineralization from organic wastes at different composting stages during their incubation with soil. Agriculture, Ecosystems and Environment 69: 175–189.

Chabbi, A.; Rumpel, C. (2009). Organic matter dynamics in agro-ecosystems – the knowledge gaps. European Journal of Soil Science 60: 153–157.

Cottenie, A. (1981). Sludge treatment and disposal in relation to heavy metals. International Conference on Heavy Metals in the Environment, Amsterdam. Commission of the European Communities, pp. 167–175.

Countinho, J.; Arrobas, M.; Rodrigues, O. (1997). Effect of composted sewage sludge amendment on soil nitrogen and phosphorus availability. Communications in Soil Science and Plant Analysis 28: 1845-1857.

Dalenberg, J.W; Jager, G. (1989). Priming effect of some organic additions to 14C-labelled soil. Soil Biology and Biochemistry 21: 443–448.

De Bertoldi, M.; Vallini, G.; Pera, A. (1983). The biology of composting a review. Waste Management and Research 1: 157–176.

Dunnivant, F.M.; Jardine, P.M.; Taylor, D.L.; McCarthy, J.F. (1992). Transport of naturally occurring dissolved organic carbon in laboratory columns containing aquifer material. Soil Science Society of America Journal 56: 437–444.

CEC. (2006). Proposal for a Directive of the European Parliament and of the Council Establishing a Framework for the Protection of Soil. Commission of the European Communities. http:// ec.europa.eu/environment/soil/pdf/com_2006_0232_en.pdf. Accessed 23 March 2010.

EEC. (1986). European Community Council Directive 86/278/EEC of 12 June 1986 on the protection of the environment, and in particular of the soil, when sewage sludge is used in agriculture.

EEC. (1991). European Community Council Directive 91/271/EEC of 21 March 1991 concerning urban wastewater treatment.

EEC. (1999). European Community Council Directive 99/31/EEC of 26 April 1999 on the landfill of waste.

Freibauer, A.; Rounsevell, M.D.A.; Smith, P.; Verhagen, J. (2004). Carbon sequestration in the agricultural soils of Europe. Geoderma 122: 1–23.

Giller, K.E.; Witter, E.; McGrath, S.P.; (1998). Toxicity of heavy metals to microorganisms and microbial processes in agricultural soils: A review. Soil Biology and Biochemistry 30: 1389–1414.

Giusquiani, P.L.; Pagliai, M.; Gigliotti, G.; Businelli, D.; Benetti, A. (1995). Urban waste compost: Effects on physical, chemical, and biochemical soil properties. Journal of Environmental Quality 24: 175-182.

Hart, P.B.S.; Rayner, J.H.; Jenkinson, D.S. (1986). Influence of pool substitution on the interpretation of fertilizer experiments with 15N. Journal of Soil Science 37: 389–403.

Horswell, J.; Speir, T.W.; van Schaik, A.P. (2003). Bio-indicators to assess impacts of heavy metals in land-applied sewage sludge. Soil Biology and Biochemistry 35: 1501–1505.

Hue, N.V.; Liu, J. (1995). Predicting compost stability. Compost Science and Utilization 3: 8–15.

Iglesias-Jiménez, E.; Pérez-Garcia, V. (1989). Evaluation of city refuse compost maturity: a review. Biological Wastes 27: 115–142.

John, B.; Yamashita, T.; Ludwig, B.; Flessa, H. (2005). Storage of organic carbon in aggregate and density fractions of silty soils under different types of land use. Geoderma 128: 63-79.

Jones, S.K.; Rees, R.M.; Kosmas, D.; Ball, B.C.; Skiba, U.M. (2006) Carbon sequestration in temperate grassland; management and climatic controls. Soil Use and Management 22: 132–142.

Lal, R. (2004a) Soil carbon sequestration impacts on global climate change and food security. Science 304: 1623–1626.

Lal, R. (2004b). Soil carbon sequestration to mitigate climate change. Geoderma 123: 1-22.

Lashermes, G.; Nicolardot, B.; Parnaudeau, V.; Thuriés, L.; Chaussod, R.; Guillotin, M.L.; Linéres, M.; Mary, B.; Metzger, L.; Morvan, T.; Tricaud, A.; Villete, C.; Houot, S. (2009). Indicator of potential residual carbon in soils after exogenous organic matter application. European Journal of Soil Science 60: 297–310.

Logan, T.J. (1992). Reclamation of chemically degraded soils. In: Advances in Soil Science 17 (Stewart, B.A.; Lal, R. Eds.), pages 15-35. Ed. Springer-Verlag, New York (EEUU).

Luczkiewicz, A. (2006). Soil and groundwater contamination as a result of sewage sludge land application. Polish Journal of Environmental Studies 15: 869–876.

Navarro-Pedreño, J.; Gómez, I.; Moral, R.; Mataix, J. (1996). Nitrogen nutrition of tomato derived from the use of sewage sludge and almonds residue as fertilizers. In: Progress in Nitrogen Cycling Studies (Van Cleemput et al., eds.), pages 243–246. Ed. Kluwer Academic Publishers, Dordrecht (Netherlands).

O'Connor, G.A.O.; Elliott, H.A.; Basta, N.T.; Bastian, R.K.; Pierzynski, G.M.; Sims, R.C.; Smith, J.E. (2005). Sustainable land application: an overview. Journal Environmental Quality 34: 7–17.

Ramlal, E,; Yemshanov D.; Fox, G.; McKenney, D. (2009) A bioeconomic model of afforestation in Southern Ontario: Integration of fiber, carbon and municipal biosolids values. Journal of Environmental Management 90: 1833–1843.

Robin, P.; Ablain, F.; Yulipriyanto, H.; Pourcher, A.M.; Morvan, T.; Cluzeau, D.; Morand, P. (2008) Evolution of non-dissolved particulate organic matter during composting of sludge with straw. Bioresource Technology 99: 7636–7643.

Roldán, A.; Albadalejo, J.; Thornes, J.B. (1996). Aggregate stability changes in a semiarid soil after treatments with different organic amendments. Arid Soil Research and Rehabilitation 10: 139–148.

Rothamsted Research. (2006). Guide to the Classical and other Long-term Experiments, Datasets and Sample Archive. Ed. Lawes Agricultural Trust Co. Ltd.

Rousk, J.; Brookes, P.C.; Bååth, E. (2009). Contrasting soil pH effects on fungal and bacterial growth suggest functional redundancy in carbon mineralization. Applied and Environmental Microbiology 75: 1589–1596.

Sánchez-Monedero, M.A.; Mondini, C.; de Nobili, M.; Leita, L.; Roig, A. (2004). Land application of biosolids. Soil response to different stabilization degree of the treated organic matter. Waste Management 24: 325–332.

Smith, P.; Powlson, D.S.; Glendining, M.J.; Smith, J.U. (1997). Potential for carbon sequestration in European soils: preliminary estimates for five scenarios using results from long-term experiments. Global Change Biology 3, 67–79.

Soriano-Disla, J.M.; Navarro-Pedreño, J.; Gómez-Lucas, I. 2010. Contribution of a sewage sludge application to the short-term carbon sequestration across a wide range of agricultural soils. Environmental Earth Sciences (DOI 10.1007/s12665-010-0474-x).

Singh, R.P.; Agrawal, M. (2008) Potential benefits and risks of land application of sewage sludge. Waste Management 28: 347–358.

Speir, T.W.; van Schaik, A.P.; Percival, H.J.; Close, M.E.; Pang, L. (2003). Heavy metals in soil, plants and groundwater following high-rate sewage sludge application to land. Water, Air and Soil Pollution 150: 319–358.

Speir, T.W.; van Schaik, A.P.; Hunter, L.C.; Ryburn, J.L.; Percival, H.J. (2007). Attempts to derive EC50 values for heavy metals from land applied Cu-, Ni-, and Zn-spiked sewage sludge. Soil Biology and Biochemistry 39: 539–549.

Tian, G.; Granato, T.C.; Cox, E.; Pietz, R.I.; Carlson, C.R.; Abedin, Z. (2009). Soil carbon sequestration resulting from long-term application of biosolids for land reclamation. Journal Environmental Quality 38: 61–74.

Vleeshouwers, L.M.; Verhagen, A. (2002). Carbon emission and sequestration by agricultural land use: a model study for Europe. Global Change Biology 8: 519–530.

Vogeler, I.; Green, S.R.; Mills, T.; Clothier, B.E. (2006). Modelling nitrate and bromide leaching from sewage sludge. Soil and Tillage Research 89: 117–184.
Wang, X.; Chen, T.; Ge, Y.; Jia, Y. (2008). Studies on land application of sewage sludge and its limiting factors. Journal of Hazardous Materials 160: 554–558.

In: Sewage Sludge Management
Editors: A. A. Zorpas and V. J. Inglezakis

ISBN: 978-1-61324-393-0
© 2012 Nova Science Publishers, Inc.

Chapter 6

SUSTAINABLE USE OF SLUDGE AND TREATED WASTEWATER IN THE AGRICULTURAL FIELD

Ioannis Papadopoulos, Costas N. Costa and Petros G. Savva

Department of Environmental Management, Cyprus University of Technology,
Limasol, Cyprus

1. INTRODUCTION

The Near East region has increased considerably over the past few years and the need for food and water is continually growing [1]. Traditionally, this situation has been met by simply increasing water supply and/or expanding agriculture. Today, this solution is nearing its limits [2]. In a number of countries in the Region, the actual consumption of water is fast approaching the limits of resources available [2]. Moreover, agricultural land is becoming rare [2]. For most countries of the Region water thus became the main factor limiting their development and consequently a major economic, social and political challenge [3]. Therefore, the use of non-conventional water resources and the overall management of water in an efficient and effective way, became pressing problems in most of the Near East countries [3]. Treatment of wastewater and its use for irrigation is an attractive option, particularly in arid and semi-arid areas of the Region as it represents an additional, renewable, reliable source of water and fertilizer as well [4]. Solid wastes are less used and more difficult to be applied. High precaution is needed due to potential high concentration in heavy metals or in parasite eggs [5].

The use of wastewater and solid wastes, besides positive effects may also have adverse impacts on public health and the environment, largely depending on wastewater and sludge characteristics, the degree of purification and the method and location of use [6]. Soil, groundwater and surface water pollution are among the most important disadvantages of the wastewater use. However, scientifically sound planning and effective management of the irrigation or fertilization regimes can minimize these disadvantages to the level of environmental insignificance effects [7]. Because of this, it is important to provide farmers with the information needed to help them improve management of treated wastewater used for irrigation and solid wastes used for fertilization. It is now feasible since considerable information and experience have been acquired at country and regional levels from successfully implemented reuse projects. In this work an attempt was done to:

- Consolidate the knowledge and experience gained on reuse in countries of the Region, and
- Provide sets of agronomic practices, in an integrated farm management form.

In 1989, WHO published the "Health Guidelines for the Use of Wastewater in Agriculture and Aquaculture" [8]. In the same year, UNEP and WHO, jointly published the "Guidelines for the Safe Use of Wastewater and Excreta in Agriculture and Aquaculture" [9], with emphasis on environmental and public health protection. In 1991 UNEP and FAO, jointly published the "Environmental guidelines for wastewater reuse in the Mediterranean Region" [10]. These were followed by an FAO publication on "Wastewater Treatment and Use in Agriculture" in 1992 [11]. These guidelines have been supporting many developing countries to implement or upgrade environmentally sound and safe wastewater use systems adapted to their own technical, socioeconomic and cultural conditions. The FAO Regional Office in Cairo also produced in 1995 a publication on "Wastewater management for agriculture protection in the Near East Region" [12] and in 1991-93 seven technical bulletins intended to help the countries of the Region for best use of wastewater in agriculture. This FAO publication on "Treated Wastewater Users Manual" intends to help irrigation operators and the extentionists who are in close contact with the farmers, irrigationists, other users in the forestry and landscaping.

1.1 Objectives

Wastewater reclamation and reuse has emerged as a realistic option for new sources of water to meet shortage and cover increasing water needs in the Near East countries, but also to meet wastewater disposal regulations in the countries of the Region aimed at protecting environment and public health. In addition, from the environmental viewpoint, reclamation and reuse of treated municipal wastewater for irrigation could be probably the most safe and feasible disposal approach.

The transfer of the available research data, management experiences and know-how gained during the last years from several countries of the Region is the main objective of this manual. Different from other manuals prepared by FAO and other organizations on "Treatment and Reuse" this manual is addressed to the final users of wastewater for irrigation, basically the farmers, and those who are in direct contact with the use of treated wastewater. In this respect, this manual provides an integrated management solution to potential problems, which enable optimum agronomic results on safe and environmentally sound-grounds. Because of this, the present manual:

- Is written in a simple way
- Literature in the text is limited
- Legal, institutional and similar aspects are very concisely presented

- Agronomic aspects of interest for the farmer like irrigation, fertilization, selection of crops, are extensively covered.

Political, boundary and inter-boundary aspects are not included in the manual. Aspects of planning, design, operation and maintenance of wastewater treatment plants are also not covered. This manual assumes that treated wastewater of a certain quality is already available for irrigation, and therefore, Best Management Practices (BMP) are proposed to achieve best possible results on sustainable basis.

We do hope that this manual will contribute positively in water reuse and that it will secure sustainability in a way to combine environmental and public protection with economic results.

1.2. Overall Water Demand in the Near East

While the Near East covers 14% of the total area of the world and contains 10% of its population, its water resources are only about 2% of the total renewable water resources of the world [13].

Reuse of municipal wastewater is not a new concept [10]. With the increase in water demand as population grows and the improvement of standard of living, wastewater reuse is getting an increasing role in the planning and development of additional water supplies. This is particularly important for the Near East countries since they are mostly arid or semi-arid. They have low rainfall, mostly seasonal and with erratic distribution [13]. Moreover, quality of water is steadily deteriorating.

Use of wastewater for beneficial purposes such as irrigation has been practiced, although without control, in many countries of the Region (Table 1) [14]. In developed countries of the Region, there are growing problems of providing adequate water supply and new approaches i.e. desalination and wastewater reuse are adopted to solve partially the problem of water scarcity, in addition to coping with the pressing environmental problem of wastewater

Table 1. Near East ccountries using the largest quantities of desalinated water and treated wastewater (Aquastat, FAO, 1997)

Country	Use of non-conventional sources of water								
	Desalinated water	As % of total	As % of withdrawal	Treated waste-water	As % of Total	As % of withdrawal	Total non-conventional	As % of total	As % of withdrawal
	10^6 m^3/yr	Withdr	of country	10^6 m^3/yr	Withdr	of country	10^6m^3/yr	withdr.	of country
Saudi Arabia	714	41.3	4.20	217	18.1	1.28	931	31.8	5.48
United Arab Emirates	385	22.3	18.26	108	9.0	5.12	493	16.8	23.38
Kuwait	231	13.4	42.94	52	4.3	9.67	283	9.7	52.61
Syria	-	-	-	370	30.8	2.57	370	12.6	2.57
Egypt	25	1.5	0.05	200	16.7	0.36	225	7.7	0.41
Other 24 countries	372	21.5	0.09	253	21.1	0.06	625	21.4	0.15
Total Near East	1 727	100.0	0.34	1 200	100.0	0.23	2 927	100.0	0.57

disposal. In several countries of the Near East, the need for new water resources is becoming even more acute and pressing [9-11]. Because of this, wastewater treatment and its use are becoming popular. However, protecting public health and the environment are concerns associated with reuse, which have not been seriously considered in a lot of countries of the Region. The health and overall risks should be within acceptable levels, safeguarding public health and protecting the environment.

1.3 Benefits and Limitations of Wastewater Reuse

Wastewater and other waters of low quality are significant in the overall water resources management. By releasing fresh water sources for potable water supply and other priority uses, reuse makes a contribution to water and energy conservation and improves quality of life. Wastewater can have positive agronomic results [15]. Moreover, wastewater use schemes, properly planned and managed, can have positive environmental and health impact, besides providing increased agricultural yields [15]. However, reuse of wastewater may also have adverse effects on the environment and health.

Currently, the quantity of treated wastewater used for irrigation in Near East countries is estimated to be **1200** million m3/year.

2. SUITABILITY OF WASTEWATER FOR IRRIGATION

Wastewater is unique in composition. Both chemical and biological constituents which occur in this water must be taken into consideration. In this chapter the problems are shortly presented and more emphasis is attached to the solutions and some integrated management approaches are presented in order to alleviate an/or overcome these problems.

2.1 Chemical and Physical Characteristics of Wastewater

The constituents of concern in wastewater are given in Appendix I [18]. However, for proper management approaches, the main constituents of concern to farmers in Near East countries are:

- Suspended solids since filtration may be needed particularly with micro-irrigation systems.
- Nutrients in order to adjust fertilization.
- Salinity in order to estimate leaching fraction and select appropriate cropping pattern, and
- Pathogens for precautionary measures, selecting cropping pattern and choosing the appropriate irrigation system.

2.1.1 Wastewater Treatment

The main objective of the treatment is to produce treated wastewater suitable for irrigation within acceptable level of risk for human health and the environment. In this respect, the most appropriate wastewater treatment is that which provides the chemical and microbiological quality required for a certain specific use at low cost and minimal operational and maintenance requirements. Wastewater treatment plants, both conventional and natural biological treatment, reduce organic and suspended solids, remove wastewater constituents that may be toxic to crops as well as pathogens which are of main concern to protect farmers and public health in general [19].

The different degrees of conventional treatment are [19]:

- Preliminary. Removal of coarse solids and other large materials from the raw wastewater.
- Primary. Removal of settleable organic and inorganic solids as well as floating materials.
- Secondary. Removes the residual organics and suspended solids from primary treated wastewater.
- Tertiary and/or advanced. Removal of specific wastewater constituents, like nutrients and heavy metals, which are not removed by secondary treatment. Disinfection, usually with chlorine, is used to reduce microbiological constituents.

The most widely used natural biological treatment is the wastewater stabilization system which may be designed to achieve different degree of wastewater purification [11]. The system consists of primary facultative ponds (anaerobic ponds), secondary facultative ponds (biological treatment) and maturation ponds which provide tertiary treatment and further pathogen reduction.

2.1.2 Treatment and Wastewater Quality Considerations

In an integrated approach of treatment and use of the wastewater for irrigation the assurance of treatment reliability and reduction of monitoring are highly desirable since these are out of the farmer's control. In the planning and implementation of new wastewater reclamation and reuse projects the intended wastewater reuse application should govern the degree of wastewater treatment required and the reliability of wastewater treatment process and operation. However, in existing treatment plants the quality of treated effluent is already known and the farmers are obliged to modify their management to the water quality provided to them. In general, in countries of the Near East, little intervention if any at all is done concerning the chemical composition of treated wastewater [18]. In this respect, the management approaches proposed to the farmers to alleviate eventual problems, are mostly presented and discussed.

2.1.3 Wastewater Quality Criteria for Irrigation

The chemical and physical quality characteristics are the same as applied to any irrigation water. In this respect, the general guidelines presented in Table 2 can be used to evaluate treated wastewater for irrigation purposes in terms of the chemical constituents such as the dissolved salts, relative sodium content and toxic ions. The procedure remains the same as with other waters [20].

Salinity. In most countries the water used for municipal supply is of the best quality available and it is usually of low salinity. However, under water scarcity conditions salinity may be a problem. The quantity and kind of salts present are important to assess the suitability of treated wastewater for irrigation. Potential problems are related to the total salt content, to the type of salt or to excessive concentration of one or more elements (Ayers and Westcot, 1985).

To overcome the problem of salinity, at farmer's level, more emphasis must be given to the following approaches:

a)   Select crops tolerant to the wastewater salinity and still be on the profitable site. Table 3 may help farmers to select appropriate cropping pattern depending on the salinity of wastewater and salt tolerance of crops. With salinity less than 3 dS/m, with good management, most fruits and vegetables can be produced. As salinity raises the selection of crops is becoming difficult and, except for certain vegetables, the choice is mostly restricted to fodder crops.

b)   Select salt tolerant crops with the ability to absorb high amounts of salts without particular toxicity effects (salt harvesting crops). In case of irrigation with treated wastewater of high salinity, particularly in areas with limited rain and natural leaching, for long term sustainable reuse salt harvesting crops may help reduce salinity build-up in the soil. Some recommended crops are sudax, sorghum, bermuda grass and barley.

Table 2. Guidelines for interpretations of water quality for irrigation (FAO 1985) [20]

Potential Irrigation Problems	Degree of restriction on use			
	Units	None	Slight to Moderate	Severe
Salinity				
EC_w[1] or	dS/m	<0.7	0.7-3.0	>3.0
TDS	mg/l	<450	450 - 2000	>2000
Infiltration				
SAR^2=0- 3 EC_w =	dS/m	>0.7	0.7-0.2	<0.2
3 -6		> **1.2**	1.2-0.3	<0.3
6-12		> 1.9	1.9-0.5	<0.5

Potential Irrigation Problems	Units	Degree of restriction on use			
		None		Slight to Moderate	Severe
12-20		>2.9		2.9-1.3	< 1.3
20-40		>5.0		5.0-2.9	<2.9
Specific ion Toxicity					
Sodium (Na)					
Surface Irrigation	SAR	< 3		3 - 9	> 9
Sprinkler Irrigation	meq/l	< 3		> 3	
Chloride (CI)					
Surface Irrigation	meq/l		< 4	4-10	> 10
Sprinkler Irrigation	meq/l		< 3	> 3	
Boron (B)	mg/l		<0.7	0.7-3.0	>3.0
Miscellaneous effects					
Nitrogen (NO$_3$-N)[3]	mg/l		< 5	5-30	>30
Bicarbonate (HCO$_3$)	meq/l		< 1.5	1.5-8.5	>8.5
pH		Normal range 6.5 - 8.4 pH			

ECW: electrical conductivity in deciSiemens per meter at 25° C.
SAR: sodium adsorption ratio.
NO3 – N: nitrate nitrogen reported in terms of elemental nitrogen. NH4-N and organic-N should be also tested with wastewater.

c)   Select irrigation system with uniform application, high efficiency and providing the possibility for more frequent irrigation. With pressurised irrigation systems particularly with drip and minisprinklers, the salinity permissible levels of irrigated crops could be higher. With such systems the guidelines concerning crop tolerance to salinity (Maas, 1977 [21]), are only rough estimates. Better yields can be achieved with those critical levels when appropriate management with modern irrigation systems are used (Goldberg et al., 1971 [22]; Papadopoulos et al. 1987 [23]).

d)   Scheduling of irrigation. The amount of irrigation water and the frequency of water application are crucial factors to control salinity. With micro-irrigation systems, irrigation could be more frequent and soil salinity in the vicinity of the irrigated plant could be maintained at a lower level.

Table 3. Electrical conductivity of irrigation water

Electrical Conductivity of irrigation water (dS/m, and mg/l)*					
<2	2-3	3-4	4-5	5-7	>7
<1280	1280-1920	1920-2560	2560-3200	3200-4480	>4480
Citrus	Fig	Sorghum	Soybean	Safflower	Cotton
Apples	Olives	Groundnut	Date palm	Wheat	Barley
Peach	Broccoli	Rice	Harding grass	Sugar beet	Wheat grass
Grapes	Tomato	Beets	Trefoil	Rye grass	
Strawberry	Cucumber	Tall fescue	Artichokes	Barley grass	
Potato	Cantaloupe			Bermuda grass	
Pepper	Watermelon			Sudax	
Carrot	Spinach				
Onion	Vetch				
Beans	Sudan grass				
Corn	Alfalfa				

e)   Leaching is a common approach by farmers but not the best probable solution in the case of water scarcity, inadequate drainage or shallow water table. In a long term the total quantity of salt applied in the soil with the wastewater (salt in) and the rate at which salt is removed by leaching and crop uptake (salt out) should be approximately the same. Which approach is to be accepted for the "salt out" is very important in selecting cropping pattern and management for the effective use of wastewater for irrigation (Papadopoulos, 1991 [24]). Salt harvesting crops of economical value like sudax and sorghum give good results. Cultivating a salt harvesting-crop every year or periodically is recommended.

f)   Soil polymers and/or other soil conditioners although effective under certain conditions for a certain period for open field crops are not recommended. Their half-life is usually short and their price is high.

g)   Drainage. One of the measures necessary to prevent irrigation-induced waterlogging and salinization in arid and semi-arid regions is the installation of drainage facilities. Drainage, in combination with adequate irrigation scheduling, allows for the leaching of excess salts from the plant root zone.

Alkalinity. Dispersion of soil colloidal phase, stability of aggregates, soil structure and permeability for water, are very sensitive to the type of exchangeable ions present in irrigation water. The increase in soil alkalinity, which may occur with treated wastewater due to high Na concentration even though leaching is allowed, reduces soil permeability for water, particularly at the soil surface, since soil clays tend to disperse and swell from the increased level of exchangeable Na. However, at certain sodium adsorption ratio (SAR) the infiltration rate increases or decreases with the salinity level. Therefore, SAR and ECW should be used in combination to evaluate eventual potential problems (Rhoades, 1977 [25]).

The Following Management Solutions Are Recommended

a)   Chemical amendments. The use of calcium source amendment such as gypsum is widely accepted for amelioration of soils with high percentage of Na in the Cation Exchange Capacity (CEC) or whenever water high in SAR is used for irrigation. Na in soil is exchanged by Ca from gypsum and dispersion of the colloidal phase is reduced. Application of gypsum should be repeated periodically depending on the Na content in water and the CEC of the soil. The farmers are advised to seek professional help to estimate the amount and frequency of the gypsum application required.

b)   Adapted Irrigation system. Crust formation at the soil surface is the result of irrigating with water high in SAR. The degree of the problem, however, is not the same with all irrigation systems. In general, the surface irrigation systems with water high in SAR create thick surface crust. Similar results are obtained with sprinklers of high discharge capacity. This way, the soil permeability for water as well as soil aeration and emergence of the seeds are affected.

c)   With low capacity minisprinklers and drippers of low discharge rate the formation of surface soil crusting is reduced, the duration of irrigation is extended and sufficient time is allowed for water penetration into the soil (Papadopoulos and Stylianou, 1988a [26]).

Organic matter. The alkalinity problem could also be solved by addition of organic matter like straw, other plant materials and organic manure.

Specific ion toxicity. The most toxic ions generally occurring in treated wastewater are sodium (Na), chloride (CI), and boron (B) which is causing most prevalent toxicity cases. Boron is one of the essential elements to plants; hence, at concentrations exceeding 0.5 mg B/l it could be toxic to sensitive crops (Table 4).

Measures Recommended to Overcome Ion-Toxicity

- With sensitive crops, B-toxicity is difficult to correct without changing the crop or the water supply (Ayers, 1977 [27]). For water with certain level of boron, one should select crops which can tolerate that concentration.
- Leaching may help to maintain boron concentration in the soil at levels comparable to those of the water used for irrigation (Bernstein and Francois, 1973 [28]). Depending on the soil, a certain leaching fraction is usually added to the irrigation requirement.
- The frequent irrigation dilutes boron in the soil solution,
- With the use of micro-irrigation systems, water application could be more uniform and the frequency of irrigation could also be controlled.

Table 4. Sensitive, semitolerant and tolerant fruits and vegetables

Sensitive (1mg/l)	Semitolerant (2 mg/l)	Tolerant (3 mg/l)
Citrus	Bean	Carrot
Avocado	Bell pepper	Lettuce
Apricot	Tomato	Cabbage
Peach	Corn	Onion
Cherry	Olives	Sugar beet
Grapes	Radish	Date palm
Apple	Pumpkin	Asparagus
Pear	Wheat	Turnip
Plum	Potato	
Strawberries	Sunflower	

Chloride (CI) and sodium (Na) are less toxic than boron. In arid and semiarid regions due to the relatively high Na and CI contents of the domestic water, wastewater may have high concentration of these elements. With proper irrigation management, (irrigation system, frequency of irrigation, leaching) toxicity effects can be reduced significantly, presenting no real constraint for reuse.

Trace elements and heavy metals. They are the main problem with treated wastewater reuse in countries with heavy industry. The metals that may be present in sewage can pose significant health hazards to humans and animals and also may affect the irrigated crops. Cadmium (Cd), copper (Cu), molybdenum (Mo), nickel (Ni) and zinc (Zn) could cause serious health risk to humans and animals. These metals in most cases are accumulated in the crop, and could adversely affect humans or domestic animals, feeding on these crops.

**Table 5. Maximum permissible cumulative loading with metals in kg/ha
for agricultural land per year (Adapted from Biswas, 1987 [29])**

Country	Cd	Cu	Cr	Pb	Hg	Ni	Zn
France	5.4	210	360	210	2.7	60	750
Germany, FR	8.4	210	210	210	5.7	60	750
Netherlands	2.0	120	100	100	2.0	20	400
UK	5.0	280	1000	1000	2.0	70	560

Table 6. Recommended limits for trace elements in reclaimed water use for irrigation[a] [30]

Constituent	Long-term Use[b] (mg/l)	Short- term Use[c] (mg/l)
Aluminium	5.0	20.0
Arsenic	0.10	2.0
Beryllium	0.10	0.5
Boron	0.75	2.0
Cadmium	0.01	0.05
Chromium	0.1	1.0
Cobalt	0.05	5.0
Copper	0.2	5.0
Fluoride	1.0	15.0
Iron	5.0	20.0
Lead	5.0	10.0
Lithium	2.5	2.5
Manganese	0.2	10.0
Molybdenum	0.01	0.05
Nickel	0.2	2.0
Selenium	0.02	0.02
Vanadium	0.1	1.0
Zinc	2.0	10.0

Adapted from: National Academy of Sciences - National Academy of Engineering (1973)
For water used continuously on all soils
For water used for a period of up to 20 years on fine - textured neutral or alkaline soils
Considerations and management aspects for trace elements and heavy metals

Because of this, many developed countries have set maximum permissible cumulative loading of metals to agricultural lands. Heavy metals are discussed in more details in connection with sludge. Biswas (1987 [29]) reported permissible heavy metal loading in some European countries (Table 5).

Recommended maximum concentrations of metals in irrigation water are presented in Table 6.

The question, for the Near East countries, is whether heavy metals represent a serious health and/or environmental problem. In general, heavy metals and trace elements should not be considered as a pressing or a serious problem for two main reasons:

- The concentration of heavy metals in municipal wastewater is low due to low heavy industry activities, and
- The soils of the Region are mostly with high CaCO3 and of pH above 7 [18], which inactivate the heavy metals and reduce their mobility and availability to crops. In such conditions the metals are rendered unavailable and higher loading and concentration than those presented in Tables 6 and 7 could be accepted.

CONCLUSION

1. Heavy metals in treated wastewater under calcareous soil conditions should not be considered as real problem and no particular management is required.
2. Under acid conditions (few cases in the Region) (pH<7) heavy metals could be a problem and the following measures at farmers level are recommended:
   - Liming (use of calcium carbonate). In this way soil pH is increased and thus the solubility of heavy metals is reduced.
- Avoid using acid fertilizers.
- Select crops tolerant to certain heavy metals.
   - Select crops having no biomagnification characteristics (accumulation of certain heavy metals by specific crops and/or parts of the crop).

Farmers should be encouraged to consult professionals before making the final decision on the measures needed.

Crop Nutrients in Treated Wastewater

The fertilizing potential of treated wastewater due to nutrients might be an asset for crops but might also be a source of pollution for the environment depending mainly on the farmer's management. Nutrients are a unique characteristic of treated wastewater of particular interest to the farmers. Because of this, a Code of Good Fertilization Practice is developed and introduced in this manual. It is intended to give new dimensions to the use of treated wastewater for irrigation in a rational, profitable and environmentally sound way.

Nutritional Value of Treated Wastewater

The suspended, colloidal and dissolved solids present in wastewater contain macro- and micronutrients, which are essential for crop nutrition. However, the nutrient content of the wastewater may exceed the plant needs and thus pose a potential source for underground water pollution. It may also cause problems related to excessive vegetative growth, delayed or uneven maturity, or reduced quality of the irrigated crops. An estimate of the nutrients present in the treated effluent as part of the overall fertilization program of the irrigated crops is necessary. In this respect wastewater analysis is required at least once at the beginning of the growing season.

Nutrients in wastewater occurring in quantities important to agriculture and landscape management include nitrogen, phosphorus and occasionally potassium, zinc, boron, and sulphur. Other macro and micronutrients may also be present. In addition, the organic matter in the wastewater beside its long-term effect on soil fertility, it can also contribute to the soil structure stability.

For correct estimation of the nutrient capacity of wastewater, based on its chemical analysis, Table 7 could be used.

Table 7. Amount of nutrients with certain amount of irrigation water in kg/ha/year [18]

Irrigation water m³/ha/year	Concentration of a nutrient in wastewater (mg/l)								
	5	10	15	20	25	30	35	40	50
	Nutrients added (kg/ha/year)								
1000	5	10	15	20	25	30	35	40	50
2000	10	20	30	40	50	60	70	80	100
3000	15	30	45	60	75	90	105	120	150
4000	20	40	60	80	100	120	140	160	200
5000	25	50	75	100	125	150	175	200	250
6000	30	60	90	120	150	180	210	240	300
7000	35	70	105	140	175	210	245	280	350
8000	40	80	120	160	200	240	280	320	400
9000	45	90	135	180	225	270	315	360	450
10000	50	100	150	200	250	300	350	400	500

With certain crops no additional fertilizers are needed. Similarly, in places where fertilizers are of concern, wastewater, if available, could give the answer for a high yield of good quality.

The Nitrogen content of municipal wastewater following secondary treatment ranges from 20 to 60 mg/l. The N in treated wastewater could be in excess of the crop needs. Knowledge about the N concentration in the wastewater and the proper management of the load of NPK are essential to overcome problems associated with eventual high N concentration.

Phosphorus in wastewater from secondary treatment systems varies from 6 to 15 mg/l (15-35 mg/l P2O5) unless removal is accomplished during treatment. Evaluation of P in the treated wastewater should be made in conjunction with soil testing for fertilization planning.

Potassium concentration in wastewater is not known to cause adverse effects on plants or the environment. It is an essential macronutrient and affects positively soil fertility, crop yield and quality. The range of K in secondary treated wastewater is 10 to 30 mg K/l (12-36 mg/l K2O). This amount must be taken into consideration in formulating the fertilization programme according to crop needs.

Other nutrients. Most of wastewaters usually contain adequate sulphur, zinc, copper and other crop micronutrients. Attention must be given to boron. Treated wastewater contains enough boron to correct any boron deficiencies. Of greater concern, however, is that this element is usually found in excess in certain wastewater creating problems of phytotoxicity. To overcome the B problem the same measures as for highly soluble salts are recommended (selection of crop, leaching, scheduling of irrigation, irrigation system). Table 5 may help the farmers to select the crop.

In general farmers should remember that:

- Fruit trees are more sensitive to B than vegetables
- In case of relatively high B concentration in wastewater to prefer annual crops.

Table 8. Fertilization potential through wastewater (FAO/RNEA, 1992) [31]

	N	P	κ
Nutrient concentration (mg/l)	40	10	30
Yearly nutrient added (kg/ha) through			
Application of 10000 m^3 water/ha	400	100	300

Nutrient Load through Irrigation with Effluents

The concentration of N and P in the secondary treated wastewater could vary substantially depending on the source of the primary wastewater and the treatment process. From conventional treatment plants, the N and P concentrations are usually higher than with the aerated lagoons and oxidation ponds. In general, N and P are reduced during treatment but K remains approximately at levels found in untreated wastewater.

The amounts of N, P and K applied per hectare with 1000mm irrigation by a typical wastewater are given in Table 8. Evidently, the load of nutrients depends on the overall amount of water applied. It is assumed that for high nutrient efficiency, irrigation should be based on crop water requirements.

Such fertilizer application rates supply all or more of the N normally required by a number of agricultural crops and also much of the P and K. In this respect, careful consideration should be given to each crop separately for estimating eventual supplementary fertilizer requirements. In some cases some nutrients in wastewater may be in excess of that required for a balanced crop growth and may potentially stimulate excessive growth of the vegetative parts of the crops rather than the flowers and seeds. This may be a problem for crops as sunflower, cotton and some fruits. In case of excess of nutrients an appropriate cropping pattern and/or mixing of the treated wastewater with fresh water to reduce the nutrient application are methods suggested to solve the problem.

Control of the Problem of Eventual Excess N in Wastewater

a) Estimate the Concentration of N

The chemical analysis for elemental N is required. Based on this analysis the farmer could calculate the amount of N added to the soil through the quantity of wastewater used. This amount should be subtracted from the fertilizer amount needed by crops. For easy estimations the farmer should remember that:

1 ppm (Part Per Million) = 1 g/m^3 in Irrigation Water = 1 mg/l

Therefore, the amount of N and other nutrients applied to the soil with wastewater depends on the amount of irrigation water. The farmers should be aware how to irrigate.

b) Select the Crop Based on the N Level

Selection of crop depending on N in treated wastewater is needed for two purposes:

• Making the best possible use of N from wastewater. If nitrogen present in wastewater is not adequate, supplemental fertilizer nitrogen is needed for satisfactory crop yields. From the standpoint of long-term application of wastewater, N input levels should be adjusted to compensate for N removal by the harvested portion of the crop plus expected losses from the system by volatilization and leaching. Total plant uptake of N may greatly exceed crop N removal, particularly in fruit crops. The N, P, K requirement of main crops is given in the following chapters.

• Avoiding nitrate pollution. Some crops are highly effective in removing nitrogen from soil, which may eventually move down in the form of NO3-N deeper in soil and contaminate underground water. Grasses such as Sudan grass, Bermuda grass, Sudax, and Rhode grass remove N efficiently from the soil. These crops are effective in removing nitrates for the following reasons:

• They have abilities to accumulate nitrate
• Several cuts are possible in one season so that more growth of the crop can be achieved
• Their nitrate content does not decrease with age, and
• They are deep-rooted crops

c) Scheduling of Irrigation

Since nutrients are always present in treated wastewater, then any amount of irrigation water above the crop water requirement would create a problem. The problem could be an environmental or an agronomic one or both. The farmer should remember that with wastewater irrigation it is even more important to follow proper scheduling of irrigation than that with water of good quality.

Amount of water. Crops must be irrigated according to their crop water requirements. It could be stressed that since the amount of irrigation water is different from place to place due to climatic conditions, the nutrients in wastewater could be in excess or inadequate for the same crop under the same soil fertility conditions in different places. Similarly, wastewater of the same quality could have adverse environmental impact in one place but be very safe in another place. Because of this, it is difficult to give absolute numbers for fertilization with wastewater, while with fresh water it is different case.

Frequency of irrigation. The farmers should be also aware that with crops at full growth stage the amount of water per irrigation must be always the same to reach a certain soil depth where the active roots are concentrated. However, since the absolute amount of irrigation water varies with climatic conditions the frequency of irrigation should be changed. The amount of water per irrigation must remain the same.

d) Irrigation System

To avoid pollution from nitrates the irrigation system should provide a uniform water application. Evidently, the higher the efficiency of irrigation system, the higher N uptake efficiency by crops is obtained and the less is the potential for nitrate losses and pollution.

Drip and micro-irrigation systems are superior. With furrow irrigation the farmers are advised to create short furrows especially if laser leveling was out of the farmer's reach.

e) Mixing Treated Wastewater with Fresh Water, if Available

Other Problems

Attention should be given to treated wastewater constituents which cause clogging of the irrigation systems. Clogging problems with sprinkler, minisprinkler and drip irrigation systems might be a serious problem. Growths (slimes, bacteria, etc.) in the sprinkler head, emitter orifice or supply line, cause plugging as do heavy concentrations of algae and suspended solids. The most serious clogging problems occur with drip systems. Filtration may be required just before use. This makes management of drip irrigation system using treated wastewater difficult.

Solutions Suggested to the Problem Are:

- To avoid problems due to suspended algae which are accumulated on the water surface and those problems due to sludge accumulation in the bottom of the reservoir, pumping of water should be taken at a depth of about one meter from the water surface.
- Filtering. Depending on the concentration of suspended solids, algae and other impurities, gravel, sand or other filters are required with micro-irrigation systems.
- Selection of the irrigation method. In case of impurities and in absent of filtering system, micro-irrigation systems should be avoided. Depending on the crop, sprinklers could be a better choice. Even surface irrigation might be preferred.

2.2 Biological Quality Criteria

Water quality criteria and guidelines are essential foundations for successful implementation of any reclamation project with wastewater. Microbiological water quality is most important for field workers as well as for the public who might be exposed to reclaimed water directly or indirectly. Reuse, depending on biological quality, could be restricted or unrestricted. In this manual, the biological quality criteria are not presented. It is assumed that each country has some guidelines and/or regulations with which the farmers are obliged to conform with. The farmers, however, should be aware of these guidelines as well as with the quality of wastewater provided to them in order to apply appropriate management within acceptable health and environmental risk. For the countries, which they don't have yet their national guidelines the WHO guidelines are proposed (Table 9). It should be stressed, although this is not farmer's responsibility, that every country should have its own guidelines and code of practices. In this line the guidelines used in Cyprus are presented as an example (Appendix II).

Table 9. Recommended Microbiological Quality Guidelines for Wastewater Use in Agriculture (WHO 1989) [8][1]

Category reuse conditions	Exposed group	Intestinal nematodes[2] (arithmetic mean no. of eggs per litre[3])	Faecal coliforms (geometric mean no. per 100 ml[3])	Wastewater treatment expected to achieve the required microbiological quality
A. Irrigation of crops likely to be eaten uncooked, sports fields, public parks[4]	Workers, Consumers, Public	≤ 1	≤ 1000[4]	
B. Irrigation of cereal crops, industrial crops, fodder crops, pasture and trees[5]	Workers	≤ 1	No standard recommended	Retention in stabilization ponds for 8-10 days or equivalent helminth and faecal coliform removal
C. Localized irrigation of crops in category B if exposure of workers and the public does not occur	None	Not applicable	Not applicable	Pre-treatment as required by the irrigation technology, but not less than primary

In specific cases, local epidemiological, socio-cultural and environmental factors should be taken into account, and the guidelines modified accordingly.

Ascaris and *Trichuris* species and hookworms.

During the irrigation period.

A more stringent guideline (< 200 faecal coliforms per 100 ml) is appropriate for public lawns, such as hotel lawns, with which the public may come into direct contact.

In the case of fruit trees, irrigation should cease two weeks before fruit is picked, and no fruit should be picked off the ground. Sprinkler irrigation should not be used.

2.3 Good Code of Practice

In every country a "Good Code of Practice" is even more important for the farmers than the quality guidelines for wastewater reuse. The Code of Practice is consisted of suggestions and recommendations, supplementing the guidelines or standards, intended to help farmers to better manage the use of these waters in best possible way.

The Code of Practice refers to the management of the wastewater for irrigation, the irrigation systems and methods and cautions related to harvesting and other details. The Code could be different from one country to another. An example of a Code, which supplements the guidelines, is presented in Appendix III. This code of practice is addressed to both treatment operators (part A and C) and farmers (part B) to indicate them how to manage wastewater for irrigation, which crops to select, which irrigation systems to use depending on the crop and the quality of the wastewater. Emphasis is also attached to the precautionary measures, which will ensure to the farmer safe reuse and good economic results. In this way, any farmer will be in the position to protect himself, the nearby population, the consumers and the environment.

2.4 Prerequisites for Efficient and Effective Reuse

Sound program for fertilization and irrigation are the main and most important parameters for long term sustainable irrigation with wastewater. Soil and wastewater analyses are needed in order to decide about the fertilization program. Fertilization with wastewater is different than with fresh water.

Code of Practice for Fertilization of Irrigated Crops with Wastewater: Step by Step Approach

Fertilization could be practiced empirically (arbitrarily) or on sound scientific way. The empirical application of fertilizers is associated with severe limitations, which lead to low recovery of fertilizers by the crop. In general, empirical fertilization is based on farmer's experience and on broad recommendations. The scientific approach, takes into consideration crop nutrition, soil fertility, nutrient content of wastewater, expected yield and provides the individual grower dependable information regarding the fertilizer needs of his field.

Crop Nutrient and Fertilizer Requirement

The selection of the proper rate of plant nutrients is influenced by the knowledge of the nutrient requirement of the crop, the nutrient supplying power of both soil and water, the efficiency of nutrient uptake under different irrigation, fertilization methods, and the expected yield. These factors taken into consideration for the same crop, different fertilization programme could be recommended for different places [18].

When soil and irrigation water does not furnish adequate quantities of elements necessary for normal growth of plants, it is essential that the required amounts be applied. This necessitates selecting a method that will permit the determination of those deficient elements and furthermore help to predict the amount of nutrients needed to supplement the crop requirement.

a. Removal of Nutrients by Crops from Soil

The quantities of nutrients removed by a crop from the soil are good information, which can be used to optimize soil fertility level needed for various crops. Part of nutrients removed by crop is used for vegetative growth (canopy) and the other for fruit production. The amount of nutrients removed by crop must be present in soil irrespective of the fact that part of them may remain or return to the soil by the end of the growing season as canopy and roots. An important key is to have enough nutrients in the right proportions in the soil to supply crop needs during the entire growing season. In crops from which just the fruits are removed much lower quantities of nutrients are lost than if the entire above ground portion were harvested. The approximate amounts of N, P, K removed from soil by various crops are given in Table 10. Uptake will vary considerably, depending on a number of factors, including yield level, nutrient supply of the soil, fertilization and irrigation method. The data in Table 10 indicate the comparative uptake among crops. In addition, they are an indication of the rate at which the reserve or "storehouse" nutrient in the soil is depleted by certain crops [18].

The amounts of nutrients that are indicated in Table 10 are needed to be available in soil. However, not all of the nutrients should necessarily come from fertilizer. Part of them is supplied from the soil and part from wastewater. In this respect, estimating the amount of nutrients which may be available to the crop from soil (nutrient supplying power of soil) and wastewater (nutrient supplying capacity of wastewater), are important. These amounts are subtracted from the overall amount, which should be supplied by fertilizer applications.

Table 10. Nutrients required by selected crops for canopy formation and fruit production [18]

Crop	N	P	K	P_2O_5	K_2O
Potato					
Canopy (kg/ha)	86	7	120	16	144
Tubers (kg/ton)	3.20	0.54	4.50	1.24	5.40
Tomato					
Canopy (kg/ha)	95	12	108	27	130
Fruits (kg/ton)	1.80	0.17	3.13	0.38	3.75
Eggplant					
Canopy (kg/ha)	105	13	113	30	135
Fruits (kg/ton)	1.96	0.17	3.2	0.40	3.8
Pepper					
Canopy (kg/ha)	90	6	90	14	108
Fruits (kg/ton)	2.0	0.26	1.83	0.6	2.2
Strawberries					
Canopy (kg/ha)	85	5	88	12	106
Fruits (kg/ton)	1.17	0.22	1.53	0.5	1.84
Lettuce (kg/ha)	115	14	160	32	192
Mango					
Canopy (kg/ha)	70	6	108	14	130
Fruits (kg/ton)	1.35	0.19	1.65	0.44	1.98
Banana					
Canopy (kg/ha)	250	26	800	60	1000
Fruits (kg/ton)	2.0	0.22	5.0	0.5	6.0
Citrus					
Canopy (kg/ha)	85	8	90	18	108
Fruits (kg/ton)	1.44	0.19	1.53	0.44	1.84

b. Nutrients Coming from Soil

A soil test can help for estimating the nutrient-supplying power of a soil for P, K, Ca, Mg and other nutrients. Soluble N leaches readily during the rainy season or with the irrigation water, so there may be little carry-over. Because of this, soil testing is not practiced as a tool for soil N fertility. Nevertheless, nitrification may build up the supply of nitrate in soil as the season advances, in case that organic matter is adequate. The soil test measures a part of the total nutrient present in the soil, assumed or experimentally found to be well correlated with the available nutrient. The values of soil analysis are of little use in themselves. To employ such a measurement in predicting nutrient needs of crops should use those chemical methods, which have been tested and calibrated against nutrient uptake rate experiments. Moreover, no analysis could give good results if soil samples are not representative. The results of the soil test could be reliable only if they are based on representative samples. The problems of representative samples, accurate analyses, correct interpretation, and environmental factors, which influence crop response all, enter in. The soil test helps to reduce guesswork in fertilizer practices.

In order to estimate the nutrient supplying capacity of a soil the following parameters are needed:

- Depth of the crop rooting system
- Percent (%) of soil occupied by the root system under different irrigation systems
- Soil Bulk Density (Bd).

These parameters are needed in order to:

1. Calculate the weight of soil of a certain area to a depth where the active root zone of the crop is developed and
2. Estimate the reserves or storehouse available nutrients for the crop in a soil.

Rooting System

The amount of available nutrients in soil is estimated up to that depth of soil where roots are active. The appearance, growth and depth to which roots penetrate in soils are in part species properties but prevailing soil conditions usually exert a pronounced influence. If there is a clay pan, hard pan, compacted layer, or other dense formation, the normal depth of roots is not possible. The rooting depths of selected vegetables and fruit trees under good soil conditions and good management are given in Tables 11 and 12.

Table 11. Rooting depths of selected vegetables (Adapted from FAO, 1990) [32]

Shallow (20-30 cm)	Moderately deep (30-50 cm)	Deep (>50 cm)
Broccoli	Bean	Artichoke
Brussels sprouts	Beet	Asparagus
Cabbage	Carrot	Pumpkin

Shallow (20-30 cm)	Moderately deep (30-50 cm)	Deep (>50 cm)
Cauliflower	Cucumber	Sweet potato
Celery	Eggplant	Tomato
Chinese cabbage	Muskmelon	Watermelon
Garlic	Pea	
Leek	Pepper	
Lettuce	Squash	
Onion	Turnip	
Potato		
Radish		
Spinach		

Table 12. Rooting depths of selected fruit trees (Adapted from FAO, 1990)[32]

Fruit trees	Rooting depths (cm)
Strawberries	15-25
Avocado	120-150
Citrus	120-150
Mango	130-180

Knowledge of the rooting habits of the crops is helpful for the farmers for determining in satisfactory way soil exploitation and revealing the depth to which the reserve nutrients in the soil could be available and contribute to the overall nutrition of the crop. In addition the same information helps to give an idea of the most effective placement of fertilizer and irrigation water.

Soil Occupied by Roots
The mass of soil occupied by roots in m^3 is estimated with the following formula:

Soil Mass Occupied by Roots = Area (m^2) x Root Depth (m) x Bd (tons/m^3)

Example:

Estimate the weight of one hectare at a rooting depth of 0.4 m and with soil   Bd of 1.2

tons/ha.

Soil Weight = 10000 m^2 x 0.4 m x 1.2 tons/m^3 = 4,800 tons/ha

Soil Nutrient Supplying Capacity
    The amount of a nutrient in g/ha is estimated with the following formula:

Amount of Nutrient = Soil Weight (tons/ha) x Available Nutrient (g/ton)

Example:

Estimate the amount of available nutrient in the previous soil where soil analysis indicates that the available nutrient is 10 ppm.

Note: 1 ppm = 1 mg/kg = 1 g/ton

Amount of nutrient = (4,800 tons/ha) x 10 g/ton = 48,000 g/ha = 48 kg/ha

    To avoid calculations, Table 13 could be used to estimate the amount of a nutrient, which could be provided by soil.

How to use Table 13. The two parameters needed are the value of the available nutrient in g/ton (ppm) and the crop rooting depth.

Example:

Given: Soil analysis: P = 50 mg/kg; K = 90 mg/kg
                Crop rooting depth 40 cm.

Reading from the table: P = 240 kg/ha
                                           K = 432 kg/ha

Fraction of Soil Ooccupied by Roots
    The values shown in Table 13 refer to the whole volume of soil up to a certain depth. However, the volume of soil occupied by roots is usually less and depends on the crop, spacing of planting and irrigation system. For drip irrigated vegetables like tomato, pepper, eggplant, the wetted soil volume is usually 30-40% of total soil volume. With wider spacing, which is the case for melon, watermelon, pumpkins the volume of soil occupied by roots could be even less than 20%. For minisprinklers (like minisprinkler irrigated potato) the volume of soil occupied up to a certain depth might be 70-80%. The fraction of soil occupied by roots must be taken into account whenever the amount of available nutrients is calculated; otherwise the available amounts could be overestimated and less fertilizer could be applied and plants could show deficiency symptoms.

Table 13. Available nutrients in soil in kg/ha as determined by chemical analyses at various soil depths. It is assumed that the soil Bulk Density is 1.2 [18]

Soil depth (cm)	Soil chemical analysis (mg/kg)										
	10	20	30	40	50	60	70	80	90	100	150
10	12	24	36	48	60	72	84	96	108	120	180
20	24	48	72	96	120	144	168	192	216	240	360
30	36	72	108	144	180	216	252	288	324	360	540
40	48	96	144	192	240	288	336	384	432	480	720
50	60	120	180	240	300	360	420	480	540	600	900
60	72	144	216	288	360	432	504	576	648	720	1080
70	84	168	252	336	420	504	588	672	756	840	1260
80	96	192	288	384	480	576	672	768	864	960	1440
90	108	216	324	432	540	648	756	864	972	1080	1620
100	120	240	360	480	600	720	840	960	1080	1200	1800

Wastewater Nutrient Supplying Capacity

This capacity depends on wastewater applied per ha per year and the content of a certain nutrient in the wastewater as indicated in Table 8. This amount together with that provided by the soil is subtracted from the overall amount of nutrients required by crop.

The amount of nutrients needed by the crop and the amount of nutrients, which should be applied, are not equivalent. Not all of the nutrients applied by fertilizers or by wastewater are used by crop. The actual amount applied by fertilizers is usually higher than the amount required by the crop. The uptake of nutrients depends, among others, on the irrigation system.

Irrigation System (Method) and Efficiency of N, P, K Uptake

The potential nutrient uptake efficiency by the same crop is different with various irrigation systems. In general, the higher the water use efficiency of a certain irrigation system the higher is the nutrient uptake efficiency. For a well designed irrigation system and with good scheduling of irrigation, the potential N, P and K uptake by the same crop is given in Table 14.

Table 14. Fertilizer-N, P and K uptake in % as influenced by the irrigation system (FAO/RNEA, 1992) [31]

Irrigation system*	Nitrogen	Phosphorus	Potassium
Furrow	40-60	10-20	60-75
Sprinkler	60-70	15-25	70-80
Microirrigation	75-85	25-35	80-90

* The values refer to good designed and operated irrigation systems.

With surface irrigation methods, the fertilizer use efficiency is the lowest. In particular, due to leaching, the N use efficiency can be considerably low. Therefore, in order to estimate the overall amount of nutrients, which must be present in soil to meet crop nutrient requirements for certain yield, the efficiency of the fertilizer nutrient uptake by the crop is needed.

Nitrogen, Phosphorus and Potassium Requirement of Different Crops

Taking into account the nutrient requirement for certain yield, the nutrient capacity of soil and wastewater, and the efficiency of nutrient uptake by crop under different irrigation systems, the following formula can be used to estimate the amount of N, P and K required being readily available in soil:

$$\text{Nutrient Requirement (kg/ha)} = \frac{NR-(SAN+WN)}{IS} \times 100$$

Where: NR = Nutrient Requirement for certain yield (kg/ha)
           SAN = Soil Available Nutrients (Soil Supplying Capacity) (kg/ha)
           WN = Nutrient capacity of the wastewater (kg/ha)
           IS = Nutrient uptake efficiency of irrigation system (%).

Example:

Crop, is drip-irrigated tomato with rooting depth of 40 cm, Drippers are wetting 35% of the soil. Soil analyses indicated 20 ppm P and 50 ppm K and wastewater analyses 25, 8 and 35 ppm for N, P and K, respectively. Find the N, P, K required for yield of 100 and 125 tons/ha. Nutrient uptake efficiency of the irrigation system is 80, 30, 85, for N, P, K, respectively.

1. *Calculate the weight of soil of one hectare to a depth of 40 cm*

Wt of soil (tons/ha/0.4 m depth) $= 10000 \ (m^2) \times 0.4 \ (m) \times Bd \ (tons/m^3)$
                                     $= 4000 \ m^3 \times 1.20 \ tons/m^3 = 4800 \ ton$

2. *Calculate available P and K from soil*

P (kg/ha) = (20 g/ton) x 4800 tons/ha = 86000 g/ha = 86
K (kg/ha) = (50g/ton) x 4800 tons/ha = 240000 g/ha = 240

3. *Actual P and K available in the 35% soil occupied by roots*

P = 96x35/100 = 33.6 kg P/ha
K = 240x35/100 = 84 kg K/ha

4. *Amount of N, P, and K from wastewater*

In order to estimate the overall supplying nutrient capacity of water the crop water requirement are needed to be determined. If crop water requirement data are not available in the region of reuse then the Fractions of Evaporation from pans recommended for selected vegetables given in Table 15 and for fruit trees in Table 16 could be used. It should be stressed that these values are based on long-term experiments and practical verification at farm level.

Table 15. Crop Water Requirements of selected vegetables expressed as fraction of Epan evaporation [18]

Crop*	Days after planting					
	0-15	15-30	30-45	45-60	Thereafter	Last month
Potato	0.4	0.7	0.9	0.8	0.7	
Carrots	0.3	0.5	0.9	0.9	0.8	0.8
Tomato	0.3	0.5	0.8	0.9	1.0	0.8
Cucumber	0.3	0.5	0.7	1.0	1.0	0.8
Lettuce	0.3	0.8	0.9	0.9		
Cabbage	0.3	0.5	0.8	0.9	0.9	
French Beans	0.3	0.4	0.5	0.7	0.8	
Onion	0.3	0.5	0.9	0.9		
Celery	0.3	0.6	0.9	0.9		
Eggplant	0.3	0.5	0.8	0.9	1.0	0.8
Bell Pepper	0.3	0.5	0.7	0.8	0.8	0.7
Watermelon	0.3	0.5	0.9	0.9	0.8	
Melon	0.3	0.5	0.9	0.9	0.8	

* Under less convenient conditions for growth the changes in the amount of water may go every three weeks interval.

Table 16. Crop Water Requirements of selected fruit trees expressed as fraction of Epan evaporation [18]

Citrus	For fully developed trees 0.55
Avocado	For fully developed trees 0.55
Olives	At full development 0.33

Table 16. (Continued)

Mango	-//-
Table grapes	The value increases gradually from 0.2 to 0.5-0.6.
Banana	Increases gradually from 0.6 to 1.0

Example:

Calculate the amount of irrigation water for tomato in an area with the following meteorological data:

Area "x"										
Month	March		April		May		June		July	
	1-15	15-31	1-15	15-28	1-15	15-31	1-15	15-30	1-15	15-30
Effect. Rain (mm)	30	20	15	-	-	-	-	-	-	-
Epan (mm)	63	64	69	68	90	90	93	94	108	128
Epan-Ef. Rain (mm)	33	44	54	68	90	90	93	94	108	128
Fraction of Epan	0.3	0.5	0.8	0.9	0.9	0.9	0.9	0.9	0.8	0.8
Irreg. water (mm)	10	22	43	61	81	81	84	85	86	102
m^3/ha	100	220	430	610	810	810	840	850	860	1020
Total amount of water is 7390 m^3/ha										

The above calculations indicate that the calculated amount of water requirement is 7390 m^3/ha of irrigation water.

Amount of N, P and K from wastewater with the 7390 m^3/ha of irrigation water is:

N = 25 g/m^3 x 7390 m^3/ha = 184750 g/ha = 184.8 kg/ha
P = 8 g/m^3 x 7390 m^3/ha = 59120 g/ha = 59.1 kg/ha
K = 35g/m^3 x 7390 m^3/ha = 258650 g/ha = 258.7 kg/ha

5. Amount of N, P, K needed to be applied as fertilizer

	Kg/ha for 100 tons/ha			Kg/ha for 125 tons/ha		
	N	P	K	N	P	K
Drip irrigation						
Required	340	94	367	396	108	459
From soil	-	33.6	84	-	33.6	84
Wastewater	184.8	59.1	258.7	184.8	59.1	258.7
Nutrients with fertilizer	**155.2**	**1.3**	**24.3**	**211.2**	**15.3**	**116.1**
Furrow irrigation						
Required	497	189	416	579	216	520
From soil	-	33.6	84	-	33.6	84
Wastewater	184.8	59.1	258.7	184.8	59.1	258.7
Nutrients with fertilizer	**312.2**	**96.3**	**73.3**	**394.2**	**123.3**	**177.3**

From the table is evident that:

- For drip irrigation no additional P and very little K are required for the 100 tons/ha.
- With furrow irrigation more fertilizer is required for the same target yield.
- With fresh water the amount supplied by the wastewater should be subtracted,
- With wastewater if the same fertilization is applied as with the fresh water it will lead to over-fertilization and pollution.

2.5 Monitoring and Evaluation

Monitoring and evaluation, particularly from the biological point of view, is not responsibility of the farmers. Therefore, the precautionary measures which the farmers must take at any time and the additional measures which they should take in case that the wastewater provided to them is not of the anticipated quality should be given to them by the appropriate authorities in each country. No unique measures for all countries could be recommended.

For the farmers (it will be discussed in following chapters), it is important to monitor effects on soil and crops, from the chemical and particularly crop nutrient point, in order to adjust the management accordingly.

3. IRRIGATION WITH WASTEWATER

Irrigation plays a vital role in increasing crop yields and stabilizing production. In arid and semi-arid regions, irrigation is essential for economically viable agriculture, while in semi-humid areas it is often required for certain crops on a supplementary basis.

3.1 Irrigation Methods

Surface (Traditional) Methods

- Flood irrigation (by border or basin), wetting almost all the land surface
- Hose-basin irrigation. The water is delivered by hose
- Furrow irrigation, wetting only part of the ground surface

These methods account for about 95% of the world's irrigation and dominate the world's irrigation. They are of low cost, simple to understand and implement. It is relatively effective technology and even though its efficiency is low, it is suitable for many developing countries particularly if water is not the limiting factor for agricultural production.

Modern Irrigation Systems

Sprinklers (Sprinklers of high capacity, ordinary minisprinklers, and sprayers). Crops and soil are wetted the same way as rain.
Drip (point or localised irrigation system). The main characteristics of the system are:

- High application efficiency. It is probably the best method for irrigation in places where scarcity of water is a problem.
- Appropriate method to cope with problems associated with saline irrigation water and soil alkalinity.
- This method is safe and in general might be the most promising for irrigation with wastewater, particularly if purification is such to prevent extensive clogging.
- Both contact of wastewater with farmers and the irrigated crops is minimized.
- No aerosols are formed and therefore no pollution of the atmosphere and of the area nearby to the irrigated fields occurs.

Subsurface Irrigation

This system is not yet used with wastewater, but it might be useful for irrigating with wastewater of lower quality and higher health risk. Subsurface irrigation together with trickle irrigation can give the best health protection.

Bubbler Irrigation

It is a localized irrigation technique with regulated flow developed lately; performs better than trickles and minisprinklers as far as clogging is concerned.

Selection of the Irrigation System

Selection of the appropriate irrigation system depends on the quality of wastewater, crop, tradition, background, skill and ability of the farmers to manage the different methods, and the potential risk to the health of farmers, public and environment. Table 17 presents an analysis of the main factors in relation to four widely practiced irrigation methods, namely, border, furrow, sprinkler and drip.

Clogging problems with the sprinkler, minisprinkler drip and subsurface irrigation systems might be a serious problem. Growths (slimes, bacteria, etc.) in the sprinkler head, emitter orifice or supply line, cause plugging. Salts and suspended solids may also create clogging. The most serious clogging occurs with drip irrigation, which although considered ideal as far as health protection and plant contamination is concerned, might be difficult to be used for irrigation with wastewater high in SS.

The decision on irrigation system selection is mainly financial. However, the health risk associated with the different systems should be seriously taken into account. The irrigation system is one of the health control measures, along with crop selection, wastewater treatment and human exposure control.

Table 17. Evaluation of common irrigation methods for their suitability to use brackish water (Kandiah, 1990 [33])

Parameters of Evaluation	Furrow Irrigation	Border Irrigation	Sprinkler Irrigation	Drip Irrigation
Foliar wetting and consequent leaf damage resulting in poor yield	No foliar injury as the crop is planted on the ridge	Some bottom leaves may be affected, but the damage is not so serious to reduce yield	Severe leaf damage can occur resulting in significant yield loss	No foliar injury occurs under this method of irrigation
Salt accumulation in the root zone with repeated applications	Salts tend to accumulate the ridge which could harm the crop	Salts move vertically downwards and are not likely to accumulate in the root zone	Salt movement is downwards and root zone is not likely to accumulate salts	Salt movements are radial along the direction of water movement. A salt wedge is formed between drip points
Ability to maintain high soil water potential	Plants may be subject to stress between irrigations	Plants may be subject to water stress between irrigations	Not possible to maintain high soil water potential throughout the growing season	Possible to maintain high soil water potential throughout the growing season and minimize the effect of salinity

Table 17. (Continued)

Parameters of Evaluation	Furrow Irrigation	Border Irrigation	Sprinkler Irrigation	Drip Irrigation
Suitability to handle brackish water without significant yield loss	Fair to medium. With good management and drainage it is possible to get acceptable yield	Fair to medium. Good irrigation and drainage practices can produce acceptable levels of yield	Poor to fair. Most crops suffer from leaf damage and low yield	Excellent to good. Almost all crops can be grown with little reduction in yield

Table 18. Factors affecting the choice of each irrigation method, and the special measures required when wastewater is used, particularly when it does not meet the WHO guidelines [9]

Irrigation method	Factors affecting choice	Special measures for wastewater
Border (flooding) irrigation	Lowest cost, exact leveling not required	Thorough protection for field workers, crop handlers and consumers
Furrow irrigation	low cost, leveling may be needed	Protection for field workers, possibly for crop handlers and consumes
Sprinkler irrigation	Medium water use efficiency, leveling not required	Some Category B crops, especially tree fruit, should not be grown. Minimum distance 50 - 100 m from houses and roads. Anaerobic wastes should not be used, due to odor nuisance*
Subsurface and localized irrigation	High cost, high water use efficiency, higher yields	Filtration to prevent emitters clogging

*Crop categories are presented in following chapters.

Recommendations

The farmers should be aware that for a treated wastewater of the WHO guideline quality (Table 14 [31]), all methods are appropriate to be used from the point of disease transmission control, provided that also meet the agricultural criteria. If the wastewater does not meet the health criteria then:

- Spray irrigation (minisprinklers, sprayers, spitters etc.) is limited to only fodder, fibre, and seed crops.
- Spray irrigation with wastewater when used for irrigating lawns or areas of unlimited access, to be practiced during night,
- Sprinkler and particularly spray irrigation is not recommended under windy conditions. There is a potential for the pathogens to be carried away in the spray mist or in the formed aerosols with the wind drift and cause health hazard to the workers, farm population and the nearby residential areas.

- Blending of sewage water with normal water to achieve prescribed guidelines is desirable. If normal irrigation water is available but not fully adequate in quantity, a blend of treated sewage water with normal water will produce mixed water that can be used with minimum adverse effects on crop production and health.

Irrigation methods must also be examined in relation to the extent that they are practiced in an area or country, the background of the farmers on certain methods and the extent of contamination they may induce on the crops, particularly the edible parts.

3.2 Amount of Water and Scheduling of Irrigation

For all practical purposes, water needed by crop plants is equal to their evapotranspiration requirement. An extensive review [34] of this subject and guidelines are given in FAO Irrigation and Drainage paper No. 24 (1977). Computer software, called CROPWAT [18] is available in FAO to determine water requirements of crops from climatic data. It could be also used the simple and practical way proposed in this manual which is lonely based on Epan evaporation.

Table 19 presents water requirements of some selected crops. It should be kept in mind that the actual amount to be applied has to be adjusted for effective rainfall, leaching requirement, application losses and other factors.

The real water requirement for a particular crop and place is estimated as indicated in the previous section or based on meteorological data.

3.3 Strategy to Protect Human Health and Environment

Human health and environment could be protected through four groups of measures (Blumenthal et al., 1989, [35]):

Table 19. Water requirements of some selected crops [18]

Crop	Water requirement (mm/growing period)
Alfalfa	800 -1600
Banana	1200 -2200
Been	300 - 500
Cabbage	380 - 500
Citrus	900 -1200
Cotton	700 -1300
Groundnut	500 - 800
Maize	500 - 800
Potato	500 - 700
Rice	350 - 700
Sanflower	800 -1200
Sorghum	450 - 650
Wheat	450 - 650

• Waste treatment
• restriction of the crops grown
• irrigation methods, and
• control human exposure to the waste, and hygiene

Full treatment prevents excreted pathogens from even reaching the field. However, the farmers in most of the cases have to cope with wastewater of a certain quality. Because of this, for the farmer's crop restriction, irrigation system and human exposure control which act later in the pathway, are more important. A combination of agrotechnical measures to be selected, depending on the local socio-cultural, institutional and economic conditions, may provide health protection.

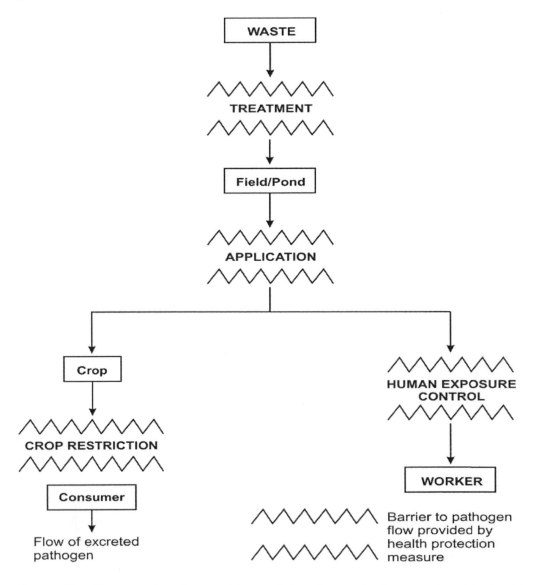

Figure 1. Flow Diagram to show the potential transmission of excreted pathogens and   points at which different health protection measures can interrupt the pathogen flow (Blumenthal et. al., 1989 [35]).

Crop Selection for Health Protection

Wastewater which has been treated to the WHO quality guidelines for unrestricted use (<1000 faecal coliforms per 100 ml and < 1 viable nematode egg per L) can be used to irrigate all crops, without further health protection measures. If the WHO quality guidelines are not fully met, it may still be possible to irrigate selected crops without risk to the consumer. Crops can be grouped into three broad categories with regard to the degree to which health protection measures are required (Shuval et. al., 1986 [36]).

Category A- Protection Needed Only for Field Workers:

- Crops not for human consumption (cotton, sisal).
- Crops normally processed by heat or drying before human consumption (grains, oilseeds, sugar beet).
- Vegetables and fruits grown exclusively for canning or other processing that effectively destroys pathogens.
- Fodder crops sun-dried and harvested before consumption by animals.
- Landscape irrigation in fenced areas without public access (nurseries, forests, and greenbelts).

Category B- Further Measures May Be Needed:

- Pasturelands, green fodder crops.
- Crops for human consumption that do not come into direct conduct with wastewater, on condition that none must be picked off the ground and that spray irrigation must not be used (tree crops, vineyards, etc.).
- Crops for human consumption normally eaten only after cooking (potatoes, eggplant, beetroots).
- Crops for human consumption, the peel of which is not eaten (melons, watermelons, citrus, bananas, nuts, groundnuts).
- Any crop if sprinkler irrigation is used.

Category C- Treatment to WHO "Unrestricted" Guidelines Is Essential:

- Any crops often eaten uncooked and grown in close conduct with wastewater effluent (fresh vegetables such as lettuce or carrots, or spray-irrigated fruit).
- Landscape irrigation with public access (parks, lawns, golf courses).

Adopting crop restriction as a means of health and environment protection requires strong institutional framework and capacity to monitor and control compliance with the regulations. Farmers must be advised why such crop restriction is necessary and be assisted in developing a cropping pattern.

A useful arrangement of crops with declining potential to transmit pathogens if irrigated with treated wastewater, irrespective of the irrigation method and wastewater quality used, is the following [18]:

1) Vegetables eaten raw.
2) Vegetables eaten cooked.
3) Ornamentals raised for sale in greenhouses.
4) Trees producing fruits eaten raw without peeling.
5) Lawns in amenity areas of unlimited access to the public.
6) Trees producing fruits eaten raw after peeling.
7) Table grapes.
8) Lawns and other trees in amenity areas of limited access.
9) Fodder crops.
10) Trees producing nuts and other similar trees.
11) Industrial crops.

The crop restriction is the strategy to provide protection to the consuming public. However it does not provide protection to farm workers and their families who remain at high risk since they are still exposed to pathogens in the waste on the soil and on the crop. Crop restriction is, therefore, not adequate for the farmers in its own; it should be complemented by other measures including their personal and family hygiene.

Potential Crop Restrictions Bases on Wastewater Chemical Composition
Apart from effluent quality criteria related to health it may be needed for the farmers to consider further selection of crops in terms of their tolerance to the chemical composition of the treated wastewater.

Wastewater Application *(Irrigation)*
Irrigation and selection of irrigation methods have been already discussed.

Control of Human Exposure to the Wastes and Hygiene
Controlling the risk that public is running to be infested by waterborne diseases when treated sewage wastewater is used for irrigation is of high importance. In this respect, the groups of persons running such a risk and the ways such groups are exposed to the risk should be identified and examined. The following groups may be recognized (Kypris, 1989 [37]):

• Farm workers. The probability for them of wetting their hands, clothes, or other parts of their body from leaks or otherwise is certainly the highest. Therefore, farmers should be aware about the risk and handle wastewater with care.
• Workers handling or packing polluted crops. If proper care was not taken at the treatment stage and proper irrigation practice were not followed by the farmers, pathogens may be present on the crops at such concentrations, as to pollute the hands, or clothes of such workers.
• Consumers. This group is actually the general public, comprising children, elderly people and others of low resistance to pathogens, being the most sensitive group. Farmers should feel responsible for this group and manage wastewater in such a way to avoid crop

contamination. Crops polluted with pathogens, particularly those consumed raw, allow the chance, if not properly washed and cleaned, that some people consuming them, to ingest enough pathogens to become sick. Risks to consumers can be reduced by thorough cooking and by high standards of hygiene. Local residents should be kept fully informed about the location of all fields where wastewater is used. In this way, they may avoid entering them and also prevent their children from doing so.

- The general public using amenity areas irrigated with wastewater. Particularly if lawn is irrigated, on which children use to play, or other people come in direct contact with. If high concentrations of pathogens remain on such lawn at the time of application, the risks of pathogens being transmitted are high. For this wastewater application farmers are not responsible.
- People passing by, or living in the neighborhood of areas irrigated with treated wastewater.

It is apparent that to control health risk, farmers should know first of all the biological quality of wastewater.

4. ENVIRONMENTAL ASPECTS ASSOCIATED WITH WASTEWATER USED FOR IRRIGATION

Environmental Benefits

When wastewater is used properly for agricultural purposes, than being disposed in any other way, improvement of the environment could be achieved. The following are some of the environmental benefits:

- Avoidance of discharge to surface waters, preventing occurrence of unpleasant esthetic situations, anaerobic conditions in rivers and eutrophication of lakes and reservoirs. Conservation of water resources will provide benefits to uses such as water supply and recreation.
- Saving groundwater resources in areas where over-utilization of these resources in agriculture is causing problems of water level depletion and salt intrusion.
- Possibility of soil conservation and improvement by humus build-up on agricultural land and the prevention of land erosion.

Potential Negative Environmental Effects

Wastewater use for irrigation may have also negative effects on the environment and human health. The principal environmental hazards associated with wastewater are:

- The introduction of chemicals into susceptible ecosystems (mainly soil and water sources) and
- The spread of pathogens.

4.1 Effects on Soil

These effects are of particular importance for the farmers since they may reduce soil productivity, fertility and yield. Soil should remain at a good level of chemical and physical characteristics in order to enable long term sustainable and profitable agriculture. The main soil problems are:

- Salinization.
- Alkalinity and water permeability.
- Accumulation of potentially toxic elements.
- Accumulation of nutrients.

The above aspects, along with management to alleviate and/or overcome the problems, have been discussed in previous chapters.

4.2   Effects on Groundwater

Although effects on groundwater are under certain conditions more important than effects on soil, the farmers take no sufficient care. Pollution of groundwater with constituents present in wastewater is very common.

The following management aspects to reduce and/or overcome the problem are recommended:

- Irrigation (amount of water) based on crop water requirement with minimum leaching if needed.
- Scheduling of irrigation based on water requirement, soil water holding capacity and wastewater quality.
- Select crops, which may absorb potentially hazardous constituents present in wastewater.
- In case of saline waters introduce in the cropping pattern salt harvesting crops.
- To avoid contamination by NO_3-N the amount of water could be restricted to that amount which supplies the N required by the crop and if N exceeds crop requirement then:

  - Select crop with higher N requirement.
  - Select the irrigation system, which provides the highest possible application uniformity.
  - Blending of water with fresh water
  - Keep operation and maintenance of irrigation systems at acceptable level.

4.3 Effects on Surface Water Bodies

Eutrophication, Growth of Algae

The high N concentration in wastewater together with P is of particular interest when wastewater is collected in dams before irrigation since they may create favorable conditions for eutrophication. Under such conditions, blooming of green algae is very common and the problems associated, particularly clogging of modern irrigation systems, is difficult to overcome. The latter is one of the main concerns for the farmers.

The problem of eutrophication and oxygen depletion due to the nutrients in the wastewater is particularly important when the effluent is discharged in water bodies (rivers, lakes, and sea). The N is the limiting factor for algae growth in the sea, while N and P are the limiting factors in the landlocked salt-water areas and in dams where the wastewater is collected before irrigation.

4.4 Effects on Crops: Phototoxicity Problem and Management

Besides the overall effect of certain characteristics of wastewater to the irrigated crops like salinity, the wastewater potentially may create plant toxicity due to high concentration of certain elements like B and some heavy metals. Necrotic spots on the leaves recognize the symptoms of B toxicity in sensitive crops. Toxicity aspects are discussed in more details in relation with sludge use in agriculture.

4.5 Human and Animal Heath Problems

The protection measures needed by farmers to protect themselves and the consumers have been already introduced and discussed. Maximum values permitted for certain elements for animal drinking purposes are given in Table 20. This is important since in most cases it is difficult to control domestic animals from drinking wastewater. Farmers being aware about the values in Table 27 may protect their domestic animals from using them regularly.

Table 20. Guidelines for levels of toxic substances in livestock drinking water1 [30]

Constituent (Symbol)	Upper Limit (mg/l)
Aluminium (Al)	5.0
Arsenic (As)	0.2
Beryllium (Be)[2]	0.1
Boron (B)	5.0
Cadmium (Cd)	0.05
Chromium (Cr)	1.0
Cobalt (Co)	1.0
Copper (Cu)	0.5
Fluoride (F)	2.0
Iron (Fe)	not needed
Lead (Pb)[3]	0.1
Manganese (Mn)[4]	0.05
Mercury (Hg)	0.01
Nitrate + Nitrite (NO_3-N+NO_2-N)	100.0

Table 20. (Continued)

Constituent (Symbol)	Upper Limit (mg/l)
Nitrite (NO$_2$-N)	10.0
Selenium (Se)	0.05
Vanadium (V)	0.10
Zinc (Zn)	24.0

[1] Adapted from National Academy of Sciences (1972, [30]).

[2] Insufficient data for livestock. Value for marine aquatic life is used here.

[3] Lead is accumulative and problems may begin at a threshold value of 0.05 mg/l.

[4] Insufficient data for livestock. Value for human drinking water used.

5. CONTROL MEASURES

5.1 General Considerations

To the planners concerned with wastewater reuse, wastewater treatment appears as a more straightforward and "visible" measure for health protection, seconded only by crop restriction. Both measures, however, are relatively difficult to implement fully: the first limited by cost and problems of operation and maintenance, and the second by lack of appropriate markets for specific products or by legal and/or institutional constraints.

The application of isolated measures, while not economical, may have only partial effects in terms of health protection. Crop restriction for instance if applied may protect the consumers but does not provide protection to farm workers and their families.

5.2 Wastewater Reuse Guidelines

Reuse guidelines may help protect public health and the environment. Currently, there are three possibilities for the countries of the Near East Region.

1. National guidelines are available. The farmers should follow and respect them.
2. No National guidelines are available. The WHO guidelines are proposed as a solution (Table 14, [31]).
3. National guidelines are not available but for certain reasons very strict guidelines are desirable. The Cyprus guidelines could be used (Table 15, [18]).

5.3 Monitoring and Control of Wastewater Quality

As mentioned already, the general monitoring and wastewater quality control are not responsibility of the farmers but of the treatment and other authorities. The quality depends on how well the treatment is, on the maintenance and operation of the treatment system and on the background of the people operating the wastewater treatment plant.

However, training the farmers to follow, at least visually or with simple tests, the quality of wastewater could be very helpful. The farmers should be in position to judge whether the wastewater has been appropriately treated. Changes in the color of wastewater or extensive growth of algae are indicators of metals and nutrients in the wastewater at higher levels. Odor may indicate insufficient treatment. The farmers should request training on these aspects prior the allocation of wastewater for irrigation.

5.4 Control of Storage, Transport and Distribution Facilities

These aspects also do not reflect responsibility of the farmers but the farmers should be informed and aware as far as it concerns storage and distribution of wastewater in their farms. This may avoid accidental use of wastewater or accidental damage of the system.

Problems Associated with Monitoring and Evaluation

Monitoring and evaluation is an essential requirement of any project that uses treated wastewater. It should be continual activity, which should constitute an integral part of the operation and management of any such project. Without regular and effective monitoring and evaluation system:

- It is not only unlikely that the benefits anticipated from the reuse project will continue to accrue on a long-term sustainable basis, but also
- There is a very real possibility that the project may in the future contribute to development of serious health and environmental hazards.

Thus, a carefully planned and effective monitoring and evaluation system is an essential pre-requisite for the success of any wastewater reuse project.

Integrated quality monitoring of treated effluent reuse for Orrigation (FAO, 1995, [12]. In crop irrigation with treated wastewater, a critical issue is the impact of its constituents on the soil and/or crops and furthermore to animals and humans fed with such crops.

Proposed chemical quality parameters, which might be regularly or periodically monitored in wastewater by farmers or for the farmers through official authorities, are:

EC_W = Electrical conductivity expressed in units of dS/m referenced at 25°C. It is one of the most commonly measured parameters, particularly in arid and semiarid regions to estimate the total amount of soluble salts in water. Salinity is probably the most important single parameter, which determines cropping pattern and management of fields irrigated with wastewater.

Cations and Anions

Ca, Mg, Na, CO_3, HCO_3, SO_4, CI. Some of these ions may be monitored only at the beginning and thereafter periodically since they don't change too much. Some other ions like

B must be tested at least at the beginning of each irrigation season since in cases where detergents with B are widely used, B in wastewater might be the main limiting factor for its reuse for irrigation.

$$SAR = \frac{Na}{(Ca + Mg)/2}$$

The SAR is a calculated value and an indicator of an influence that Na+ ion has on soil properties.

Heavy Metals and Trace Elements

Although heavy metals (Al, As, Ba, Cd, Cr, Cu, F, Fe, Pb, Li, Mn, Hg, Ni, Se, Hg) may not be a problem for wastewater reuse, it is recommended that all listed elements be determined at least once before initial irrigation. Periodical monitoring of those (if any) found in concentrations that may affect the soil-plant system, is recommended.

Plant Nutrients

NO_3-N, NH_4-N, P and K is advisable to be monitored for three main reasons:

• To estimate additional fertilizers needed to be supplied for optimizing yield and quality of the crops.
• To help deciding the appropriate cropping pattern for the best possible water and nutrient use efficiency.
• To protect surface and underground water pollution by NO_3-N.

6. SOCIO-CULTURAL, LEGAL AND INSTITUTIONAL ASPECTS

This chapter is concise and up to the point of direct interest for the farmers. It is more relevant for the extension agronomist. The extension agronomists should be aware particularly on legal and institutional aspects existing in their countries in order to control and also give the best possible advice to the farmers to overcome problems that are common with wastewater use.

6.1 Social Acceptability and Public Information

Reuse of wastewater in some countries presents a new approach. Therefore, an important requirement of the safe and cost efficient use of municipal wastewater is the education of all participants in the process of reuse. Training is necessary for the personnel managing and maintaining treatment facilities, and the farmers using the wastewater.

Wastewater, as a resource, is often not utilized, or even considered for reuse, for five main reasons:

- Lack of information about its benefits
- Fear of health risk involved
- Cultural bias
- Lack of a method for comprehensive economic analysis of reuse projects and
- Poor experience with wastewater reuse where it has been practiced under uncontrolled conditions.

Therefore, special care should be taken to provide information and education programmes that will involve people from all community levels and not only farmers. The second step should be a training programme for the users (farmers), because misuse of wastewater can lead to repugnance.

A well - organized public information campaign should be planned as a means of making the public aware of the issue. Its primary objective will be to raise collective consciousness and to present wastewater reuse as a reliable substitution technique, It should also make the potential users aware of the facts related to wastewater reuse. Farmers and the public in general should be aware not only of benefits, which will result from reuse but also of environmental and health hazards connected with wastewater use. The information campaign should result in removal of cultural and psychological bias linked with wastewater.

6.2 Training and Human Resource Development

Lack of skills and knowledge can cause failure in project implementation and, in the case of wastewater reuse projects, can potentially increase environmental and public health risk. Because of this, training programmes should be an integral part of projects. Programmes of training should include technical, environmental, health and socioeconomic aspects. The educational input must provide the farmers with an understanding of the details of techniques and their associated hazards as well as of the precautions to be taken so that the operations take place within acceptable safety levels and a reasonable cost.

The timing of a training programme is of crucial importance since persons should be sufficiently trained to be able to work properly when it is required. Thus, farmers should already have undergone any training necessary before use of wastewater starts. Farmers should request for training. The aspects of training could be more or less those covered in this manual.

The provision for training is required not only before use of wastewater but from time to time thereafter, since refreshing and/or upgrading of skills and training of new farmers should be a continual process.

6.3 Institutional Aspects, Monitoring and Control Measures

Wastewater reclamation and reuse is viewed increasingly as a means to augment existing and future water resources against the growing demand for water. Reclaimed water is a reliable source even in drought years, existing right in the urban environment, which is capable of replacing potable water for non-potable water uses. However, municipal wastewater is of unique composition associated with environmental and health risks.

Considering the risks associated with wastewater reuse an adequate institutional framework should always be created to control, supervise and advise on reuse schemes, in order to ensure safe reuse. At a national level, the use of wastewater is an activity, which involves the responsibilities of several ministries or agencies. The institutional framework should be well defined and the distribution of responsibilities clearly specified. Considering that usually a great number of institutions is involved, smooth operation of individual interactions is not always easy to achieve. The form and operational characteristics of the institutional framework it varies in different countries and should be designed to suit local conditions. The farmers should be aware about the responsibilities of each institution in order to address their questions and/or problems faced. Without this background a general confusion will be developed.

6.4 Regulatory Considerations and Legal Issues

Public health and environment are the main concerns in reuse projects. In this respect quality guidelines as well as requirements for treatment, sampling and monitoring are essential in each country. In a number of countries, strict control measures are imposed on reuse. In this way farmers are obliged to act within the framework of these regulations in order to safeguard public health and the environment. However, how well the farmers respect the regulations is questionable. Strong monitoring is essential. Moreover, legal authorization to enforce compliance with the regulations might be needed.

Most countries where wastewater reuse is practiced have public health and environmental regulations, which govern treatment and reuse practices. These regulations as guidelines or code of practices are intended to protect the health of both the consumers and the workers. They may prohibit wastewater irrigation within specified periods prior to harvesting, require appropriate clothing and provide for preventive health care of workers. The guidelines for wastewater reuse, although variable in different countries are usually very stringent because of water pollution control requirements. The technology involved in the treatment process together with appropriate operation of the system available should achieve the guidelines, but regulatory agencies must be ready to monitor wastewater quality and enforce appropriate regulations. The health of the farmers and the consumers must be safeguarded.

In most of the Near East Countries, the technology to produce an intended effluent quality is often not available or if available not properly maintained. Regulatory agencies are seldom able to enforce the guidelines. Wastewater reuse is, therefore, often uncontrolled and workers and consumers are usually at risk. To resolve the legal problem of unenforceable guidelines, the first step is to set realistic criteria reflecting prevalent risks. It is important that the quality criteria be such as to promote rather than to restrict effluent reuse.

7. SLUDGE

The Problem

Efforts to reduce the pollution of rivers, lakes, and oceans by treating sewage are generating a rapidly growing amount of sludge, solid material that is removed from a wastewater to produce a clean wastewater.

7.1 Sludge Composition

The composition and therefore the suitability of sewage sludge for agricultural use (USDA, 1980, [38]) depends on the:

- Type of wastewater treatment (primary, secondary, digestion, etc.)
- Chemical used for flocculation and
- Sludge source (industrial or domestic).

Four are the main aspects of concern in sludge reuse:

- Heavy metals content of sludge varies with the level of industrial contribution. Under certain conditions heavy metals (zinc, copper, and nickel) can kill plants. Cadmium can be absorbed by plants in concentrations, which may be harmful for humans. Even domestic wastewater may yield sludge containing enough metals to warrant limiting continuous application. Therefore, heavy metal analyses are needed to assess the suitability of sludge as a fertilizer source or as a soil conditioner [39].
- Pathogens. Viruses, helminthes, protozoa and bacteria. In most Near East Countries pathogens are the main factor of concern. In particular eggs of some parasites with long life and very resistant to high temperatures may be present even after drying sludge in thin layers or processing it to compost. Farmers should be aware on the risk and manage sludge with care. Precautionary measures as wearing cloves and boots should be considered as obligatory measure [11].
- Salinity. In certain cases salinity may be high and the overall loading of agricultural land with sludge per application is calculated on salinity and not the N content of sludge.
- Nutrients. Sludge may contain high levels of certain nutrients which may damage crops (B and other metals) and environment (N). These constituents should be taken into account, particularly in sensitive areas with water resources, when the overall amount of sludge is calculated per unit area per year.

7.2 Sludge as a Fertilizer and Soil Conditioner

Sludge applied at a rate to supply the nitrogen requirement of the crop will supply most of the plant nutrients except potassium. However, it is unlikely that sewage sludge will be used to supply the total nutrient requirements of crops because of the large amounts that would have to be applied. Other factors may restrict the amount of sludge per application per year [40].

The maximum value of sludge is realized when it is employed in combination with inorganic fertilizers; In this way, sludge partly meets the crop's nutrient requirements and also serve as a valuable organic soil conditioner for maintaining soil productivity. Sludge and sludge compost are known to improve soil physical properties, as evidenced by enhanced aggregation, increased soil aeration, lower bulk density, less surface crusting, and increased water infiltration, water content, and water retention.

Sludge added to sandy soils will increase the moisture available to the plant and reduce need for irrigation. In heavy textured clay soils, the added organic matter will increase permeability to water and air, and increase water infiltration into the profile, thereby minimizing surface runoff. The soils also will have a greater water storage capacity. Addition of sludge to clay soils has also been shown to reduce compaction (i.e., lower the bulk density) and increase root development and depth.

The fertilizer benefit to the crop from nitrogen and other nutrients contained in the sludge may be approximated by appropriate calculations. Nitrogen is usually the limiting factor and the most common used for calculations. Several facts must be established to make such calculations.

a. N-Crop Requirement
The crop requirement for nitrogen should be estimated taking into account:

•   Soil fertility. The amount of nitrogen supplied from the soil (including previously applied compost, crop residues, manure, and chemical fertilizers) should be estimated and
•   Predicted yield levels (Table 17, [18])

b. The Amount of Nitrogen
Available to the crops during the initial growing season from the applied sludge can be estimated as follows:

% Available N = 0.1 x % Organic N

The inorganic N in the sludge immediately available to the plant is accounted for in the 10% mineralization rate prediction. Alternatively, the percent mineralizable nitrogen may be determined more accurately by an incubation technique, using the specific soil to which the sludge will be added.

Mineralization of organic N from earlier sludge or compost applications will supply a considerable portion of the N requirement along with that which is available from the current application. The second year mineralization of sludge compost is about 5% of the remaining organic N and it is estimated that 2% of the remaining organic N will mineralize per year after the second year. Table 21, contains the available N levels for a designated application of sludge containing 1.0% organic N. If a farmer uses sludge as the only N source and the crop requires 100 kg N/ha, he would apply 100 metric tons/ha the first year, about 60 metric tons/ha the second, third and fourth years, and 30 metric tons/ha thereafter until a

Table 21. Available N in kg from a single application of sludge at indicated rates*

Sludge Applic. dry metric Tons/ha	Total N Applied (kg)	1styr	Available N (kg)	
			2nd yr	Subsequent yrs
20	200	20	9	3
40	400	40	18	7
100	1000	100	45	17
200	2000	200	89	34

* Mineralization rates are 10%, 5% and 2% for the first, second, and subsequent years, respectively.

mineralization equilibrium would be established; at that time, the amount of available N would equal the total N applied. The user might also consider supplementing nitrogen needs with fertilizer, depending on availability and cost. The user should be aware that, in addition to N availability, heavy metal and salt accumulation will also be factors in the determination of a beneficial cumulative loading rate for sludge or sludge compost.

Sludge can supply besides the organic matter and N, other macronutrients P, K, Ca, Mg and micronutrients Fe, Cu, Zn, Mn, Mo, and B.

Uses of high rates (>50% by volume) is wasteful of the nutrients in the sludge and can cause salt toxicity. If such a situation happens then leaching is recommended.

7.3 Constraints on Uses

Pathogens. Because of its origin, sewage sludge may not be acceptable by the farmers and public from the standpoint of aesthetic or health aspects. Aesthetic reservations are generally offensive odors and appearance of sludge. This is one of the reasons why composting of sludge is favored. In addition If composting is properly done it destroys or reduces to insignificant levels all primary pathogens present in sewage sludge. Once destroyed, viruses, helminthes, protozoa, and most bacteria will not repopulate the compost, since they cannot grow external to their hosts. Salmonella, one of the most common organisms causing food poisoning, can regrow to a limited extent in the finished compost, but it does not compete well with other microorganisms present [18].

Heavy metals. Sewage sludge contains large amounts of heavy metals, which may reduce the value of sludge as a fertilizer for either direct application to land or for composting. Excessive amounts of these metals are often found in sludge where industrial effluent is discharged into the sanitary sewers without pre-treatment. Application of high metal sludge on land results in soil enrichment in heavy metals. Soil enrichment by zinc, copper, and nickel can cause direct phytotoxic effects manifested as decreased growth and yield, especially where soil pH is low (pH 5.5) and rates of application are high. Heavy metals may also accumulate in plant tissues and enter the food chain through direct ingestion by humans or indirectly through animals [18].

The element of greatest concern to human health where sewage sludge and sludge compost are applied to land is cadmium (Cd), since it is readily absorbed by most crops and is not generally phytotoxic at the concentrations normally encountered. Therefore, Cd can accumulate in plants and enter the food chain more readily than, for example, lead (Pb) or mercury (Hg), which are not readily absorbed and translocated to the edible portion of crops.

Plant species, as well as varieties, have been found to differ markedly in their ability to absorb and translocate heavy metals, to accumulate them within edible organs of the plant, and to resist their phytotoxic effects. In this respect, the farmers should remember the following:

• Leafy vegetables are usually sensitive to the toxic effects of metals and accumulate them.
• Cereal grains, corn, and soybeans are less sensitive and
• Grasses are relatively tolerant.

Uptake studies with corn, soybean, and cereal grains have shown that heavy metals accumulate less in the edible grain than in the leaves; similar results are found for edible roots, as radish, turnip, carrot, and potato, and fruits, as tomato, squash etc.

The availability to and uptake of heavy metals by plants are influenced by certain chemical and physical properties of soil, especially pH, organic matter content, cation exchange capacity (CEC), and texture (i.e., the proportions of sand, silt, and clay). Phytotoxicity and plant availability of sludge - borne metals are higher in acid soils than in those with neutral or alkaline pH. Maintaining soil pH in the range of 6.5 or above by liming reduces the availability of heavy metals to plants. Evidently, under calcareous conditions heavy metals are rendered unavailable and the problem is becoming insignificant. Application of organic amendments such as manure and crop residues can also decrease the availability of heavy metals to plants. The CEC is a measure of the soil's capacity to retain cations; higher CEC is usually associated with higher clay and organic matter contents. Heavy metals are generally less available to plants in soils of high CEC (e.g., organic matter rich soils or clay loams) compared with soils of low CEC (e.g., loamy sands).

Table 22 shows the recommended maximum cumulative sludge metal loadings for privately owned agricultural land according to the soil cation exchange capacity (USDA recommendations). Soils in the 0 to 5 CEC range are sands through sandy loams; the 5 to 15 range includes sandy loams, loams, silt loams; and > 15 includes silty clay loams and clays. Higher metal loadings would be considered reasonably allowable on heavier textured soils. Cadmium loading on land should not exceed 2 kg/ha/year for dewatered sludge or sludge compost and should not exceed the total cumulative loading shown in Table 22. When sludge is applied, the soil should be limed to pH 6.5 and maintained at 6.2 or higher. Sludge and sludge compost should not be applied to land used to grow tobacco as this crop allows high transfer of Cd to humans; sludge and compost used on land used to grow leafy vegetables should be low in Cd and Cd/Zn ratio to minimize any effects on humans.

Table 22. Recommended maximum cumulative metal loadings from sludge or sludge compost applications to privately owned land [18]

Metal	*Soil*	*cation exchange capacity (meq/100g)*[1]	
	0 - 5	5 - 1 5	> 15
	(Maximum metal addition, kg/ha)		
Zn	250	500	1000
Cu	125	250	500
Ni	50	100	200
Cd	5	10	20
Pb	500	1000	2000

[1]CEC determined prior to sludge application using I N neutral ammonium acetate and is expressed here as a weighted average for a depth of 50 cm.

8. ECONOMIC ASPECTS

8.1 General Considerations

This chapter is addressed also to the farmers and not to the decision-makers who they need economic justification in order to proceed in reuse. From the farmers the main point is the profit expected from the use of wastewater for irrigation. He is not interested on the cost of collection, treatment, distribution and other aspects. His interest is whether with certain quality and amount of wastewater, given to him with some price or free, and with restriction for certain crops, could practice profitable irrigated agriculture.

8.2 Benefits

The benefits could be briefly summarized as follows:

• Saving in fresh water
• Saving in fertilizers
• Increasing yield
• Creating job opportunities

These aspects have been already covered. In summary, for the farmers the main benefits are resulted from the point that wastewater is reliable water source even in very drought years and that the nutritional value of wastewater may lead to high yield of good quality without or with limited amount of fertilizer.

REFERENCES

[1]   Popul. Environ. (2006) 28 :113-131, "A SRES-based gridded global population dataset for 1990-2100", Magnus Bengtsson, Yanjum Shen,Taikan Oki.

[2]   Allan, J.A.; The Middle East water question: Hydropolitics and the Global Economy, I.B. Tauris & Co. Ltd: London, 2002, p. 9.

[3]   Scheumann, W. ; Schiffler, M., Water in the Middle East: Potential for Conflicts and Prospects for Cooperation; Springer: Berlin, 1998.

[4]   Aboukhaled, A., International Journal of Water Resources Development, 1992, vol. 8, Issue 3, p. 204-215.

[5]   Turkdogan, M.K.; Kilicel, F.; Kara, K.; Tuncer, I.; Uygan, I., Environmental Toxicology and Pharmacology, 2002, vol. 13, p. 175-179.

[6]   Arceivala, S.J.; Asolekar, S.R., Wastewater treatment for pollution control and reuse, Tata McGraw-Hill: New Dheli, 2008.

[7]   Newell, F.H., Irrigation Management: The operation, maintenance of works for bringing water to agricultural lands; Kessinger Publishing, 2009.

[8]   World Health Organization Scientific Group, Health Guidelines for the use of wastewater in Agriculture and Aquaculture, Technical Report No. 778, June 1989, Geneva.

[9]   Mara, D.D.; Cairncross, S., Guidelines for the use of wastewater and excreta in agriculture and aquaculture, WHO/UNEP, Geneva, 1989.

[10]   Tedeschi, S.; Pescod, M.B., Environmental guidelines for municipal wastewater reuse in the Mediterranean region, Mediterranean Action Plan – Priority Actions Programme, Regional Activity Centre, UNEP/FAO, 1991, Split, Yugoslavia.

[11]   Pescod, M.B., Wastewater treatment and use in agriculture, Irrigation and drainage paper No. 47, FAO, 1992, Rome, Italy.

[12]   Papadopoulos, I., Wastewater management for agriculture protection in the Near East region, Technical Bulletin, FAO regional office for the Near East, 1995, Cairo, Egypt.

[13]   Bashour, I.; Nimah, M., Fertigation potentials in the Near East region, IPI regional workshop on potassium and fertigation development in West Asia and North Africa, 24-28 November 2004, Rabat, Morocco.

[14]   Food and Agriculture Organization, Irrigation in the Near East region in figures, FAO Water Report 9, 1997, Rome, Italy.

[15]   Bielorai, H.; Vaisman, I.; Feigin, A. Drip irrigation of cotton with treated municipal effluents: I. Yield response. Journal of Environmental Quality, v.13, p.231-234, 1984.

[16]   Tsagarakis, K. P.; Dialynas, G. E.; Angelakis, A. N., Agricultural Water Management, Vol. 66, Issue 1, 2004, p. 35-47.

[17]   Al-Salem, S.S., Eastern Mediterranean Health Journal, Vol. 7, No. 6, 2001, p. 1056-1060.

[18]   Papadopoulos, I., Users Manual for Irrigation with treated wastewater, FAO Regional Office for the Near East, 2003, Cairo, Egypt.

[19]   Tchobanoglous, G.; Burton F.L.; Stensel, H.D., Wastewater engineering: treatment and reuse, McGraw-Hill, NY, 2003.

[20]   Ayers, R.S.; Westcot, D.W., Water quality for agriculture: irrigation and drainage, FAO, Paper 29, Rev. 1, 1985, p.174.

[21] Maas, E.V.; Hoffman, G.J., Jour. of the Irrig. and Drain. Div., ASCE, Vol. 103 (2), 1977, pp. 115-134.

[22] Goldberg, S.D.; Rinot, M.; Karn, N., Soil Sci. Soc. Am. Proc., vol. 35, p. 127-130 (1971).

[23] Papadopoulos, I., Nitrogen fertigation of greenhouse-grown strawberries. Fertilizer Research, vol. 13, 1987, pp. 269-276.

[24] Papadopoulos, I.; Stylianou, Y., Agric. Water Manag., vol. 19, p. 67-75,1991.

[25] Rhoades, J.D., Proc. Water Management for Irrigation and Drainage, ASCE, Reno, NV, 1977, pp. 85-116.

[26] Papadopoulos, I.; Stylianou, Y, Journal of Environment Quality, vol. 17, 1988a, pp. 574-580.

[27] Ayers, R.S., Jour. of the Irrig. and Drain. Div., ASCE., Vol. 103, No. IR2, 1977, p. 140.

[28] Bernstein, L.; Francois, L.E., Soil Science, Vol. 115, No. 1, 1973, pp. 73-86.

[29] Biswas A.K.; Arar, A., Role of wastewater reuse in planning and management, In A.K. Biswas and A. Arar (Eds) Treatment and reuse of sewage effluent for irrigation, Butterrsworth Scientific Guildford, 1987, U.K.

[30] National Academy of Sciences Water quality criteria 1972: a report of the committee on water quality criteria. US Environmental Protection Agency, pp. 232–353.

[31] FAO/RNEA, Wastewaters a crop nutrient source, 1992, Tech. Bul. No. 5, p. 15.

[32] FAO, Report on the expert consultation on revision of FAO methodologies for crop water requirements, 1990, Annex V, Rome, Italy.

[33] Kandiah A., Environmental impacts of irrigation development with special reference to saline water use in Water, Soil and Crop Management Relating to the Use of Saline Water, 1990, AGL/MISC/16, FAO, Rome, Italy.

[34] Doorenbos, J.; Pruitt, W.O., Crop Water Requirements, Irrigation and drainage paper No. 24, FAO, 1977, Rome, Italy.

[35] Blumenthal, U.J.; Strauss, M.; Mara, D.D.; Cairncross, S., Wat. Sci. Tech., Vol. 21, 1989, p.p. 567-577.

[36] Shuval, H.I.; Adin, A.; Fattal, B.; Rawitz, E.; Yekutiel, P., Wastewater Irrigation in Developing Countries: Health Effects and Technical Solutions, Technical Paper No. 51, 1986, The World Bank, Washington, DC.

[37] Kypris, D., Considerations of the quality standards for the reuse of treated effluents. In "Proceedings of Wastewater Reclamation and Reuse", 1989, Cairo, Egypt.

[38] USDA, Willson, G.B.; Parr, J.F.; Epstein, E.; Marsh, P.B.; Chaney, R.L.; Colacicco, D.; Burge, W.D.; Sikora, L.J.; Tester, C.F.; Hornick, S., Manual for Composting Sewage Sludge by the Beltsville Aerated-Pile Method, United States Department of Agriculture, 1980.

[39] Soler-Rovira, E.; Soler-Soler, J.; Soler-Rovira, J.; Polo, A., Fertilizer Research, 1996, Vol. 43, pp.173-177.

[40] Perera, K.L.G.; Uyasatian, U;, Dilokwanich, S., Environment and Natural Resources Journal, Vol. 5, No.1, 2007, pp. 11-21.

GLOSSARY

Alkalinity	The capacity of water to neutralize acids; a property imparted by carbonates, bicarbonates, hydroxides, and occasionally borates, silicates, and phosphates. It is expressed in milligrams of equivalent calcium carbonate per liter.
Alkalinity	The capacity of water to neutralize acids; a property imparted by carbonates, bicarbonates, hydroxides, and occasionally borates, silicates, and phosphates. It is expressed in milligrams of equivalent calcium carbonate per liter.
Aerated lagoon	This is an adaptation of the waste stabilization pond in which oxygen is added by mechanical aerators.
Available water	The portion of water in a soil that can be readily absorbed by plant roots. Considered, by most workers to be that water held in the soil against a pressure of up to approximately 15 bars. See field capacity, permanent wilting point, and soil moisture tension.
BOD$_5$	1. Biochemical oxygen demand. The quantity of oxygen used in the biochemical oxidation of organic matter in a specified time, at a specified temperature, and under specified conditions. 2. A standard test used in assessing wastewater strength.
Cation Exchange Capacity (CEC)	The sum of exchangeable cations that a soil can adsorb expressed in milliequivalents per 100 grams of soil or in millimoles of positive charge per kilogram of soil. CEC is directly related to a soil's ability to retain cations against leaching. CEC is also used in calculating exchangeable sodium percentage (ESP) - a measure of excessive sodium hazard in the soil.
COD	Chemical oxygen demand. A quantitative measure of the amount of oxygen required for the chemical oxidation of carbonaceous (organic) material in wastewater using dichromate or permanganate salts as oxidants in a two - hour test.
Coliform bacteria	Are group of bacteria of the enteric tract of mammals used as an indicator of faecal pollution.
Denitrification	The biological conversion of nitrate or nitrite to gaseous N_2 or N_2O.

Effluent	Partially or completely treated wastewater flowing out of a treatment plant, reservoir, or basin.
Electrical conductivity	A measure of salinity expressed in milliohms per centimeter (mmho/cm) or decisiemens per meter (dS/m) at 25°C. Empirically related to total dissolved solids (in mg/L) divided by 640.
Evapotranspiration (ET)	The combined loss of water from a given area and during a specified period of time by evaporation from the soil surface and by transpiration from plants. Et_0 is reference ET devised as the ET from an extended surface of 3 to 6-inch tall green grass cover of uniform height, actively growing, completely shading the ground and not short of water. E_p or E_{pan} is evaporation from a standard evaporation pan.
Exchangeable sodium percentage (ESP)	The ratio (as percent) of exchangeable sodium to the remaining exchangeable cations in the soil. See SAR.
Field capacity (FC)	The percentage of water (either weight or volume) remaining in a soil 2 and 3 days after having been saturated and after free drainage has practically ceased. This term is obsolete in technical work. For many soils, FC is in the range of 1/10 to 1/3 bar water potential. See soil moisture tension.
Hydraulic conductivity	The rate of water flow in soil per unit gradient of hydraulic head or potential.
Immobilization	The conversion of an element from the inorganic to the organic from in microbial or plant tissues. Often used to describe the conversion of nitrate or ammonium into organic forms in soil microorganisms.
Infiltration	(1) The downward entry of water into soil. (2) The flow or movement of water through the pores of a soil or other porous medium. (3)The quantity of groundwater that leaks into a pipe through joints, porous walls, or breaks. (4) The entrance of water from the ground into a gallery.

Infiltration rate	(1) A soil characteristic describing the maximum at which water can enter the soil under specified conditions, including the presence of excess water. It has the dimensions of velocity, i.e., inch/hr or cm/sec. Formerly, the infiltration capacity. (2) The rate, usually expressed in cubic feet per second or million gallons per day per mile of waterway, at which groundwater enters an infiltration ditch or gallery, drain, sewer, or other underground conduit.
Land application	The recycling, treatment, or disposal of wastewater or wastewater solids to the land under controlled conditions.
Land disposal	Application of raw or treated wastewater, sludge, or solid waste to soils and/or substrata without production of usable agricultural products (See also land treatment).
Leaching fraction (LF)	The fraction of water applied to soil that leaches below a depth of interest such as the rooting depth.
Leaching requirements (LR) Mineralization	The leaching fraction required maintaining average root zone salinity below a phytotoxic threshold value. The conversion of an element from an organic to an inorganic from (e.g., the conversion of organic nitrogen in wastewater to ammonium nitrogen by microbial decomposition).
Oxidation pond	Natural or artificially accelerated transfer of oxygen (e.g., algae pond, and lagoon) effects a relatively shallow pond or basin in which biological oxidation.
Permeability	The ease with which gas, liquids or plant roots penetrate or pass through a soil horizon.
pH	The degree of acidity or alkalinity, defined as the negative logarithm of hydrogen ion activity of water.
Primary treatment	(1) The first major treatment in a wastewater treatment facility, usually sedimentation but not biological oxidation, (2) The removal of a substantial amount of suspended matter but little or no colloidal and dissolved matter. (3) Wastewater treatment processes usually consisting of clarification with or without chemical treatment to accomplish solid-liquid separation. See also secondary treatment, tertiary treatment.
Rapid infiltration	A type of land treatment in which water is applied to relatively porous soil at rates far in excess of normal crop irrigation.

Secondary treatment	(1) Generally, a level of treatment that produces removal efficiencies for BOD and suspended solids of 85%. (2) Sometimes used interchangeably with concept of biological wastewater treatment, particularly the activated sludge process. Commonly applied to treatment that consists chiefly of a biological process followed by clarification with separate sludge collection and handling.
Sludge	Is the solid matter (often having high water content) that is formed both when sewage is allowed to stand so that the denser solids settle out, and as a product of various treatments processes.
Sodium adsorption ratio (R_{na} or SAR)	A measure of the amount of sodium relative to the amount of calcium and magnesium in water or in soil saturation extract. It is defined as follows: $$R_{na} \text{ or SAR} = \frac{Na}{(Ca+Mg)/2}$$ Where the quantities Na, Ca and Mg are expressed in milliequivalent/liter. The SAR can be used to predict the exchangeable sodium percentage of a soil equilibrated with a given solution.
Soil structure tension (or pressure)	The soil water content expressed as an equivalent negative pressure. It is equal to the equivalent pressure that must be applied to soil water to bring it to equilibrium, through a permeable membrane, with a pool of water of the same composition. Usually expressed in bars or atmospheres.
Soil structure	The combination or arrangement of primary soil particles into secondary particles, aggregates, or pees. These secondary units are classified by soil morphologist on the basis of size, shape, and degree of distinctness.
Reclaimed wastewater	Wastewater that, as a result of treatment, is suitable for a beneficial use.
Soil texture	The relative proportion in a soil of sand, silt, and clay-sized mineral particles.
Soil total water potential	The amounts of work which must be done to transport a unit of pure water to the soil water isothermally and at atmospheric pressure. Comprised mainly of osmotic, gravitational, and capillary components.
Soil water content	The amount of water lost from the soil upon drying to constant weight at 105°C, expressed as g water per g dry soil or cm^3 water per cm^3 bulk soil. In the field, water content is often expressed on a percent dry weight basis. This can lead to ambiguity when it is not stated whether a weight or volume basis is being used.
Tertiary treatment	See advanced wastewater treatment.
Total dissolved solids (TDS)	The sum of all dissolved solids in water or wastewater and an expression of water salinity in mg/L. Empirically related to electrical conductivity (EC) multiplied by 640.

Wastewater irrigation	Land application of wastewater with the primary purpose of maximizing crop production per unit of water applied. Often used in a broader sense to mean land treatment and disposal of wastewater where maximum crop production is a secondary objective.
Wastewater reclamation	The process of treating wastewater to produce water for beneficial uses, its transportation to the place of used, and its actual use.
Waste reuse	The additional use of once-used water.

APPENDIXES

APPENDIX I

Fertilizer Requirements of Certain Crops

Potato

Yield (tons/ha)	Irrigation method	kg/ha				
		N	P	K	P_2O_5	K_2O
40	Sprinkler	251	95	353	218	423
	Furrow	256	143	400	327	480
50	Sprinkler	289	113	406	260	487
	Furrow	409	170	460	390	552
60	Sprinkler	327	131	459	301	550
	Furrow	463	197	520	452	624
70	Sprinkler	364	149	512	342	614
	Furrow	516	224	580	513	696
Each additional 10 tons	Sprinkler	38	18	53	41	64
	Furrow	53	29	60	62	72

Eggplant

Yield (tons/ha)	Irrigation method	kg/ha				
		N	P	K	P_2O_5	K_2O
50	Drip	255	105	326	241	392
	Furrow	367	210	370	482	445
75	Drip	313	136	419	312	503
	Furrow	456	272	476	625	572
100	Drip	349	167	512	383	615
	Furrow	545	334	582	766	699
125	Drip	410	198	605	454	726
	Furrow	633	396	688	908	825
Each additional 10 tons	Drip	24	12	38	28	45
	Furrow	35	24	43	56	51

Pepper

Yield (tons/ha)	Irrigation method	kg/ha				
		N	P	K	P_2O_5	K_2O
30	Drip	185	46	182	106	218
	Furrow	269	92	207	212	248
40	Drip	210	55	204	126	245
	Furrow	305	110	233	252	280
50	Drip	235	64	228	146	273
	Furrow	341	127	259	292	311
60	Drip	260	72	250	166	300
	Furrow	377	145	285	332	342
Each additional 10 tons	Drip	25	9	23	20	28
	Furrow	36	18	26	40	31

Mango

Yield (tons/ha)	Irrigation method	kg/ha				
		N	P	K	P_2O_5	K_2O
20	Microirrigation	139	30	177	68	212
	Surface	194	52	202	119	242
25	Microirrigation	148	32	187	74	224
	Surface	208	57	213	130	256
30	Microirrigation	158	35	197	80	236
	Surface	221	62	225	141	270
35	Microirrigation	168	38	207	86	248
	Surface	235	66	237	152	284
40	Microirrigation	177	40	217	92	260
	Surface	248	71	249	163	298
45	Microirrigation	187	43	227	98	272
	Surface	262	76	260	174	312
50	Microirrigation	197	45	237	104	284
	Surface	275	81	272	185	326
Each additional ton	Microirrigation	2.0	0.5	2.0	1.2	2.4
	Surface	2.5	0.96	2.3	2.2	2.8

Banana

Yield (tons/ha)	Irrigation method	kg/ha				
		N	P	K	P$_2$O$_5$	K$_2$O
40	Microirrigation	388	117	1176	268	1411
	Sprinkler	440	175	1333	401	1600
50	Microirrigation	412	123	1235	282	1482
	Sprinkler	467	185	1400	424	1680
60	Microirrigation	435	130	1294	298	1553
	Sprinkler	493	195	1467	446	1760
70	Microirrigation	459	137	1353	314	1624
	Sprinkler	520	205	1533	469	1840
80	Microirrigation	482	147	1412	337	1694
	Sprinkler	547	220	1600	504	1920
90	Microirrigation	506	157	1470	359	1764
	Sprinkler	573	235	1667	538	2000
100	Microirrigation	529	167	1529	382	1835
	Sprinkler	600	250	1733	572	2080
Each additional 10 tons	Microirrigation	23	7	59	16	71
	Sprinkler	27	11	67	25	80

Strawberries

Yield (tons/ha)	Irrigation method	kg/ha				
		N	P	K	P$_2$O$_5$	K$_2$O
20	Drip	136	32	149	73	179
	Furrow	216	64	170	146	204
30	Drip	150	39	168	90	202
	Furrow	240	78	192	180	230
40	Drip	165	47	188	107	225
	Furrow	263	93	214	214	257
50	Drip	179	54	207	123	248
	Furrow	287	108	234	246	283
Each additional 10 tons	Drip	1.46	0.73	1.91	1.67	2.29
	Furrow	2.34	1.47	2.19	3.34	3.63

In: Sewage Sludge Management
Editors: A. A. Zorpas and V. J. Inglezakis

ISBN: 978-1-61324-393-0
© 2012 Nova Science Publishers, Inc.

Chapter 7

LAND APPLICATION OF SEWAGE EFFLUENTS AND SLUDGE: SELECTED ABSTRACTS FROM 1897-1973

Antonis A. Zorpas[*]

Institute of Environmental Technology and Sustainable Development,
Paralimni, Cyprus

ABSTRACT

This Chapter is addressing a trip in the literature from the year 1897 until 1973. Present the most common practise for the treatment of sewage sludge and sewage effluents. The literature on applying sewage effluents and sludges to the land has been annotated quite thoroughly through the year 1965. The period from 1968 into 1973 represents a very active period during which the literature reflects the current attractiveness of land application approaches as an advanced waste treatment approach. This report includes several selected abstracts which have been compiled as part on land application of sewage effluents and sludges. These abstracts have been arrayed in chronological groupings and identified as to emphasis on effluent or sludge. The report also includes an author index and a subject matter (descriptor term) index to facilitate reference to specific abstracts or to abstracts addressing narrower subject matter areas.

Application of wastewater to the land as an alternative or as an adjunct to more conventional methods of wastewater management is receiving increased attention as a result of recent environmental concerns and legislation. Byplay generated through the brash claims of strong proponents and the counter statements of adament opponents regarding the viability of land application approaches confuses one seeking reliable information to support objective decisions for selection of a wastewater management alternative.

---

[*] Website: www.envitech.org Tel: +357-23743440, Fax: +357-23743441, P.O.Box 34073, 5309, Paralimni, Cyprus. Email: antoniszorpas@envitech.org, antoniszorpas@yahoo.com.

Land application of sewage predates all of the more conventional approaches to treatment and direct discharge to watercourses. It is of interest to note that some of the land application systems in operation at the present time were constructed and placed in operation prior to the development of such waste treatment concepts as the rotary sprinkler for trickling filters, the imhoff tank, or the activated sludge process.

This chapter differ from the others in this book. Is divided in several sections and presents the title of the paper, the authors where is applicable, the name of the journal, and the year of publication.

1. ABSTRACTS PRIOR TO 1941

001 S   THE UTILIZATION OF SEWAGE SLUDGE AS FERTILIZER: A REPORT OF THE COMMITTEE ON SEWAGE DISPOSAL APHA

Anonymous
Sewage Works J., 9:861-912, 1937.
Descriptors: Canada, United States, Economic feasibility, Operation and maintenance, Costs.

A comprehensive report resulting from a survey of sludge disposal practices in the United States and Canada. Data were furnished by State sanitary engineers, sewage works operators, and municipal officials. Discussed in connection with use of sludge as fertilizer are soil fertilization, fertilizer value, use in mixed fertilizers, tonnage of sludge available, types of sludge, and comparisons of sludge from different treatment processes. Results of sludge utilization on many different crops are presented and general conclusions based on the results of the survey are given.

002 M IRRIGATION WITH SEWAGE

Anonymous
Eng. News-Rec, 121 821, 1938.
Descriptors: Texas, Raw sewage. Evapotranspiration, Flood Irrigation, Climatic data.

A storage lake and land irrigation solved the sewage disposal problem at Kingsville, Texas. In this region of low rainfall and high evaporation, suitable streams for the disposal of sewage are practically nonexistent. The distribution and disposal system is described. The storage lake makes the system quite flexible. Disposal operations continue even when irrigation is not in progress.

003 M SEWERAGE AND SEWAGE PURIFICATION, 2nd ED.

Baker, M. M.
New York, D. Van Nostrand Company, 1905. 153 p.
Descriptors: Raw sewage, Economic feasibility, Soil physical properties, Flood irrigation.

The author devotes five pages to the subject of "Broad Irrigation or Sewage Farming." He states that the biggest problem is the need for land, about an acre per 100 people. Both price and availability of land are obstacles, as well as soil type. Soil should be fairly light and porous--wet, cloying soil can be used sometimes, and cultivation with sewage can improve its characteristics. The use of sewage for irrigation also has political and economic problems and the method of sewage irrigation is best suited for arid regions.

004 M EFFECTS ON ANIMAL HEALTH OF FEEDING SEWAGE

Crawford, A. B., and A. H Frank
Civil Eng., 10:495-496, 1940.
Descriptors: Pathogenic bacteria, Animal diseases, Coliforms, Water reuse, Raw sewage.

A study was conducted at the U. S Department of Agriculture's Beltsville Research Center in which swine and cattle remained in good condition after a severe six-month feeding period including raw sewage, treatment plant effluent, and sludge. This study concluded that virulent bacteria were not present in sufficient concentration in the incoming sewage, effluent, or sludge of this sewage treatment plant to cause disease in susceptible animals. These test animals were subjected to a more severe exposure to effluent than would normally be expected to occur.

005 S ADAPTABILITY OF SEWAGE SLUDGE AS A FERTILIZER

DeTurk, E. E.
Sewage Works J., 7:597-610, July 1935.
Descriptors: Fertilizer value, Anaerobic digestion, Economic feasibility, Soil chemical properties Reviews.

A significant historical review of sewage utilization is made. Comparisons of chemical composition and fertilizer value are made between sludge produced by the activated sludge process and digested or imhoff sludge. Digested sludge is also compared to arm manure. Methods of sludge utilization are discussed. Future development of sludge-processing methods may well result in the production of sludge with increased fertilizer value. Even though sludge may be low in fertilizer value, its organic content should not be overlooked as a soil-conditioning agent.

006 M SEWAGE DISPOSALS BY IRRIGATION

Eddy, Harrison, and Leonard Metcalf
In: American Sewerage Practice Vol. III, 3rd ed., New York, McGraw-Hill, 1935. Chap. 9, p. 233-252.
Descriptors: Reviews, Soil physical properties, Soil chemical properties, Rates of application.

The physical, chemical and biological processes of the soil in treating sewage effluent have been widely used. Initial settling is preferred to prevent clogging of the soil and rates of 18 feet per year have been recorded. Examples of sewage farms in Europe and America are given. In 1910, Berlin had over 43,000 acres in use for sewage disposal. Pasadena had its farm in use since 1887.

007 M SIMPLE APPLICATIONS OF FUNDAMENTAL PRINCIPLES
OF SEWAGE TREATMENT

Gillespie, C. G.
Sewage Works J., 1:68-69, 1928.
Descriptors: California, Primary effluent, Rates of application, Soil physical properties,
Costs.

The author describes two cases where expensive disposal alternates were avoided by
conversion to land application. At Carmel, California a flow of 0.05 mgd from a septic tank
was diverted to 3,000 sq ft of sand beds. The coarse sand (0.3 mm effective size) is raked
after inundation for a week and no pounding develops. At Vacaville, California 0. 125 mgd of
septic tank effluent was passed over a field of heavy clay prior to pounding. The 8 acre field
was graded to a 1 percent slope and a thick stand of grass grew immediately. Only about 10
percent of the effluent applied over 8 months appeared as runoff.

008 M SEWAGE IRRICATION IN TEXAS

Goodwin, Earl H.
Pub. Works, 66:23-24, March 1935.
Descriptors: San Antonio, Texas, Primary effluent, Soil types, Flood irrigation, Grain
crops.

Since 1900, San Antonio has utilized a part of its sewage effluent to irrigate up to 3500
acres. At present, about 47 plants in Texas use irrigation as a means of treatment and disposal.
Spray, border, and furrow methods are used to irrigate grains, grasses, cotton, alfalfa, nuts,
and citrus. Porous sandy soils seem to be most suitable. Careful supervision is required so that
soil type and crops form a compatible combination for the efficient utilization of the effluent.

009 S SEWAGE SLUDGE AS A FERTILIZER

Harper, Horace J.
Sewage Works J., 3:683-687, 1931.
Descriptors: Milwaukee, Wisconsin, Anaerobic digestion, Fertilizer value, Nutrient
removal, Microbial degradation.

Fertilizing value of sewage sludge depends upon its source and the sewage treatment
process. The nitrogen content may be quite different for sludges obtained from Imhoff tank
and activated sludge processes. Activated sludge may contain 5 to 6 percent total nitrogen,
with total phosphoric acid content of approximately 3 percent. Sludge from Imhoff tanks
seldom contains more than 2.5 percent nitrogen and 2 percent phosphoric acid.

Cites the city of Milwaukee's experience with "Milorganite" produced from activated
sludge and several experiments concerned with fertilizer value of sludge. Liquid sludge has
considerably more fertilizer value than dried sludge since much of the nutrient content drains
away in the drying process.

010 M SEWAGE IRRIGATION AS A METHOD OF DISPOSAL

Harrell, Riley B.
In: Proc. 21st Tex. Water Works Sewerage Short Sch., December 1939.
P. 121-123. Abstr., Sewage Works J., 12:1019, 1940.
Descriptors: Texas, Primary effluent, Flood irrigation, Fiber crops.
The author describes sewage irrigation as practiced by the city of Munday, Texas. Use of row crops is recommended for better weed control. Cotton is most satisfactory crop. In 1934, 24 acres of irrigated land produced 23 bales of cotton. Dry land produced less than one-third bale per acre of poorer quality cotton. Serious problem of effluent disposal has been solved by irrigation.

011 M SEWAGE DISPOSAL PRACTICE IN EUROPE

Hyde, C. G.
West. Constr. News, 4:345-352, July 1929.
Descriptors: Europe, Primary effluent, Furrow irrigation, Reviews.
The author reports on inspection tour of 28 sewage treatment plants in Great Britain and Germany. The treatment processes observed are described. A number of plants employed "broad irrigation" as a means of treatment. The sewage of Berlin, Paris, and Milan is used to irrigate crops.

012 M TREATMENT ON NATURAL SOIL

Imhoff, Karl, and Gordon Fair
In: Sewage Treatment, New York, John Wiley and Sons, Inc., 1940.
Chap. 7, p. 108-118.
Descriptors: Economic feasibility, Rates of application, Soil types, Organic loading, United States.
When land is available cheaply, land treatment of sewage can be economical. Rates of application vary from 750 gpd per acre to 24000 gpd per acre, depending on the soil type, cover crop, and land use. Use of sand beds with secondary effluent can take loadings of 800000 gpd per acre.

013 M UTILIZATION OF DOMESTIC SEWAGE AND INDUSTRIAL WASTES
       BY BROAD IRRIGATION

Kreuz, C. A.
Sewage Works J., 8:348-349, March 1936.
Descriptors: Forage crops, Flood irrigation, Costs, Odor, Reviews.
The author discussed the use of sewage for agricultural purposes. He suggested that, in order to produce maximum yield of agricultural products the sewage from as many as 80 inhabitants can be distributed on an acre, if the grass is cut three or four times, and the sewage evenly distributed over the year. The fertilizing constituents can be used to greatest advantage when the sewage is artificially sprinkled but this method has the highest operating cost and requires the largest amount of land.

Two methods of irrigation are discussed. In ordinary broad irrigation, land is divided into areas about 1.25 acres in size. The disadvantages are: the use of agricultural machinery is difficult; smoothing of the surface is expensive, too much of the land is used for roads, dikes and ditches; cultivation of cereals has not proved satisfactory: and odors during the summer cannot be entirely avoided.

In another method of irrigation, the water is held on plots of land 6 to 30 acres in size by dikes about 20 inches high. These plots are covered with sewage to a depth of 8 to 12 inches and allowed to stand from 3 to 6 days in summer and about a week in winter. The water remaining after that time is drained off through ditches. The distribution of land with at least a two percent slope has also proved very satisfactory.

014 S USE OF SLUDGE AS FERTILIZER

Maloy, Thomas P.
Sewage Works J., 3: 458-487, July 1931.
Descriptors: Anaerobic digestion, Soil physical properties, Fertilizer value.

Sludge does not have the high fertilizing value of commercial fertilizers. Experiments indicated that it is very effective in improving the physical condition of the soil: "Where the ground was hard and compact before, it is now soft and resilient. It has a certain amount of spring to it." "The whole texture of the soil was improved and made more pliable."

015 M SEWAGE TREATMENT FOR OBTAINING PARK IRRIGATING WATER

McQueen, Frank
Pub. Works, 64:16-17, 1934. Abstr., Sewage Works J., 6:145-146.
Descriptors: San Francisco, California, Greenbelts, Sprinkler irrigation, Odor, Costs.

Golden Gate Park (1,013 acres), San Francisco, was originally irrigated with sewage from an outfall sewer traversing the park. This was soon discontinued because of objectionable odors. The purchase of potable water proved too costly. A new sewage treatment plant, completed in 1932 and employing the activated-sludge process, supplies 1 mgd for irrigation of the Park. The water is clear, odorless, and completely satisfactory for the purpose. Odors and suspended matter are so completely removed that the excess plant effluent is used for lakes and waterfalls.

016 M SEWERAGE AND SEWAGE DISPOSAL

Metcalf, Leonard, and Harrison P. Eddy
New York, McGraw-Hill, 1922. 598 p.
Descriptors: Europe, Clogging (Soil), Reviews, Flood irrigation, United States.

Part of this textbook describes sewage irrigation in Britain, France, Germany and the U.S. Although sewage does have valuable fertilizer value, it is most valuable for its water content, as sewage is 99.9 per cent water. Unfortunately, sewage also contains fat and soap which are harmful to land, causing clogging of the soil. This condition can be overcome with the use of screening and sedimentation before application to the land. Sewage irrigation in the U. S has not been used on a large scale, but numerous small-scale operations have been used for many years around the country, especially in Southern California. Although satisfactory, none of the operations has been an unqualified success. One city, San Antonio, Texas, is described as having a system which pumps the sewage 12 miles from the city to a 6,700-acre privately owned tract.

017 S SLUDGE DISPOSALS AT A SEWAGE IRRIGATION FARM

Mitchell, George A.
Eng. News-Rec., 107:57, July 1931.
Descriptors: Vineland, New Jersey, Soil types, Sludge disposal, Odor.
The author describes a method of disposing of sludge-bearing sewage as practiced at Vineland, New Jersey. The sludge is pumped onto a field prepared with deep furrows. Immediately upon drying, the sludge is plowed under, and the field prepared to receive another sludge treatment. Once the sludge is plowed under, odour and insect nuisances are eliminated. The surface soil proved to be an excellent purifier.

018 M MUNICIPAL SEWAGE IRRIGATION

Mitchell, George A.
Eng. News-Rec., 119:63-66, July 1937.
Descriptors: Vineland, New Jersey, Flood irrigation, Soil physical properties, Costs, Operation and maintenance.
A sewage irrigation farm in use since 1928 at Vineland, New Jersey, provides disposal facilities for a population of 8,000 and aids crop production in poor soil. A detailed description of the operating methods is given. Distribution system details include land slope and flow, land preparation, and crops grown. Revenue and cost data are included.

019 S THE VALUE OF RAW SEWAGE SLUDGE AS FERTILIZER

Muller, J. F.
Soil Sci., 28 :423-432, 1929.
Descriptors: Fertilizer value, Soil physical properties, Soil types, Flood irrigation, Nitrification.
Several samples of dried fresh sewage sludge were analyzed, and pot experiments carried out to determine the fertilizer value of such material. The analyses showed considerable potential plant food is present. The carbon-nitrogen ratio was narrowed to below eight by the addition of available nitrogen, markedly increasing the fertilizer value of the sludge. A phosphate supplement appears to be necessary for good plant growth; and a potash supplement, in small quantities, seems desirable. The dried sludge alone, with no mineral supplements, when applied to turf grown on sand gave a good stand of grass and prevented its dying off. Dried fresh sludge applied to a sandy soil materially increased its water-holding capacity, a most desirable result on soils of this type. Although the experiments reported did not indicate a need for using lime with the sludge, lime almost certainly would be required after several years continuous sludge application.

020 M SEWAGE IRRIGATION

Rafter, George W.
U. S. Geol. Surv. Water-Supply Pap. 3, Washington, D.C., 1897. 100 p.
Descriptors: United States, Reviews, Human diseases, Public health regulations, Operation and maintenance.

The author presents the method of sewage irrigation as a method which is not new. His purpose is to point out to American farmers and to municipal authorities the benefits of this use of sewage. It is especially useful in the arid West and has been used in Cheyenne, Wyoming since 1883, and is in use at about 60 installations in the U. S. Problems of disease are described as minimal with proper care, and crops fit for human consumption can be grown. Descriptions are given of various methods of sewage irrigation along with descriptions of foreign operations in Germany, France and England.

021 S SEWAGE SLUDGE AS FERTILIZERS

Rudolfs, Willem
Soil Sci., 26:455-458. 1928.
Descriptors: Anaerobic digestion, Aerobic digestion, Fertilizer value.

The nitrogen content of sewage sludge varies with the type of treatment. Aerobically and anaerobically treated sewage sludges contain about 5 and 2.25 percent nitrogen, respectively. Only one sewage treatment plant markets its sludge on a large scale; at a number of places the sewage sludge is given away or sold for a small nominal sum. An estimated 150,000 to 200,000 tons of nitrogen are lost annually. Analysis of sludges from a number of sewage plants indicates that 8,000 to 10,000 tons nitrogen per year could be saved.

022 S SLUDGE DISPOSAL AND FUTURE TRENDS

Rudolfs, Willem, and E. J. Cleary
Sewage Works J., 5:409-428, may 1933.
Descriptors: Reviews, Economic feasibility, Human diseases, Sludge disposal, Fertilizer value.

Present methods of sludge handling and disposal are discussed. The need for further development is stressed. Sludge as a fertilizer is discussed from the standpoint of fertilizing value, preparation of fertilizer, and economic and hygienic considerations. Fertilizer production may be profitable only for the large plants. Small plants usually must dispose of sludge material locally as a market develops. Dewatering and incineration offer promise as a means of ultimate disposal.

023 S SEWAGE SLUDGE AS FERTILIZERS

Skinner, John F.
Sewage Works J., 4:279-282, March 1932.
Descriptors: New York, Operation and maintenance, Costs, Vegetable crops, Fertilizer value.

Sludge from four Imhoff tanks at Rochester, New York is utilized as fertilizer for gardens, shrubs, lawns, and local farm crops. Details of the operation are described, and sales data are shown. Liquid sludge has much greater nutrient value than dried sludge. Consideration is given to dispensing liquid sludge thus eliminating cost of constructing and operating drying beds.

024 S PUBLIC HEALTH SIGNIFICANCE OF SEWAGE SLUDGE
      WHEN USED   AS A FERTILIZER

Tanner. Fred. W.
Sewage Works J., July 1935.
Descriptors: Sludge disposal. Pathogenic bacteria. Vegetable crops, Human diseases, Public health regulations.

The application of sewage sludge to soil on which vegetables, which ……..be eaten raw, are grown should be practiced with caution. …. .longevity of pathogenic bacteria in sludge would probably be greatly, influenced by the nature of the sludge and the conditions under which it is stored and handled, sufficient data have been recorded to indicate the presence of viable Bacterium typhosum cells in sludge. At best, the sludge should be added to the soil in the late fall, winter, or early spring. Wolman's advice, probably sound, is that sludge not be added to growing crops. Sanitary districts and others concerned with the sale of sewage sludge to farmers might well consider the health hazards involved.

2. ABSTRACTS 1941 THROUGH 1950

025 S LIQUID SLUDGE – THE VITAMIN B FERTILIZER

Damoose, N
Sewage Works Eng., 12:308-312, 1941.
Descriptors: Michigan, Greenbelts, Odor, Public acceptance, Sludge disposal.

The sale of liquid sludge has been promoted at Battle Creek, Michigan. After successful trials on the plant grounds, a truck was fitted with a 600-gallon tank for delivery of liquid digested sludge to the public. Use of this material on lawns produced "astounding" results. Odor and appearance nuisances were overcome by wetting down with sprinklers following application. In addition to the plant food value, the author believes vitamin B plays an important role and discusses this in some detail.

026 S EXPERIENCE OF CHICAGO, ILL., IN THE PREPARATION
      OF FERTILIZER

Dundas, William A., and C. P. McLaughlin
Amer. Soc. Civil Eng., J. Sanit. Eng. Div., 69: 80-102, 1943.
Descriptors: Sludge disposal, Chicago, Illinois, Fertilizer value, Odor, Design data.

In the sludge disposal of The Sanitary District of Chicago (III) a method has been worked out by a chain of experiments to dewater and heat dry activated sludge for use as fertilizer material or for incineration. The principles of design are indicated from the heat standpoint together with the control of odors. A typical heat balance is presented, with a discussion of the relative merits of various fuels. At the Southwest Works a combination of power generation with heat drying has been successfully used.

027 M LAND DISPOSAL OF SEWAGE

Gray, H. F., and W. O'Connell, Jr.
Sewage Works J., 16: 729-740, April 1944.
Descriptors: California, Design data, Organic loading, Flood irrigation, Rates of application.
Land application of sewage effluents can be done with different objectives. If the objective is an agricultural operation, the loading rates should be 2,000 to 3,000 gpd per acre depending on the crop. The area for irrigation should be planted to one crop and a second area for excess effluent should be set aside. Three installations using municipal effluent in California are described.

028 M AGRICULTURAL UTILIZATION OF DOMESTIC SEWAGE IN EUROPE AND U.S.A.

Halamek, Ferdinand
Bull, Czechoslovak Academy Agriculture (Prague), 22:396-402, 1948.
Biol. Abstr., 23:3092, 1949.
Descriptors: Europe, United States, Wastewater disposal, Operation and maintenance, Legal aspects.
The present status of sewage irrigation and the utilization of fertilizing ingredients in sewage and sewage sludge in Europe and U.S.A are presented. Germany uses sewage irrigation and often overlooks the hygienic problem. In Britain the use of sewage as a fertilizer is decreasing. In the U.S.A. sewage irrigation is practiced only in the south western states and, there, hygienic regulations are severe. Directions are given for proper management of sewage irrigation.

029 M EFFECT OF SEWAGE WATER ON THE YIELD AND QUALITY OF GRASSLAND

Maloch, M.
Sbornik Ceskoslovenske Akademie Aemedelskych (Progue), 19: 57-107, 1946, Abstr., Soils Fert., 13:364(2021), 1950.
Descriptors: Salinity, Forage crops, Water reuse, Plant growth.
Application of sewage water for 3 years to grassland raised the hay yield by 132.9 percent and the yield of crude protein by nearly 300 percent. Additions of superphosphate, Ca

and K salts to the sewage waters gave even higher increases. There was evidence of residual action of N from the sewage, and its effect in increasing the resistance of the grasses to summer drought was very marked.

030 M THE AGRICULTURAL USE OF SEWAGE

Muller, Wilhelm
Wasser Boden (Hamburg), p. 124, 1949. Abstr., Sewage Ind. Wastes, 22: 589, April 1950.
Descriptors: Germany, Public health regulations, Water management (applied), Pathogenic bacteria, Economic feasibility.

During recent years, the agricultural use of sewage has often been discussed in Germany, with no final answer being reached on this important question. Public health requires hygienic sewage disposal but local authorities alone cannot solve the problem. It is a task for the state organization, especially in densely settled countries.

Water conservation has become more important and water use must be regulated. Within this water planning, sewage has its own part. In particular, the agricultural use of sewage by different technical means may appreciably increase a country's productivity. For this purpose, sewage must be fully treated and freed of pathogenic organisms.

The humus matter and the manure value of sewage should be used in agriculture to compensate for the organic matter exported from the country as food. Sewage disposal into the ocean only "manures" the sea water. Treated sewage should be discharged into rivers only in such volume as is necessary to manure the rivers for fishing industries.

031 S UTILIZATION OF SEWAGE SLUDGE AS FERTILIZER

Pearse, Langdon, and A. H. Niles, et al.
Fed. Sewage Works Ass., Champaign, Ill., Manual of Practice No. 2, 1946. 120 p.
Descriptors: Recycling nutrients, Bibliographies, Costs, Fertilizer value, Human diseases.

The major topics in the manual are: Fertilizer Requirements of Soils; Fertilizer Characteristics of Sewage Sludge; Effects of Sewage Treatment Practices on Fertilizing Elements in Sludge; Processing Sludge for Use as Fertilizer; Application of Sludge as Fertilizer; Hygienic Aspects of Sludge Utilization as Fertilizer; Prices, Marketing, and Economic Considerations. A closing chapter includes conclusions and recommendations relative to sludge utilization. The bibliography contains 251 entries.

032 S DISPOSAL OF LIQUID SLUDGE AT KANKAKEE, ILLINOIS

Schriner, Phillip J.
Sewage Works J., 14:876-878, April 1942.
Descriptors: Costs, Illinois, Sludge disposal, Operation and maintenance, Recycling nutrients.

Cost figures show that liquid sludge disposal by tank truck delivery constitutes a considerable saving over operation and maintenance of sludge drying beds. It also eliminates the severe operating difficulties of sludge drying beds in the winter months. The use of liquid sludge to fertilize lawns will be promoted, and the practice should show a considerable profit to the city over past experience with sludge drying beds.

3. ABSTRACTS 1951 THROUGH 1955

033 S SEWAGE SLUDGE FOR SOIL IMPROVEMENT

Anderson, Myron S.
U.S. Dept. Agr., Wash., D.C., Circ. No. 972. 1955. 27 p.
Descriptors: Sludge disposal, Nitrification, Bibliographies, Anaerobic digestion, United States.
Chemical compositions of sludges variously prepared in different parts of the United States are reported. Only 18 to 25 percent of the nitrogen present in digested sludges is normally nitrified during a 16-week period. Activated sludges show nitrification values of 50 to 60 percent for a similar period. The bibliography has 41 entries.

034 S HYGIENE OF IRRIGATION AND THE USE OF SEWAGE RESIDUES

Anonymous
Staedteh. (Hamburg), 6:259-260, 1955. Water Pollut. Abstr., 29:244(1348), 1956.
Descriptors: Public health regulations, Sludge disposal, Sprinkler irrigation, Water quality standards, Europe.
A draft is given of a proposed standard (DIN 19650) dealing with requirements for water used for irrigation and with the use of sewage and sludge in agriculture.

035 M THE SEWAGE UTILIZATION PLANT AT MEMMINGEN

Bachmann, G.
Wasserwirt. – Wassertech. (Berlin) 4:191, 1954, Water Pollut. Abstr., 29:28(166), 1956.
Descriptors: Sludge disposal, Sprinkler irrigation, Forage crops, Memmingen, Germany.
The author describes the sewage works of Memmingen where sewage, after sedimentation for 1.5 hours, is used as artificial rain. The yield of hay on watered land has been increased by 52 percent.

036 I STUDIES OF WASTE WATER RECLAMATION AND UTILIZATION

Bush, A. F., and S. F. Mulford
Calif. State Water Pollut. Contr. Board, Sacramento, Pabl. No. 9, 1954. 82 p.
Descriptors: California, Groundwater recharge, Groundwater contamination, Soil types, Odor.
This report covers the determination of (a) relationship of underground water pollution to methods and rates of spreading and percolation rates of reclaimed waste waters; (b) the effects on underground water pollution of increasing the percolation rates of reclaimed waste

waters by means of cropping and vegetation, additives to top layers of soil, forced irrigation, or other means; (c) degree of contamination and/or pollution of a variety of truck crops on maturity, where the reclaimed waste waters: from various sources, including sewage and industrial wastes, have been applied to the crops and spreading areas by flooding, spraying, or other means; and (d) the extent of odor and other nuisances which may result from this procedure.

It is recommended that waste water be considered a water resource and that further studies be made of percolation rates and degree of treatment required to handle the pollution load, also to remove salts (sodium and boron) from sewage and industrial waste. Useful references and a bibliography are included.

037 S REGULATIONS FOR IRRIGATION AND THE USE OF SEWAGE SLUDGE

Falkenhain, H. S.
Wasserwirt. – Wassertech. (Berlin) 3: 293-294, 1953. Water Pollut. Abstr., 28: 213(1805), 1955.
Descriptors: Wastewater treatment, Human diseases, Sludge disposal, Pathogenic bacteria, Germany.

Proposed German regulations for irrigation and for the use of sewage sludge are discussed. The author considers the divergent views in the literature on the subject of preliminary treatment of sewage and trade waste waters for use in agriculture. Special importance attaches to the destruction of parasitic worms and pathogenic bacteria. During sedimentation the numbers of these are reduced, but the danger of infection is not removed.

038 I SEWAGE EFFLUENT RECLAMATION FOR INDUSTRIALS AND AGRICULTURAL USE

Greenberg, Art. old E., and Jerome F. Thomas
Sewage In. Wastes, 26:751-770, June 1954.
Descriptors: California, Water quality standards, Groundwater contamination, Groundwater recharge, Spreading basins.

Planned reclamation is designed to produce a usable water from sewage such reclaimed waters may be used by industry or agriculture "directly" "indirectly." The latter involves replenishing groundwater basins from ... industrial, agricultural, or domestic supplies are drown.

Experiments performed by the University of California's Sanitary Engineering Research Laboratory are discussed, and the conclusions are summarized:

1. A bacteriologically safe water can be produced from settled or more highly treated sewage if the liquid passes through at least four feet of soil.
2. A water of chemical quality satisfactory for most uses can be produced from settle sewage or final effluents provided high concentration of undesirable wastes are not included in the raw sewage.
3. To obtain relatively high rates of percolation highly treated sewage plant effluent must be used for ...

4. The optimum method of operation is to spread continuously for a month, preferably with liquid containing large amounts of organic matter, then to allow the basin to rest until it is air dried. Thereafter cultivation of the dry soil is desirable. Following this preliminary treatment, continuous application of a final effluent may be carried on for as …as six months. Resting and cultivation then be repeated.
5. Mosquitoes in spreading basins will create a nuisance and health hazard unless control measures are adopted. If algal odors are pronounced, the control of algae also may be necessary.
6. Further investigation is needed of sewage percolation in different soils and of phenomena ass ……. with the movement of water into such soils to generalize the conclusions reached as a result of this study with Hanford fine sandy loam.

039 S THE AGRICULTURE UTILIZATION OF SEWAGE

Grubinger. H
Bodenkulter. Abstr. Soils Fert., 18:64(327). 1955
Descriptors: Recycling nutrients.
The author discusses the technical features of the purification of sewage and sprinkler irrigation with sewage. Irrigation with 1500-3000 cu.m. per hectare of purified sewage annually supplies 120-240 kg/ha N, 30-60 kg/ha P_2O_5; and 52-104 kg/ha K_2O. Irrigation with such waters and application of the sludge are particularly suited for nitrogen fertilizing, especially of grassland.

040 S DISPOSAL OF WET DIGESTED SLUDGE AT MARSHALL, MISSOURI

Hayob, Henry
Sewage Ind. Wastes. 26:93-95, January 1954.
Descriptors: Missouri, Sludge disposal. Soil types, Anaerobic digestion.
Sludge drying beds were eliminated, and liquid sludge is now hauled by tank truck to be spread directly onto farming land. No mention is made of crops grown, soil type, or crop yields. Several problems were solved with this new type of disposal.

041 S FIFTEEN YEARS OF THE UTHLEBEN SEWAGE COOPERATIVE

Holler, K.
Wasserwirt. –Wassertech. (Berlin), 2:397, 1952. Abstr., Sewage Ind.
Wastes, 26:118, January 1954.
Descriptors: Sprinkler irrigation, Grain crops, Raw sewage, Waste-water disposal.
Progress and effect of 15 years of sewage farming and technical installations are described. Of about 613 hectares (1,515 acres) 440 hectares received spray irrigation and 173 hectares received trench irrigation. The present maximum sewage flow is about 4,800 cu. m. per day (1.27 mgd). The raw sewage flows through a grit removal unit and thence into a 5,000 cu. m. equalizing lagoon functioning at the same time as a settling basin. This is drained once a year for sludge removal. About 50 percent of the wet sludge is used directly with cattle manure on the fields. The other 50 percent is dried on beds and composted with garbage. The sewage is pumped to the land through movable pipes. The importance of proper timing of

irrigation is emphasized and illustrated for grains (oats, wheat), beets, rape seed, and pastures. Comparisons of crop yields with non-irrigated surrounding areas are given.

042 S HYGIENIC EVALUATION OF THE AGRICULTURAL UTILIZATION OF SEWAGE

Kruez, C. A.
Gesundh. Ing. (Munich), 76:206-211, 1955.
Descriptors: Public health regulations, England, United States, Germany, Soviet Union.

The author discusses the hygienic problems arising from the agricultural utilization of sewage and the amount of agricultural use in the United States, Great Britain, the Soviet Union, and Germany. He considers the dangers to health, precautions necessary in the use of sewage and sludge, the fertilizer and humus-forming effects of sewage and sludge, and the effect of treatment on the fertilizing constituents.

043 M LAND TREATMENT OF SEWAGE

Reploh, H
Kommunalwirtschaft (Germany), 8:401, 1955. Water Pollut. Abstr., 29:352 (1932), 1956.
Descriptors: Sprinkler irrigation, Odor, Human diseases. Groundwater recharge.

The author discusses the advantages of agricultural utilization of sewage, methods of preventing odor nuisance and spreading of bacteria, and the importance of this method of disposal to the groundwater supply.

044 M AGRICULTURAL VALUES OF SEWAGE

Skulte, P
Sewage Ind. Wastes. 25:1297 1305, November 1953.
Descriptors: Planning. Groundwater Recycling nutrients. Primary effluent. Europe.

Actual experiences have demonstrated that sanitary sewages and many industrial wastes can be successfully used for agricultural, industrial, and other purposes. The sewage as used for irrigation is usually given primary treatment. After that, the processes of natural soil filtration and biochemical stabilization can produce an oxidized and well-treated percolation water. Reclamation programs planned only for groundwater replenishment are usually unsound economic ventures. Therefore, the emphasis should be on the agricultural values in sewage, greatly reducing the immense fertilizer bill and improving soil conditions.

045 M EFFECTS OF SEWAGE ON CATTLE AND GARBAGE ON HOGS

Snyder, W.
Sewage Ind. Wastes, 23: 1235-1242, October 1951.
Descriptors: Human diseases, Pathogenic bacteria, Wastewater treatment, Animal diseases.

In making use of sewage effluents, the degree of treatment required must be governed by its subsequent reuse if the greatest social and economic advantages are to be realized. Since there are many diseases common to both man and animals, the possibility of disease transmission by sewage must not be overlooked. Experiments are cited in which swine were fed with incoming sewage mixed with bran; others were fed with effluent mixed with bran. Careful post-mortem examinations revealed no evidence of disease. Cows were supplied with effluent for drinking Likewise, no evidence of disease was found. It was concluded that virulent bacteria were not present in sufficient concentration in the sludge and effluent of the treatment plant at Beltsville to cause disease in susceptible animals. The feeding of raw garbage to hogs can lead to trichinoisis in the animals and transfer to man if the meat is not properly cooked. Garbage can be cooked to destroy the infection before feeding to the animals.

046 M EFFECT OF SEWAGE IRRIGATION UPON SOILS

Steel, W., and E. J. M. Bery.
Sewage Ind. Wastes, 26: 1325-1339, November 1954.
Descriptors: Soil chemical properties, Organic loading, Soil physical properties, Wastewater disposal, Texas.

The relatively small changes in the soils studied indicate that sewage irrigation is neither especially beneficial nor injurious to soils. Sewage irrigation encourages accumulation of chlorides. Leaching of the chlorides was readily accomplished and there should be no injurious accumulations under normal operation. A slight increase in organic matter (humus) can be expected. Pore space is increased by sewage irrigation, and crumb structure shows improvement. Sewage-irrigated soils accumulate slightly more phosphorous than water-irrigated soils. Boron also will be contributed to soils by sewage.

047 I LAND DISPOSAL OF SEWAGE AND INDUSTRIAL WASTES

Stone, Ralph
Sewage Ind. Wastes, 25:406-418, April 1953.
Descriptors: Waste quality standards, Human diseases, Groundwater recharge, Rates of application, California.

The author discusses the disposal of waste effluents by irrigation of restricted crops, or within spreading areas. Soil organisms and filtration provide a "high quality, fully oxidized, pathogen-free, nonturbid water". Intermittent dosage assures an aerobic environment which is required for nuisance-free disposal. Several examples are cited along with data describing operations. Variable factors important to the design and operation of land disposal facilities are discussed.

Nuisance and health hazards may be controlled through proper design and technical supervision of the operation. Certain highly mineralized industrial wastes may present special problems. Land disposal methods appear to be satisfactory for domestic sewage effluent.

048 S FERTILIZER VALUE IN WASTE DISPOSAL METHODS

Van Kleeck, W.
Amer. J. Pub. Health, 44:349-354, March 1954.
Descriptors: Sludge disposal, Recycling nutrients, Connecticut.
Sprinkler irrigation, Odor.

Dried or partially dewatered sewage sludge makes an excellent soil conditioner and a good though incomplete, fertilizer. Heat-dried raw activated sludge is the best sludge product, both chemically and hygienically, although some odor may be encountered in its use. The reasons why more sludge is not used are discussed, as are the advantages and disadvantages of broad irrigation and sewage farming. Garbage composting, both alone and in connection with sewage sludge, is receiving more attention although current practice ... much to be desired.

4. ABSTRACTS 1956 THROUGH 1960

049 S FERTILIZING CHARACTERISTICS OF SEWAGE SLUDGE

Anderson, Myron S.
Sewage Ind. Wastes, 31: 678-682, June 1959.
Descriptors: Climatic data, Recycling nutrients, anaerobic digestion, Soil types, United States.
The chemical composition and fertilizer value of sewage sludge depend in part on the method of treatment from which the sludge is obtained. Undigested sludge from secondary treatment has the greatest fertilizer value. Digested sludge from primary treatment has a lower fertilizer value but may be beneficial as a soil conditioner or mulch. The nitrogen content of digested sludge is appreciably lower. Heat treatment is recommended for sanitary purposes before sludge is sold as a fertilizer.
It is important that prospective users be told what to expect from sludge applied to soil. The plant responses will vary with type of soil, kind of plants grown, and climatic conditions.

050 S SEWAGE SLUDGE AS SOIL CONDITIONER

Anonymous
Water Sewage Works, 105:489, 1958, Pub. Health Eng. Abstr., 39:S:46.
Descriptors: Greenbelts, Economic feasibility, Sludge disposal, Surveys.
The results of a national survey are presented. Reports the increased sale and demand for sewage sludge to be applied to soil in recent years. Many examples and case histories are cited. Sludge analyses, sales trends, price, etc, are tabulated for a large number of cities where sludge is sold. Sludge is recommended for lawns, parks, and flower gardens. It deepens the colour of grass and stimulates a luxurious growth that is noticeable for longer than one season. It should be applied late in March and again in September, if desired. It provides needed humus as well as a moderate amount of nitrogen to flower garden soils.

051 S SEWAGE SLUDGE AS SOIL CONDITIONER

Anonymous
Water Sewage Works, 106:424, 1959.
Descriptors: Sludge disposal, Recycling nutrients, Public health regulations, Anaerobic digestion, United States.

This article discusses the use of sewage sludge by farmers and gardeners in the United States, methods of processing the sludge and its fertilizer value, how to use the sludge and the safety of using sludge.

052 S SALE OF "RAPIDGRO" GRAND RAPIDS $150,000 REVENUE
FROM DRIED SLUDGE

Delano, E. H.
Wastes Eng., 28:30-31, 1957.
Descriptors: Michigan, Costs, Operation and maintenance, Surveys.
The city of Grand Rapids has produced an organic soil builder from its sewage treatment plant since 1932. In the past 14 years, its dried sludge sales have totalled $150,975 against operating costs of $87,771. Analyses of the sludge are given. The cost and other manufacturing aspects of "RAPIDGRO" production are discussed in detail.

053 S SLUDGE UTILIZATION AND DISPOSAL

Fleming, J. R
Sewage Ind. Wastes, 31:1342-1346, November 1959.
Descriptors: Sludge disposal, Recycling nutrients, Southeast U. S., Surveys, Fertilizer Value.
Presents a resume of general methods of sludge disposal and summarizes methods used in 28 towns in Alabama, Arkansas, Florida, Kentucky, South Carolina, and Tennessee. Soil conditioning and fertilizer value of digested sludge are discussed, as are problems associated with the agricultural use of sludge.

054 M SEWAGE SALINITY PREVENTS USE OF EFFLUENT FOR GOLF
COURSE IRRIGATION

Guymon, Boyd E.
Wastes Eng., 28:80-83, 1957.
Descriptors: Sprinkler irrigation, Golf courses, Salinity, California, Costs.
The salinity of the treated sewage of the city of Coronado, California, was found to be too high to permit its use for irrigating a proposed 18- hole public golf course on land bordering San Diego Bay. The annual cost of irrigating the tract with the municipal supply was estimated to be $26.000. The author presents salinity data for both the public water supply and the sewage, covering a typical 24-hour day. A method of separating the merging flows of two main outfall sewers, one high in salinity and the other acceptable, is illustrated and briefly explained.

055 M UTILIZATION OF SEWAGE FOR CROP IRRIGATION IN ISRAEL

Heukelekian, H.
Sewage Ind. Wastes, 29:868-874, August 1957.
Descriptors: Odor, Economic feasibility, Activated sludge effluents, Sprinkler irrigation, Israel.

Distinction is drawn between the disposal of sewage on land and the utilization of sewage for crop production. Failures and difficulties encountered with the former need not arise when utilization is the main objective. With land disposal systems, the tendency is to apply more sewage to a limited area than can percolate through the soil. Water logging of the soil and odors result. When application is geared to the needs of the soil and crop, these problems do not arise. Under proper climatic and soil conditions and with proper control, irrigation with sewage effluents need not create nuisance conditions or health hazards. In certain areas of the world, crop irrigation is an economic necessity for the development of agriculture and, at the same time, it can solve the sewage disposal problem.

Sewage effluents are proposed as a resource to augment present water supplies and allow further expansion of irrigated agriculture. Factors affecting the successful utilization of sewage for crop production are discussed. Two alternatives for the treatment of irrigation sewage are suggested:

1. The sewage could be subjected to secondary treatment and thorough disinfection and the effluent used for unrestricted crop irrigation.
2. The sewage could be given minimum treatment and the effluent utilized without disinfection for irrigation of crops not eaten raw by human beings.

Experiments are described which compared sewage irrigation with municipal water supply. Significantly greater yields on sewage-irrigated plots were attributed to regular supply of nutrients throughout growing season instead of just at beginning.

056 S SLUDGE DISPOSAL PRACTICES IN THE PACIFIC NORTHWEST

Leaver, R. E.
Sewage Ind. Wastes, 28:323-328, March 1956.
Descriptors: Washington, Recycling nutrients, Economic feasibility, Wastewater disposal Costs.

Summarizes the sludge disposal practices at several communities in the Pacific Northwest, principally the State of Washington. Four communities are cited as typifying current practice in the disposal of liquid sludge. Some of it is sold, and some is free to private haulers. Other examples are cited in which communities' dispose of air-dried digested sludge and filter-dried sludge. Income from sales is given. Fertilizing ingredients are compared in tabular form for various types of sludge, manures, and other organic materials. The Washington State Department of Health, "General Guide for the Utilization and Disposal of Sewage Sludge," is presented as an appendix to the report.

057 M THE WHY AND HOW OF SEWAGE EFFLUENT RECLAMATION
     WATER AND SEWAGE

McGauhey, P. H.
Water Sewage Works, 104:265-270, 1957.
Descriptors: Groundwater recharge, California, Economic feasibility, Flood irrigation.

The author discusses the need for sewage reclamation in California and the amount of water available from sewage. The total volume of water which could be reclaimed would be only eight percent of that required for crop irrigation but would be of more value for industry and irrigation purposes in towns. Sewage effluent can be used to recharge groundwater supplies by spreading on the soil. The bacteria present are removed during the first four feet of travel through the soil. Ground-water may also be recharged by direct injection of effluent into water bearing strata. Bacteria do not travel more than 100 feet in moving groundwater. The author considers the present use of effluents and suggests methods of increasing future use.

058 M DIRECT UTILIZATION OF WASTE WATERS

Merz, Robert C.
In: Proc. 11th Ind. Waste Conf., Purdue Univ., Lafayette., 1956.
p. 541-551
Descriptors: Golf courses, Recreational facilities, Climatic data, Economic feasibility, Water reuse.

A survey shows that more than 150 industries in 38 states reclaim industrial wastes, and about 15 in nine states employ sewage effluent. The primary reason is that there are significant saving. Specific examples are cited.

Land and climate are the primary factors affecting agricultural utilization of wastewater. Several successful operations are described. Restrictions are the amount of water to be disposed of, quality of the effluent, and health regulations.

Other direct utilization is employed for recreational areas (golf courses, decorative lakes, parks, etc.) and groundwater recharge. The author states, "The investigation made thus far indicate strongly that the reclamation of sewage effluents is a sound practice for industry, agriculture, and other uses."

059 M WATER RECLAMATION AND REFUSE DISPOSAL

Merz, Robert C.
Water Sewage Works, 105:306-307, July 1958.
Descriptors: Sprinkler irrigation, Water reuse, California, Surveys, Costs, Groundwater recharge.

This report summarizes the reports of five speakers at the forum. Settled sewage from Santa Ana irrigates 2,250 acres growing alfalfa and sugar beets, and on pre- irrigation, bases, lima beans and chilli peppers. San Bernardino sells effluent for irrigation but plans to improve the treatment process and use the effluent for recharging the groundwater. Golden Gate Park uses 0.75 mgd from the San Francisco treatment plant.

Contemplated plant improvements will make 5 mgd available for irrigation. Oceanside will use oxidation ponds for final treatment to prepare its sewage effluent for commercial and irrigation use. A Marine Corps base in the Imperial Valley pays $55 per acre-ft for water piped in from 15 miles away. Reclaiming its sewage plant effluent by pond treatment and using it for irrigation and flushing will effect an annual savings of $24,000.

060 S UTILIZATION OF LIQUID SLUDGE

Merz, Rober C.
Water Sewage Works, 106:489-493, 1959.
Descriptors: Land reclamation, California, Costs, Sludge disposal, Recycling nutrients.

A method for reducing the cost of treatment plant operation plus reclamation of waste land is presented. The city of San Diego has found land disposal a satisfactory and practical method of utilizing the liquid digested sludge produced at its sewage treatment plant. Sludge-drying facilities have been shut down, and all sludge is now utilized for soil improvement. The value of organic matter in the soil for maintaining and storing moisture is well known.

The San Diego liquid sludge disposal operation has shown that (a) waste land can be reclaimed for agriculture with liquid sludge of reasonable solids content at less cost than with dried sludge; (b) sludge loadings as high as 100 tons dry solids per acre can be applied without impairing crop growth; (c) sludge loadings as low as 25 tons dry solids per acre will produce crop growth comparable to that achieved by the use of commercial fertilizer at conventional application rates; (d) an initial sludging with 50 tons dry solids per acre will grow a second superior crop due to residual fertilization not used up by the first; (e) it is possible to avoid serious sludge handling or nuisance problems.

061 M IRRIGATION WITH SEWAGE IN AUSTRALIA

Muller, W.
Wasser Boden (Hamburg), 7:12, 1955; Zb. Bakt. I, Ref., 1956, 159:503. Water Pollut. Abstr., 20:202(1108), 1956.
Descriptors: Water reuse, Wastewater disposal, Australia, Forage crops, Fruit crops.

The author gives an account of the conditions under which sewage is used for irrigation in Australia. Only settled and biologically treated sewage may be used. Surface irrigation is preferred. The amounts vary from 350 to 7,500 mm. per year. Land for pasture use and fruit growing is irrigated.

062 S DIE INFEKTION VON GEMUSEPFLANZEN DURCH DIE BEREGNUNG
MIT HAEUSLICHEM ABWASSER (INFECTION OF VEGETABLES
BY APPLICATION OF DOMESTIC SEWAGE AS ARTIFICIAL RAIN)

Muller, Gertrud
Staedteh. (Hamburg), 8:30-32, 1957.
Descriptors: Sprinkler irrigation, Vegetable crops, Flood irrigation, Pathogenic bacteria, Germany.

The author describes experiments in which plots of land, on which carrots, cabbages, potatoes, and gooseberry bushes were growing, were watered with settled sewage. The soil, vegetables, and fruits were tested for the presence of Bact. coli and salmonella at intervals up to 40 days after application of sewage. The amounts of sewage used were small, but Salmonella were detected in the soil and on the potato tubers after 40 days, on carrots after 10 days, and on cabbage leaves and gooseberries after 5 days.

063 S UTILIZATION OF SEWAGE SLUDGE FOR THE PRODUCTION
OF TOPSOIL

Scanlon, A. J.
Sewage Ind. Wastes, 29:944-950, August 1957.
Descriptors: Odor, Recycling nutrients, Design data.
In New York City, the idea of recovering organic sewage solids and returning them to the soil as fertilizer has been advocated since the early screening plants were built in Brooklyn and Queens. No significant progress was made in this direction until 1935, when the practice of separate sludge digestion was established at the Coney Island sewage treatment plant. Digestion not only produces an odourless sludge acceptable as a fertilizer or soil conditioner, but also yields a valuable product in the form of methane gas which is used for heat and power.

064 S LAND IN SEWAGE PURIFICATION

Stone, A. R.
J Inst. Sewage Purif. (London), Pt. 4:417-424, 1960.
Descriptors: England, Sludge disposal, Recycling nutrients.
The author cites British experience with land disposal and discusses reasons behind several monumental failures. The agricultural disposal of sludge is dealt with in some detail and is recommended as an integral part of land disposal of sewage. Experience at Nottingham is described in detail. Where sludge disposal and irrigation with effluent have been studied.

065 M SOIL IMPROVEMENT SOCIETY ORGANIZED TO PRODUCE
AND TO UTILIZE COMPOST OF REFUSE

Tietjen, Cord, and Hans Joachim Banse
Calif. Vector Views, 7:1-7, January 1960.
Descriptors: Wastewater disposal, Sludge disposal, Water reuse, Germany.
The article demonstrates how cooperation between people who produce wastes and those who can use the wastes leads to increased efficiency in waste disposal.

066 M ANNOTATED BIBLIOGRAPHY ON ARTIFICIAL RECHARGE
OF GROUNDWATER THROUGH 1954

Todd, David K.
U. S Geol. Surv. Water-Supply Pap. 1477, 1959. 115 p.
Descriptors: Groundwater recharge, Wastewater disposal, Water reuse, Bibliographies, United States.
The author lists the various methods used in artificial recharge operations and considers the factors that are important in the selection of the method. The bibliography lists the literature pertaining to artificial recharge of groundwater up to and including the year 1954. Each reference is abstracted, authors are listed alphabetically, and an index based on subject and locality is included. The bibliography section covers pages 5 through 107 of the report.

067 S DO'S AND DON'TS OF USING SLUDGE FOR SOIL CONDITIONING
AND FERTILIZING

Van Kleeck, W.
Wastes Eng., 29:256-257, 274, May 1958.
Descriptors: Public health regulations, Sludge disposal, anaerobic digestion, Plant response, Connecticut.
Various methods of sludge disposal are discussed with emphasis on use as a soil conditioner. The characteristics of both raw and various forms of digested sludge which affect the soil and crops are pointed out. As a guide to the attitude of public health authorities or the use of sludge on soils, the author quotes the policy of the Connecticut State Department of Health.

068 M A REPORT OF PRESENT ACTIVITIES IN ISRAEL

Yehuda, Peter
Water Sewage Works, 105:493, 1958.
Descriptors: Israel, Spreading basins, Flood irrigation, Forage crops, Groundwater recharge.
The primary effluent from a number of sewage treatment plants is pumped to a 55- acre sand dune at the rate of 2,000 to 2,500 gpd/acre. Sandy loam underlies the dune at depths of 13 to 23 feet. The annual layer formed during the first year of cultivation, helping to stabilize the shifting sands.
A natural depression is used as an experimental percolation area. Sewage effluent is applied at the rate of 85,000 to 100,000 gpd/acre, ten times the agricultural irrigation rate. This raises the fresh water table and prevents salt water intrusion.

5. ABSTRACTS 1961 THROUGH 1965

069 S WASTES DISTRIBUTED TO LAND

Anonymous
Compost. Sci., 3(1): 14, March 1962.
Descriptors: Sludge disposal, Soil types, Public health regulations.
A member of the Board of Health reports that digested liquid sludge, excess whey, and spent sulfite liquor are the organic wastes that may be deposited directly on the land.

070 S ADVANCES IN SLUDGE DISPOSAL IN THE PERIOD FROM OCTOBER
1, 1954 TO FEBRUARY 1, 1960.

Anonymous
Amer. Soc. Civil Eng., J. Sanit. Eng. Div., 88(SA2): 13-51, March 1962.
Descriptors: Sludge disposal, Soil types, Soil chemical properties, Costs, Reviews.

This report relates to matters of sludge disposal: digestion; gas utilization; elutriation; vacuum filtration; drying; incineration; use as fertilizer and soil conditioner; disposal at sea; by-products; and special processes. Continued efforts are being made to reduce the operational problems and costs of this phase of sewage treatment.

071 S SLUDGE TO SEA OF LAND (Management Memoranda)

Anonymous
Water Waste Treat. J. (London). 8(5):240-242, 1961.
Descriptors: Sludge disposal, Costs.
This article compares costs of ocean disposals.

072 S LIQUID SLUDGE DISPOSAL AT MIAMI, FLORIDA

Backmeyer, D. P.
Compost Sci., 6(2): 10-11, Summer 1965.
Descriptors: Florida, Operation and maintenance, Rates of application, Sludge disposal.
All digested sludge at Miami plant is now successfully applied in liquid form, eliminating expense of dewatering operation.

073 M AIR FORCE ACADEMY SEWAGE TREATMENT PLANT DESIGNED
          FOR EFFLUENT REUSE

Bauer, J. H.
Pub. Works, 92:120-122, June 1961.
Descriptors: Design data, Planning, Water reuse, Colorado.
Because of limited rainfall and the planned landscaping of the site, the need for a large amount of irrigation water was foreseen. It was planned to utilize the sewage plant effluent to meet this need. Only the excess effluent was to be released to the creek. Engineering design data and operating results are also given.

074 S COMBINED DISPOSAL OF SEWAGE SLUDGE AND REFUSE

Black, Ralph J.
Compost Sci., 3(4):16-17, December 1962.
Descriptors: Sludge disposal, Raw sewage, Economic feasibility, United States.
Combined treatment of refuse and sewage sludge is one approach to solid waste disposal problem that should receive serious attention.

075 M METHODS OF TREATING AGRICULTURAL LAND –A GOOD
          METHOD FOR THE TREATMENT AND APPLICATION OF SEWAGE

Blumel, F.
Wasser Abwasse (Leipzig). 35-45, 1965.
Descriptors: Rates of application, Raw sewage, Microbial degradation, Agriculture.

Existing methods for use of sewage in agriculture and plans for its use in modern farming and soil improvement are discussed, and the author reviews the application of sewage on land without preliminary treatment, after mechanical treatment, and after biological treatment, and compares these methods with those used abroad. Special reference is made to the use of sewage effluents and trade wastewaters for irrigation.

076 M CHEMICAL CHANGES IN SOIL IRRIGATED WITH MUNICIPAL SEWAGE

Bocko, J., and L. Szerszen
Zesz. Nauk. Wyzsz. SZK. Roln. Wroclawiu Melio. (Wroclaw, Pol.). 7:71-82, 1962.
Descriptors: Raw sewage, Soil chemical properties, Soil physical properties, Rates of application, Poland.
No significant changes were observed in humus and nutrient content in the soil after sprinkling with sewage at low rates of application; but alkalinity increased slightly, especially in the deeper soil layers. In filtration fields, because of the high sewage load, organic substances accumulated in the soil and the sorption capacity increased. A constant supply of sewage resulted in a decrease in pH. No accumulation of alkalis occurred in light-soil filtration fields.

077 S EXPERIMENTS IN EAST HERTFORDSHIRE ON THE USE OF LIQUID DIGESTEL SLUDGE AS A MANURE FOR CERTAIN FARM CROPS

Coker, E. G.
In: Proc. Inst. Sewage Parif. (London), 1965. p. 419-426.
Descriptors: Sludge disposal, Crop response, Hertfordshire, England, Rates of application, Heavy metals.
The author summarizes the results of studies carried out by the Hertfordshire Institute of Agriculture on the use of liquid digested sludge as fertilizer for various crops. In all the trials the sludge was applied thinly to the soil. It was found that the liquid sludge gave an increase in dry-matter content similar to that produced by an equivalent amount of nitrogenous fertilizer and grass responded much more rapidly to application of the liquid sludge than to air-dried sludge, suggesting that the nutrients in liquid sludge are more rapidly available. In discussion, the possible risk of transmitting pathogens to grazing stock and the effect of toxic metallions in the sludge were considered.

078 M IRRIGAITON POTENTIAL OF SEWAGE EFFLUENTS

Cormack, R. M. M.
J. Inst. Sewage Purif. (London), Pt. 3:256-257, 1964.
Descriptors: South Africa, Flood irrigation, Economic feasibility.
Scarcity of water in Southern Africa makes the use of sewage effluent for irrigation attractive and worthy of consideration. Whether reclaimed water is used for one purpose or another, matters little in the final analysis. Every gallon of reclaimed water represents a saving from other sources of supply.

The use of reclaimed sewage effluent for agricultural and horticultural purposes represents not only sound water economy but also good fertilizer economy. It is estimated that the fertilizer value in sewage effluent from the Aisleby Works at Bulawayo is about seven cents per 1,000 gallons.

079 S UTILIZATION AND DISPOSAL OF SEWAGE SLUDGE

Eberhardt, H., and H. Ermer
Staedteh.. (Hamburg), 13: 175-179, 1962.
Descriptors: Sludge disposal, Economic feasibility, Soil types, Rates of application.

The authors discuss the various methods of sewage sludge utilization or disposal and the conditions under which each can be applied. Methods are described for the agricultural use of wet or dry sludge, for drying and composting, and for disposal on land or at sea. Methods for incinerating sludge and for producing gas are also discussed.

080 S UBER DIE WIRKUNG VON KLARSCHLAMM AUF BODEN UND
MIKROORGANISMEN (THE EFFECT OF SEWAGE SLUDGE ON SOILS
AND MICRO-ORGANISMS)

Glathe, H., and A. A. M. Makawi
Z. Pflanzenernaehr. Dueng. Bodenk. (Berlin), 101:109-121, 1963.
Descriptors: Sludge disposal, Coliforms, Microbial degradation, Soil types, Raw sewage.

In pot experiments with sewage sludge applied to loam and sandy soil, fresh sludge had a superior effect in increasing total counts of micro-organisms and cellulose decomposers. Though there was little difference in the effect of autoclaved and ethylene-oxide-sterilized sludge in increasing microbial populations, the former material had the greatest effect in promoting azotobacter. Application of fresh, sterilized, and autoclaved sewage sludge (in this decreasing order of efficiency) increased CO_2 production in soil and promoted the production of NO_3 and NH_4. Fresh sewage contained large numbers of coliform bacteria which rapidly increased in soil for eight days after application and persisted even after thirty days.

081 M IRRIGATED LANDS OF THE WORLD

Highsmith, Richard M., Jr.
Georg. Review, 55:382-389, 1965.
Descriptors: Surveys, Reviews, United States, Sprinkler irrigation, Flood irrigation.

A comprehensive review of the areas of the world where irrigation agriculture is practiced. Data were collected from several sources, and total irrigated acres for each country are presented in tabular from. A world map is included, showing the geographical location of irrigated areas.

The impact of irrigation on agriculture is realized when the total irrigated acres are reviewed. I n 1961, approximately 37.7 million acres were irrigated in the United States. World totals were estimated to be something over 431 million acres.

082 M STATUS OF LAND TREATMENT FOR LIQUID WASTE--FUNCTIONAL DESIGN

Hill, R. D., T. W. Bendixen, and G. G. Robeck.
Presented at Water Pollut. Contr. Fed., Bal, Harbour, Florida, October 1, 1964. 28 p. (Unpublished).
Descriptors: Wastewater disposal, Rates of application, Soil types, Reviews, Surveys.

The distribution of liquid waste-land disposal systems by type of waste, application technique, and geographic location has been presented. The over 2,000 reported installations, spread over many states with vastly different climatic conditions, have been used for the disposal of different wastes, and have had great diversity in the waste application techniques used.

A review of the literature revealed that the median application rates at seepage ponds, ridge and furrow, and spray systems were 1.5, 0.74, and 0.22 inches per day, respectively. Sand soils were used almost exclusively at seepage pond installations, while soils with poorer infiltration and percolation rates were usually used at ridge and furrow, and spray sites. Cover vegetation reported in the literature is described and its use and soil systems reviewed.

083 S EFFECT OF DIFFERENT PUTREFYING WASTE-WATER SLUDGE APPLICATIONS ON THE WATER-STORAGE CAPACITY AND YIELD OF A SANDY SOIL

Husemann, C., and D. Pannier
Z. Kulturtech. (Berlin), 3: 193-204, 1962. Abstr., Soils Fert., 27:327(2374), 1964.
Descriptors: Rates of application, Vegetable crops, Crop response, Soil physical properties.

Applications of sewage sludge markedly increased the water-holding capacity of the soil and increased yields of lettuce. Effectiveness of the sludge depended on its source, composition, consistency, and preliminary treatment.

084 M TREATMENT OF SEWAGE AS AN AID TO ECONOMICAL DISPOSAL

Isaac, P. C.G.
Compost Sci., 3:7-11, 1962.
Descriptors: Costs, Wastewater disposal.
Disposing of sewage by irrigation and land application can assist in reducing disposal costs.

085 I DISPOSAL OF HIGH ORGANIC CONTENT WASTES ON LAND

Scott, Ralph H.
J. Water Pollut. Contr. Fed., 34:932-950, September 1962.
Descriptors: Sludge disposal, Groundwater contamination, Pulp and paper wastes, Dairy wastes.

The author describes the practical aspects of strong waste application to land and cites experience gained from the practice. Disposal of liquid digested sludge, cheese whey, and spent sulfite liquor is discussed. Examples and costs figures are given. Careful planning is needed to safeguard groundwater quality, especially where spent sulfite liquor is involved.

086 S USING SLUDGE MEANS CLEAN STREAMS FOR PENNA. TOWN

Sine, Richard L.
Compost Sci., p. 40-41, Winter 1963.
Descriptors: Pennsylvania, Sludge disposal.
Sewage sludge as a soil conditioner to improve municipal property is used.

087 M CONTROL OF WATER POLLUTION BY WASTEWATER UTILIZATION:
THE ROLE OF THE WATER POLLUTION CONTROL FEDERATION

Steffen, A. J.
Water Sewage Works, 111: 384-385, 1964.
Descriptors: Water reuse, Water quality standards, Reviews, Conferences.
The Water Pollution Control Federation has stressed the great importance of wastewater reuse in its Statement of Policy, Point No. 9: "That wastewater represents an increasing fraction of the nation's total water resource and is of such value that it might well be reclaimed for beneficial reuse through the restoration of an appropriate degree of quality." The concern of WPCF is evidenced by the many papers and discussions on this subject that are presented at Association and Federation publications and by the various medals and awards presented for research in this field.

088 S TEST OF SEWAGE SLUDGE FOR FERTILITY AND TOXICITY IN SOILS

Vlamis, J., and D. E. Williams
Compost Sci., 2(1): 26-30, 1961.
Descriptors: Salinity, Sludge disposal, California, Reviews.
California researchers report on studies comparing the growth of plants receiving applications of sludge and chemical fertilizers. Both fertility and toxicity are considered.

089 S USING TREATED SEWAGE EFFLUENT FOR CROP IRRICATION

Weiss, Rudolph H.
Compost S. 2(3): 33-34, 1961.
Descriptors: Texas, Design data, Sludge disposal, Flood irrigation.
The city of Kerrville, Texas solves a stream pollution problem and provides water for growing crops at the same time. The treatment plant design and operation are given. The wet digested sludge is diluted with sewage effluent and discharged onto the land in conjunction with the regular irrigation program, thereby eliminating the need for sludge drying beds and the tedious labor involved. The operation won Kerrville an award for the most efficient disposal of sewage sludge in the State of Texas.

090 M EFFECT OF ORGANIC MATTER CONTENT OF THE SOIL
ON INIFILTRATION

Wischmeier, W. H., and J. V. Mannering
J. Soil Water Conserv., 20:150-152, 1965.
Descriptors: Soil physical properties, Soil types, Soil chemical properties.
Measurements of soil physical properties were obtained from 44 different soils and related to runoff. Soil texture classes included sandy loam, loam, silt loam, clay loam, silty clay loam, and silty clay. Organic matter contents ranged from 1 to 4 percent, and slopes from 4 to 14 percent.

The organic matter content of the soil was the measured variable most closely correlated with runoff. Results of linear regression analyses are discussed.

The study indicated that the entry of rain into the soil was influenced much more by the organic matter content and by management practice than by texture and topography.

6. ABSTRACTS 1966 THROUGH 1970

091 S WHERE DOES THE GRASS GROW GREENEST? WHERE SEWAGE
SLUDGE AND FERTILIZER ARE MIXED WITH SEED

Anonymous
Water Pollut. Contr. (London), 65(4): 26-28, August 1966.
Descriptors: Canada, Forage crops, Recycling nutrients.
In order to find a use for sewage sludge as fertilizer at Kitchener, Ontario, dried, liquid or filtered sludge, together with superphosphate, was spread over or mixed with the topsoil of test areas on a filled-in municipal waste tip, and grass seed was sown. The best growth was produced when the sludge was uses. The filtered wet sludge used with superphosphate produced too strong a mixture.

092 S INCREASED INCOME FROM SLUDGE SPRAYING

Anonymous
Water Waste Treat. J., 12:32, 1968.
Descriptors: Sludge disposal, Recycling nutrients.
East Kilbridge has completed its second year of land disposal of digested sewage sludge, selling to farmers in a 15-mile area. Undigested sludge is disposed of at sea.

093 S HOW TO SAVE TAXES AT THE SEWAGE PLANT

Anonymous
Compost Sci., 11(1):21, January-February 1970.
Descriptors: Pennsylvania, Public acceptance, Sludge disposal.
Allentown, Pennsylvania, is getting citizens to use its sewage sludge as a soil conditioner.

094 M REUSE OF MUNICIPAL WASTE WATER

Amramy, A
Civil Eng., 38:58-61, May 1968.
Descriptors: Israel, Water quality standards, Nitrate contamination, Groundwater recharge, Water reuse.
This author feels that only one type of wastewater use is complete in itself-groundwater recharge. The work that he has done in Israel has shown that this can be successfully accomplished, although final calcium and nitrate concentrations in the groundwater were too high to be acceptable as drinking water.

095 M WASTEWATER RECLAMATION BY GROUNDWATER RECHARGE
ON LONG ISLAND

Baffa, John, and Nicholas Bartilucci
J. Water Pollut. Contr. Fed., 39 :431-445, March 1967.
Descriptors: Activated sludge effluents, Tertiary treatment, New York, Rates of application, Groundwater recharge.
With the decrease in groundwater tables and other available supplies of fresh water, many communities are faced with both a water shortage and the danger of salt water intrusion into existing supplies. Research for new supplies has indicated that the use of sewage effluent for groundwater recharge is feasible and bacteriologically safe. Rates of infiltration as high as 3 ft/day have been recorded using activated sludge effluent which has been chemically coagulated and then filtered through sand and carbon. Recharge basins have been found to give the equivalent of tertiary treatment.

096 S LIQUID SLUDGE AS A FARM FERTILIZER

Conn, R. L.
Compost Sci., 11(3) :24-25, May-June 1970.
Descriptors: Chicago, Illinois, Rates of application, Economic feasibility, Anaerobic digestion, Recycling nutrients.
The University of Illinois' Depts. of Agronomy and Civil Engineering are working on a three-year project to study the fertilizer value and application of digested liquid sludge to farmland. The research plots used in the study will try to determine the method and maximum rate of sludge application possible before bad effects appear.

097 S RECLAIMING: LAND WITH CHICAGO SEWAGE SLUDGE

Dalton, F. E.
Composts Sci., 8(2): 5-8, December 1968.
Descriptors: Chicago, Illinois, Planning, Land reclamations. Sludge disposal.
This article describes present methods of sludge disposal and plans for construction of a ten-acre small-scale farm irrigation system for crop utilization of digested sludge.

098 S ULTIMATE DISPOSAL OF WASTE WATER CONCENTRATES
TO THE ENVIRONMENT

Dean, Robert B.
Environ. Sci. Technol., 2:1079-1086, December 1968.
Descriptors: Wastewater treatment, Sludge disposal.
Advanced treatment of wastewater is basically the separation of a valuable product--water--from its pollutants. The residue of substances which remains normally has no positive economic value, and must be degraded or disposed of. In this respect, waste treatment is similar to other industrial processing: separation the wheat from the chaff, copper from its gangue, or wood pulp from lignin. These low value residues, like those from wastewater treatment, are usually an economic liability and may become a pollutant. It is the assignment of the Ultimate Disposal Research Activity of the Federal Water Pollution Control Administration to devise methods for treating these residues so that they will not pollute the environment.

099 S USING SEWAGE SLUDGE ON FARMLAND

Evans, James O.
Compost Sci., 9(2):16-17, June 1968.
Descriptors: Pennsylvania, Recycling nutrients, Sludge disposal.
Digested sewage sludge gets big demand after Pennsylvania farmers take a look at results of applications to fields.

100 M UTILIZATION OF SEWAGE FOR AGRICULTURAL PURPOSES

Feinmesser, A., and S. Z Hershkovitz
Water Sewage Works, 114:181-184, May 1967.
Descriptors: Detergents, Odor, Salinity, Heavy metals.
The use of sewage for irrigation is important for three reasons: (1) re-utilization of a source of water, (2) utilization of the fertilizer present in sewage; (3) inexpensive and efficient solution of the problems of sanitary disposal and prevention of possible sanitary nuisances by sewage. Problems occur from the salts, toxic materials and boron present in sewage, the latter often from detergents.

101 S THE EFFECT OF SEWAGE, OVERLYING LIQUOR AND COMPOSTING
ON THE VIABILITY OF PARASITE, REPRODUCTIVE STATES

Forstner, M. J.
Wasser Abwasser Forschung (Munich), 3:176-184, 1970.
Descriptors: Human diseases, Sludge disposal. Anaerobic digestion, Sprinkler irrigation, Europe.
Detailed parasitological tests have been carried out to study the distribution, viability and rate of secondary infection caused by worm ova in cattle and agriculture originating from

fields irrigated with silage or sewage as artificial rain. Results showed that sewage sludge containing parasites and ova was rendered harmless after composting fro two or three months or after pasturization.

102 M CHEMICAL INTERACTIONS OF WASTEWATER IN A SOIL ENVIRONMENT

Hajek, B. F.
J. Water Pollut. Contr. Fed., 41 :1775-1786, October 1969.
Descriptors: Wastewater disposal, Soil chemical properties, Soil physical properties, Alabama.
The effects of wastewater disposal to soil will be scrutinized increasingly to predict the assimilative capacity of soil. One phase of wastewater disposal to soil, that of chemically contaminated wastewater has been discussed to acquaint environmental engineers and scientists with experimental methods presently available for making such a prediction. These methods are wastewater chemical characterization, and chemical interactions of soil-waste systems.

103 M UTILIZATION OF SEWAGE FOR AGRICULTURAL PURPOSES

Hershkovits, S. Z, and A. Feinmesser
Water Sewage Works, 114:181-184, May 1967.
Descriptors: Oxidation lagoons, Economic feasibility, Wastewater treatment, Israel.
The re-utilization of sewage is included as part of the water potential in Israel and is estimated at about 40 bil gal. per annum, or about 10 percent of the total potential of the country. The utilization of an additional source of water, the utilization of sewage for agricultural purposes is important for the re-utilization of an additional source of water, the utilization of fertilizers present in the sewage, and as an inexpensive and efficient solution of the problems of sanitary disposal. The use of sewage for irrigation in an efficient manner can increase crop yields. Oxidation ponds are the most suitable purification means in Israel.

104 S DIGESTED SLUDGE DISPOSAL ON CROP LAND

Hinesly, T. D., and Ben Sosewitz
J. Water Pollut. Contr. Fed., 41:822-830, May 1969.
Descriptors: Odor, Chicago, Illinois, Groundwater contamination, Costs, Heavy metals.
The Metropolitan Sanitary District of Greater Chicago concluded that digestion followed by land disposal could cope with Chicago's 1,000 tons (907,000kg)/day of sludge successfully at a cost of $20 to $23/ton ($0.022 to $0.25/kg). Research designed to determine groundwater contamination, the effect of heavy metals, and crop irrigation parameters was conducted by the University of Illinois and the District. Kenaf and corn were used as the experimental vegetables. An 8-acre (3.2-ha) and a 30-acre (12.2-ha) plot were used for the tests. Crops responded favorably to digested sludge; odors and flies were not problems; and nitrate rates in drainage waters increased.

105 S A SOLUTION TO THE SLUDGE PROBLEM AT THE SEWAGE WORKS
OF MUNICH

Karnovsky, F.
Muernchner Beitr. Abwasser-Fisch.- Flussbiol. (Munich), 13:211-225, 1966.
Descriptors: Sludge disposal, Munich, Germany.
The author discusses the utilization and disposal of domestic and industrial sludges in
relation to recent developments at the Munich sewage works. Details are given of sludge
drying in beds followed by removal with sludge dredgers.

106 S THE UTILIZATION OF SEWAGE SLUDGE IN MUNICH

Karnovsky, F.
Gas-Wasserfach (Munich), 107(34):962-964, August 1966.
Descriptors: Sludge disposal, Munich, Germany, Recycling nutrients.
Further reference is made to the disposal and utilization of sewage sludge in connection
with the successful operation of the sludge digestive plant at Munich-Graslappen and to the
increasing use of sewage sludge in agriculture.

107 S AGRICULTURAL UTILIZATION OF SEWAGE EFFLUENT
AND SLUDGE--AN ANNOTATED BIBLIOGRAPHY

Law, J. P
U. S Dept. Interior, Washington, D. C., Fea. Water Pollut. Contr. Admn., Report No.
CWR-2, January 1968, 89 p.
Descriptors: Review, Bibliographies, Recreational facilities, Public health regulations.
An excellent reference source, this book contains over 200 references and reviews of
articles pertaining to effluent utilization for agriculture, industry, recreational purposes and
other uses. Included are sections on sanitary aspects of wastewater utilization, effects of
effluents on soil properties, and pollution abatement. Period covered is from before 1951 to
1965.

108 S SEWAGE SLUDGE DISPOSAL- BACK TO THE LAND

Lewin, V. H.
Effluent Water Treat. J (London), 8:21-23, January 1968.
Descriptors: Human diseases, Odor, Forage crops, Public acceptance, England.
The author discusses the ways in which modern technical advances have made the
disposal of liquid sludge on farm land an inexpensive and useful proposition and suggests that
disposal of liquid sludge on agricultural land by tanker no longer be a health hazard or
nuisance it once was. At Oxford, vacuum-dried sludge was acceptable to very few farmers
and only for brief periods prior to ploughing but it was found that tanker distribution of
digested activated sludge to grazing land was beneficial and could be practiced for 9 months
of the year. By 1967 demand exceeded supply and more than 8 mil. gal. were disposed of in
this way.

109 S RECONSTRUCTION OF OFFERTON SEWAGE WORKS OF HAZEL
        GROVE AND BRAMHALL UDC

Morton, J., and F. Summerfield
Water Pollut. Contr. (London). 68:85-92, 1969.
Descriptors: England, Costs, Sludge disposal.
The cost of sludge disposal at the Offerton Sewage Works in England which served about
20000 people is discussed and compared to the use of drying beds.

110 S PIPELINES TO TRANSPORT ORGANIC WASTES

Olds, Jerome
Compost Sci., 7(3): 3-5, December 1967.
Descriptors: Wastewater treatment, Planning.
Developing pipeline systems to convey wastes to compost plants, sewage treatment
plants, utilization sites.

111 M ULTIMATE DISPOSALS OF WASTES TO SOIL

Routson, R. C., and R. E .Wildung
Eng. Progress Symp. Series, 65(97): 19-25, 1969.
Descriptors: Soil chemical properties, Soil microbiology, Wastewater disposal.
Soil is a dynamic system which is capable of reacting with a broad spectrum of
extraneous components. The nature of this reaction is such that soil may function as a
medium for either waste storage or for ultimate waste disposal. This paper will elaborate upon
those aspects of the soil system which are relevant to the problems of ultimate waste disposal.
Since soil represents a relatively thin surface layer and is generally an integral part of man's
environment, soil disposal alone should not be considered permanent. Furthermore,
concentration mechanisms in the soil often negate the usefulness of soil as medium for
disposal by dispersion or dilution. However, both transformational and recyclization and
reuse mechanisms occur as normal functions of most soil systems. Thus, the latter two
ultimate disposal methods will be emphasized in the following discussion of nature of the
soil, soil properties amenable to waste disposal, and general soil waste interactions.

112 M WHAT HAPPENS IN SOIL-DISPOSAL OF WASTES

Routson, R. C., and R. E .Wildung
Ind. Water Eng., 7:25-27, October 1970.
Descriptors: Soil chemical properties, Soil contamination, Soil physical properties, Soil
microbiology, Reviews.
The soil is a combination of chemical, physical and biological aspects which work upon
substances to change them into organic or inorganic forms. This article discusses some of
these actions upon carbon, sulphur, phosphorus, and nitrogen compounds, explaining some of
the limiting conditions for the soil to process these materials.

113 S DIE ABWASSERSCHLAMMVERWERTUNG AUF
LANDWIRTSCAFTLICHEN NUTZFLACHEN (THE UTILIZATION
OF SEWAGE SLUDGE ON AGRICULTURAL LAND)

Schaffer, G.
Z. Acker-Pflanzenbau (Berlin), 126:73-99, January 1967.
Descriptors: Aerobic digestion, Racycling nutrients, Sludge disposal, Europe.
This paper is based on the experimental work of 19 authors on the value of sewage sludge to agriculture and contains details of the problems and difficulties which are experienced owing to the different properties present in sewage sludge which effect sludge decomposition both in the digestion tank and in the soil during composting. The value of sludge as a commercial product in view of its low nutrient content and future aspects regarding crop rotation are discussed.

114 M SOIL SYSTEMS FOR LIQUID WASTE TREATMENT AND DISPOSAL:
EVNIRONMENTAL FACTORS

Schwartz, W. A., and T. W Bendixen
J. Water Pollut. Contr. Fed., 42:624-630, April 1970.
Descriptors: Climatic data, Ohio, Denitrification, Soil types, Phosphorus removal.
Studies were performed at the Robert A. Taft Water Research Center in Cincinnati and at Loveland, Ohio, on the effects of several environmental factors on soil treatment efficiency. The effects of depth to groundwater, climate, and vegetation on the removal of COD, N, P, and MBAS and the maintenance of hydraulic acceptance were studied. A minimum depth to groundwater of 2 ft for satisfactory COD removal and 4 ft for satisfactory nitrification was found. In general 5 ft represents an adequate depth and 1 ft is completely inadequate. Biological activity is the main victim of cold weather, however, this can be partially overcome by achieving biological maturity in the system prior to the onset of cold weather.

Cultivation of vegetation served to extend hydraulic longevity by a factor of about 2. Trickling filter effluent was dosed to 6 inch buried soil lysimeters at a rate of 1.6in./day. Removals of total nitrogen and total phosphorus were 85 and 99 percent, respectively.

115 M SOIL RESPONSE TO SEWAGE EFFLUENT IRRIGATION

Thomas, R. E., and J. P. Law, Jr.
In: Proc. Symp. Munic. Sewage Effluent for irrigation, Wilson, C. W, and F. E Beckett (ed.), Ruston, Louisiana, Louisiana Polytechnic Inst., July 30, 1968. p. 5-12.
Descriptors: Reviews, Furrow irrigation, Overland flow, Wastewater treatment, Soil chemical properties.
A review of soil systems that have been used for the treatment of wastewater. The soil systems investigated by the authors were grouped into 4 systems that included (1) infiltration basins, (2) ridge and furrow, (3) spray irrigation, and (4) spray-runoff. These systems were being used specially for wastewater treatment rather than irrigation of economic crops. Each system has its design features that provide advantages for particular operating conditions and needs. Many industrial and most domestic wastewaters are suitable for irrigation use under good management practices.

116 S EXPERIENCES WITH THE DISPOSAL OF SEWAGE SLUDGE
          IN AGRICULTURE

Triebel, W.
Korresp. Abwass. (Germany), 10:11-16, 1966.
Descriptors: Public health regulations, Soil microbiology, Germany, Vegetable crops,
Evapotranspiration.

In a detailed report on the activities of the Niersverband, which is responsible for the
protection of German waters in the 1348-km^2 catchment area between the Rhine and the
Maas, special reference is made to the disposal and utilization of sewage sludge in
agriculture. Details are also given of the sewage-treatment facilities at the group sewage
works of the Niersverband. The increased use of sewage sludge in agriculture is illustrated in
tables, which show that since 1960 the group sewage works have supplied on average 820
farms, resulting in increased yields of truck crops, especially beet, and pastures. Studies
showed also that wet sludge, deposited on grassland, had lasting effects in reducing and thus
regulating the soil evaporation. Special reference is made to the advantageous sorption
capacity of digested sludge which greatly improves dry soil, contrary to dried sludge which,
owing to irreversible hydrophobia (caused by the drying process) has adverse effects on the
sorptive soil structure. Compared with artificial fertilizers the use of sludge in agriculture
presents more work; this, however, is compensated by the valuable properties in the humus,
restoring the exploited soil. Existing parasites and micro-organisms are destroyed by
pasteurization plants which have recently been installed, operating at a temperature of 65 ^{0}C
and for a period of 15 min, to comply with health regulations.

117 S PLANT EXPERIMENTS WITH SEWAGE SLUDGE FROM HELSINGFORS

Vlitasalo, Likka
Grundfoerbattring (Uppsala, Swed.), 22(1-2): 22-23, January 1969.
Descriptors: Rates of application, Grain crops, Spreading basins.

Root crops and oats were cultured in soil enriched with sewage sludge. The sludge was
treated in dry beds, by vacuum filtration or by precipitation with $FeSO_4$. Dosage for dry bed
sludge were given as 80 tons/acre, content of dry matter was not mentioned. Maximum crops
were harvested when additions of N and K were given.

7. ABSTRACTS 1971 THROUGH 1973

118 S CHICAGO RECLAIMING STRIP MINES WITH SEWAGE SLUDGE

Anonymous
Civil Eng. , 42:98-102, September 1972.
Descriptors: Chicago, Illinois, Costs, Sludge disposal, Land reclamation, Spray disposal.

This article covers incineration, dewatering, drying. These were other alternatives studied
by Chicago engineers. Then several years ago, MSD decide--following an intensive research
program--the most attractive alternative to improving existing sludge-disposal operations was

land reclamation. The plan was to digest sludge and transport it to rural land beyond city limits. Spraying digested sludge on abandoned strip-mined land, or on other areas with poor soils (e.g., low grade pastures) would have several unique advantage: it would eliminate land, air, and water pollution stemming from sludge processing in the urban area; cost less than other sludge disposal methods; solve the problem of hunting for lagoon space in urban areas; and it would make beneficial use of the organic materials, nutrients, and water making up sludge.

119 S USING ORGANIC WASTES TO BENEFIT FARMS AND RURAL AREAS

Anonymous
Compost Sci., 14(1):7-9, January-February 1973.
Descriptors: Pennsylvania, Sludge disposal, Recycling nutrients, Wastewater disposal, Legal aspects.

For years, waste disposal methods in cities and farms simply tried to get rid of those wastes as cheaply and quickly as possible. In many areas, dumps burn almost continually; water supplies are fouled; state regulations make many old installations illegal. In the future, there will be a much closer look at recycling wastes--specifically how to use wastes so they benefit the land, the people in rural areas, so they aid crop growth, and even produce methane gas for heating, lighting and powering trucks and tractors. Some of the research and projects now under way which give some information on how organic wastes can be used--or are already being used--to benefit Pennsylvania farms and Pennsylvania farmers are listed.

120 S INCORPORATION OF SEWAGE SLUDGE IN SOIL TO MAXIMIZE BENEFITS AND MINIMIZE HAZARDS TO THE ENVIRONMENT

Agricultural Research Service
U. S. Dept. Agr. , Beltsville, Maryland, 1972, 118 p.
Descriptors: Sludge disposal, Pathogenic bacteria, Maryland, Groundwater contamination, Soil chemical properties.

The principal objective of the project is to conduct a pilot study on land of the ARS at Beltsvillle, Maryland, on incorporation of sewage sludge in agricultural land by a trenching method. Work includes different depths and spacing of trenches, different sludges and different management of the land after incorporation of sludge. Evaluations will include physical and logistical problems in the operation, effects on quality of groundwater, plant growth and nutrient uptake, survival and movement of pathogens, and effect on soil properties. Research is also being conducted in green- houses and laboratories on the chemistry, microbiology and behaviour in soil of different kinds of sewage sludges.

121 S ENRICHMENT OF TRACE ELEMETS FROM SEWAGE SLUDGE FERTILIZER IN SOILS AND PLANTS

Anderson, A., and K. O Nilsson
AMBIO (Oslo, Norway), 1(5): 176-179, September / October 1972.
Descriptors: Sweden, Recycling nutrients, Soil chemical properties.

This paper reports on trace element analyses of soil and vegetation from a field trial where one of the treatments has received sewage sludge as a source of plant nutrients, five received no form of organic substance, but received varying amounts of chemical fertilizers, and nine received organic substance other than sewage sludge and chemical fertilizers.

122 M EFFECTS OF TREATMETN PLANT EFFLUENT ON SOIL PROPERTIES

Day, A. D., et al.
J. Water Pollut. Contr. Fed., 44:372-375, March 1972.
Descriptors: Soil chemical properties, Soil types, Salinity, Tucson, Arizona.

Recognizing that water supplies are becoming harder and harder to find, sewage effluent is becoming more and more popular as a source of irrigation water. Long-term studies of the effects of effluent on soil condition are lacking; this is the purpose of this study. Comparing Grabe silt irrigated with effluent to soil which has been fertilized at recommended rates, shows that the effluent-irrigated soil has higher concentrations of soluble salts, nitrates, phosphates, calcium and magnesium that the control. Organic material was greater in the control. No declines in crop yield or adverse effects on the soil were recorded.

123 S THE GIVE AWAY MARKETEERS

Gaffmey, M. P.
May 1972. 16 p.
Descriptors: Hertfordshire, England, Soil chemical properties, Heavy metals.

A description of the land application of digested sludge by the West Hertfordshire Main Drainage Authority, England. Sludge, called Hydig, is distributed over a 300-square-mile radius by the district without charge. Control of heavy metals is mentioned, as well as the development of a "zinc-equivalent", to express the relative build-up of metals.

124 S AGRICULTURAL BENEFITS AND ENVIRONMENTAL CHANGES RESULTING FROM THE USE OF DIGESTED SEWAGE ON FIELD CROPS, AN INTERIM REPORT ON A SOLID WASTE DEMONSTRATION PROJECT

Hinesly, T. D., O. C. Braids, and J. E. Molina
Metropolitan Sanit. District of Greater Chicago, II. U. S Environmental Protection Agency, Washington, D. C., 1971. 73 p.
Descriptors: Chicago, Illinois, Coliforms, Anaerobic digestion, Rates of application, Aerobic digestion.

The Metropolitan Sanitary District of Greater Chicago worked with the University of Illinois to test the use of digested sewage sludge on farm land. Crop response, chemical pollution of runoff and the possibility of contamination by fecal coliform were studied. It was found that freshly digested sludge had serious agricultural disadvantages, inhibiting seed germination. In some cases, germination fell to zero, while other tests showed germination of other seed types at only 30 percent of normal. Aging or aeration of the sludge corrected this problem. Similarly, fecal coliform counts were high in freshly digested sludge and fell rapidly upon aging. Sludge was found to be infiltrated very slowly into the ground, and application rates must be kept low to avoid runoff of sludge from slopes. Crops raised on sewage sludge produced better yields than the control crops and did not show any absorption of toxic heavy metals found in the sludge.

125 S PROBLEMS OF SLUDGE TREATMENT AND DISPOSAL

Horner, R. W.
In: Proc. Inst. Civil Eng. (London), 49:87-90, 1971.
Descriptors: Wastewater treatment, Sludge disposal, Review, United States.

Mr. Horner described sewage treatment as predominantly a concentration process in which each stage and type of treatment produced its own particular concentrated pollutant. The main problem, however, was the treatment and disposal of the sludges produced by the primary, secondary, and tertiary treatment processes. A very wide range of processes had been employed, with varying degrees of success, in their treatment and disposal.

126 M THE USE OF THE SOIL AND ITS ASSOCIATED BIOSYSTEMS TO RECYCLE URBAN INDUSTRIAL AND AGRICULTURAL WASTES

Kardos, Louis T.
APWA Reporter, May 1973. 13 p.
Descriptors: Sprinkler irrigation, Soil chemical properties, Penn State, Pennsylvania, Soil physical properties, Soil microbiology.

Author's thesis is that use of the land and its associated biosystems, rather than surface waters, as an extensive recycling system enables one to utilize the strong physical chemical and biological buffering capacity of the soil in relieving various kinds of water pollution crises. Pertinent aspects of the Penn State living-filter project are discussed. Present research at Penn State, in which final digested sludge from the treatment plant is injected into the wastewater pipe line and applied through the sprinklers, indicates that the same acres being used for the wastewater disposal can take care of the sludge disposal in a totally integrated operation without unbalancing the living-filter system.

127 S EXPERIENCES WITH THE SLUDGE PROGRAM IN THE DENVER AREA

Korbitz, W.
Compost Sci., 12(5)3-5, September-October 1971.
Descriptors: Reviews, Sludge disposal, Planning, Colorado.

The report shows the Metro Denver District staff is firmly convinced that "future sewage sludge must be recycled to the land".

128 S THE SOIL AS A BIOLOGIGAL FILTER

Miller, R. H
In: Recycling Treated Municipal Wastewater and Sludge through Forest and Cropland, Sopper, W. E., and L. T. Kardos (ed). Univ. Park, Penn State Univ. Press, 1973. p. 71-94.
Descriptors: Soil microbiology, Biodegradation, Heavy metals, Nitrification.

The primary functions of the soil microbial component of the biological filter are as follows: (1) as a "decomposer" population metabolizing biodegradable organic materials to CO_2 and H_2O. The rate of this process often determines the loading rates and capacity of the soil for waste renovation. As part of this microbial reaction soil humic materials accumulate which are significant in modifying soil physical and chemical properties, (2) to degrade or detoxify potentially toxic or unwanted organic compounds, e.g., ABS, pesticides, NTA, phenols, etc., (3) to modify the adsorption and mobility of cations and anions including phosphorus and heavy metals within the soil profile, (4) to modify the adsorption of nutrient elements and heavy metals by plants associated with the soil filter. The mechanisms involved are oxidation reduction, mineralization-immobilization, chelation and solubilisation, (5) as the responsible agents for the nitrogen transformations necessary for the proper functioning of soil in waste renovation, e.g., immobilization-mineralization, nitrification and denitrification, and (6) in the elimination of pathogenic microorganisms.

129 S DIVISION S-3--SOIL MICROBIOLOGY AND BIOCHEMISTRY-- AERATION-INDUCED CHANGES IN LIQUID DIGESTED SEWAGE SLUDGE

Molina, J. A. E., O. Braids, T. D. Hinesly, and J. B Cropper Soil Sci. Soc. Amer. Proc., 35:60-63, 1971.
Descriptors: Grain crops, Anaerobic conditions, Anaerobic digestion, Plant growth, Chicago, Illinois.
Explored the question of germination of corn and soybean seeds in the presence of liquid anaerobically digested sludge. Exposure of the sludge to the air resulted in ammonia losses, and inhibition of germination was not detected after a few days.

130 S RECYCLING MUNICIPAL SLUDGES AND EFFLUENTS ON LAND

National Association of State Universities and Land-Grant Colleges Washington, D., July 1973. 244 p.
Descriptors: Reviews, Sludge disposal, Wastewater disposal, United States, Recycling nutrients.
This report contains 26 topical papers and the information gathered at a Research Needs Workshop co-sponsored by the National Association of State Universities and Land-Grant Colleges, the Environmental Protection Agency, and the United States Department of Agriculture.

131 M RECYCLING TREATED MUNICIPAL WASTEWATER AND SLUDGE THROUGH FOREST AND CROPLAND

Sopper, W. E., and L. T. Kardos (ed.) Univ. Park, Penn State Univ. Press, 1973. 479 p.
Descriptors: Reviews, Wastewater treatment, Pennsylvania, Silviculture.
Abstracts of 33 papers delivered at the symposium. Current knowledge related to the potential of using land areas for the disposal of treated municipal wastewater and sludge is reviewed and discussed. There is an attempt to determine technological gaps and research needs.

132 S PROTECTION OF PUBLIC HEALTH

Sorber, C. A.
In: Proc. Conf. Land Disposal of Municipal Effluents and Sludges, U. S. Environmental Protection Agency, Washington, D. C., Report No. EPA-902/9-73-001, March 1973. p. 201-209.
Descriptors: Groundwater contamination, Pathogenic bacteria, Vegetation damage, Aerosols.
Public health considerations in land application include aerosol travel and infectivity, bacterial contamination of vegetables, and groundwater pollution from toxic chemicals and pathogens. A discussion is included of each of these topics plus current research in the field.

133 M PROBLEM DEFINITION STUDY: EVALUATION OF HEALTH
AN HYGIENE ASPECTS OF LAND DISPOSAL OF WASTEWATER
AT MILITARY INSTALLATIONS

Sorber, C. A .S. A. Schaub, and K. J. Guter
U. S .Army Medical Environ. Eng. Res. Unit, Aberdeen Proving Ground, Md., August 1972 32 p.
Descriptors: Viruses, Groundwater contamination, Aerosols, Bibliographies.
Many of the detrimental health and hygiene aspects of land application would be significantly reduced by proper wastewater pretreatment (secondary treatment, filtration, complete disinfection). By choosing a land disposal site that has from 5 to 10 feet of continuous fine soil, biological contamination of groundwater should be avoided. The probability of inhaling pathogenic aerosols near a spray irrigation site is significant. Chemical components of sewage may enhance the viability of bacteria, virus and protozoans in aerosols. Pathogenic microorganisms may survive longer in sewage aerosols and in soil than common indicator organisms such as coliform organisms. Mosquito breeding is enhanced as a result of ponding in land disposal areas. In areas where land application is the first step in a water recycle program, total dissolved solids (sodium and nitrate ion build up) in the groundwater supply can be a problem.

134 S MODEL OF ORGANIC WASTES RECYCLING

Stickelberger, D.
IRCWD News, 3:2-6, March 1972.
Descriptors: Recycling nutrients, Sludge disposal.
Excerpt from a speech proposing a thermodynamic model for organic waste recycling. Waste material cannot be destroyed or even removed. The refuse which has seemingly been disposed of always reappears in one form or another, either in the air, water, soil or all three "elements" at once. The author maintains that "the disrupted ecological system must be brought into balance and aligned with initial production conditions through buffered feedback loops of end products". He suggests that his model will be specially useful in developing countries.

135 S PROCESS CONTROL AT SHEFFILD'S SLUDGE FILTRATION PLANT

Swanick, K. H.
Filtr. Separ., 8:137-139, March/April 1971.
Descriptors: Aerobic digestion, Sludge disposal, Dewatering, Design data, England.

The sewage sludge filter pressing plant at Sheffield's Blackburn Meadows Sewage Works is one of the largest installations of its kind in the world. He plant consists essentially of sludge storage tanks, milk of lime tanks, variable rate sludge and lime pumps, flash mixers conditioned sludge storage tanks and six air compressors with associated pairs of rams, which feed the 36 filter presses.

136 S UTILIZATION OF MUNICIPAL SOLID WASTE COMPOST: RESEARCH
RESULTS AT MUSCLE SHOALS, ALABAMA

Terman, G. L., and D. A. Mays
Compost Sci., 14(1):18-21, January-February 1973.
Descriptors: Costs, Forage crops, Vegetable crops, Alabama.

Experiments conducted show that large tonnages of municipal compost can be applied on grassland or cropland and result in positive yield responses. Economic comparisons are then made between the actual value of dry compost in terms of crop response and the estimated costs of $8 to $30 per ton of garbage (double these costs per ton of compost) for producing compost from municipal wastes. Carlson and Menzies concluded that benefits to agriculture from applying waste to land are generally minimal, but the benefits to urban governments are substantial. Thus, composting of organic wastes would be considered largely as a useful alternative method of disposal.

137 S   PALZO RECLAMATION PROJECT, VIENNA RANGER DISTRICT,
SHAWNEE NATIONAL FOREST, WILLIAMSON COUNTY, ILLINOIS: FINAL

Environmental Statement
U. S Department of Agriculture
Forest Service, 1972. 102 P.
Descriptors: Anaerobic digestion, Land reclamation, Illinois, Heavy metals, Nitrate contamination.

The pilot project described in this statement deals with the application of anaerobically digested municipal waste on strip-mined land. The expected environmental impact will be to significantly reduce the water pollution problem now associated with the tract as well as to control extensive erosion through establishment of vegetative cover. The long-range environmental impact could be the development of techniques for reclaiming thousands of acres of strip-mined land, resulting in the general improvement of water quality in areas similar to Southern Illinois. This land may be the ideal place for the deposition of large volumes of waste which are presently a growing problem in large metropolitan areas. Adverse environmental effects which cannot be avoided may include slight increases in concentrations of cadmium, chromium, nickel, and nitrates in the runoff.

138 S UTILIZATION OF MUNICIPAL ORGANIC WASTES
        AS AGRICULTURAL FERTILIZERS

Vlamis, J., and D. E. Williams
Compost Sci., 13(1):26-28, January-February 1972.
Descriptors: Recycling nutrients.
Sewage sludge and garbage compost found to be good sources of plant nutrients.

In: Sewage Sludge Management
Editors: A. A. Zorpas and V. J. Inglezakis

ISBN: 978-1-61324-393-0
© 2012 Nova Science Publishers, Inc.

Chapter 8

SEWAGE SLUDGE COMPOST EVALUATION AND UTILIZATION

Antonis A. Zorpas[*]

Institute of Environmental Technology and Sustainable Development,
Paralimni, Cyprus

ABSTRACT

Sewage sludge is a wastewater industrial sub-product with high organic matter and nutritional contents traditionally used as an agricultural soil fertilizer and to promote biomass production. Generally any Compost as an organic such as N, P, K, Ca and Mg and many of the micronutrients required mater resource improves the soil's physical properties. In additions, it provides various levels of macronutrients for plant growth. The largest potential user of compost is the agricultural industry. The evaluation of the final product usually includes different test and parameters. From the use of the final product depend the evaluation procedures and the parameters, which must be concerned. The evaluation and the utilization of the composted material affected from the composting procedure, from the raw material and from the final use of compost. The chapter is providing general review and the methodology for different parameters that must be taking into consideration before the final utilizations of the compost.

1. GENERAL INTRODUCTION AND LITERATURE REVIEW

Utilization of compost depends on product quality and consistency. The parameters of concern vary depending on the compost feedstock source and intended use. A different type of compost provides significant and important characteristics.

---

[*] Website: www.envitech.org Tel: +357-23743440, Fax: +357-23743441, P.O.Box 34073, 5309, Paralimni, Cyprus.

For the Biosolids compost the most important characteristics are the concentrations of pathogens, the presents of the heavy metals, the soluble salts, the odor, the stability, the pH and finally the particle size.

In contrast to Biosolids the MSW compost has different characteristics. Those involves pathogens, the present of heavy metals, the soluble salts, the concentration of boron, the stability, the odor, the maturity, the pH, the EC, the inert (plastic, metals, glass), the humics, and the particle size.

On the other the most important characteristics from yard waste compost are the stability maturity, odor, pH and houmus.

Also the compost produces from sewage sludge present similar characteristics with the other compost but among the most important characteristics are the present of heavy metals, the chemical extraction of metals, the metals leachability, the houmics, the pH and the Electronic Conductivity, the pathogens, the phytotoxicity, the maturity and the stability.

The globalisation of trade, global climatic changes, the diminution of rainfall and the expansion of desertification are some of the serious problems that agricultural areas of most Mediterranean countries have to deal with (Zacharias and Koussouris, 2000; Manios 2004). Agricultural income has declined significantly (National Statistical Service of Greece, 2002). The availability of water resources restricts agricultural production, and desertification is gradually expanding in Mediterranean countries like Cyprus, Greece (Craete), Malta, Italy, Egypt etc. Additionally, despite a chronic deficiency of soil organic matter, recycling of biodegradable organic wastes on land is only poorly developed and is not standard agricultural practice in many countries.

It appears that although solutions can be found in these problems, they are not easy to implement. The effort to increase agricultural income may come from the expansion of integrated production systems and organic farming. This approach will provide certified agricultural products, which on one hand will be beneficial to consumer's health, and at the same time attract higher financial premiums. However, this will require gradual replacement of chemical fertilizers that are currently extensively used in agriculture production, with organic soil amendments. Additionally, organic soil amendments increase the soil organic matter content and water reserve and reduce runoff and soil erosion from the typical hilly, terrain characteristic of the Mediterranean region (Bazzoffi et al., 1998; Navas et al., 1998; Edwards et al., 2000; Querejeta et al., 2001). Moreover, increasing the organic matter content of soil has the additional benefit of reducing the problems associated with the use of brackish water for crop irrigation, which is frequently the case in these areas (Tsikalas and Manios, 1986; Tomar et al., 2003; Manios, 2004).

Zorpas (Zorpas et al. 1999c, 2000, 2003, 2008a, 2008b,) refer to the production of high quality compost from sewage sludge, waste paper, organic fraction of municipal solid waste and clinoptilolite. Also according to the same Researcher is mention that the compost quality is improved by adding organic fraction of municipal solid waste, sewage sludge (anaerobically and stabilized) and natural zeolite.

According to Manios (2004) the raw materials that can be used to produce high quality composts are mainly the residues of local cultivations and agricultural industries. A secondary source, both in amount and quality, is the organic fraction of municipal solid waste (OFMSW) (when source separated) and sewage sludge, when mixed with green waste or other bulky organic materials (Manios and Siminis, 1988; Manios and Dialynas, 1995, 1997; Manios et al., 2002b; Vrilakis et al., 1999). However, given the bulky nature of these materials and the large transportation cost, composting should take place as close as possible to the place of waste production to ensure this is an economically viable and affordable option for growers (Manios et al., 2001a). Large quantities of the above residues are currently disposed of in landfills, resulting in rapid exhaustion of landfill capacity, or they are burned on-site creating an important fire hazard that has been responsible for the complete destruction of large forestry areas.

The European Landfill Directive (EU 1999/31) requires the phased reduction of landfill disposal of such biodegradable organic materials. Therefore, recycling of biodegradable agroindustrial residues and the organic fraction of municipal solid waste (after source separation) could also reduce the problems relating to the increasing production of wastes and the difficulties of locating new landfill sites, especially in island communities. Recycling of organic residues by composting and land application would contribute to the sustainable development of the Mediterranean islands. Prime quality compost is suitable for the production of high quality agricultural products, and the secondary quality compost has utility for the improvement of landscape and forests, an essential element for the development of agro-tourism and tourism in general.

Soil incorporation of composted municipal solid waste (MSW) usually results in a positive effect on the growth and yield of a wide variety of crops and the restoration of ecologic and economic functions of land. Agricultural uses of MSW have shown promise for a variety of field crops (e.g., maize, sorghum, forage grasses) and vegetables for human consumption (e.g., lettuce, cabbage, beans, potatoes, cucumbers). Responses by plant systems have ranged from none to over a twofold increase in yield. Specific responses are crop and site dependent. In most cases, yields were highest when composts were applied with fertilizer management programs. In some cases, elevated trace metal uptake was noted with lead and boron of greatest concern. Where long-term monitoring has been possible, benefits persist and actually accrue when sound soil/crop management practices are followed. Levels of toxic elements in plants for human consumption are either not well known or thresholds were not reached, as little mention was made in the literature. Container-grown ornamental crops are not for human consumption and this concern does not apply. However, they are high value commodities, and therefore, safe, satisfactory growth is important. Because composts can replace. peats and barks which are becoming expensive, this application shows great promise. Silvicultural or forestry applications also represent a nonconsumptive, potentially large use of compost. These uses range from low volume/high value nursery and Christmas tree applications to high volume/low value uses in forest regeneration. Land reclamation whether for minespoil, landfill *cover* or soil stabilization as in the establishment and rehabilitation of road shoulders represents another potential beneficial use. However, tonnages recycled in this way will be less than in agriculture and silviculture.

Compost utilization will occur in response to a number of factors including their benefits to soil-plant systems, especially those having economic value. As important, however, is the realization by policymakers that cornposting and compost use in agro or natural ecosystems are an integral part of the entire waste management system. Thus, a systems perspective that includes full cost accounting of waste collection, handling, and processing must incorporate marketing, distributing, and recycling in a life cycle analysis that reflects external costs and societal benefits for composting-based solid waste systems to be competitive.

Factors which affect the use of composted material include feedstock properties, regulations, product uniformity, contaminant levels (weed seeds, pathogens, pesticide residuals, trace metals, etc.) and economic considerations relating to distribution and utilization benefits.

1.1. Land Application of Sewage Sludge and Sewage Sludge Compost

Sewage sludge is a wastewater industrial sub-product with high organic matter and nutritional contents traditionally used as an agricultural soil fertilizer and to promote biomass production.

Land application of sewage sludge has been extensively used as an effective dispersive method throughout Canada, the United States and Europe for more than 40 years. Many studies have demonstrated the positive effect of land application of sewage sludge or sludge compost on corn and forage yields and soils (Catroux et al., 1981; Davis et al., 1985; Hornick et al., 1984; Tiffany et al., 2000; Warman, 1986). In the few instances where a nil or negative response to these organic amendments have been observed, either a high C:N ratio, excess metals, high soluble salts or extremely high application rates were responsible for the reduced yields or negative effects to soils or crops. The primary plant nutrient associated with sewage sludge is N; however, sludges also contribute significant amounts of other macro and micronutrients (Carlton-Smith and Coker, 1985; Shober et al., 2003; Sims,1990; Soon et al., 1978a,b; Warman, 1986; Zebarth et al., 2000). Nitrogen availability from sewage sludge and sludge compost is reported to range from 0% to 56% (Carlton-Smith and Coker, 1985; Cripps et al.,1992; Hutchings, 1984; Magdoff and Amadon, 1980; Serna and Pomares, 1992; Sims, 1990; Warman, 1986; Zebarth et al., 2000). Sikora and Enkiri (1999) suggest that biosolids compost N can replace one-third of the fertilizer N required by fescue without decreasing yield. The main factors that influence N availability from sewage sludge are its inorganic N content (Hutchings, 1984), digestion process (aerobic vs anaerobic) (Amundson and Jarrell, 1983; Hutchings, 1984; Serna and Pomares, 1992), C:N ratio (Sims, 1990), pH, the method and timing of application (Cripps et al., 1992), and soil type and properties (Hutchings, 1984; Magdoff and Amadon, 1980). Phosphorus content in sewage sludge and sludge compost and P availability to crops varies even more widely than N (Sikora et al., 1983; McCoy et al., 1986). To some extent, the soil conditions and characteristics of biosolids which influence N mineralization and plant availability apply to P, but not to K, which is not organically-bound and not subject to mineralization reactions. However, the use of chemical flocculants in some wastewater treatment processes can decrease P mineralization from sludge's and composts while P availability to crops is controlled by soil Fe, Al and Ca content. Excessive applications of sewage sludge beyond crop requirements and the soils absorptive capacity or applications made in the fall or winter may result in groundwater contamination by nitrates, loss of N through denitrification, toxic nitrate concentrations in animals (especially from grass forages), and surface water contamination by P. Therefore, there are still questions about the release of essential and non-essential nutrients from sewage sludge and sludge composts, and the short and long-term economic benefits to the farmer and the community from applying sludges to the land.

Utilization of compost in crop management provides considerable advantages, as it reduces the input of expensive chemically-synthesised nitrogen and phosphorus fertilizers and contributes to prevent land degradation (Albiach et al., 2001). During the composting process of sludge from wastewaters microbiological stabilization, partial dehydration and enrichment in nutrient content occur (Davis, 1989; Sa´nchez-Monedero et al., 2004). Nevertheless, depending on the conditions during this process, sewage sludge might contain pathogenic microorganisms (e.g., Salmonellas and enteroviruses) (Sa´nchez-Monedero et al., 2004; Gale, 2005) and heavy metals that can affect both soil pollution and the uptake of these metals by roots, especially in horticultural crops. Many relevant publications have already stated some beneficial effects of the application of composted sewage sludge from wastewater to several kinds of soils, including agricultural soils, under different weather conditions and using compost from different sources highly variable in nutritional composition (Kvarnstrom et al., 2000; Engelhart et al., 2000). Amongst the drawbacks of this practice, there is a deep concern about the heavy metal and other trace pollutants plant uptake from the soil through the roots and the potential risk for human health (During and Gath, 2002). This concern is especially high regarding vegetables and other non-woody plants that are grown in close contact to the soil surface and that are usually included in human diets processed, pre-cooked or even fresh, as it is the case of cauliflower. Thus, the evaluation and feasibility of the application of composted sewage sludge on horticultural plots are becoming more and more relevant, as water treatment policies are becoming widespread and composting-process techniques are affordable. A review of the literature available showed previous experiments using other non-woody economically relevant species as experimental models, such as barley (Moreno et al., 1996; Costa et al., 1992), lettuce (Moreno et al., 1997) and petunia (Smith, 1992).

1.2. Sewage Sludge – Compost and Metals

Composting provides a simple and a cost effective alternative treatment method for sewage sludge by decomposing organic matter, producing a stabilized residue and disinfecting pathogens (Fang et al., 1999). The composted product can also be used as a fertilizer or soil conditioner because of its large content of stabilized organic matter. However, the high content of heavy metals in sewage sludge compost has proven to be a limiting factor in the land application of sewage sludge compost (Wong et al., 1997). The addition of natural zeolite, clinoptilolite, during sewage sludge composting has been proven to be a promising way to reduce the heavy metals content (Zorpas et al., 2000, 2008b), since zeolite (clinoptilolite), has the ability to take up heavy metals. Zeolite utilization has become popular in the last decade, due to its cation exchange and molecular sieving properties (Ouiki and Kavannagh, 1997; Zorpas, 1999a).

A major limitation of sewage sludge compost on land is the potential high heavy metal content due to the metal content of the original sludge. Zeolites may be useful as metal scavengers in metal-rich sludges. Natural zeolites such as Clinoptilolite (Cli) has the ability to take-up and remove these metals by utilizing ion exchange. The sludge is classified as solid waste that requires special methods of disposal, because of its noxious properties. However, much of the sludge originating from urban wastewater treatment is contaminated with heavy metals (Wozniak and Huang, 1982; Hasit and Christensen, 1987; Sims and Skline, 1991; Langenbach et al., 1994; Karvelas et al., 2003; Wei and Liu, 2005). Theses metals, may leach from sludge and enter the ecosystem, the food chain and finally the human body. Also, the total concentration of heavy metals cannot provide useful information about the risk of bioavailability, toxicity and capacity for remobilization of heavy metals in environment (Fernandez et al., 2000; Kunito et al., 2001; Liu et al., 2007).

Several studies has carrying out by Zorpas (1999a, 1999b, 1999c, 2000, 2003, 2008a, 2008b) which proves that the use of zeolites in sewage sludge compost has the ability to remove and uptake heavy metals before the application of the final product to the agricultural.

1.3. Compost Metals

The concerns over metal accumulation from compost application are phytotoxicity, increased human exposure to metals, and the ecological health of soil. Of the metals to be of concern are As, Cd, Pb, Cr, Cu, Ni and Zn. While all of the metals are phytotoxic if present in elevated concentrations in soil, the first four metals are highly toxic to human health. Since soil is central to the transfer of the metals (As, Cd, Pb, Cr) to the plants, animal and humans, limiting their excess accumulations in soils is fundamental to the compost utilization and management. This has led to the establishment of metal concentration limits for composts in some countries discussed earlier.

Closely examining the metal inputs and outputs is key to what soil metal concentration might be in the long term. The inputs of metals into soils can be from diverse sources including aerial deposition particularly near smelters and land application of metalcontaminated organic amendments (compost, animal wastes, sewage sludge) and fertilizers (Fig. 1). The outputs are plant uptake, leaching/runoff, volatilization (Hg and As), and removal of surface soil by excavation (Fig. 1). If the inputs were equals to outputs, metal concentration in soil would remain the same, which is the principle of no net degradation approach in establishing metal concentration limits by some countries. Whether the metal concentration limits of composts are developed based on no net degradation or no observable adverse level, knowledge of metal concentrations and application rate of compost and metal utilization or uptake by plant are critical since they largely determine the net input of metals in the soils where leaching and runoff loss of metals are minimal.

1.4. Compost Metal Concentrations

Compost metal concentrations vary depending largely on the metal concentrations in the feedstock. Metal concentrations in the finished composts tend to be higher than in the initial composting mix as a result of compost weight loss by degradation of organic matter (Leita

Figure 1. Inputs and Outputs of Metal in Soils

Table 1. The concentrations of Metals in some biosolids and MSW compost

Metal	MSW compost			Biosolids compost		
	Range	Mean	n	Range	Mean	n
Cd	0.6 – 11	3.7 (2.5)	12	2.0 – 16	6.7 (4.4)	8
Cr	0.1 – 762	70.0 (48)	13	45.0 – 578	163.0 (174)	8
Cu	36.0 – 762	308.0 (194)	13	180.0 – 890	401.0 (217)	8
Ni	5.1 – 80	45.0 (45)	14	15.0 – 350	74.0 (112)	8
Pb	24.0 – 603	281.0 (172)	14	14.0 – 863	347.0 (321)	8
Zn	38.0 – 1420	715.0 (396)	14	116.0 – 1801	898.0 (462)	8

Values within the parenthesis are standard deviations.

and Nobili, 1991). The concentration of metals in the finished compost may be higher or lower than the feedstock depending on the type of bulking agent used. It should be lower if sawdust or straw is used to compost with biosolids (Table 3), but it may be higher if mixed municipal waste is used. While straw or sawdust is typically low in metal concentration, mixed municipal solid wastes may contain metal (e.g., Pb) concentration even higher than biosolids. Table 1 shows the range and mean concentrations of some metals for some municipal and biosolids composts in the US. The wide variations in the concentration of the metals among the MSW or biosolids composts reflect the difference in metal concentration in the feedstock, bulking agent and/or composting method used (Epstein, 1996). The data show that biosolids compost on the average has higher metal concentrations than MSW compost. The concentrations of the metals in these types of composts generally are far greater than yard waste debris or animal manure compost with the exception of compost from manure of animal fed with feeds enriched with Cu and/or Zn as additives (Mullins et al., 1982, L'Herrous et al., 1997, Hsu and Lo 2000). The concentration of Cu and Zn in some swine manure composts reached as high as 1380 and 2840 mg kg^{-1}, respectively.

The concentrations of Cd, Cu, Ni, Cr, and Pb in the MSW and biosolids composts in the US are highly correlated with the average annual consumption of the metals by the US over a span of 20 years (1970 to 1990) for Cd, Cr, Cu, Pb, Ni, and Zn. The average annual consumption rates for the metals were calculated using data from USGS for Cd (Buckingham and Plachy, 2004), Cu, Cr, Pb, Ni and Zn (Porter and Edelstein 2003; Goonan and Papp 2003; Francesco and Plachy 2004; Goonan and Kuck 2004; DiFrancesco and Smith 2003). A fraction of the metals used in the manufacturing of various products ends up in the waste streams. The deviation of Zn from the other metals in the relationship signals a far greater amount of Zn entering the waste streams than the other metals at the same amount of consumption annually. This perhaps is related to the extensive use of Zn in agriculture, and consumer goods including anti-corrosion coatings on steel, construction material, brass, pharmaceuticals and cosmetics, tires, micronutrient for humans, animal, and plants. Zinc would have a greater probability to enter the waste streams. It is anticipated that the magnitude and direction of flows of the metals in the whole ecosystem vary with metal and detailed analysis of the structure of the flows of metal as was done for Pb by Socolow et al. (Socolow and Thomas 1997) should be encouraged to quantify the amount of the metals which could end up in the waste stream annually. The magnitude of each flow in a region will likely be affected by its social and economical conditions, regulations and by season, as well. The structure of the flows could help explain the regional differences in the concentrations of metals in the MSW and biosolids composts.

1.5. Forms and Mobilization of Compost Metals

Because of the impact of compost metals on metal accumulation in soil and in plant, investigations were made to the forms and solubility of metals particularly in the MSW, in sludge compost and biosolids composts, and in the metal-enriched swine manure composts. Compost itself is comprised of organic and inorganic components, and metals in the compost are likely to distribute in the two components in magnitudes depending on the affinities of the metals with organic and inorganic constituents, and redox conditions of the compost.

The typical method to study forms of metals in compost is using sequential extractionsCompost is sequentially extracted with KNO3 (0.5 M), H2O, NaOH (0.5 M), EDTA (0.05 M), and HNO3 (4 M) to separates compost metals into operationally defined fractions as exchangeable, water soluble, bound to organic, precipitated as inorganic (carbonate and sulfide precipitates), and residual metals (Petruzzelli 1989; Sims and Kline 1991; He et al., 1992). Being an excellent chelator of the metals with a stability constant ranging from 16.46 for complex Cd to 23.4 for complex with Cr(III), it is not surprising that a large fraction of the metals in compost can be extracted with EDTA. Another extraction scheme by sequentially extracting compost with $MgCl_2$ (1 M) (exchangeable), sodium acetate (1 M) (carbonate bound), hydroxyamine hydrochloride [$NH_2(OH)HCl$] (0.04 M) (bound to Fe and Mn oxides), HNO_3 (0.02 M)–$H2O2$ (bound to organic matter), and HF-H3BO3 or HNO_3-$HClO_4$ (residual) was also extensively used (Tisdell and Breslin, 1995; Hsu and Lo 2000; Pueyo et al., 2003; Zhang et al., 2004). Using the latter extraction scheme, Tisdell and Breslin (Tisdell and Breslin, 1995) showed a comparatively high exchangeability of Cd in MSW composts (> 10%) and the exchangeability of some MSW compost metals decreased in an order as follows: Cd > Zn > Ni > Cu > Cr > Pb.

During 1998 Zorpas (Zorpas et al., 1998; Zorpas 1999) develop a methods that was base on Tessier Scheme (Tessier et al., 1979) in order to extract the deferent forms of metal. In order to study the forms of metals in the sludge samples, a sequential chemical extraction procedure were used for the partitioning of these metals into five fractions. According to this scheme, heavy metals are associated with five fractions:

- ✓ the exchangeable fraction which is likely to be affected by changes in water ionic composition (e.g. in estuarine waters), as well as sorption-desorption processes, (extracted by 1M CH_3COONa for 1 h at room temperature and at pH 8.2)
- ✓ the carbonate fraction that would be susceptible to changes of pH, (extracted by 1M CH_3COONa for 5 h at room temperature and at pH 5.0)
- ✓ the reducible fraction that consists of iron and manganese oxides which are thermodynamically unstable under anoxic conditions, (extracted by 0.04M $NH_2OH.HCl$ in 25 % v/v CH_3COOH for 6 h at 96 °C)
- ✓ the organic fraction, that can be degraded leading to a release of soluble metals under oxidizing conditions, (extracted by 0.02 M HNO_3 and 30 % W/V H_2O_2 for 2 h at 85 °C and pH 2.0, followed by the addition of 3.2 M CH_3COONH_4 in 20 % w/v HNO_3 with 30 % W/V H_2O_2 for 3 h at 85 °C, diluted with distilled water and let for 30 min at room temperature)
- ✓ the residual (extracted by 40 % w/v HF, c.HNO_3 and 0.2 M NH_4NO_3 at pH 3.0, for 1 h and at room temperature) that contains mainly primary and secondary minerals,

which may hold metals within their crystal structure. These metals are not expected to be released in solution over a reasonable time span under the conditions normally encountered in nature.

The same methods were used by the same author (Zorpas et al., 2000, 2008a) in order to observed the behaviour of heavy metals in sludge compost with or with out zeolite. The methods were used before, during and after the composting of sewage sludge. The results indicated that a significant ($p < 0.05$) amount of Cr and Cu are bound to the organic and residual fractions with less than 2% in the exchangeable and the carbonate fractions. Almost 60 % of Mn is bound to the reducible fraction. Iron and Pb were found to be bound in the reducible and residual fractions (about 95% for the Fe and 83% for the Pb). Only Ni and Zn were found in substantial proportions in all phases.

Comparing the results with the raw materials and the So cured compost (Figures 2-8), it is evident that Cd, Fe and Zn are not affected during the composting process. Additionally, Cr seems to be transformed from the reducible fraction and the organic fraction to the residual fraction. The organic fraction of Cu changes to other fractions, especially residual and exchangeable. The residual fraction increased from 27.53 % to 45.37 % and exchangeable from 1.55 % to 10.22 %. Approximately 72 % of the Pb was found to be bound in the residual fraction. A substantial percentage of Mn was removed from the reducible, carbonate and organic fractions and transferred to the residual fraction. As it is observed in Figures 2-8, zeolite has the ability to take up all the metals content bound in the exchangeable and carbonate fraction.

The changes, which are observed in metal partitioning of the sewage sludge compost, are the results of the thermophilic phase of composting, which is the first step of composting and affects the exchangeable fraction. During the first step of composting we have changes in the pH value (acetic acid formation) and in ammonia content (affected by the action of the proteolytic bacterial and by the temperature), (Zorpas et al., 1999a). Also the pH changes, which occur in the beginning of the composting process, affect the exchangeable and carbonate fractions. The pH changes may be due to acid formation during the decomposition of organic matter contained in the sludge.

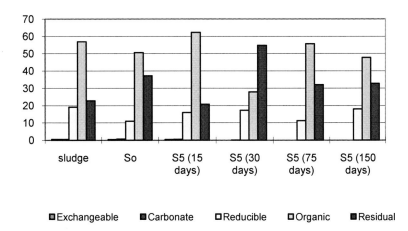

Figure 2. Cr concentration before during and after the composting process in %.

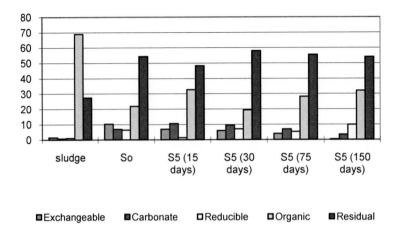

Figure 3. Cu concentration before during and after the composting process in %.

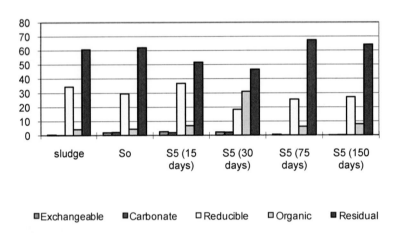

Figure 4. Fe concentration before during and after the composting process in %.

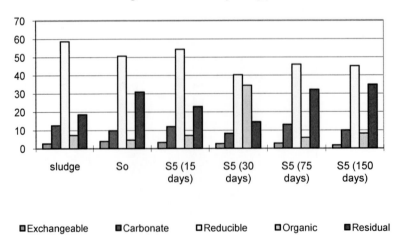

Figure 5. Mn concentration before during and after the composting process in %.

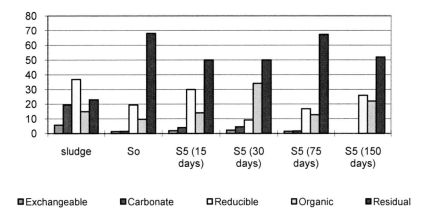

Figure 6. Ni concentration before during and after the composting process in %.

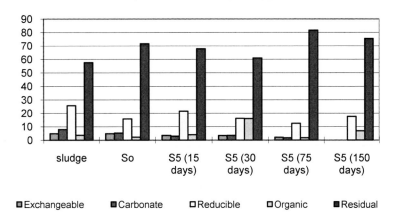

Figure 7. Pb concentration before during and after the composting process in %.

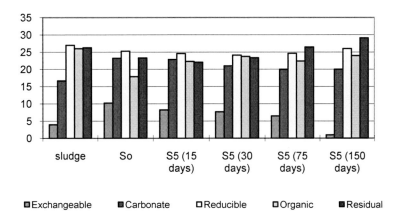

Figure 8. Zn concentration before during and after the composting process in %.

Table 2. Uptake Coefficients For Cu, Cr, Ni, And Zn For Several Crops (216)

Crop	Cu	Cr	Ni	Zn
Lettuce	0.09	0.042	0.17	0.90
Carrot	0.05	0.028	0.15	0.37
Tomato	0.06	0.016	0.09	0.11
Grass	0.07	0.041	0.20	0.33

A high exchangeability of Cd and Zn was also found in biosolids compost (Henry and Harrison, 1992). The sequential extractions also show high percentage of compost Cu in organic, high compost Cr in the residual fraction, and high compost Zn in the Mn and Fe oxides components of the compost. A high percentage of Pb is in organic, and Fe and Mn oxide fractions. The highly inorganic nature of Zn in MSW compost is reflected by more than 70% of total Zn in inorganic fractions (Exchangeable + carbonate bound + Fe and Mn oxides bound). Hsu and Lo (Hsu and Lo, 2000) also found the same high percentage of Zn in swine manure compost in the inorganic fractions, although the percentage of Cu in the swine manure compost was not as high as in the MSW compost. Differences among composts in the percentage of the metals associated in various fractions are to be anticipated as compost may vary in the contents of lime, Fe and Mn oxide, organic matter, maturity, and particle size. Metals tend to be more concentrated in the fine particle size (<1 mm) with an average enrichment of 1.30 ± 0.11(SD) across Pb, Cu, Zn, Cd, Ni, and Cr (Petruzzelli et al., 1979).

The uptake slope of roots was comparatively high for Cd and Zn, intermediate for Cu and lowest for Cr, Pb, and Ni. This was true also for grain. Except for Ni, this order is similar to that for the uptake of the metals by grass and several vegetable crops from biosolid compost (Henry and Harrison, 1992) averaged across compost rates (Table 2).

By comparing the metal availability of biosolids before and after compost, Epstein (Epstein et al., 1978) showed composting reduced Cd uptake by corn leave and grain. Composts typically have pH levels near or above neutral (Petruzzelli et al., 1989; Giusquiani 1992; Canet et al., 2000) and increasing proportions of compost in soil increase soil pH levels (Lutzow et al., 2002; Zheljazkov and Warman 2004) at least within a period of time after compost addition. Increased soil pH often decreases accumulation by plants Zn and Cd from a variety of sources (Kuo et al., 2004), and this was true also for compost metal (Sims and Kline 1991). Part of the composting effect on the reduction of Zn and Cd availability to plants has been attributed, at least in part, to soil pH (Simeoni et al., 1984). Whether or not simply including soil total metal content along with pH is sufficient to predict metal accumulation in plants across diverse soils remains to be seen. Such a combination was found sufficient to predict Cd accumulation by wheat grain or other crops (McBride, 2002; Adams et al., 2004). However, it excludes the role of metal buffering capacity of soil and the organic matter added along with the metals in biosolids or compost. The metal buffering capacity of organics added along with metals is an important factor determining the long-term availability of the added metals (Bergkvist and Jarvis, 2004). For other metals such as Cu, the combination may not work as the increased mobilization of Cu at high pH levels increased, rather than decreased, Cu uptake by plants (Sims and Kline 1991).

1.6 Heavy Metals Leachability before, during and after Composting of Sewage Sludge with Natural Clinoptilolite

Because of the impact of compost metals on metal accumulation in soil and in plant, investigations were made to the leachability of metals during, before and after composting of sewage sludge. During 2010 Zorpas (Zorpas, 2010), apply a simple a method that was base on Generalized Acid Neutralization Capacity (GANC) in order to estimate the leachability of metals in the 3 phases of composting. The aim of this study is to determine the leachability of the heavy metals (Cd, Cr, Cu, Fe, Mn, Ni, Zn, Pb), in the raw sludge, during the composting process and in the final compost products when natural zeolite is used in order to take up the heave metals. The Generalized Acid Neutralization Capacity (GANC) test procedure was used for the estimation of the leachability of metals from the sludge and compost samples, (Sims and Skline, 1991; Mendoza, 2006). This test is a single batch procedure that utilizes a series of sludge samples extracted with increasingly acidic leachant. A known quantity (1 g) of sludge sample is placed in a series of 100 ml polyethylene bottles. 20 ml of liquid is added to each bottle. A declining amount of distilled water is added to each bottle followed by an increasing amount of 2N acetic acid. This process produces a series of bottles containing increasing equivalents of acid per kilogram of samples but the same total liquid volume (20 ml). The mixtures were tumbled in rotating extractors for 48 hours and then were let to stand for 15 minutes. Following, in the supernatant the pH was measured and the metal concentrations were determined by atomic absorption spectroscopy. Leachants strength starts out at the 0 equivalents of acetic acid and is increased until pH is below 5 for three consecutive equivalents. For this study the prepared sambles was Ao (no zeolite and 100% sludge), A1 (5% Cli + 95 % sludge), A3 (15% Cli + 85% of Sludge), A5 (25% Cli + 75% sludge) and A6 (30%CLi + 70% sludge).

Figure 9 resents the metals leachability before the composting process (in raw materials). The GANC test results showed that by increasing the leachate pH, the heavy metals concentration in leachates was decreased. It was observed that at pH greater than 5 the metal amount in leachate was decreased. This phenomenon can be explained by the fact that, the metal load was not bound to the exchangeable and carbonate fractions, (Zorpas et all, 1999). Figures 10a-g) present the metals leachability during the composting process for A5 compost sample. It is obvious that the zeolite has a different selectivity during the first 30 days of composting. This is due to the fact that, the pH during the composting process is initially low, due to the acid formation, then increases and in the final stage of the composting process remains constant. Also, due to the high temperature and the amount of moisture the mobile forms of the metals move faster.

However, it is obvious that zeolite seems to be constant for most of the metals after 15 days. Figure 11shows that by increasing the leachate pH, a decrease to the heavy metal concentration is caused. According to Figure 4 zeolite bounds a significant ($p<0.05$) amount of heavy metals at pH lower than 4.5. Specifically zeolite bounds 100 % of Cd in A5 and A6 compost sample, 98 % of Cr, 92 % - 95 % of Cu (A5 and A6 respectively), almost 90 % of Fe and Zn, 50 % - 60 % of Mn (A5 and A6 respectively), almost 50 % of Ni, and 82 % - 88 % of Pb (A5 and A6 respectively).

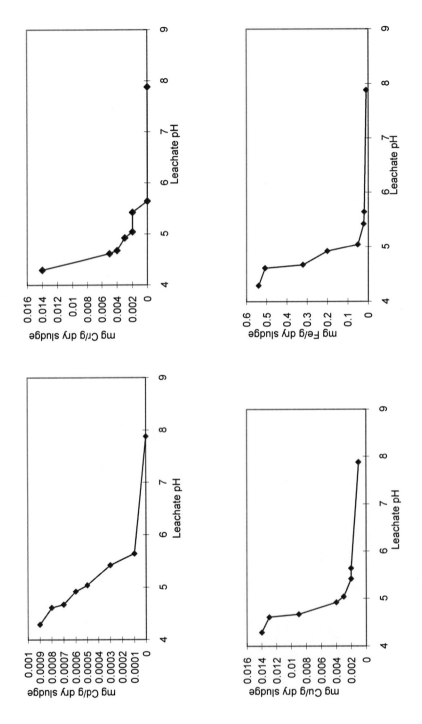

Figure 9. (Continued).

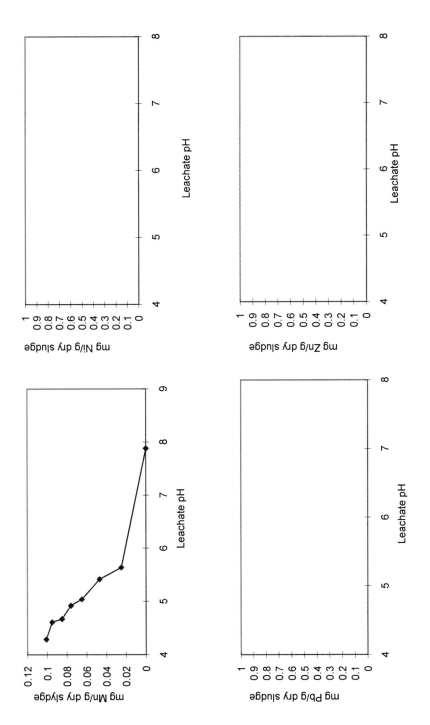

Figure 9. GANC metals releases to leachates from primary sludge samples as a function of pH of leachate.

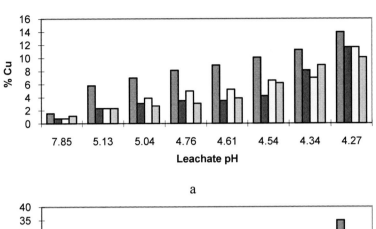

a

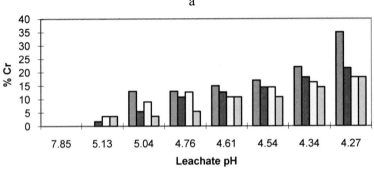

b

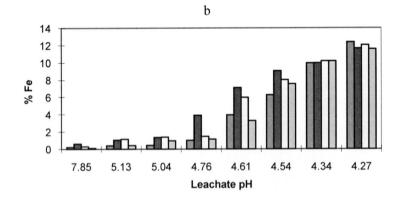

c

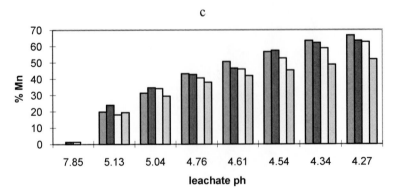

d

Figure 10. (Continued).

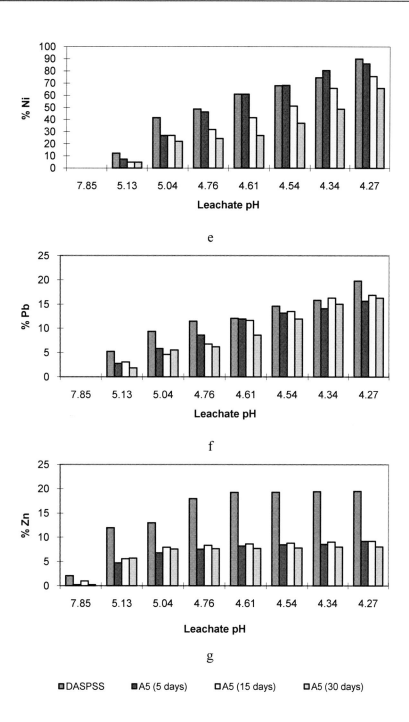

Figure 10. GANC metals releases to leachates from A5 compost sample as a function of pH of leachate, during the first step of composting.

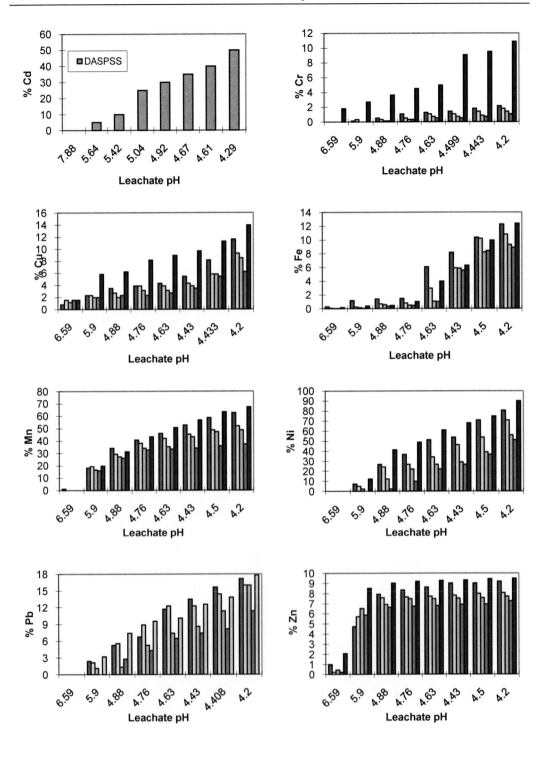

Figure 11. GANC metals releases to leachates from A1, A3, A5 and A6 compost sample as a function of pH of leachate after from 150 days of maturity.

When comparing the metal content of the Ao, sewage sludge compost with no zeolite and Sewage Sludge sample, it may be observed that the concentration of chromium, nickel, manganese, lead and zinc appeared increased while that of copper and iron decreased. Composting can concentrate or dilute the heavy metals present in sewage sludge. This change in metal concentration depends on the metal loss through leaching and on the overall concentration of metals due to organic matter destruction.

2. EVALUATION CRITERIA PARAMETERS – QUALITY ASPECTS

2.1. Organic Mater

Organic matter is one of the important components of soil contributing to its metal retention capability as indicated earlier. Its decomposition in soil over time could lessen soil capability to retain metal and increases metal availability in the soil. This is the basic justification in proposing 'Time Bomb' concept by McBride (1995). A rapid increase in flux of CO2 within the first several weeks after compost had been applied on the forest floor (Borken et al., 2004) or incorporated in the soil (Hadas and Portnoy, 1994) was followed by a very slow degradation rate of 4 to $5x10^{-4}$ d^{-1} or lower, if the compost is well mature (Sikora and Yakovchenko, 1996). Since loss of compost C also involves SOC loss via leaching as discussed earlier, a long-term study under the field condition is needed to determine compost C accumulation or depletion rate is. This, in conjunction with field measurement of accumulation of compost metal to plants would allow assessment of the loss of compost C on the compost metal availability in soil. Loss of soil organic C or compost C may not necessarily reduce sorption of compost metal by soil as degradation of SOC increases polymerization of residual C and metal retention capacity. It is likely that the long-term availability of compost metal will be controlled by whether or not metal sorption in the compost amended soil is changed as a result of C degradation (Bergkvist and Jarvis, 2004). A typical methods of the determination of Organic Mater in Sludge is Referred to the Soil Analysis Standard Methods and Zorpas (1999). The methods referred that (%) organic matter of sludge can determine by incinerate a know quantity of samples in 500 $^{\circ}C$ (until we have a stable weight) while the % Total Organic Carbon (TOC) = % O.M. / 1.724.

2.2. Humic Substances in Compost

A reduction of carbohydrate, hemicellulose, and cellulose during the composting process is accompanied by increased humification. The humification is an index of compost maturity. It is a critical component of composting as humic substances from composting are beneficial to soil physical and chemical properties and plant productivity.

Accompanying with increase in humification is increases of alkyl C, aromatic C, and carboxyl (-COOH), phenolic (-OH), and cabonyl (-CO) groups. The aromatic and phenolic C containing groups increased by 23 and 16%, respectively, following composting of municipal solid waste (Chefetz et al., 1998). The other major changes in the characteristics of compost are the marked increase in the content of humic acids and consequently the cation exchange capacity of compost (Inbar et al., 1989), which appear to follow the first-order kinetics (Inbar et al., 1989). Because of increased aromatic nature of composting product, composting process influenced not the elemental composition of finished compost but the functional groups (Sanchez-Monedero, 2001).

Questions arise as to the origin of humic substances from the composting. Adani et al. (1999) indicated that no net humic substances were formed during composting. This is in line with the findings that humic substances in the compost retain structural characteristics of lignin in the wood waste or crop residue used as bulking agent, and increased aromaticity of humic acids result as the protective coating materials (polysaccharides, polypeptides, and lipids) are degraded (Chefetz et al., 1998). The humic substances formed is dissimilar to soil humic substances (Sanchez-Monedero, 2001) as they have comparatively lower concentrations of total acidity, and carboxyl and phenolic functional groups than soil humic acids (Inbar et al., 1989). However, other investigators (Gonzalez-Vila et al., 1999) found that compost humic substances are similar to soil humic substances. The question is difficult to answer, as the mechanism of humic substances production in soil and in compost is not exactly clear. If polymerization of quinones from degradation of lignin by microbial activity or polymerization of polyphenols synthesized by microorganisms is the mechanism involved in the formation of humic substances in soil and in the compost, a similarity in humic substance between soil and compost should be anticipated. However, while compost may be mature in the context of humic substances production reaching the plateau during the composting process, it will undergo further degradation once incorporated into the soil. The re-synthesis of microbial by-products over time may make the humic substances from compost before soil incorporation different from that of soil humic substances.

Humic substances were extracted according to Schnitzer (1982), with some modifications by Zorpas (1999a, 1999c). According to this methods the total humics can determine as follow: A know quantity of sample is used by HCl 0.05N in order to removed the carbonates salts and after is extracted by NaOH 0.1 N for 24 h. The concentration of humics may reported in mg/g dry mater or %. Foulvic and Humic Acids are estimated by fractionate the extraction of total humics with HCl 2 Í (reaction time 24 h). The insoluble part is the humic acid while the soluble is the fouvlic acids. The determination is done by dry the sambles in the temperature of 40 °C and the results is referred in mg/g dry sambles. The ratio E4/E6 specifies the characterization of the humics substances. The E4/E6 ratio show the characterisation of humic materials (by spectrometry methods after the extraction of a know quantity of a sample in 1:10 ratio with $NaHCO_3$ 0.05 N. Then we used spectroscopy in order to determine the absorption in 465 nm and 665 nm). As the E4/E6 ratio is bellow 5 the samples are characterised as Humic Acid, (whereas if the ratio is above 5 the sample is characterised as Fulvic Acid). As the E4/E6 lead to zero the molecular weight of the substrate is to hight. The humication index (H.I) = % Total Humics / % Organic Mater (Zorpas 1999a). Knowing the ratios of humic acids (HA) to fulvic acids (FA) help to predict the possible reactions of the organic matter fraction with soil components because FAs are more soluble and reactive than HAs. This information can compliment current knowledge about compost application, making it easier to develop compost application programs that improve soil characteristics such as water and nutrient retention and soil structure.

2.3. Compost C and N

The loss of C ranged from 46 to 62% as compared to 19 to 42% N loss during composting, depending on type of composting system, waste stream (e.g., lignin content), and composting conditions (e.g., temperature, moisture) (Eghball et al., 1987). The majority of C loss is from carbohydrates, hemicellulose, and cellulose as they constitute the majority of plant C and are comparatively more easily degradable than lignin. With a much greater loss of C than N during composting especially during the active decomposition by thermophilic microorganisms, the C:N ratio of organic materials in the compost mix declines also in a first-order faction (Inbar et al., 1989) and reached a steady state after 100 days of composting of cattle manure (Inbar et al., 1989, Chefetz et al., 1996). The ratio of C: N ratio of the finished compost [(C:N)final] to the C: N ratio of composting mix initially [(C:N)initial] is about 0.6 to 0.75 (Jimenez and Garcia, 1989). This indicates that the C:N ratio of the initial composting mix will be reduced by 25 to 40% under normal composting conditions. The C:N ratio of compost should be about 20 to prevent N immobilization and to facilitate the release of mineral N for crop use once the compost is added to soil (Allison, 1973). This suggests that a C:N ratio of about 30:1 (20 to 40) (B.C. Agriculture and Food. 1996; USDA 2000) before composting is commenced is desirable. Initial C:N ratio over 40:1 is not conducive to the degradation of organic matter due to N immobilization at least initially (Allison, 1973). In contrast, low C:N ratios of feedstock, or overabundance of N, tend to cause the accumulation of NH_4- N as $(NH_4)2CO_3$, which at high pH levels dissociates into NH_3 and CO_2 according to following reaction: $(NH_4)2CO_3 + 2 (OH^-) <=> 2NH_3 \uparrow + CO_2 \uparrow + H_2O$.

Rise in pH (>8.5), coupled with elevated temperature in the composting pile during the thermophilic phase of degradation, promote the volatilization of odorous NH_3. Martin and Dewes (1992) found that of the total N loss during composting, a majority of it was lost by NH_3 volatilization during the thermophilic phase and the loss could be as high as 55% of total N (Hammouda and Adams 1986). The extent of NH3 volatilization is influenced by temperature, pH, C:N ratio, and turning (Hammouda and Adams 1986; Martin and Dewes 1992, Hansen et al., 1993). Covering the compost pile with a bio-filter comprising of wood waste (e.g., wood chips or sawdust or finished compost) can be practiced to reduce volatilization of NH_3, organic acids and associated bad odor from the composting pile. Control of compost pH to less than 8 is desirable as hydrolysis of NH_3 turns it to NH^{4+}, thereby reducing its partial pressure and volatilization potential.

As the thermophlic phase of degradation reaches the peak, the readily degradable organic C and O_2 in the areas of active degradation could be reduced to levels insufficient for generation of heat to compensate for the heat loss through radiation and convection. The composting pile begins to cool. Aeration at this state of composting could bring the temperature back up if sufficient readily decomposable organic C is still present. The cooling brings the composting back to mesophilic degradation and begins the stage of curing. The slow mesophilic curing process may take several weeks to months depending the feedstock, method of composting, and the extent of the degradation of degradable organic C during the thermophilic decomposition phase. This curing process carried out predominately by fungi and actinomycetes is critical as it provides time for further degradation of organic acids, huminification, and nitrification that transforms NH_4-N to NO_3-N [Nitrification: $NH_3 + O_2 \rightarrow NO^{3-} + H_2O + H^+$], raising the ratio of NO_3-N to NH_4-N. The ratio of NH_4-N to NO_3-N, the ratio of humic acid to fulvic acid, and dissolved organic C content, and self heating all have

Table 3. Methodology for the determination of the forms of Nitrogen in Compost

Ammonia, NH_4^+-N	Nitrite, NO_2^- - N	Nitrate, NO_3^- - N
2 g of sample 10 ml dis-distilled water extracted for 2h 2 ml Nesller reagent and 0.25 ml Rochelet salt	2 g of sample 50 ml dis-distilled water extracted for 2h 1 ml mixed reagent (2g Sulfanilamide + 100 mg N1-napthylenediamine- dicloride in 50 ml of water)	2 g of sample 10 ml dis-distilled water extracted for 2h 0.02 ml HCl 1N
Measured in 415 nm (after from 10 minutes)	Measured in 543 nm (after from 10 minutes)	Measured in 220 nm and 275 nm ($ABS=2*ABS_{220}-ABS_{275}$) - immediately

been used as indexes of compost maturity (Zorpas 1999, Zorpas et al 2008b). The curing process also allows colonization of compost by certain beneficial fungi for the suppression of pathogens such as *pythium, Rizoctonia*, and *Fusarium* when the compost is applied to the field or used in the potting mix for container crop production.

As the primary purposes of composting are to stabilize the organic wastes to facilitate recycling of nutrients in the organic wastes, and to reduce the volume of wastes going to the landfills, many types of organic wastes have been used as feedstock for composting. The wastes include sewage sludge, animal manures, yard waste, crop residues, municipal solid waste, fish scraps and mortality, and food waste and food process wastes. The materials vary widely in C:N ratios. Since the C:N ratio of the composting mixture initially should be about 30:1 (Epstain, 1997, Zorpas 1999a), co-composting is feasible for some of the materials, if they are available in or near the composting facility to reduce the distance and cost of transportation. Straw from grain crop, peat moss, sawdust, wood chips and shredded and ground papers, municipal solid waste typically have high C:N ratio, and they can be blended with animal manures or biosolids for co-composting. Density of the bulking agent is related to porosity/aeration of compost and degradability of bulking agent has some bearing on N conservation (Liao et al., 1995). Passive aeration would require low bulk density to facilitate the diffusion of O_2 from the atmosphere into the interior of compost pile. The selection of a bulking agent for composting is mostly dictated by its cost and availability.

Table 3 present according to Zorpas (1998, 1999a) the methods for the determination of the different forms of Nitrogen while the Total Nitrogen is estimated according to the Kjeldahl methods as described by Adams (1990).

2.4. Quality of Compost - Physical Factors Affecting Compost Quality

The sustainability of composting is largely determined by its end use, health risk and consumer acceptance of the product. In essence, the compost quality or standards are driven by market development and protection of human health. To protect human health in the US, the US federal guidelines (USEPA. 1993) limit E. Coli to less than 3 E. coli g^{-1}, and fecal coliforms to less than 1000 $MPNg^{-1}$, Salmonella to less than 3 MPN per 4g of total solids. Currently in the US, the finished composts are mostly for landscaping purposes, although

Table 4. Ranking of popular uses of compost in the United States (Hogg et al., 2000)

Type of compost used	Use rank	Estimated use quantity (mT)
Landscaping	1	4,000,000
Landfill cover	2	2,000,000
Gardening and horticulture	3	1,000,000
Commercial farming	4	1,000,000

Table 5. The compost class (type) based on the content of foreign matter of 2 mm in the compost (CCC 1999)

	Type AA	Type A	Type B
Foreign matter content, %	<0.01	<0.5	<1.5
Foreign matter, maximum dimension, mm	12.5	12.5	25.0
Organic matter, %	50.0	40.0	30.0

considerable amounts are also used in agriculture and horticulture to improve soil quality and the growth of field and garden crops, and turfgrass (Table 4). When quality standards are oriented towards marketing (or end use), compost quality standards differ from state to state within the U.S. and from country to country.

When compost quality is determined by end use, it is ranked by its physical, chemical, and biological characteristics. In the use of physical characteristic including particle size, texture, and the content of non-composable debris (stone, plastic and glass) to define compost quality, there is considerable subjectivity involved. This is reflected by the wide difference in the percent of non-decomposable debris in defining compost quality among countries (Table 6). This is true also among the states within the U.S. In the case of Canada, three standards were established based on the percentage of foreign matter (Table 5).

For Type AA and A in the Canadian Standards, the compost is considered to be high quality, provided that their heavy metal contents do not exceed their standards, which will be further discussed later. For Type B, its use is restricted.

Table 6. Maximum foreign matter particles allowed in composts in various countries (Brinton 2000)

Country	Stone	Man-made foreign matter
Australia	< 5%* of > 5mm size	< 0.5% of > 2mm size
Austria	< 3% of > 11mm size	< 2% of > 2mm size
Germany	< 5% of > 5mm size	< 0.5% of > 2mm size
Netherlands	< 3% of < 5mm size	< 0.5% of > 2mm size
Switzerland	< 5% of > 5mm size	< 0.5% of > 2mm size
United Kingdom	<5% of > 2mm size	Max. 0.1% plastic < 1% of > 2mm size < 0.5% if plastic

* on a dry weight basis.

Table 7. Methods for predicting compost stability/maturity

Methods	Parameter
Physical analyses	Temperature, color
Chemical analyses	NO_3-N, NH_4-N, waster-soluble C, C:N ratio, cation exchange capacity, Humic and fulvic acid
Microbiological assays	Respiration (CO_2 evolution; O_2 consumption)
Plant assays	Cress germination test in water extract, ryegrass growth in compost containing mixtures
Spectroscopy analyses	Solid state CPMAS 13C-NMR, infrared-FTIR

2.5. Compost Maturity Tests – Phytotoxicity Test

For compost to be used not for mulching but for row or container crops, its high stability or maturity is desirable as instable or immature compost is often odorous and phytotoxic, and interfere with seed germination due to the elevated concentration of NH_3, salt content, and/or organic acids. A number of methods to test compost stability and maturity have been proposed over the last two decades with varying degrees of sophistication. Some of the methods focus on the measurement of decomposition products including temperature, CO_2, nitrogenous products of various forms (NH_4, NO_3, C:N ratio), organic acids and cation exchange capacity, while others monitor the biological response to the compost (seed germination and seedling growth). The rate of the release of the decomposition products (CO_2) and the rise of temperature of composting mix are related to compost stability, whereas the biological response to compost is related to compost maturity. All these methods are listed on Table 7.

Calculating the humification index by using only the carbon concentration in humic and fulvic acids can introduce mistaken values. Indeed, measuring carbon concentration alone does not take molecular structure into account. The UV method for measure the state of maturity of a compost, takes into account the non-humified faction and the fact that the compost is young and not yet in the humification phase, making possible to approach the global chemical structure of molecules as a whole, in particular their aromatic structure.

This method avoids the problem of distorted values for carbon concentration in humic acids (Domeizel et al., 2004). The use of spectroscopy analyses allows the determination of functional groups in the compost to explain the change to the compost composition and functional groups with time (Inbar et al., 1989, 1990). With increasing composting, increases in the aromaticity, alkyl C and carboxyl groups concentration, and decreases in polysaccharides and aliphatic chain results. This leads to increased cation exchange capacity of the composting mix, which is important in terms of retention of cations following its use as soil amendment or potting mix.

As heat is released during microbial degradation of organic matter, temperature of composting mix is a good indicator of its stability. The Dewar Test (Brinton 2000) is often employed for the determination of compost self-heating. When the increase above ambient temperature from self-heating is less than 10 $^{\circ}$C, the compost is classified as completely stable (Table 8). As self-heating intensifies by the further rise in temperature above ambient, the compost is increasingly immature. Being simple and easy to use, Dewar Test is used by many composting facilities around the world.

**Table 8. Dewar self-heating test and CO2 loss for determining
the compost maturity (USDA 2000)**

Heating rise over ambient	CO_2 loss mg C g-1 Cd-1	Rating	Description of stability
0 to 10	0 to 2	V	Completely stable, can be stored
10 to 20	2 to 8	IV	Maturing compost, can be stored
20 to 30	8 to 15	III	Material still composting, do not store
30 to 40	15 to 25	II	Immature, active composting
40 to 50	>25	I	Fresh, very active composting

Associated with a rise in microbial activity is the release of CO_2 and consumption of O_2. The respiration rate in C loss $g^{-1}d^{-1}$ at 34 °C has been used in defining the class of compost (Table 8). A low respiration rate is indicative of high compost stability. However, a low respiration rate of compost may not necessarily be a good indication that phytoxicity would not occur. This is particularly the case when microbial activity is inhibited by high metal concentration. Wu et al. (2000) found that compost samples from one composting facility showed phytotoxicity despite having a low CO_2 evolution rate. As composting proceeds, a greater C than N loss results in the reduction of C:N ratio (Eghball et al., 1997). When C:N ratio reaches below 25:1, composting is generally considered to be mature. Despite a close relationship between the change in C:N ratio and the change in CEC or in the concentration of humic acids (Inbar et al., 1989), C:N ratio alone is not a good index of maturity of composts made from a diversity of waste streams and bulking agents. The representation of compost maturity by C:N ratio can be misleading particularly when the compost contains elevated levels of NH^+-N. This form of N is part of organic N if kjeldhl N is used to reflect organic N. While C:N ratio is an important characteristic of organic residue that determines net N mineralization and N mineralization potential (Kuo et al., 1997; Kuo and Sainju, 1998), it has not been a consistently good indicator of compost N mineralization potential. For characterization of compost maturity, C:N ratio should be used in conjunction with other stability indexes. In the Canadian Standards (CCGC) (CCC, 1999), the compost is deemed mature if C:N ratio is ≤ 25 and the oxygen uptake rate is ≤ 150 mg O_2 kg^{-1} or phytotoxicity test .

Immature compost containing high VOAs is phytotoxic to seed germination and seedling growth (Murillo et al., 1995). While the phytotoxicity test is of value for compost used as whole or part of growing medium for nursery crops, it may not be as useful for compost used as a soil amendment. This is because VOAs are readily degradable by soil microbes and the amount of compost added is only a small fraction of surface soil (<1%). The dilution effect could render those composts containing medium levels of VOAs less or non-toxic. This is true also for soluble salts. For use as a soil amendment, the compost does not necessarily have to have the same high quality as used for nursery container crops, many of which are sensitive to VOAs and soluble salts. Zorpas et al., (2000) present a mature and stable compost from sewage sludge with C:N ratio 11 after from 75 days of maturity. The ratio remain stable and in 150 days.

According to Zorpas (1999a, 2008b) the evaluation of the cured compost may determinate by a the Seed germination index and the Phytotoxicity test. The test is described below.

Briefly, water extracts were obtained by mixing 10 g of compost samples with 20 ml of de-ionized and distilled water (DDW) for 2 h. The test solution was obtained by centrifuging the slurry at 5000 rpm for 30 min, and then filtered through Whatman No 1 filter paper. A 10ml of test solution was pipetted into a sterilized plastic Petri dish. Also, one blank with only DDW was carried out. Twenty seeds of lettuce group were placed in each Petri dish and incubated at 22°C in the dark for 7 days. The GI was calculated according to the following formula:

$$GI = \frac{(\%G)(\%L)}{100} \tag{1}$$

Where % G = Growth index,

$$\%L(Root\,Length) = \frac{(Root\,Length\,of\,Treatment)}{(Root\,Length\,of\,Control)} * 100 \tag{2}$$

Zorpas (2008b) mentions that, if the $0 < GI < 26$ the substrate is characterized as very phytotoxic, $27 < GI < 66$ the substrate is characterized as phytotoxic, $67 < GI < 100$ the substrate is characterized as non-phytotoxic and if the $GI > 101$ then the substrate is characterized as phyto nutrient. The test was used in several other cultivations like: Oats, Pepper, Eggplant, tomatoes etc.

2.6. Contaminants in Composts

Contaminants in feedstock can impact the quality, marketability, and use of finished composts. Overuse and persistence of some herbicides and insecticides could result in pesticide contamination of yard waste and compost. Since composting is a biologically mediated process, there is a considerable interest in knowing if the intense biological activity during the active thermophilic degradation period could induce degradation of pesticide contaminants. In the review of pesticide occurrence and degradation during composting, Buyuksonmez et al. (1999) found that a few of the targeted pesticides were found in the composts with organochlorine compounds being the most resistant to degradation during composting. Some herbicides are resistant to degradation. Clopyralid and picloram had been detected in some compost (Bezdicek et al., 2001). Compost contaminated with Clopyralid, a broadleaf herbicide, caused plant damage in Washington State in 1999, even though the damage largely disappears if the use of contaminated clippings as feedstock is delayed for a year or longer (Miltner et al., 2003). Residues from other broadleaf herbicides for lawn care, 2,4-D (2,4- dichlorophenoxy acetic acid), dicamba (2-methoxy-3,6-dichlorobenzoic acid) and MCPP [2-(4-chloro-2-methylphenoxy) propionic acid) were detected and found toxic to tomato (*Lycopersium esculentum* L.) (Bugbee and Saraceno, 1994). Other pesticides that have also been detected in composts include no atrazine, carbaryl, chlordane, etc. While diazinon (O,O-diethyl O-[2-isopropyl-6methyl-4-pyrimidinyl] insecticide was used extensively for insect control on turfgrass, its biodegradability is evident as only a trace of it (<1%) could be found after composting (Michel et al., 1997). Close to 11% of the insecticide was degraded during composting, and a majority of the insecticide was converted to a potentially leachable but less toxic hydrolyzed product. The release of diazinon via volatilization is extremely small (<0.2%).

Waste stream from biosolids or municipal solid waste can contain elevated concentration of various metals and the impact of metals on compost quality and use is a major concern throughout the world. The change of the chemical forms of metals during composting has been a subject of some intensive investigation. As the composting of municipal solid waste proceeds, the concentration of metals increases progressively as a result of mass reduction due to C loss as CO_2 (Leita and de Nobili, 1991). A rapid decline of water-soluble Pb and Zn was associated with a rapid reduction of water-soluble C during composting.

The composts that use sewage sludge or biosolids or mixed or source separated municipal solid waste as waste streams in general have much higher metal content than those using animal waste as a waste stream. Attempts were made by adding natural zeolite (clinoptilolite) to the composting of municipal solid waste or biosolids to increase cation exchange capacity of the compost, thereby reducing the eachability and bioavailability of metals in the compost (Zorpas et al., 2000).

2.7. Compost Metal Concentration Limits

To protect the quality of composts from excess contamination of metals, standards of metal concentration limits should be established. This is being done around the world, but the standards or guidelines developed are not uniform. Currently, a major difference in the maximum metal concentration limits exists between U.S. and Canada or European Union. It evolves from the fundamental difference in the approach on how to best protect human health.

In the development of guidelines for metal and metalloid concentration in compost in Canada, three approaches were considered:

✓ No net degradation
✓ Best achievable approach
✓ No observable adverse level (NOAEL)

The basic principal of no net degradation is that the use of compost over time will not lead to increase in metal or metalloid accumulation over their background levels in the soils. The amount of metals applied should not exceed their losses through plant uptake or leaching loss. In Canada, the average metal and metalloid concentrations in soils are their background levels in uncontaminated soils in Alberta, Ontario, and Quebec Providences plus three standard deviations. This approach was used for establishing the limits for Cd, Co, Ni, Pb, and Zn (Table 9) for two classes (Classes AA and A) of compost that are allowed for all types of applications. To encourage recycling and source separation to reduce metal concentrations in compost product, best achievable approach via source separation and reduction was used to

**Table 9. The metal and metalloid concentration limits for composts
in several countries (Brinton, 2000)**

Metal or metalloids	CAN mg kg-1	DK	FR	GER	IT	NL	SP	US*	US**
As	13	25	-	-	10	15	-	-	-
Cd	3	1.2	8	1.5	1.5	1	10	39	2.0
Cr	210	-	-	100	100	70	400	1200	100
Cu	100	-	-	100	300	120	450	1500	100
Pb	150	120	800	150	140	120	300	300	150
Hg	0.8	1.2	8	1	1.5	0.7	7	17	0.5
Ni	62	45	200	50	50	20	120	420	50
Zn	500	-	-	400	500	280	1100	2800	400

US* (biosolids rule); ** Woods End QSAP standards.

establish the limits for As, Cr, Cu, and Hg. No net degradation approach was also taken by several European countries, and resulted in low concentration limits and the amount of metals or metalloids allowed for application to agricultural soils (Table 9). The metal or metalloid concentrations for Class B compost are far higher than those for Class AA and A, and this class of compost has a restricted use and should not be applied to food crops.

The limits developed based on no net degradation obviously is far more stringent than those based on risk assessment (Table 9). The risk-based concentrations were developed originally to regulate land application of biosolids under the section 503 of U.S. Clean Water Act (55). No observed adverse effect is the basis for the development of risk-based concentration. Numerical limit is established for each contaminant based on the analysis of the potential impacts of the contaminant on human health and environmental risks. For those contaminants that are below the concentrations limits, no restriction of the use of material is imposed as long as the annual applications of the contaminants do not exceed their annual loading limits.

In the US, there is no uniform guideline for metal standards in compost, although a number of states utilize Biosolids Rule to set metal concentration limits for composts now. An attempt was made by US Composting Council in cooperation with the American Association of Plant Food Control Officials (AAPFCO) to setup uniform limits. Woods End Research Laboratory in Maine has recommended metal concentration limits for their QSAP (Table 9) (Brinton, 2000) certification program, which are more in line with the European standards than with the limits in the US Biosolid Rule.

3. UTILIZATIONS OF COMPOST

Composting is the controlled biological decomposition of organic matter under aerobic conditions. Organic matter decays naturally, but slowly. Composting involves human intervention to speed up the decay process by manipulating various materials and conditions.

Composting is one form of recycling. Organic, compostable material comprises 68% of MSW (Municipal Solid Waste.) Most communities can implement some form of composting in order to reduce the amount of waste going into their landfills.

Compost can be used in many ways. It can improve soil conditions and plant growth, and reduce the potential for erosion, runoff and non-point source pollution. Compost has also been found to be useful as a medium in plant disease suppression and in biofiltration. Through these uses, compost can be used to remediate or prevent the pollution of soil and groundwater systems.

Application of composts to land presents a potential way to recover value and avoid disposal to landfill (Petersen et al., 2003). An important criteria used to determine if land application will be beneficial, or not, is the ability of the organic by-product to enhance, or at least have no eleterious effect on, soil productivity and the growth and/or yield of plants (Sims and Pierzynski, 2000). Improvements in soil chemical properties have been reported for re-use of organic by-products and include; increased soil pH (Oue'draogo et al., 2001), increased plant available potassium (Erhart and Hartl, 2003) and plant available calcium and magnesium (Jakobsen, 1996; Wen et al., 1999; Miyasaka et al., 2001). Organic amendments can also promote plant health and increase yield of certain crops. Barley yields were increased at 50 t/ha MSW (Zhang et al., 2000). The influence of organic matter on soil properties depends on its amount and composition (Unsal and Ok, 2001). Low quality compost essentially arises from an excess of heavy metals and salt and a low degree of stabilization (Murillo et al., 1995). Depending on feedstock, certain composts have been shown to contain elevated concentrations of metals including Pb, Cd, Cu, and Zn (Breslin, 1999; Zheljazkov and Warman, 2003). Also, land application of some composts can result in increased electrical conductivity (EC) due to high salt content and can restrict seedling performance (Hsiao-Lei Wang et al., 1984). Consequently, the type of compost being used for land application will affect the overall impact on soil properties and crop growth and composts organic mattercontent, nutrient values and potential levels of metals are of concern. Courtney and Mulle on 1999 was investigate the application of of two agroindustrial composts; spent mushroom compost (SMC) and forced aeration compost produced from industrial sludges (FAC), on soil properties and the growth of barley (Hordeum vulgare) and how they compared to a commercial mineral fertilizer treatment.

In some countries, before 1980's, the soil fertility was maintained mainly by the use of organic fertilizers such as farmyard, manure, compost, green manure, straw and organic wastes (Yang and Hansen 1997). At the present, low soil organic matter content on arable lands has made them become less fertile (Yang and Hansen 1997). The over use of inorganic fertilizers and intensive fieldcrop production can cause the quality of agricultural soils to decline. Reduced soil physical quality is, in turn, linked to declined crop performance and/or profitability, as well as negative environmental impacts related to the off-field movement of soil (wind/water erosion) and agrochemicals (pesticide/nutrient leaching into surface and ground waters) (Reynolds et al., 2002). The beneficial effects of the use of compost can be seen in landscaping, which can be hard and soft landscaping schemes. Hard, includes road construction and motorway edges and surfacing of landfill sites. Soft landscaping involves the utilization of compost in parks, gardens, playground and golf courses (Manser and Keeling, 1996). In horticulture and agriculture, compost can be used as an alternative to peat as a growing medium, as a source of organic matter, as a cover material to conserve moisture and suppress plant disease and also as a nutrient source (Liu, 2000).

As the quantity and quality of organic matter declines in many farming systems, farmers are now faced with finding alternative or supplementary sources of nutrients. The compound fertilizer made by compost is welcome very much at present because peasants know that too much use of inorganic fertilizers is not good to the soil and the environment (Wei et al., 2000). The variety of tropical agro-ecosystems and the diversity of organic inputs used in those systems, including trees, shrubs, cover crops and composts present a challenge for research and extension activities in soil fertility management (Palm et al., 2001).

4.1. Horticulture/Agriculture

Compost products have been traditionally used predominately in horticulture and agriculture to improve soil conditions and enhance plant growth. Compost improves the soil's physical and chemical properties that play an important role in crop production. Addition of compost provides for a better root environment. For this reason, it is often referred to as a "soil conditioner."

It is important to remember, however, only to apply finished compost to a soil system. Finished compost is cool (within 10 degrees of ambient soil temperature), is approximately 1/3 the original volume of material and has an "earthy" smell. Applying unfinished compost may be detrimental since it can remove nutrients from the soil as the decay process continues. Also, unfinished compost may contain substances that are detrimental (i.e., phytotoxic) to seedlings or sensitive plants (Zorpas 1999a)

4.2. Physical Soil Improvements

Soil structure - Compost increases soil aggregation and decreases compaction. Compost also increases soil porosity and consequently aeration. All of these conditions aid in plant growth, (Zorpas 1999a).

Water relations - Compost increases the water holding capacity of soils thereby providing for higher water availability to plants, (Zorpas 1999a).

Infiltration - Compost tilled into soils eliminates crusting and increases infiltration and permeability. This reduces the potential for runoff and erosion, (Zorpas 1999a).

4.3. Chemical Soil Improvements

Cation exchange capacity - Compost provides an abundance of organic Carbon. This increase in organic C essentially results in and increase in CEC. This is very important as CEC ultimately governs the potential fertility of the soils, (Zorpas 1999a).

Soil pH - The pH of most stable compost products ranges from 6.5 to 7.5. The addition of compost can raise or lower a soil's pH to this level. The acidity of a soil impacts metal solubility, plant uptake and movement, plant growth, soil microbes and many reactions.

Electrical conductivity (EC) - The EC measures the amount of salt in the soil solution. The salinity can affect seed germination and plant growth. Compost addition can increase the EC causing a problem for plants. This is easily remedied by leaching the compost prior to application.

Nutrients - Compost generally contains low levels of major nutrients N, P and K. It also contains many essential micronutrients needed for crop growth. The quantities of compost used often provide the necessary nitrogen when considered in conjunction with N mineralization, (Zorpas 1999a).

4.4. Plant Disease Suppression

This is a newer and non-traditional use of compost. Numerous studies have shown that organic amendments can reduce or suppress plant pathogens. The control of plant diseases is accomplished in two ways.

Thermophilic temperatures achieved through composting can destroy many plant disease organisms.

Compost products can themselves suppress plant disease organisms (Zorpas 1999a). The biological control is accomplished by:

I.   antibiosis: the production of fungal inhibitors
II.   nutrient competition
III. parasitism: direct attack of pathogens by non-pathogens

4.5. Effects of Compost on Soils and Crops

Mature compost is a brown-black crumbly material with an earthy smell and a C/N ratio of approximately 10:1. If applied to the soil, microorganisms continue to degrade the compost through a process called mineralization. This process takes place slowly in temperate climates and at increased rate under warm temperatures and moist, but not excessively wet conditions. In tropical conditions, high radiation loads and high soil temperatures eventually lead to the total disappearance of the compost leaving only the mineral nutrients behind. In temperate regions of the world, the mineralization reaction is much slower and a portion of the organic matter generally becomes stabilized as soil humus. As a result of the degradation kinetics under the two climatic regimes, annual applications may be required in tropical soils to achieve optimum benefits of the compost. In contrast, applications of compost to soils in temperate regions may result in benefits several years after their application (Dick and McCoy, 1993).

The addition of organic matter to soil, especially in the form of compost, results in increased mineralization of nitrogen and also micronutrients (Dick and McCoy, 1993). Application of organic by-products to soil increases soil fertility by improving the physical and chemical properties and augmenting microbial activity (and is a common way to dispose of waste). The yeast (*Saccharomyces cerevisiae*) is a byproduct obtained from the recovery, processing and drying of the yeast surplus generated during the alcoholic fermentation from sugarcane. It is a product rich in proteins, carbohydrates, minerals and vitamins. Yeast amendment stimulated CO_2 production, microbial biomass and induced higher enzyme activities than in mineral fertilized or control soils (Rezende et al., 2004). Improvements in the properties of cultivated soil and increased crop production from additions of compost to soil have been reported. Smith et al. (1992) reported that cabbage (*Brassica oleracea capitata*) and onion (*Allium cepa*) grown on a soil fertilized with 25% compost N + 75% NH_4NO_3-N had significantly higher yields than those fertilized with NH_4NO_3 alone (both treatments provided the same amount of total N). Similarly, Buchanan and Gliessman (Buchanan and Gliessman, 1991) reported that when inorganic fertilizer N was incorporated with compost, efficiency and improved crop response was higher than without compost. Higher yields of turf grass, onion, and lettuce, were observed on a sandy calcareous soil amended with a cumulative total of 37 or 74 Mg ha^{-1} of biosolids compost over a 2-year period compared to the unamended control (Bevacqua and Mellano, 1993). Application of moderate additions of three sugar beet vinasse compost to a calcareous loamy sand soil had a positive effect on plant nutrition, on soil chemical fertility (soil organic matter, humic substance and Kjeldahl-N contents, and on cation exchange capacity) and yield, without serious risks of salinization or sodification of coarse textured –or well drained- soils under irrigation (Madejón et al., 2001).

In Hawaii, Miyasaka et al. (2001) found that organic soil amendments increased taro yield (*Colocasia esculenta*), reduced soil erosion and improved soil structure, but due to the high cost of those organic inputs, the increased yield did not result in larger profits but in losses. In Portugal, Correia Guerrero et al. (1995) increased the production of fresh and dry matter lettuce (*Lactuca sativa*) planted after enriching the soil with dried orange pulp and peel wastes compost. The use of MSW compost had a moderate effect on the concentration of plant nutrients in corn (*Zea mays*) at physiological maturity. Application of 270 days MSW compost (considered mature) increased yield as MSW compost rate increased and resulted in corn grain yield near that obtained with recommended N fertilizer (Wolkowski, 2003). Manios (2004) show nutritional contents of some mature compost (Table 10), and strongly recommend the mixtures of the organic wastes to ensure a good final product.

Table 10. Nutrient content of mature composts (Manios 2004)

Compost	Nutrients (100 % dried matter)				
	N	P	K	Ca	Mg
Unmixed					
Extracted olive press cake (100%)	2.64	0.28	1.03	1.36	0.09
Olive tree leaves (100%)	2.87	0.19	0.69	8.53	0.64
Vine branches (100%)	3.74	0.29	1.4	3.17	0.61
Pig manure (100%)	2.74	0.51	0.63	11.66	1.45
Mixtures					
Olive tree branches with cucumber plant biomass (50%+50%, by volume)	2.43	1.39	2.59	12.59	1.82
Pig manure with olive tree leaves and vine branches (66%+17%+17% by volume)	2.85	0.29	0.82	11.55	0.92
Sewage sludge with green waste (33%+66%, by volume)	2.92	2.96	0.49	4.48	0.63

Table 11. Effects of compost application on sorghum (Sorghum bicolor) yield components (5 Mg ha⁻¹) in a Ferri-gleyic lixisol (Ferrudalfs) (Southern Burkina, Africa) (Ouédraogo et al., 2001)

Treatment	Grains yield (kg ha^{-1})	Number of grains/panicle	1000 grains weight (g)	Straw dry matter (kg ha^{-1})
Compost (5 Mg ha-1)	1689	4213	25.5	5145
No compost	1160	2035	23.2	4450

Table 12. Effects of compost application on sorghum (Sorghum bicolor) yield components (10 Mg ha-1) in a Ferric lixisol (Ferrudalfs) (Southern Burkina, Africa) (Ouédraogo et al., 2001)

Treatment	Grains yield (kg ha^{-1})	Number of grains/panicle	1000 grains weight (g)	Straw dry matter (kg ha^{-1})
Compost (10 Mg ha^{-1})	1380	4071	30.04	3285
No compost	408	871	31.32	2175

The applications of compost result in a significant increase in crop production, but the type of soil is important in determining the response of crop to compost (Tables 11 and 12).

4.6. Compost Effect on Soil Quality

Soil quality definition accepted by the Soil Science Society of America is "the fitness of a specific kind of soil, to function within its capacity and within natural or managed ecosystem boundaries, to sustain plant and animal productivity, maintain or enhance water and air quality, and support human health and habitation" (Karlen and Stott, 1994; Arlen and Martin 2002). The content and quality of soil organic matter are the most important factors maintaining the quality and fertility of soils. Land application of organic wastes and biosolids has some risks as it involves the use of complex residuals that can contain high levels of toxic constituents, for example, there may be greater risk of heavy metals entering the food chain via uptake and excess supply of P, K, and nitrates. One way of improving the quantity of applied wastes is composting, which yields an excellent product due to the accumulation of humus like substances produced from biochemical process that takeplace during composting. Thus, the application of composted materials to soils is expected to increase both the quantity and quality of soil organic matter (Zorpas 1999a). Monitoring soil quality procedures include biomass and respiration measurements but extended also to nitrogen mineralization, microbial diversity and functional groups of soil fauna (Schloter et al., 2003). Several parameters, including soil biological, chemical and physical properties, should be used to indicate soil quality, depending on scale and objectives of specific investigations (Gardi et al., 2002).

Common Indicators of Soil Quality

To make indicator measures meaningful, decision tools must provide agronomic and environmental interpretations, reflecting both farmer and societal values (Andrews et al., 2003).

Indicators for Biological Soil Properties

Biological properties are dynamic and sensitive to changes in soil conditions (Nortcliff, S. 2002), and represent different aspects of soil quality in different ecosystems (Elliot, 1997). Microbial biomass and activity are often used as indicators (Elliot 1997). Ratio of the respiration to biomass C, called metabolic quotient, is also used as a reliable indicator of the effects of environmental influences on the microbial population (Anderson, 2003). Specifically, consumption of O_2 or production of CO_2 by the treated soil is measured to indicate microbial activity (Anderson, 2003). Chloroform fumigation and incubation is often used as a measure of microbial biomass (Horwath and Paul, 1994). Specific enzymes and biomolecules, such as phosphatase, dehydrogenase, invertase, urease and arginine ammonification have also been suggested as indicators of soil biological activity and diversity (Dick 1994). Faunal populations and the rates of litter decomposition, which includes the interaction of vegetation, soil nutrient availability, micro – and macro-fauna and microbial populations, are also considered as biological indicators of soil quality.

Indicators for Chemical and Physical Soil Properties

Many chemical properties directly influence biological process (e.g. via nutrient and C supply). Physical – chemical processes determine (i) the capacity of soils to hold, supply and cycle nutrients (especially, C and N), and (ii) the movement and availability of water (Schoenholtz et al., 2000). As suggested by Karlen and Stott (1994), soil nutrient (N, P, K, Ca, etc.) availability, soil pH, cation exchange capacity (CEC) and electrical conductivity (EC) are used as chemical indicators. Soil bulk density, aggregate stability, total and/or available water content, and soil hydraulic conductivity can also be measured as indicators of physical soil quality. Soil texture, as considered by some authors, is a master soil property that influences most other properties and processes (Schoenholtz et al., 2000).

Compost Improves Soil Quality

Compost applications improve soil fertility over time (Clark 1998). The transition period between conventional and organic composted practices is often marked by a decrease in nitrogen (N) availability and in yields due to a shift in biological activity and N sources that are not immediately available for plant use (144). Soil Biological Properties: Among the benefits of composting is an increase in soil microbial activity and biological processes (Werner, 1997; Gunapala and Scow, 1998)

Wander et al. (1994) studied three farming systems: (i) animal-based (cover crops and animal manure only), (ii) legume based (cover crop only), and (iii) conventional (N fertilizer). Their results showed that the two organic systems had higher levels of microbial activity and more diverse species than the conventional system. Soil Physical Properties: Organic fertility inputs (animal and/or green manures) improve soil physical properties by lowering bulk

Table 13. Several Specification of the use of compost

Parameters [1,6]	Reported as (units of measure)	Specification turf establishment with Compost	Specification planting bed establishment with compost	Specification Compost as a landscape backfill mix component	Specification compost as a erosion control blanket Blanket media to be vegetated	Blanket media to be left un-vegetated	Specification compost as a soil blanket for sediment control Filter Berm to be Vegetated	Filter Berm to be left Un-vegetated
pH [2]	pH units	6.0 - 8.5	6.0 - 8.5	6.0 - 8.5	6.0 - 8.5	N/A	6.0 - 8.5	N/A
Soluble Salt Concentration [2] (Electrical conductivity)	dS/m (mmhos/cm)	Maximum 10	Maximum 10	Maximum 10	Maximum 5	Maximum 5	Maximum 5	N/A
Moisture Content	%, Wet weight basis	30 – 60	30 – 60	30 – 60	30 – 60	30 – 60	30 – 60	30 – 60
Organic Matter Content	%, dry weight basis	30 – 65	30 – 65	30 – 65	25 – 65	25-100	25 – 65	25 – 100
Particle Size	% Passing a selected mesh size, dry weight basis	98% pass through 3/4" screen or smaller	98% pass through 3/4" screen or smaller	98% pass through 3/4" screen or smaller	• 3" (75 mm), 100% passing • 1" (25mm), 90% to 100% passing • 3/4" (19mm), 65% to 100%passing • 1/4" (6.4 mm), 0% to 75% passing • Maximum particle length of 6" (152mm)	• 3" (75 mm), 100% passing • 1" (25mm), 90% to 100% passing • 3/4" (19mm), 65% to 100%passing • 1/4" (6.4 mm), 0% to 75% passing • Maximum particle length of 6" (152mm)	• 3" (75 mm), 100% passing • 1" (25mm), 90% to 100% passing • 3/4" (19mm), 70% to 100% passing • 1/4" (6.4mm), 30% to 75% passing Maximum: • particle size length of 6" (152mm) (no more than 60% passing 1/4" (6.4 mm) in high rainfall/flow rate situations)	• 3" (75 mm), 100% passing • 1" (25mm), 90% to 100% passing • 3/4" (19mm), 70% to 100% passing • 1/4" (6.4mm), 30% to 75% passing Maximum: • particle size length of 6" (152mm) (no more than 50% passing 1/4" (6.4 mm) in high rainfall/flow rate situations)
Stability3 Carbon Dioxide Evolution Rate	mg CO2-C per g OM per day	< 8	< 8	< 8			< 8	N/A

Table 13. (Continued)

Parameters [1,6]	Reported as (units of measure)	Specification turf establishment with Compost	Specification planting bed establishment with compost	Specification Compost as a landscape backfill mix component	Specification compost as a soil blanket for erosion control		Specification compost as a filter berm for sediment control	
					Blanket media to be vegetated	Blanket media to be left un-vegetated	Filter Berm to be Vegetated	Filter Berm to be left Un-vegetated
Maturity [3] (Bioassay) Seed Emergence and Seedling Vigor	%, relative to positive control %, relative to positive control	Minimum 80% Minimum 80%	Minimum 80% Minimum 80%	Minimum 80% Minimum 80%				
Physical Contaminants (inerts) [4]	%, dry weight basis	<1	<1	<1	< 8	N/A		
Chemical Contaminants	mg/kg (ppm)	Meet or exceed US EPA Class A standard, 40 CFR § 503.13, Tables 1 and 3 levels	Meet or exceed US EPA Class A standard, 40 CFR § 503.13, Tables 1 and 3 levels	Meet or exceed US EPA Class A standard, 40 CFR § 503.13, Tables 1 and 3 levels	<1	<1	<1	<1
Biological Contaminants [5] Select Pathogens Fecal Coliform Bacteria, or Salmonella	MPN per gram per dry weight MPN per 4 grams per dry weight	Meet or exceed US EPA Class A standard, 40 CFR § 503.32(a) levels	Meet or exceed US EPA Class A standard, 40 CFR § 503.32(a) levels	Meet or exceed US EPA Class A standard, 40 CFR § 503.32(a) levels				

Recommended test methodologies are provided in Test Methods for the Examination of Composting and Compost (TMECC, The US Composting Council)

It should be noted that the pH and soluble salt content of the amended soil mix is more relevant to the establishment and growth of a particular plant, than is the pH or soluble salt content of a specific compost (soil conditioner) used to amend the soil. Each specific plant species requires a specific pH range. Each plant also has a salinity tolerance rating, and maximum tolerable quantities are known. Most ornamental plants and turf species can tolerate a soil/media soluble salt level of 2.5 dS/m and 4 dS/m, respectively.

Seeds, young seedlings and salt sensitive species often prefer soluble salt levels at half the afore mentioned levels. When specifying the establishment of any plant or turf species, it is important to understand their pH and soluble salt requirements, and how they relate to existing soil conditions.

Stability/Maturity rating is an area of compost science that is still evolving, and as such, other various test methods could be considered. Also, never base compost quality conclusions on the result of a single stability/maturity test.

US EPA Class A standard, 40 CFR § 503.13, Tables 1 and 3 levels = Arsenic 41ppm, Cadmium 39ppm, Copper 1,500ppm, Lead 300ppm, Mercury 17ppm, Molybdenum 75ppm, Nickel 420ppm, Selenium 100ppm, Zinc 2,800ppm.

US EPA Class A standard, 40 CFR § 503.32(a) levels = Salmonella <3 MPN/4grams of total solids or Fecal Coliform <1000 MPN/gram of total solids.

Landscape architects and project (field) engineers may modify the allowable compost specification ranges based on specific field conditions and plant requirements.

density, increasing water-holding capacity, and improving infiltration rates (Werner, 1997). Lower bulk density implies greater pore space and improved aeration, creating a more favorable environment for biological activity (Werner, 1997).

Soil Chemical Properties

Compost increases soil organic matter (SOM) content (Clark et al., 1998; Zorpas 1999a; Petersen et al. 1999). Alvarez et al. (1988) found a positive correlation between SOM content and available Ca, K, Mg, Na, and P. Wong et al. (1999) found that manure compost can be an alternative to chemical fertilizer to restore the soil nutrient balance; it increases the soil macro and micronutrient for plant growth.

4.7. Potential Markets

The success of a composting operation will depend greatly on the market developed for the product. It is important to appraise the value of the compost for its potential uses; both beneficial and detrimental characteristics should be considered. A realistic evaluation of the potential market relative to the amount of compost produced is especially important. Some municipalities may find it advantageous to distribute compost to consumers at no cost, since this may be a least-cost option to the municipality.

Livestock manures and their composts, peat, topsoil, and chemical fertilizers already hold significant portions of the potential market. There are, however, possibilities for increasing the market by developing new uses. The potential market can be classified into three broad categories: (i) a very high-profit, but usually small, market for intensive plant culture practices (luxury garden market); (ii) a market for restoration of disturbed lands by mixing compost into the unproductive soil of strip mines, gravel pits, road construction sites and areas of urban or suburban development, and (iii) a market for use as a fertilizer-soil conditioner for farm crops.

Some promotional effort will be necessary to distribute a steady volume of compost in any of the above markets. Since the cost of composting is very competitive with the cost of alternative disposal practices even with no credit for value of the compost, it will rarely be essential to show a profit on marketing. However, it will be necessary to develop a market sufficient to handle the planned production.

Table 13 presents several specification of the use of compost, which depends from the final application.

4.6. Beneficial Effects as a Fertilizer and Soil Conditioner

Sludge compost applied at a rate to supply the nitrogen requirements of the crop will supply most of the plant nutrients except potassium, thus it may be necessary to apply supplemental potash. However, it is unlikely that sewage sludge composts will be used to supply the total nutrient requirements of agronomic crops because of the large amounts that would have to be applied.

The maximum value is realized when they are employed in combination with inorganic fertilizers; they partly meet the crop's nutrient requirements and also serve as a valuable organic soil conditioner for maintaining soil productivity.

Nearly all of the nitrogen in sewage sludge compost is in the organic form and must be mineralized to inorganic ammonium or nitrate before it is available for crops. Research indicates the compost from the Beltsville Aerated Pile Method mineralizes only about 10% of the organic nitrogen (N) during the first cropping period after the compost is applied. Thus, sludge compost can indeed be considered as a slow-release N fertilizer.

The application of sludge compost alone, at fertilizer rates (i.e., the N requirement of the crop), to marginal soils can produce significantly higher yields than cormnercial fertilizers applied alone at the same N level. Attributed to an improvement in soil physical properties by the compost, properties, as evidenced by enhanced aggregation, increased soil aeration, lower bulk density, less surface crusting, and increased water infiltration, water content, and water retention.

Sludge compost added to sandy soils will increase the moisture available to the plant and reduce need for irrigation. In heavy textured clay soils, the added organic matter will increase permeability to water and air, and increase water infiltration into the profile, thereby minimizing surface runoff. The soils also will have a greater water storage capacity. Addition of sludge compost to clay soils has also been shown to reduce compaction (i.e., lower the bulk density) and increase root development and depth.

The higher yields are Sludge compost is known to improve soil physical. Large quantities of the sludge compost produced at Beltsville have been mixed with subsoil and used successfully as a topsoil substitute. A number of public agencies, including the National Capital Park Service and the Maryland State Park Service, have used the compost for land reclamation and development research at Beltsville indicates that sewage sludge compost can be used to great advantage in the commercial production and establishment of turfgrasses, trees, and ornamental plants. Plants and turfgrasses produced with sludge compost were of better quality, had developed more extensive root systems, were transplanted with lower mortality, and were marketable earlier than those grown with inorganic fertilizer alone. It is likely that large amounts of sludge compost will eventually be used on golf courses and cemeteries, and for landscaping the grounds of public buildings. In addition to the above uses, sludge compost has a major potential for use in the revegetation and reclamation of lands disturbed by surface mining, by removal of topsoil, and by excavation of gravel deposits.

4.7. Significant Disadvantages for the Utilization of Compost

4.7.1. Pathogens

Because of its origin, the public from the standpoint of esthetic or health aspects may not readily accept compost made from sewage sludge. Esthetic reservations will generally be dispelled upon direct examination of the compost itself, which the observer will perceive as being free from offensive odors and having the appearance of a highly fertile soil. Reassurance on health safety may depend more on an explanation of the composting process and the high temperature disinfections involved, reinforced perhaps by approval from local health authorities. If composting is properly done, as described in this manual, it destroys or reduces to insignificant levels all primary pathogens present in sewage sludges. Once destroyed, viruses, helminths, protozoans, and most bacteria will not repopulate the compost, since they cannot grow external to their hosts. Salmonella, one of the most common organisms causing food poisoning, can regrow to a limited extent in the finished compost, but it does not compete well with other microorganisms present.

Salmonella bacteria are frequently found in the environment. They are often present in fecal material of wild and domestic birds and animals, including pets, and have been isolated from streams and vegetation of mountainous areas remote from human population centers. No significant hazard should be associated with their presence in compost so long as the compost does not come into contact with food.

4.7.2 Metals

Many sewage sludges contain large amounts of heavy metals (Zorpas et al., 1998, 1999a, 2000, 2008a, 2008b), which may reduce the value as a fertilizer for either direct application to land or for composting. Excessive amounts of these metals are often found in sludges where industrial effluent is discharged into the sanitary sewers without pretreatment.

Application of high metal sludges on land results in soil enrichment in heavy metals. Experiments on sludge application have shown that soil enrichment by zinc, copper, and nickel can cause direct phytotoxic effects manifested as decreased growth and yield, especially where soil pH is low (pH 5.5) and rates of application are high. Heavy metals may also accumulate in plant tissues and enter the food chain through direct ingestion by hwans or indirectly through animals.

The element of greatest concern to human health where sewage sludges and sludge composts are applied to land is cadmium (Cd), since it is readily absorbed by most crops and is not generally phytotoxic at the concentrations normally encountered.

Therefore, Cd can accumulate in plants and enter the food chain more readily than, for example, lead (Pb) or mercury (Hg), which are not readily absorbed and translocated to the edible portion of crops. However, since mushrooms accumulate Hg, sludge compost should not be used for mushroom culture. Most human exposure to Cd comes from food (principally grain products, vegetables, and fruits) and results in an accumulation of the element in the liver and kidneys. Humans retain approximately 3 to 5 percent of dietary Cd. If dietary Cd is substantially increased over long periods of time, Cd can accumulate to levels that might be expected to cause kidney injury. Among the sources that contribute to the level of Cd in food are (a) soils and surface waters contaminated by disposal of wastes, (b) soils inherently high in Cd because of geochemical factors, (c) industrial fallout, and (d) phosphatic fertilizers containing Cd, and (e) industrial contamination of soil and/or food.

The World Health Organization (WHO) has recommended that the maximum level of dietary Cd should not exceed 70 pg/person/day. Woirkers in the U.S. Food and Drug Administration (FDA) advance the view that any further increase in dietary intake of Cd seems undesirable, However, the extent of the hazard is debated because current data are incomplete about human exposure to Cd and risk through their lifetime. Thus, in order to limit this risk, the utilization of organic wastes on land is restricted by regulatory agencies to control the level of Cd in food chain crops. Plant species, as well as varieties, have been found to differ markedly in their ability to absorb and translocate heavy metals, to accumulate them within edible organs of the plant, and to resist their phytotoxic effects. Leafy vegetables are usually sensitive to the toxic effects of metals and accumulate them; cereal grains, corn, and soybeans are less sensitive; and rasses are relatively tolerant. Uptake studies with corn, soyfean , and cereal grains have shown that heavy metals accumulate less in the edible grain than in the leaves; similar results are found for edible roots, as radish, turnip, carrot, and potato, and fruits, as tomato, squash, etc.

The availability to and uptake of heavy metals by plants are influenced by certain chemical and physical properties of soil, especially pH, organic matter content, cation exchange capacity (CEC), and texture (i.e., the proportions of sand, silt, and clay).

Phytotoxicity and plant availability of sludge-borne metals are higher in acid soils than in those with neutral or alkaline pH. Maintaining soil pH in the range of 6.5 or above by liming reduces the availability of heavy metals to plants. Application of organic amendments such as manures and crop residues can also decrease the availability of heavy metals to plants. The CEC is a measure of the soil's capacity to retain cations; higher CEC is usually associated with higher clay and organic matter contents.

Heavy metals are generally less available to plants in soils of high CEC (e.g., organic matter rich soils or clay loams) compared with soils of low CEC'(e.g., loamy sands).Recent research (Zorpas et al., 1999a - 2008b) suggests that, on a total metal basis, heavy metals are less available to plants in composted sewage sludges than in uncomposted raw and digested sludges, although the reason for this is not yet known, and also the use of natural zeolite clinoptilolites has the ability to remove heavy metals from sludges

REFERENCES

Adani, F., Genevini, P.L., Gasperi, F., and Tambone, F. 1999, *Compost Sci. Util.,* 7, 24.

Adams, M.L., Zhao, F.J., McGrath, S.P., Nicholson, F.A., and Chambers, B.J. 2004,*J. Environ. Qual.,* 33, 532.

Allison, F.E. 1973, *Soil organic matter and its role in crop production.* Elsevier Scientific Publ. Company, Amsterdam.

Anderson, T.H. 2003, Microbial eco-physiological indicators to asses soil quality. *Agri. Ecosystems Environ.,* 98, 285

Arshad, M. A., and Martin, S. 2002, Identifying critical limits for soil quality indicators in agro-ecosystems *Agri, Ecosys. Environ.,* 88, 153.

Alvarez, C.E., Garcia, C., and Carracedo, A.E. 1988, Biol. Agric. Hort., 5, 313.

B.C. Agriculture and Food. 1996, *Composting fact sheet.* British Columbia Ministry of Agriculture and Food, Abbotsford, BC, Canada.

Bevacqua, R.F., and Mellano, V.J. 1993, *Compost Sci. Util.,* 1(3), 34.

Bergkvist, P., and Jarvis, N. 2004, *J. Environ. Qual.,* 33, 181.

Borken, W., Su, Y.J., and Beese, F. 2004, *J. Environ. Qual.,* 33, 89

Brinton, W.F. 2000, *Compost quality standards and guidelines: an international view.* Woods End Research Laboratory Inc., ME.

Buckingham, D.A., and Plachy, J. 2004, *Cadmium statistics.* USGS.

Buchanan, M, and Gliessman, S.R. 1991, *BioCycle,* 32(12), 72.

Buyuksonmez, F., Rynk, R., Hess, T.F., and Bechinski, E. 1999, *Compost Sci. Util.,* 7, 66.

Bezdicek, D., Fauci, M., Caldwell D., Finch, R., and Lang, J. 2001, *BioCycle* 42(7), 28.

Bugbee, G.J., and Saraceno, R.A. 1994, *Bull. Environ. Contam. Toxicol.,* 52, 606.

Canet, R., Pomares, F., Albiach, R., Tarazona, F., Ibanez, M.A., and Ingelmo, F. 2000, *BioCycle*, 41 (12), 72.

CCC (Composting Council of Canada). 1999, *Compost standards review.* Composting Council of Canada.

Chefetz, B., Adani, F., Genevini, P., Tambone, F., Hadar, Y., and Chen, Y. 1998, *J. Environ. Qual.*, 27, 794.

Chefetz, B., Hatcher, P.G., Hadar, Y., and Chen, Y. 1996, *J. Environ. Qual.*, 25, 776.

Clark, M.S., Horwath, W.R., Shennan, C., and Scow, K.M. 1998, *Agron. J.*, 90, 662.

Correia Guerrero, C., Carrasco de Brito, J., Lapa, N., and Santos Oliveira, J.F. 1995, Re-use of industrial orange wastes as organic fertilizers, *Bioresour. Technol.*, 53, 43.

Dick, W.A., and McCoy, E.L. 1993, *Science and engineering of composting*: Design, environmental, microbiological and utilization aspects, Hoitink, H.A. and Keener, H.M. (Eds.), Ohio Agricultural Research and Development Center. The Ohio State University, 622.

Dick, R. P.1994, Defining soil quality for a sustainable environment, J.W. Doran, D.C. Coleman, D.F. Bezdicek, and B.A. Stewart (Eds.), *Soil Sci. Soc. Am.* Publ. 35. Madison, WI, 107.

Domeizel, M., Khalil, A, and Prudent, P. 2004, UV spectroscopy: a tool for monitoring humification and for proposing an index of the maturity of compost *Bioresour. Technol.*, 94, 177.

Elliot, E.T. 1997, *Biological indicators of soil health,* C. Pankhurst, B.M. Doube, and V.V.S.R. Gupta (Eds.), CAB International, NY, 49.

Epstein, E. 1996. *The Science of composting.* Technomic Publishing Company, Inc., Lancaster, PA

Epstain Eliot. 1997. *The Science of Composting.* Technomic Pup. Co, Pennsylvania, USA.

Epstein, E., Keane, D.B., Meisinger, J.J., and Legg, J.O. 1978, *J. Environ. Qual.*, 7, pp. 217.

Eghball, B., Power, J.F. Gilley, J.E., and Doran, J.W. 1997, *J. Environ. Qual.*, 26, 189.

Gardi, C., Tomaselli, M., Parisi, V., Petraglia, A., and Santini, C. 2002, Soil quality indicators and biodiversity in northern Italian permanent grasslands *Eur. J. Soil iol.*, 38, 103.

Gonzalez-Vila, F.J., Almendros, G., and Madrid. 1999, Molecular alterations of organic fractions from urban waste in the course of composting and their further transformation in amended soil, *The Sci. Total Environ.* , 36, 215.

Goonan, T.G., and Papp, J.F. 2003, *Chromium statistics*, USGS.

Goonan, T.G., and Kuck, P.H. 2004, *Nickel statistics*, USGS.

Giusquiani, P.L., Gigliotti, G., and Businelli, D. 1992, *J. Environ. Qual.*, 21, 330.

Gunapala, N., and Scow K.M. 1998, Dynamics of soil microbial biomass and activity in conventional and organic farming systems, *Soil Biol. Biochem.*, 30, 805.

DiFrancesco, C.A., and Smith, G.R. 2003*, Lead statistics,* USGS.

Francesco, C.A., and Plachy, J. 2004, *Zinc Statistics*, USGS.

Hammouda, G.H.H., and Adams W.A. 1986, Compost: production, quality and use, M. De Bertoldi, M.P. Ferranti, P. L'Hermite and F. Zucconi (eds.), *Elsevier Applied Science,* NY, 245.

Hansen, R.C., Keener, H.H., Marugg, C., Dick, W.A., and Hoitink, H.A.J. 1993, *Science and engineering of composting: design, environmental, microbiological and utilization* aspects, H.A.J. Hoitink and H.M. Keener (Eds.), Renaissance Publ., Worthington, OH, 131.

Hadas, A., and Portnoy, R. 1994, *J. Environ. Qual.,* 23, 1184.

He, X.T., Traina, S.J., and Logan, T.J. 1992, J. Environ. Qual., 21, 318.

Henry, C.L., and Harrison, R.B. 1992, *Biogeochemistry of trace metals*, D.C. Adriano (ed.), Lewis Publishers, Ann Arbor, MI., 195.

Hsu, J.H., and Lo, S.L. 2000, *J. Environ. Qual.*, 29, 447.

Horwath, W.R., and Paul, E.A. 1994, Methods of soil analysis, part 2: icrobiological and biochemical properties, R.W. Weaver, S. Angle, P. Bottomley, D. Bezdicek, S. Smith, A. Tabatabai, and A. Wollum, (Eds.), *Soil Sci. Soc. Am.,* Madison, WI. 753.

Hogg, D., Barth, J., Favoino, E., Centemero, M., Caimi, V., Amlinger, F., Devliegner, Brinton, W., W., and Antler S. 2000, *Review of compost standards in the United States.* The Waste and Resource Action Programme. Banbury, UK.

Inbar, Y., Chen Y., and Hadar, Y. 1989, *Soil Sci. Soc. Am. J.,* 53, pp 1695.

Inbar, Y., Chen, Y., and Hadar, Y. 1990, *J. Environ. Qual.*, 54, 1316.

Jimenez, E.I., and Garcia, V.P. 1989, *Biol. Waste,* 27, 115.

Jorge Mendoza, Tatiana Garrido, Gabriela Castillo, Nilsa San Martin. Metal availability and uptake by sorghum plants grown in soils amended with sludge from different treatments. Chemosphere, Volume 65, Issue 11, December 2006, Pages 2304-2312

Karlen, D.L., and Stott, D.E. 1994, *Defining soil quality for a sustainable environment,* Doran, J.W., Coleman, D.C., Bezdicek, D.F., and Stewart, B.A. (Eds.). Soil Sci. Soc. Am. Publ. 35. Madison, WI, 53.

Kuo, S., Huang, B., and Bembenek, R. 2004, *Soil Sci.*, 169:363.

Kuo, S., Sainju U.M., and Jellum, E.J. 1997, *Soil Sci.Soc. Am. J.*, 61, 1392.

Kuo, S., and Sainju, U.M. 1998, *Bio. Fertil. Soils*, 26,346.

Liu, J. 2000, *Composting and use of compost as a soil amendment.* Doctoral Dissertation. Natural Resources and Environmental Management Department, University of Hawaii. 138 p.

Liao, P.H., Vizcarra, A.T., Chen, A., and Lo, K.V. 1995, *Compost Sci. Util* 3, 80.

L'Herrous, L., Le Rous, S., Apprious, P., Martinez, J. 1997, *Envrion. Pollu.,* 97

Leita, l., and de Nobili, M. 1991, *J. Environ. Qual.,* 20, 73.

Lutzow von, M., Leifeld, J., Kainz, M., Kogel-Knabner, I., and Much, J.C. 2002, *Geoderma,* 105, 243.

Martins, O., and Dewes T. 1992, *Bioresour. Technol.*, 42, 103.

Manios, T. 2004, *Environ. International,* 29, 1079.

Madejón, E., López, R., Murillo, J.M., and Cabrera, F. 2001, *Agri, Ecosyst.Environ.*, 84, 55.

Miyasaka, S., Hollyer, J., and Cox, L.J. 2001, Impacts of organic inputs on taro production returns. *Soil and crop management.* CTAHR, University of Hawaii.

Mullins, G.L., Martens, D.C., Miller, W.P., Kornegay, E.T., Hallock, D.L. 1982, *J. Environ. Qual.*, 11, 316.

Murillo, J.M., Cabrera, F., Lopez, R., and Martin-Olmedo, P. 1995, *Agri. Ecosys. Environ.* 54, 127.

Manser, A.R.G., and Keeling, A.A. 1996, *Practical handbook of processing and recycling municipal waste*, A.G.R. Manser (Ed.), Lewis Publ., Boca Raton, FL, 279.

Miltner, E., Bary A.D., and Cogger, C. 2003, *Compost Sci. Util.*, 11, 289.

Michel, F.C. Jr., Reddy, C.A., and Forney, L.J. 1997, Fate of carbon-14 diazinon during the composting of yard trimmings. *J. Environ. Qual.*, 26, 200.

McBride, M.B. 1995, *J. Environ. Qual.*, 24, 5.

McBride, M. 2002, *Soil Sci.*, 167, 62.

Nortcliff, S. 2002, *Agric. Ecosys. Environ.*, 88, 161.

Ouédraogo, E., Mando, A., and Zombré, N.P. 2001, *Agri., Ecosys. Environ.*, 84, 259.

Palm, C. A., Gachenco, C. N., Delve, R. J. Cadish, G., and Giller, K. E. 2001, *Agri., Ecosys. Environ.*, 83, 27

Petersen, C., Drinkwater, L.E., and Wagoner, P. 1999, T*he Rodale Institute Farming Systems Trial: The first fifteen years,* Kutztown, PA.

Petruzzelli, G., Lubrano, L., and Guidi, G. 1989, *Plant Soil,* 116, 23.

Porter, K.E., and Edelstein, D.L. 2003, *Copper Statistics.* USGS.

Pueyo, M., Sastre, J., Hernandez, M., Vidal, M., Lopez-Sanchez, and Rauret, G. 2003, *J. Environ. Qual.*, 32, 2054.

Rezende, L.A., Assis. L.C., and Nahas, E. 2004, *Bioresour. Technol.,* 94, 159.

Reynolds, W.D., Elrich, D.E., Young, E.G., Amoozegar, A., Booltink, H.W.G., and Bouma, J. 2002. Methods of soil analysis, part 4: physical methods, J.H. Dane and G.C. Topp, (Eds.) *Soil Sci. Soc.* Am. No. 5. Madion, WI, 797.

Sanchez-Monedero, M.A. 2001, *Biodegradation,* 13, 361.

Schloter, M., Dilly, O., and Munch, J.C. 2003, *Agri. Ecosys. Environ.*, 98, 255

Schoenholtz, S.H., van Miegroet, H., and Burger, J.A. 2000, *Forest Ecol. Manage.*,138, 357

Sikora, L.J., and Yakovchenko V. 1996, *Soil Sci. Soc. Am. J.*, 60, 1401.

Sims, J.T., Skline, J.S., 1991. Chemical fraction and plant uptake of heavy metals in soil amended with co-composted sewage sludge. J. Environ. Qual. 20, 387-395.

Simeoni, L.A., Barbarick K.A., and Sabey, B.B. 1984, *J. Environ. Qual.,* 13, 264.

Socolow, R., and Thomas, V., 1997, 1(1), 13.

Smith, S.R., Hall, J.E., and Hadley, P. 1992, *Acta Hort.*, 302, 203.

Tesier, A., Campell, P.G.C, Bisson, M., 1979. *Sequential Extraction Procedure for the Speciation of Particulate Trace Metals. Analytical Chemistry* 51(7), 844-851.

Tisdell, S.E., and Breslin V.T. 1995, *J. Environ. Qual.*, 24, 827.

USDA. 2000, Composting. Part 637, *National Engineering Handbook*, NRCS, U.S. Department of Agriculture, Washington, D.C.

USEPA. 1993. *Fed. Regist.*, 58, 9248.

Wander, M.M., Traina, S.J., Stinner, B.R., and Peters, S.E. 1994, Soil Sci. Soc. Am. J. 58, 1130.

Werner, M.R. 1997, *Appl. Soil Ecol.*, 5, 151

Wei, Y.S., Fan, Y.B., Wang, M.J., and Wang, J.S. 2000, *Resources conservation and recycling,* 30, 277.

Wolkowski, R.P. 2003, *J. Environ. Qual.*, 32, 1844.

Wong, J.W.C., Ma, K.K., Fang, K.M., and Cheung, C. 1999, *Bioresour. Technol.*, 67, 43.

Wu, L., Ma, L.Q., and Martinez, G.A. 2000, *J. Environ. Qual.*, 29, 424.

Yang, H. S., and Hansen, B.H. 1997, European J. Agron., 7, 211.

Zhang, M.K., He, Z.L., Stoffella, P.J., Calvert, D.V., Yang, X.E., Xia, Y.P., Wilson, S.B. 2004, *J. Environ. Qual.*, 33, 373.

Zheljazkov, D.Z., and Warman P.R. 2004, *J. Environ. Qual.*, 33, 542.

Zorpas A. Antonis, A.G.Vlyssides, M.Loizidou, 1998, *Fresenious Environmental Bulletin*, Vol. 7, pp. 502-508.

Zorpas A. Antonis 1999a, D*evelopment of a Methodology for the composting of Sewage Sludge Using Zeolites,* PhD, National Technical University of Athens , Greece

Zorpas A. Antonis, Apostolos G. Vlyssides and Maria Loizidou. 1999b. Dewater Anaerobically Stabilized Primary Sewage Sludge Composting. Metal Leachability and Uptake by Natural Clinoptilolite. Communications in Soil Science and Plant Analysis, Vol. 30, No 11/12, pp 1603-1614

Zorpas A.A, Stamatis V, Zorpas A. G., Vlyssides A.G, Loizidou M. 1999c, *Fressenious Environmnetal Bulletin,* Vol. 8, 3-4, pp 154-162.

Zorpas, A.A., Constantinides, T., Vlyssides, A.G., Haralambous, I., and Loizidous, M. 2000, *Bioresour. Technol.,* 72, 113

Zorpas Antonis, Dimitris Arapoglou, Karlis Panagiotis, 2003, *Waste Management,* Vol. 23, pp. 27-35

Zorpas A. Antonis, Maria Loizidou, 2008a, *Bioresource Technology.* Vol 99 (16) November, 7545-7552

Zorpas A. Antonis, 2008b *Dynamic Soil – Dynamic Plants,* Global Science Book 2 (2) pp 103-109.

Zorpas Antonis, 2010. Heavy Metals Leachability Before, During and After Composting of Sewage Sludge with Natural Clinoptilolite. Dessalination and Water Treatmnt, In Press

In: Sewage Sludge Management   ISBN: 978-1-61324-393-0
Editors: A. A. Zorpas and V. J. Inglezakis   © 2012 Nova Science Publishers, Inc.

Chapter 9

ANAEROBIC TREATMENT OF SEWAGE SLUDGE

Christos Venetis

Waste Management Engineer,
Associate of EPTA Ltd. Environmental Engineers – Consultants,
Waste Treatment Technologies Department, Athens, Greece

Abstract

Waste water treatment plants (WWTP) generate sludge as a by-product of biological, chemical and physical processes implemented during wastewater treatment. Sludge disposal represents 50% of the current operating costs of a WTTP. Anaerobic digestion of sewage sludge occurs with the purpose of its stabilization and is the most common method used, although in small WWTP aerobic stabilization is a competitive option as well. Nevertheless digestion is usually the most favorable option from an economical and ecological point of view for plants serving population equivalents higher than 25.000 ± 10.000 and it constitutes an important element of any modern WWTP. Anaerobic digestion is a complex treatment method where several parameters related to microbiology, plant equipment, process design and biogas utilization interact with each other and have to be taken into consideration in order to be able to treat sewage sludge efficiently and within an economic viable concept. This is often difficult since those parameters are interrelated and changes in one condition may directly or indirectly affect others.

The main component of a sewage sludge digestion system is the anaerobic reactor which can have different configurations from an equipment (standard or high rate), temperature (meso- or thermophilic) and design (mostly egg or cylidrical shaped) point of view. During the operation of high rate digesters certain operations such as mixing, inoculation, heating and uniform feeding are essential in order to achieve sufficient degradation rates of the organic compounds and a steady production of biogas which will allow its effective utilization. Apart from treating sewage sludge the existing infrastructure can be also used for treating source separated bio-waste in the form of co-digestion, in order to enhance biogas production rates and increase the economic viability of the plant.

The abovementioned aspects are described in this chapter so as to provide the reader with the basic concept of sewage sludge anaerobic digestion.

1. INTRODUCTION

Waste water treatment plants (WWTP) generate sludge as a by-product of biological, chemical and physical processes used during wastewater treatment. The quantity, composition and characteristics of the sludge depend on the original wastewater contamination level and the treatment process implemented. Current daily amounts vary from 50 to 90 g per population equivalent and the water content ranges from 95-99% again depending on the process used in the WWTP. Sewage sludge contains also mineral and organic components while toxic substances and pathogens are also present. Because of those characteristics it cannot be directly utilized or disposed before it has undergone a certain treatment. Anaerobic digestion is an important treatment method especially with a view to agricultural use or other land applications.

There are three main sludge streams produced in a WWTP:

1.   Primary sludge

During the mechanical pre-treatment step primary sludge occurs through the physical process of sedimentation. It possesses a total solids concentration between 2-7%, it is highly putrescible and generates an unpleasant odor if it is stored untreated.

2.   Secondary sludge

After the primary settling step follows the biological process step during which secondary sludge and activated sludge are produced as a consequence of the microbiological activity. The latter is recirculated to the aeration tank in order to maintain the desired level of micro-organisms concentration while the secondary sludge stream is combined with primary sludge.

3.   Tertiary

Through the use of precipitants tertiary sludge is produced. Precipitants are either used in a distinct treatment step or during the sedimentation or the biological process.

A typical WWTP process with an anaerobic digestion step is presented in the next figure [17].

The sum of all sludge streams produced in a WWTP prior to their stabilization is referred to as raw sludge and its disposal is a very important issue representing 50% of its operating costs [3]. Anaerobic digestion of sewage sludge occurs with the purpose of its stabilization and is the most common method used, although in small WWTP aerobic stabilization is a

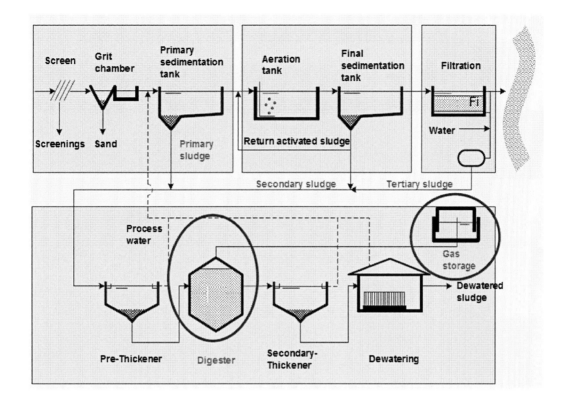

Figure 1. WWTP with an integrated anaerobic digestion step.

competitive option as well. Nevertheless digestion is usually the most favorable option from an economical and ecological point of view for plants serving population equivalents of 25.000 ± 10.000 [2] and it constitutes an important element of any modern WWTP.

Sludge anaerobic stabilisation can be characterized through the following main objectives:

- Reduction of the organic dry substance by 50% [2]
- Reduction of the total dry substance by 33% [2]
- Enhancement of the sludge characteristics in respect to its dewaterability
- Reduction of pathogenic germs
- Production of energy through biogas exploitation.
- Reduction of odours

From an ecological and economical point of view the production of biogas is a very important aspect of sewage sludge digestion since it represents a usable source of energy. In addition through the reduction of the organic dry mass the amount of sludge to be disposed is reduced and this again has a positive impact to the final disposal costs.

Figure 2. Largest sewage sludge digesters in the world in the WWTP of Emschermündung Germany (3x16.700 m^3).

2. PARAMETERS IN ANAEROBIC DIGESTION

Anaerobic digestion consists of a series of chemical reactions that lead to the decomposition of organic materials in the absence of oxygen. The full process occurs in four steps:

1. Hydrolysis

During this stage complex organic substances are broken down to into their constituent parts by hydrolytic bacteria making them thus more easily usable by the acidogenic bacteria. Proteins are converted to amino-acids, fats into long-chain fatty acids and carbohydrates into sugar. The hydrolysis rate depends on substrate availability, bacterial population, temperature and pH and is a slow process generally limiting the overall rate of the anaerobic digestion process.

2. Acidogenesis

Hydrolysis is followed by the acid-forming phase of acidogenesis. In this stage, acidogenic bacteria turn the products of hydrolysis into simple organic compounds, mostly short chain (volatile) acids, ketones and alcohols. Depending on the reactor operating conditions the percentage of the hydrolysis products that is transformed directly to

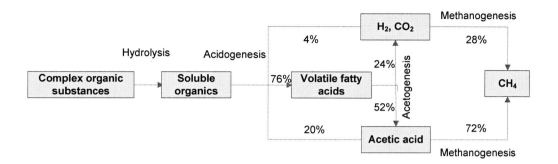

Figure 3. The Anaerobic Digestion Process.

methanogenic substances such i.e., H_2, CO_2 and acetate varies. The remaining is transformed to intermediate products such as volatile fatty acids (VFAs) which are further degraded by acetogen bacteria during the next step.

1. Acetogenesis

This stage is often considered with acidogenesis to be part of a single acid forming stage. Acetogenesis occurs through carbohydrate fermentation, through which acetate is the main product. The result is a combination of acetate, CO_2 and H_2.

2. Methanogenesis

In the last stage methanogenic bacteria convert the organic acids into methane and CO_2 and only then are organic materials removed as the produced biogas desorbs from the liquid face. The methanogenic bacteria are very sensitive to pH and temperature changes.

During the process is has to be ensured that all bacteria populations are working under optimal conditions. As a consequence the affecting process parameters have to be monitored and controlled in order to enhance microbiologic activity. The most important parameters are briefly presented in the next sections.

Temperature

Temperature is one of the most important parameters affecting the process of anaerobic digestion in respect to bacterial growth rates and affects the relationship between solid retention time and digester performance. Methane production is favoured in mesophilic (30-40 °C) and thermophilic (45-60 °C) temperature ranges. Although the selection of the initial design operating temperature is important, maintaining a stable temperature is of greater importance. Especially methanogenic bacteria are very sensitive to temperature changes. An increased temperature has several benefits including enhanced biological and chemical reaction rates.

pH

The growth of the micro-organisms is closely linked to optimal pH-values. Each group of bacteria has a different optimal pH range. For the methanogenic bacteria the optimal value lies between 6,5-7,2 while fermentative bacteria can function in a range between 4-8,5. Generally pH values during anaerobic digestion range from 5,5-8,5. Volatile acids produced in the acid forming phase tend to reduce the pH. This reduction is countered by methane formers which produce alkalinity in the form of carbon dioxide, ammonia and bicarbonate. Falling pH is an indication for acid accumulation which can result to process failure clearly depicted in the gas production.

Solids and Hydraulic Retention Time

In order to ensure efficient degradation of the volatile solids the sewage sludge has to remain for sufficient time inside the digester. The parameters describing the retention time are the solids retention time (SRT), which is the time the solids are held into the digester and the hydraulic retention time (HRT) which is the time the liquid sludge is held into the digester. In anaerobic digestion systems without recycle SRT and HRT are equal.

The reactions inside the digester are directly related to the SRT. Increasing the SRT is translated to an increase of the reactions and vice versa. When sludge is withdrawn from the system accordingly the bacterial population is reduced and the existing bacterial growth rate has to compensate for the losses in order to keep the population steady and avoid a process failure.

Inhibitory and Toxic Materials

A variety of inorganic and organic wastes can cause toxicity or inhibition in anaerobic digesters. Inhibitory substances affect the bacterial kinetic parameters in a reversible way while toxic substances cause the bacterial population to decline. Any substance can be inhibitory or toxic to the process at a certain concentration level. Ammonia, heavy metals, light metal cations, oxygen, short chain organic acids and sulfides are of main concern.

Heavy Metals

Industrial wastewater treated with municipal wastewater is the main source accounting for the heavy metal content in sewage sludge. Industrial contaminants include zinc, copper, chromium, nickel, cadmium and lead. High concentrations of metals in sludge affect disposal options and costs. Certain traces of heavy metals can have a stimulating effect on the activity of the micro organisms. The limits between stimulation, inhibition and toxicity depend on the type of heavy metals, their concentration and on the chemical and physical parameters present inside the reactor [4].

Oxygen

The presence of oxygen inside the digester causes the reaction rate of the methanogenic bacteria to decline while a large share of the acidogenic bacteria can also grow in the presence of oxygen.

Sulfides

Sulfides are either present to the wastewater stream, produced within the digester through the reduction of sulfates and other sulfur containing inorganic compounds or are the degradation products of sulfur containing organic substances. Sulfur compounds are present in many industrial processes and when industrial wastewater is treated together with municipal waste water then increased sulfur concentrations are expected to be found in the sewage sludge. Inhibition occurs through the competition of sulfate reducing and methanogenic bacteria which use the same substrates. Additionally non-dissociated hydrogen sulphide is toxic for both methanogens and sulphate reducers [3]. If during the planning phase of an anaerobic digester it is known that high contents of sulfur compounds are to be expected then a two phase digestion can offer significant advantages. However pilot tests have to be performed first before this concept is put to realisation [4].

Ammonia

Ammonia originates from the degradation of nitrogenous matter such as proteins and urea. Ammonium ion (NH_4^+) and free ammonia are the two principal forms of inorganic ammonia nitrogen in aqueous solution. Ammonium ions are used by bacteria as a nutrient source for nitrogen [5] while free ammonia is considered to be the main reason for inhibition since it is freely membrane-permeable.

Organics

Short chain organic acids are a by-product of the non-methanogenic phase of the process. In a stable running digestion system the concentration of organic acids is kept low through their degradation by the methanogenic bacteria. If the acid formation exceeds the methane-formation rate then this has an inhibiting impact to methanogens who are not able to remove volatile organic acids fast enough. As a result an acid accumulation is observed and the pH decreases to a low value. The undissociated part of the acids which depends on the pH level is responsible for the inhibition effect [4].

Light metal cations

Light metal cations including sodium, potassium, calcium and magnesium are present in the influent of anaerobic digesters. They are either released with the degradation of organic materials or added directly through addition of alkaline materials for pH control. Another source is the addition of lime for phosphorous removal. Moderate concentrations can have a stimulating effect on bacterial growth while exceeding certain limits can cause inhibition or toxicity.

3. METHODS FOR THE ANAEROBIC DIGESTION OF SEWAGE SLUDGE

Standard Rate (Cold) Digester

This method is the simplest and oldest form of AD and is carried out as a single stage process. The retention time is around 30 to 60 days. The cold digestion requires a large digester volume and for that reason it is not being implemented in new facilities any more

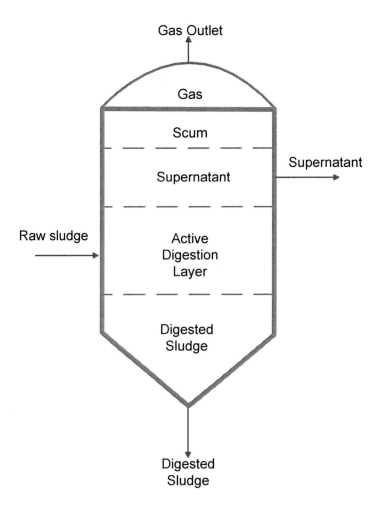

Figure 4. Standard rate digester.

since the construction cost is too high. In the past this method was used for small plants (2.000 eq. population). The biogas production is not constant hindering thus its effective exploitation.

High Rate Digester

The high rate digester is characterized through heating, mixing and uniform feeding of the sludge. Mixing and heating contribute towards creating a uniform environment inside the digester which leads to a more efficient and stable process and a lower tank volume. Mixing is achieved through mechanical stirring, gas recirculation, pumping or draft-tube mixers. For heating external heating exchangers are mostly utilized.

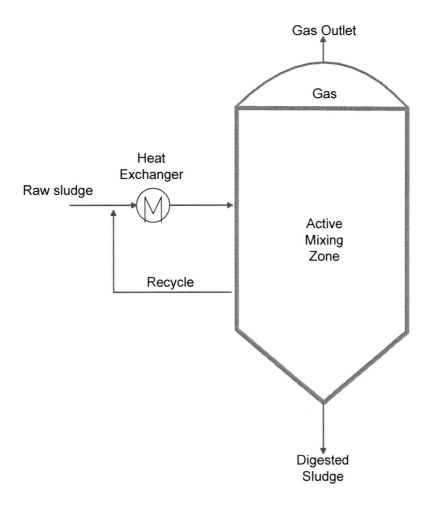

Figure 5. High rate digester.

Mesophilic Digestion

The most common used method for the digestion of sewage sludge is the mesophilic digestion which operates at temperatures between 30-40 °C. Exceeding a temperature of 42 °C can lead to a breakdown of the gas production while temperatures below 30 °C cause the process to slow down. Compared to thermophilic digestion it requires less energy and is less affected by temperature changes but requires also a larger digestion volume.

Thermophilic Digestion

Thermophilic digestion operates between 45-60 °C. Apart from the fact that through this method a better sanitation of sludge is achieved, a faster degradation which results to higher biogas yields is also ensured. However the latter is still not clearly clarified and there are studies that claim that the reaction kinetics remain the same [1]. Nevertheless the sanitation advantage cannot be overlooked. Another advantage compared to mesophilic digestion is the improved dewaterability of the sludge. Thermophilic digesters are more sensitive to temperature changes. While during mesophilic digestion temperature variations of 2-3 °C can be of minor importance in thermophilic digestions a variation even of 1 °C can lead to problems such as external odours and reduction of the sludge dewaterability [7]. Other disadvantages are the higher energy requirements and a lower quality supernatant.

Two-Stage Digestion

Two stage digestion usually involves the use of a high rate digester coupled in series with a second tank, which is usually neither heated or mixed and is mainly used for storage, settling of digested solids and decanting of supernatant liquor. The gas produced in the second tank accounts to less than 10% of the total production [6]. In some cases the second tank is similar to the first one with heating and mixing equipment in order to provide standby digester capacity or to achieve further stabilization prior to further treatment. This practice is rarely used in modern WWTP due to cost reasons combined with small operational benefits.

Two-Step (Phase) Digestion

During two-step digestion the process is divided in two tanks whereas in the first tank the steps of hydrolysis and acid formation occur and the second tank optimizes methanogenesis. In each tank optimum conditions for each bacteria population are present and as a result the microbial activity is enhanced.

Uncoupling hydrolysis and acidification from the methanogenic phase allows the reduction of the reactor volume and respectively the necessary hydraulic retention time while the metabolic rate (biogas production rate) remains constant or even increases. Furthermore this uncoupling allows a separate treatment of the low quality gas (predominantly CO_2) produced during hydrolysis and acidification [9].

Either stage can be operated either in mesophilic or thermophilic conditions. There are two modes of temperature phased digestion: thermophili-mesophilic or mesophilic-thermophilic. In the thermophilic-mesophilic mode the retention time in the first tank is 2-3 days while in the second tank 6-12 days. A thermophilic-mesophilic process with 2+10 retention time achieves the same degradation rates compared to a one step process operated at 35°C and a retention time of 20 days. Extending the retention time in the mesophilic step to 18 days will result to an increase of the degradation rate by 4% compared to a one-step thermophilic process with the same retention time [1]. In Germany where temperature phased digestion was first developed ten large scale plants operate with this configuration where the first step operates under 55 °C and aprox. 3 days retention time and the second step between 35-37 °C and retention time of minimum 12 days.

The advantages and disadvantages of one and two-step digestion are summarized in Table 1 [17]:

Table 1. Comparison between one and two-step digestion

	Advantages	Disadvantages
One-step process	Simple operation	larger specific reactor volume
		negligible higher energy consumption
Two-step process	Separation of bacteria populations	higher equipment investment
	Decoupling of hydrolysis and methanogenesis results in shorter retention time in the methan reactor	not suitable for small plants

4. TYPES OF DIGESTERS FOR SEWAGE SLUDGE

The advantages of anaerobic stabilization are only opposed by the investment cost of the reactor. For that reason optimal and low cost solutions have to be applied when it comes to the size and configuration of an anaerobic digester in order to increase the economic viability of the plant.

Sizing

The first important parameter when designing an anaerobic digestion process is the required tank volume. There are various methods used which include per capita sizing, solids loading, solids retention time, volumetric loading and volatile solids destruction. Digester sizes can vary from 70 m^3 (WWTP Boxholm, Sweden) to 16.670 m^3 (WWTP Emschermündung, Germany).

Today sludge digesters are mostly dimensioned by using the solid retention time as a design parameter. Usually the retention time is equal to 20 days [2]. Newly constructed digesters are also designed with lower retention time taking into consideration that the stabilization process shows an asymptotic trend and after 15 days 95% of the degradation rate is achieved [2]. Other factors such as the final disposal route of the digested sludge and the facility size have to be taken into account as well.

Digester Forms

The digester form depends strongly on the desired volume and the mixing system in order to:

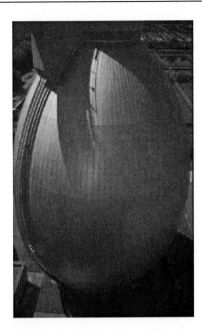

Figure 6. Sewage sludge anaerobic digester in the WWTP of Emschermündung Germany.

- Ensure intensive contact between biomass and fresh feedstock
- Enhance gas recovery
- Prevent the formation of a floating scum layer, dead spaces and accumulations

In respect to the aesthetic nuisance of the digester it has to be integrated in the surrounding environment. The most common used types of anaerobic digesters are the cylindrical and the egg shaped.

Cylindrical Digesters

Cylindrical digesters are very commonly used in the United States and are typically made out of concrete. Their diameter reaches from 6-38 m with a height to diameter ratio of 1:4 [2,16]. This type of digester is susceptible to scum accumulation and foam flotation. Although this can be somehow prevented through intensive energy mixing emptying cleaning of the digester is required every 3-5 years. The construction cost of a cylindrical digester is low but it has to be considered that a second back-up digester is needed when the primary digester is being emptied for cleaning. One of the advantages of this design is a relative large volume for gas storage.

Egg Shaped Digesters

The first egg-shaped digesters were constructed in Germany during 1955/56. Egg shaped digesters offer a higher mixing efficiency than cylindrical digesters and have a smaller footprint. The absence of corners and edges prevents the creation of dead spaces inside the digester. In addition they do not accumulate scum and grit and have lowered operational and maintenance costs although their construction costs are higher. They are made out of steel or concrete [8]. The disadvantages of egg shaped digesters include lack of storage capacity for biogas. In addition their high profile can cause aesthetic nuisances. They are mostly favored for a digestion volume of more than 5.000 m^3 [2].

Construction Materials

Digesters are constructed by steel or concrete. Both materials are suitable and have been used for many years in the construction of sludge digesters. In Germany about 90% of the digesters are from concrete while for smaller digesters (max. 1500 m^3) usually steel has prevailed.

5. CO-DIGESTION OF ORGANICS WASTE AND SEWAGE SLUDGE

Co-digestion is the simultaneous digestion of a homogenous – mixture of two or more substrates. In the case of sludge digesters this is a very common practice with the addition of organic waste from food industry or households as a co-substrate. The basic concept of co-digestion is shown in the next figure:

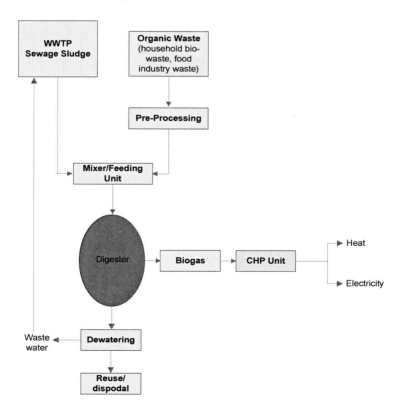

Figure 7. Co-digestion process.

Apart from the obvious cost benefits when exploiting existing digestion capacities the co-digestion process is linked with other benefits as well:

- higher solid substances degradation rates
- enhancement of biogas production (by a factor of 2 with a mixing ration of sludge to bio-waste of 1 to 0,5 and by a factor of 2 with a mixing ration of sludge to bio-waste of 1 to 1 [13]

All WWTP equipped with a digester are potential bio-waste receptors and thus decentralized solutions with low transportation costs are possible. The existing infrastructure offers a quick realization of the concept, no time consuming authorizations are needed and the familiarity of the WWTP personnel with the anaerobic digestion process can be of great benefit.

Before a co-substrate is co-digested with sewage sludge it has to fulfil certain conditions such as:

- Degradable under anaerobic conditions
- Easily mixable with sludge
- Pumpaple
- Free from impurities

Additional equipment that has to be considered in order to accept a co-substrate includes:

1. Weighbridge
2. Reception area
3. Pre-processing equipment
4. Wheel loader for feeding the pre-processing equipment
5. Exhaust air treatment equipment
6. Expansion of the gas utilization unit

The weighbridge is only necessary only when solid organic waste is delivered. Liquid organic wastes can be registered through their volume and are delivered in a temporary storage tank. Their pre-processing should include removal of solid particles such as stones, plastic, bones etc. A screen can be used for that purpose. A feeding unit should be available in order to ensure uniform feeding of the digester. Depending on the origin of the liquid waste mixing, heating, inoculation and hygienisation should be taken into consideration.

For solid bio-waste such as separate collected kitchen waste a flat reception bunker is required which should be within an enclosed building in order to avoid odour problems. By means of a wheel loader the waste is forwarded to the pre-processing line.

German WWTP that applied co-digestion have reported that the concept was integrated without any significant problems. The existing infrastructure was easily adapted to the new requirements and there was no impact to the sludge quality. Moreover no floating layers were caused by the addition of the co-substrate. However the WWTP faced odour problems and for that reason an enclosed reception bunker is required. Another problem observed were accumulations in pipes and pumps. In case the co-substrate has a high content on fats then the dewaterability of the sludge can be negatively affected.

6. OPERATION OF SEWAGE SLUDGE DIGESTERS AND EQUIPMENT

Digester Start Up

Before a newly constructed digester can operate a sufficient concentration of active biomass has to be available. Therefore the organic loading rate during the start up phase has to be gradually increased. Accordingly operating conditions such as intensive mixing, constant temperature and feeding have to be provided. The start up phase consists of the following steps:

Initial Filling
During this step the digester is fully filled with process water or waste water (free from impurities). This is necessary in order to avoid the creation of an explosive gas-air mixture inside the tank and to test that the recirculation system is operating without any problems.

Heating
Following the inital filling the temperature is increased depending on the operational parameters of the digester and the process (mesophilic or thermophilic).

Inoculation
After heating is complete, seeding sludge from another digester which operates in the same temperature levels as the start-up digester is used as inoculum. High quantities of inoculum will decrease the start up time. In case no seeding sludge is available then raw sludge fed in small batches is used. A constant monitoring of pH, organic acids concentration and biogas composition is essential during this step.

Raw Sludge Feeding
The raw sludge feeding rate should be kept to a minimum during the start up phase and gradually increased until a stable operating environment has been accomplished. This is depicted in the gas quality where increased CO_2 levels are an indicator that the process has to be optimized. Depending on the quality and quantity of the incoculum 20 days to 6 months are required in order to complete the start up phase and have the digester running in its full operating capacity.

Sludge Piping System

The sludge piping system has to be able to support the following operations:

- inoculation, heating, circulating
- removal of sand and emptying the digester
- removal of digested sludge and water

Feeding, Inoculation

An important issue that has to be taken into consideration during digester feeding is the intensive mixing of raw sludge with the partly digested material. Therefore before entering the digester raw sludge is inoculated in a mixing chamber before the heat exchanger. The mixing ratio should be at least 1:1 while with a ratio of 1:4 the biological process is accelerated [12]. In addition turbulent flow inside the recirculation system ensures that bacteria and fresh feedstock are efficiently blended.

Ideally feeding should be performed in a continuous way in order to keep a constant daily organic loading rate and to avoid temperature fluctuations. In small facilities where this is not possible the total amount of raw sludge should be divided into small batches.

Mixing and Circulating

Intensive mixing ensures constant process parameters throughout the reactor volume. Proper inoculation, heat diffusion and biogas removal are enhanced through effective mixing as well. In addition scum layer formation and grit accumulation are prevented. Through the rise of gas bubbles and the thermal convection currents created by the addition of heated sludge a natural self-mixing is present to a certain degree, which however is not sufficient. Therefore additional means have to be applied. Regarding the intensity and duration of mixing operations contradictory opinions exist depending on the feeding rate and process of the system.

The most common methods are:

- Mechanical stirring

The use of mechanical stirring was abandoned decades ago due to interference of impellers by rags and other impurities. Today such problems are partially solved through the use of fine grates and improvements on the impeller form. However problems like wear and bearing failures are still present and the use of mechanical stirring should be avoided for a trouble free operation of the digester. Mechanical stirring is performed through low speed flat-blade turbines or high speed propeller mixers.

- Mechanical pumping

External pump recirculation systems withdraw sludge from the digester through the external heat exchangers and inject it back near the surface to break up scum accumulation. Through this process the pumped sludge is blended with raw sludge and heated. The use of external pump recirculation as stand alone system for digester mixing is not sufficient for large volume tanks and should only be used for heating and inoculation purposes while an additional mixing system should also exist.

Gas Injection

Gas mixing systems can be classified as confined and unconfined. Confined systems collect biogas at the top of the digester, compress it and then reinject it through confined draft tubes. There are two main types of confined systems, the gas lifter and the gas piston. Confined gas mixing has a lower power requirement and is an effective measure against scum building.

Unconfined systems collect biogas at the top of the digester, compress it and then reinject it through diffusors at the bottom of the reactor. The gas bubbles released rise to the surface while carrying and moving the sludge. This system is effective against solids deposition but does not provide good top mixing, which results in scum buildup.

Heating

Heating is required during the anaerobic digestion process in order to heat the incoming raw sludge and to compensate for heat losses through walls, floor and roof of the digester. Maintaining a constant temperature is also very important for certain bacteria populations such as the methane-bacteria which are very sensitive to sudden temperature changes.

A momentarily drop of the temperature is not harmful and the activity of the bacteria is only inhibited until the original temperature is reached again. For example during digester feeding a temperature decrease can be observed which can be limited when the cold raw sludge is sufficiently mixed inside the digester. In order to limit this decrease to less than $1°C$ the raw sludge batches should be smaller than 4% of the effective digester volume. During peak periods it is rather better to increase the number of the incoming batches than their volume.

External heat exchangers have prevailed as standard heating equipment due to their better heat transfer efficiency. Another advantage of external heat exchangers is that the recirculated sludge can be blended with raw sludge for inoculation purposes. Three types of external heat are commonly used: water bath, jacketed pipe and spiral. Sludge should not be heated in a temperature of more than $60°C$ in order to avoid sludge encrustation and the flow rate should be around 1,5 m/s.

Another method that can be applied for heating is steam injection. Disadvantages of the steam injection method are the need for boiler feed water processing and sludge thinning due to condensate water.

7. PRAXIS IN BIOGAS UTILIZATION FROM SEWAGE SLUDGE ANAEROBIC DIGESTION FACILITIES

Biogas is a flammable gas with high energy content that consists mainly of methane and carbon dioxide. Fractions of water vapour, traces of H_2S and H_2 and possibly other contaminants can be found depending on the composition of the raw sludge.

The composition and properties of biogas are presented in the next table [13].

Table 2. Composition and properties of sewage sludge digester biogas

Parameter	Range	Average
CH_4 (Vol. %)	60-70%	65%
CO_2 (Vol. %)	30-40%	35%
H_2S (Vol. %)	0-0,7%	
N_2 (Vol. %)	0-0,2%	
H (Vol. %)	0-0,2%	
O_2 (Vol. %)	traces	
Density (kg/m^3_N)		~1,2
Heating value (kWh/m^3_N)		~6,4
Ignition-/Explosions limit (Vol. % in air)		5-15%

Quantity and quality of biogas depend on various factors such as:

- Quantity and characteristics of the raw sludge
- Impact of inhibiting/toxic substances
- Digestion process (retention time, engineering process)
- Operation of the digestion plant

Primary sludge is relative easily degradable with degradation rates of 60% and specific gas production of 0,5-0,6 m^3_N/kg oDS. Excess sludge on the contrary has a degradation rate of 30-40% and specific gas production of 0,25-0,3 m^3_N/kg oDS. Standard production rates for raw sludge amount to 0,35-0,45 m^3_N/kg oDS [13].

The intensity of the biogas production is also influenced from the digester's feeding pattern. Discontinuous feeding (total daily sludge amount in one batch) leads to a peak in the biogas amount and a storage capacity is needed otherwise the excess biogas has to be flared. For that reason a constant feeding of the digester is recommended. In case this is not possible due to the size of the WWTP a semi-continuous operation of the digester is the next ideal solution.

Praxis has shown that during dimensioning of a facility the biogas quantities are mostly overestimated and that in reality lower values are to be expected. A way to increase biogas quantities is to co-digest sludge with bio-waste.

Biogas utilization worldwide occurs mainly through combined heat and power (CHP) applications.

Biogas Storage

Purpose of biogas storage is to compensate fluctuations in gas quantity and quality. The storage component allows downstream equipment to work in a constant rate. The required capacity depends on the operational parameters of the digester and the biogas utilization concept and no standard design criteria for dimensioning of storage equipment exist. However the following has to be taken into consideration for the selection of biogas storage tanks:

- Volume
- Pressure
- Safety
- External loads
- Tank diameter
- Stir technology
- Costs

When the total daily amount of sludge is fed to the digester in one batch then biogas production can reach a peak value that corresponds to 150% of the average generation rate [14]. On the contrary if the digester is fed continuously or semi-continuously then a smoother biogas production takes place and a storage capacity that corresponds to 10% of the daily amount should be sufficient. In WWTP low pressure storage tanks are mostly used [13] while medium or high pressure tanks are used for higher biogas amounts. Low pressure tanks are usually made of galvanized iron, concrete and plastics and medium or high pressure tanks out of mild steel.

Biogas Upgrading

The main incentive for gas upgrading is to fulfil the requirements of gas appliances, to increase its heating value and to standardize its quality. Gas upgrading includes the following main operations when it comes to utilization in CHP engines:

Desulphurization

Hydrogen sulphide is created through the decomposition of organic and inorganic sulphide components and leads to chemical corrosion of the gas storage tanks and engines. Desulphurization includes methods such as NAOH scrubbing, water scrubbing, biological removal on a filter bed and active carbon adsorption. The simplest method is to add air or oxygen directly into the digester or in the gas storage tank whereas H_2S can be reduced by up to 95% to levels up to less than 50 ppm [15].

Siloxanes

Siloxanes are volatile contaminants not broken down during the digestion process and cause significant physical wear to gas engines. These silicon-containing compounds are widely used in cosmetics and pharmaceutical products. Problems mostly occur when silicon concentrations exceed 15 mg/m^3_N (based on silicon) [13]. They can be removed by absorption on a liquid medium, adsorption on active carbon or cryogenic condensation

Biogas Utilization

There are four main concepts for the utilization of biogas in WWTP. The production of heat / steam, cogeneration of heat and power, upgrade to vehicle fuel and fuel cells. The decision for one of the aforementioned concepts depends mainly on the local conditions, the plant size and the legislative framework. The most common applications for biogas utilization however are the use as boiler fuel for the production of heat/steam or the combine heat and power production.

Boiler

This application involves the direct use of biogas as boiler fuel and is particular attractive since conventional equipment can be used with only minor modifications. The concentration of corrosive materials should not affect the operation of the system other than that the boiler has to be constructed from corrosion resistant materials. A boiler with an efficiency of 90% can cover the total heat demand of a WWTP which includes digester and building heating and production of hot water. In most of the cases the gas production exceeds the total need and excess gas has to be flared. Boilers can also operate with oil, propane or natural gas so that when no biogas is available during the startup phase or due to a system failure the heating requirements of the plant are met.

CHP gas engines operate with a thermal energy efficiency factor of 50% and the produced heat does not always suffice. In order to cover the heat deficit and to ensure stable operation in the WWTP a boiler operated with backup fuel is held in reserve.

Combined Heat and Power (CHP)

Cogeneration includes the simultaneous production of two or more forms of energy from a single fuel source (biogas), i.e. electricity and heat. The electrical energy efficiency is around 35% and the thermal energy efficiency 50%. The overall efficiency factor is approx. 85%.

The technologies that are commercially available are internal combustion reciprocating engine-generator sets and gas turbine-generator sets.

There are two main types of gas reciprocating engines:

- dual-fuel engines (otto cycle)
- spark ignition engines (diesel cycle)

In most CHP applications internal combustion engines are used. When selecting the most appropriate system several factors such as capital and operating costs, efficiency, emissions, noise and changes in gas composition have to be taken into consideration.

CONCLUSION

In comparison to aerobic stabilization of sewage sludge, anaerobic digestion (AD) is a far more attractive option which is related to the production of biogas and its subsequent utilization which is profitable from an economical and ecological point of view. However AD is a complex process where several parameters such as temperature, pH, retention time and inhibitory/toxic materials have to be taken into consideration during the operation of a plant.

The main distinction between anaerobic digesters refers to standard and high rate digesters. The first are a simplified execution of the AD method where the material is usually neither heated nor mixed, the retention time ranges between 30-60 days and are used only for smaller WWTP. The latter are more sophisticated solutions which are characterized through heating, mixing and uniform feeding of the sludge in order to achieve high degradation and biogas production rates. Another distinction between anaerobic digestion systems refers to the operating temperature of high rate digesters, where mesophilic systems operate at temperatures between 30-40 °C and mesophilic systems at 45-60 °C. Most digesters operate under mesophilic conditions. Thermophilic systems provide faster degradation rates and increased destruction of pathogenic organisms. Another configuration of AD systems implicates the use of two tanks where the second tank is used for storage, settling of digested solids and decanting of supernatant liquor. Alternatively the second tank can operate as a high rate digester and the methanogenic phase takes place in it while in the first tank hydrolysis and acidification occur. In each tank optimum conditions for each bacteria population are present and as a result the microbial activity is increased.

There are several methods for dimensioning an anaerobic digestion tank with solid retention time being the most common practice while always taking the estimated daily sludge amount into consideration. Since the only disadvantage of anaerobic in comparison to aerobic stabilization is the increased cost, optimal solutions have to be applied in the reactor design. The most common shapes for digesters are the cylindrical and egg shaped digesters. The first have a low construction cost and offer a relative large volume for gas storage while the latter offer high mixing efficiency and a smaller footprint. Construction materials consist of concrete or steel. In Germany 90% of the digesters are made of concrete and steel has prevailed for smaller systems only.

Concerning the costs of a sewage sludge digestion plant those can be mitigated by the co-digestion of sludge and source separated bio-waste. Apart from the obvious economic profit in terms of a gate-fee for the incoming bio-waste, a higher degradation of solid substances and an increased biogas production are also achieved. Moreover the usual dispersion of WWTP inside a region makes the application of decentralized solutions possible. Depending on the exact characteristics of the co-substrate additional infrastructure and equipment will be required in order be able to treat the incoming feedstock.

Start-up of a digester includes steps such as initial filling, heating, inoculation and raw sludge feeding. Before the digester reaches its full operational capacity a period of 20 days to 6 months is required depending on the quality and quantity of the inoculum. Apart from the operational parameters that affect the anaerobic digestion process (temperature, pH, etc.) there are other aspects that refer to the configuration of the plant and the process design. The feeding pattern of the digester should be performed in a continuous way in order to keep a constant organic loading rate and avoid temperature fluctuations caused by the temperature of the incoming sludge. In case this is not possible for technical or other reasons (small plants with low sludge production) then the total sludge amount should be divided in small batches. In addition before the incoming material is fed into the tank it should be inoculated with bacterial population through its mixing with partly digested material. Mixing is also essential in order to achieve constant parameters and characteristics of the materials inside the digester. There are three mixing methods which refer to mechanical mixing, mechanical pumping and biogas injection. Finally heating is of great importance in order to keep a constant temperature. Temperature fluctuations can affect the bacterial population. Especially methanogenic bacteria are sensitive to temperature changes. External heat exchangers have prevailed as standard heating equipment due to their better heat transfer efficiency.

Apart from the digested sludge whose utilization is often limited by legislative restrictions, the main product of the anaerobic process is biogas, a flammable gas with high energy content that consists mainly of methane and carbon dioxide. Biogas can be utilized in several ways depending on its quality the incentives give in each country. Combined heat and power (CHP) is the most favourable option. The electrical energy efficiency is around 35% while the thermal energy efficiency 50%. The overall efficiency factor lies approx. at 85%.

REFERENCES

[1]   Kapp, H., *Verfahren der Schlammfaulung und Bemessung*; In Müll-Handbuch; Bilitewski, B., Schnurer, H., Zeschmar-Lahl, B.; Erich Schmidt Verlag: Berlin, Germany, 2008, Vol. 2, No. 3070, pp 1-15

[2]   Niehoff, H.-H., *Bau von Behältern für die Klärschlammfaulung*; In Müll-Handbuch; Bilitewski, B., Schnurer, H., Zeschmar-Lahl, B.; Erich Schmidt Verlag: Berlin, Germany, 2008, Vol. 2, No. 3075, pp 1-15

[3]   Appels, L., Baeyens, J., Degreve,J., Dewil, R., *Principles and potential of the anaerobic digestion of waste-activated sludge.* Progress in energy and combustion science. 2008, *Vol. 34, 755-781*

[4]   Bischofsberger, W., Dichtl, N., Rosenwinkel, K.-H. Seyfried, C.-F., Böhnke, B., *Anaerobtechnik*, Springer Verlag, Berlin, Germany,2005, pp 1-718

[5]   Gerardi, M.-H., *The Microbiology of Anaerobic Digesters*, John Wiley & Sons, Inc., New Jersey, US, 2003, pp 1-177

[6]   Chen, Y., Cheng, J.-J., Creamer, K.-S., *Inhibition of anaerobic digestion process: A review.* Bioresource Technology. 2008, Vol. 99, 4044-4064

[7]   Schumacher, *Betrieb von Schlammfaulungsanlagen*; In Müll-Handbuch; Bilitewski, B., Schnurer, H., Zeschmar-Lahl, B.; Erich Schmidt Verlag: Berlin, Germany, 2008, Vol. 2, No. 3080, pp 1-12

[8]   Turovskiy, I.-S., Mathai, P.-K., *Wastewater Sludge Processing*, John Wiley & Sons, Inc., New Jersey, US, 2006, pp 173-210

[9]   Blank, A., Hoffmann, E., *Upgrading a co-digestion plant by implementation of a hydrolysis stage;* Proceedings of 7th International Conference Orbit 2010, Organic Resources in the Carbon Economy, 2010, pp 263-270

[10]   Schmelz, K.-G., *Co-fermentation of sewage sludge and biowastes;* Proceedings of the International Conference Orbit 1999 on Biological Treatment of Waste and the Environment, Organic Recovery and Biological Treatment, 1999, pp 187-195

[11]   Schmelz, K.-G., *Co-Vergärung auf kommunalen Kläranlagen*; In Müll-Handbuch; Bilitewski, B., Schnurer, H., Zeschmar-Lahl, B.; Erich Schmidt Verlag: Berlin, Germany, 2008, Vol. 2, No. 3059, pp 1-27

[12] Roediger, M., *Ausrüstung von Schlammfaulungsanlagen*; In Müll-Handbuch; Bilitewski, B., Schnurer, H., Zeschmar-Lahl, B.; Erich Schmidt Verlag: Berlin, Germany, 2008, Vol. 2, No. 3079, pp 1-17

[13] Niehoff, H.-H., *Die Praxis der Faulgasbewirtschaftung und -nutzung*; In Müll-Handbuch; Bilitewski, B., Schnurer, H., Zeschmar-Lahl, B.; Erich Schmidt Verlag: Berlin, Germany, 2008, Vol. 2, No. 3088, pp 1-29

[14] Ronchetti, C., Bienz, P., Pridal, R., *Ökobilanz Klärgasverstromung*. Bundesamt für Energie. 2002, Switzerland, pp 1-33

[15] Wellinger, A., Lindeberg, A., *Biogas upgrading and utilization*, Task 24, Energy from Biological Conversion of Organic Wastes, 1999, 1-20

[16] Tchobanoglous, G., Burton, F.-L., Stensel, D.-H., , *Wastewater Engineering Treatment and Reuse*, McGraw-Hill, New York, US,2003, pp 1505-1531

[17] Werning, J., Klärschlammbehandlung- und Entsorgung, Schlammstabilisierung, Lehrtuhl für Siedlungswasserwirtsschaft und Siedlungsabfallwirtsschaft, RWTH Aachen, 2009

In: Sewage Sludge Management
Editors: A. A. Zorpas and V. J. Inglezakis

ISBN: 978-1-61324-393-0
© 2012 Nova Science Publishers, Inc.

Chapter 10

STABILIZATION OF MUNICIPAL SEWAGE SLUDGE BY FLY ASH

E. Papastergiadis[1], C. Papadimitriou[1], P. Samaras[1], A. Karagiannidis[2,] and A. Zouboulis[3]*

[1]Department of Food Technology,
Alexander Technological Educational Institute of Thessaloniki, Thessaloniki, Greece
[2]Laboratory of Heat Transfer and Environmental Engineering,
Department of Mechanical Engineering, Aristotle University of Thessaloniki,
Thessaloniki, Greece
[3]Division of Chemical Technology, Department of Chemistry,
Aristotle University of Thessaloniki, Thessaloniki, Greece

1. INTRODUCTION

Sewage sludge is defined as the solid, semi-solid, or liquid residue generated during the treatment of domestic sewage. The term biosolids defines "the primarily organic solid product yielded by municipal wastewater treatment processes that can be beneficially recycled" as soil amendments (EPA, 1995). Use of the term biosolids has been controversial because of the perception that it was created to improve the image of sewage sludge in a public-relations campaign by the sewage industry (Rampton 1998). Biosolids are a complex mixture that may contain organic, inorganic, and biological pollutants originating from the wastewaters of households, commercial establishments, and industrial facilities and compounds added or formed during various wastewater treatment processes, including inorganic contaminants (e.g. metals and trace elements), organic contaminants (e.g. polychlorinated biphenyls-PCBs, dioxins, pharmaceuticals, and surfactants), and pathogens (e.g. bacteria, viruses, and parasites).

* E-mail: makis@aix.meng.auth.gr.

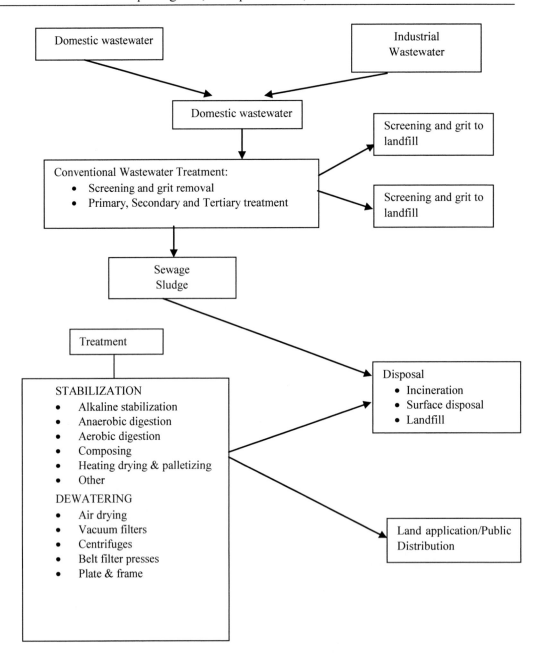

Figure 1: Sources and fate of sludge.

Sewage sludge treatment processes are intended to reduce the volume and organic content of biosolids and to reduce the presence of pathogens, but retain beneficial properties for soil-amendment and land-reclamation purposes. Figure 1 provides a simplified schematic of biosolids production sources and illustrates the way that the content of biosolids may vary depending on the wastewater streams and the variation in treatment processes. The Council of European Communities (1986) published the Sewage Sludge Directive (86/278/EEC); all members had to promulgate their own version of the directive as national regulations by

Table 3. European Union limit values for concentrations of heavy metals in biosolids for land use, according to Sewage sludge Directive

Elements	Limit values (mg/kg dm)	
	Directive 86/278/EEC	Proposed
Cd	20 – 40	10
Cr	-	1000
Cu	1000 - 1750	1000
Hg	16 - 25	10
Ni	300 - 400	300
Pb	750 - 1200	750
Zn	2500 - 4000	25000

1989. The directive included a recommended range of pollutant concentration values for seven constituents in biosolids, for member nations to use, in adopting their standards (Table 1). A comprehensive review of biosolids use and disposal practices was published by the International Association on Water Quality (IAWQ), International Water Association (IWA), the Water Environment Federation (WEF), and the European Water Pollution Control Association (EWPCA) (Matthews 1996); representative data from Europe to complement U.S. information have been assembled to provide a basis for comparison and some determination of the current and future status of biosolids management. An assessment of the status of disposal and recycling within the European Community (European Communities 2001) reviewed existing legislation and regulations and provided an analysis of stakeholder positions, motivations, and constraints, as well as solutions for reducing constraints and encouraging the use of biosolids.

Analysis of existing legislation indicated that specific requirements focus principally on the use of biosolids in agriculture, both nationally and in Europe. The EEC directives, which have the strongest influence on biosolids use, are directive 91/271/EEC on urban wastewater treatment and 86/278/EEC on the use of biosolids in agriculture (Council of the European Communities 1986). Requirements set by the latter directive are a crucial element in the management of biosolids produced in the member states and some member states have introduced provisions that go beyond the requirements of the directive. In particular, the limit values for concentrations of heavy metals in biosolids are lower than those specified in the directive in a majority of the countries.

The countries in which the limitations on heavy metal concentrations are the most stringent are Belgium (Flanders region), Denmark, Finland, the Netherlands, and Sweden. Greece, Luxembourg, Ireland, Italy, Portugal, and Spain have set limit values similar to those in the directive; values for Poland, are also lower than the European Union standards. The United Kingdom legislation differs by not providing any limit values for heavy metals in biosolids but rather specifies the maximum annual average loads of heavy metals to soil that are similar to the directive. In addition, the regulations on biosolids use include limit values for pathogens (Table 2) in France, Italy, and Luxembourg and, for organic compounds (Table 3) in Austria, Belgium-Flanders, Denmark, France, Germany, and Sweden, neither of which are included in the directive.

In all member states, regulations on the use of biosolids specify limit values for heavy metals in soil that are similar in most cases to the requirements set in the directive (Table 2). Some countries have defined limit values for several categories of soil pH or limit the maximum load of heavy metals to agricultural lands on a 10-year basis. Maximum quantities of biosolids that can be applied on land have been set between 1 metric ton by the Netherlands for grasslands and 10 metric tons by Denmark per hectare and per year. The debate on biosolids recycling and disposal differs in intensity and resolution throughout the European community.

Table 4. Limit values for pathogens in biosolids in national regulations

	SALMONELLA	Other pathogens
France	8 MPN/10g of DM	Enterovirus: 3MPCN/10g of DM Helminths eggs: 3/1010g of DM
Italy	1000 MPN/g of DM	
Luxembourg		Enterobacteria: 100/g No egg of worm likely to be contagious
Poland	Biosolids cannot be used in agriculture if contain *Salmonella*	Parasites: 10/kg DM

**Table 5. Limit values for organic compounds in biosolids (mg/kg of DM)
in national regulations**

	Dioxins and Furans (PCDD, PCDF) ng/TE/kg of DM	PCB s	AOX	LAS	DEHP	NPE	PAH	Toluene
Austria	100	0.2	500				6	
Belgium				2.6	100	50	6	
Denmark				1.3	50	10	3	
France		0.8					2 – 5	
Germany	100	0.2	500					
Sweden		0.4				100	3	5

An analysis of stakeholder groups (European Communities, 2001), including the farming community, landowners, industries, water and wastewater plants and companies, local authorities, national authorities, and citizens and consumer groups, indicated a significant diversity of opinion ranging from opposition to advocacy as shown below:

• The regulatory requirements in the Netherlands and Flanders region of Belgium have prevented almost all use of biosolids in agriculture since 1991 and 1999, respectively.

• In countries such as Denmark and the United Kingdom, new regulations are considered sufficiently strict to reduce risks to an acceptable level (Denmark), and agreement in 1998 between water and sewage operators and retailers as well as farmers' associations and government (United Kingdom) led to the joint adoption of a "safe sludge matrix" providing for additional restrictions on the use of biosolids on agricultural land as well as the categories of crops on which biosolids may not be used.

• In Sweden, a voluntary agreement was signed in 1994 between the Swedish Environmental Protection Agency, the Swedish Federation of Farmers (LRF) and the Swedish Water and Waste Water Association concerning quality assurances relating to the use of biosolids in agriculture. However, in October 1999, the LRF recommended that its members stop using biosolids because of quality concerns.

• The opinion in Germany has recently swung in favor of agricultural land application, mainly because this practice is considered economically viable and that the potential risks are sufficiently reduced by the existing legislation, which is now being reviewed.

• In Austria, France, and the Walloon region of Belgium, national (or regional) agreements have been considered, and in France, such an agreement was supported on the condition that additional quality controls and an insurance fund have to be developed. One party to the agreement (farmers' union) asked for a ban on biosolids because current methods used are not considered sufficient to address the perceived risks related to the agricultural cycling of biosolids.

• In Finland and Luxembourg, the farming community is generally hostile toward the use of biosolids for land application, mainly because of the pressure to use animal manure (e.g., Finnish Union of Agricultural Producers requested a ban on the use of biosolids for land application, and have renewed its stand against the use of biosolids in agriculture in 2001).

• In Ireland and Portugal, farmers tend to support the agricultural use of biosolids for economic and for agronomic (organic matter and phosphorus content) reasons, although biosolids use in these countries has been relatively recent.

• In Spain, Italy and Greece, available information indicates that there is little debate on use of biosolids. The analysis of stakeholders' positions indicates that the main areas of concerns on sewage sludge disposal and biosolids recycling are that the growing quantities of sewage sludge must be treated with the aim of keeping both environmental and economic costs as low as possible. Similarly, improving practices of treatment and use of biosolids is now considered essential. Moreover, within the context of uncertainties concerning the potential impacts on human health and the environment of the various disposal and recycling options, additional research is needed to increase confidence in the use of biosolids in agriculture. Some strategies suggested by the recent European Union biosolids-management assessment for reducing constraints and encouraging recycling of biosolids include the following (European Communities 2001):

- Certify the treatment process involved, the quality of biosolids, and recycling practices.
- Develop a trust fund or insurance system to cover any loss of profits, damages, or other costs related to the use of biosolids in agriculture together with legal provisions to regulate producer liability.
- Standardize science-based laws and regulations.
- Enhance mutual confidence and communication and transfer of information between stakeholders.
- Diminish uncertainty over risks to human health and environment, and extend the assessment and dissemination of information beyond heavy metals to include organic pollutants and pathogens.
- Develop codes of practice for the recycling of biosolids, the possible use of labels for quality assurance, and associated training programs and outreach activities for stakeholders.

When European Union biosolids-management practices are compared with those of the U.S., it is apparent that European and U.S. contaminant limits apply largely to heavy metals and are based on (1) the concentration of the biosolids itself; (2) the loading or total amount of metal that can be added and how quickly it can be applied; and (3) the maximum concentration of metals in soil allowed to build up after biosolids application. According to an analysis of regulations in the United States and some European countries by McGrath et al. (1994), three basic approaches to setting limits were distinguished: (1) analyzing the pathways of pollutant transfer to selected target organisms and an assessment of the likely harmful effects that metals might have on the target; (2) setting limits consistent with the lowest-observed-adverse-effect concentrations, which are actual cases of effects due to metals but not necessarily derived from studies that involved applications of biosolids; and (3) attempting to match the metal inputs to soils to the small losses of metals due to crop removal, soil erosion, and leaching (metal balance approach). These approaches were considered responsible for the widely different numerical limits for metals arising either from a policy decision to reach zero impact (metals balance) and associated low levels or from approaches that allow some increase in metal concentrations in soils based on target organisms and use of associated models and sparse toxicity data. Thus, the practice of implementing vastly different regulations for biosolids application to land in the United States and within European Union member nations create differing social, economic, technological, and environmental impacts that beg consensus resolution in the scientific, technical, and regulatory communities.

Within the European Union, the intended goal and most widely applied biosolids disposition option is agricultural use. However, the selection of an option and its implementation according to European Commission directives is affected by local or national circumstances. Thus, the degree of flexibility varies. Notably, ocean disposal has been phased out, so that the principal disposal options now include agricultural use, landfill, and incineration. As in the United States, the European Commission has developed regulatory limits (Sewage Sludge Directive 86/278/EEC) when biosolids are used in agriculture. The Sewage Sludge Directive requires member states to apply maximum limit values for certain heavy metals in the biosolids and in the soil to which it is applied; to pre-treat sewage sludge; and to restrict its use, including the frequency and quantity of application, on certain soils.

These regulations establish conditions relating to pretreatment, nutrient needs, quality of soil, protection of surface waters and groundwaters, and compliance with concentration limits of heavy metals in soil. Use of biosolids is prohibited on specified categories of land within defined periods prior to harvesting and where concentrations of heavy metals in the soil exceed specified limit values. Records must be kept and made available to the competent authorities on the quantities, composition, use, treatment, and results of analysis on biosolids, the names and addresses of recipients of biosolids, and the places where biosolids are to be used (European Union, 2000a). Accordingly, member states have performed biosolids surveys to comply with the reporting requirements, such as the U.K. Sludge Survey for 1996-1997 (Gendebien et al., 1999). Summary reports indicating biosolids quality and ultimate disposition quantities are to be submitted to the European Union every 5 years (e.g. UK Department of the Environment, 1993). A part of the implementation of the directive is that application for biosolids use is made in advance of the operation, and conditions are applied to the methods and type of biosolids used.

Consideration is given to the links between biosolids use and potential transmission of pathogens to the human food chain and into water courses or supplies through nutrient leaching. In addition, biosolids producers are obliged to provide details of biosolids composition to owners of land where biosolids will be applied. Analytical methods, sampling frequencies, monitoring procedures, and record-keeping requirements are also prescribed. Proposed revisions are included in the European Union Working Document on Sludge (European Union, 2000b), and changes in limit values are being considered for heavy metals and organic compounds on the basis of biosolids concentrations and soil characteristics.

There are three major alternatives for final disposition of sewage sludge: (1) recycling as biosolids to agricultural land as a fertilizer or soil amendment or selling or giving away to the public for use on home gardens or lawns; (2) burying in a municipal solid-waste landfill or a surface disposal site; or (3) burning in an incinerator. When assessing any of these practices, they should be evaluated holistically for risk. For instance, if all land application should cease, how would the overall risk be altered if additional landfills, surface disposal sites, and incinerators were constructed and operated to accommodate the additional volumes. In response to EPA's beneficial-use policy, the publication of risk-based regulations and the general trend toward recycling, numerous states began to encourage POTWs to use their biosolids in the late 1980s and 1990s. This policy was further aided by political and legal difficulties associated with siting and constructing incinerators and landfills.

Application of sewage sludge in agricultural and horticultural industries has always been difficult in terms of the trade-off between nourishment of nutrients and organic matters and contamination of toxic heavy metals and pathogenic agents (Chaney, 1983; Robert and Winkler, 1991; Wong et al., 1997; de Bertoldi et al., 1983). Although composting of sewage sludge would be a feasible pre-treatment step to eliminate pathogens and to stabilize organic matter through biological stabilization (de Bertoldi et al., 1983; Inbar et al., 1993), high heavy metal concentrations in the compost may be harmful to soil microbial activity and plant growth; in addition, they pose a threat to human health by bioaccumulation and biomagnification in the food web and the environment (Adriano et al., 1982). Lime treatment is often carried out as an alternative method for reducing the microbial content and the availability of heavy metals. Due to its properties, lime is able to maintain the high pH values of the mixtures throughout time, having as an effect the removal of microbial communities in sludge; changes in pH values may imply consequent changes in metal bioavailability.

However, the application of lime for the stabilization of sewage sludge depends upon a number of parameters such as availability of lime, the associated costs, the required period for stabilization, etc.

Fly ash may have increased calcium concentrations and high pH values similar to lime. Moreover, continuous worldwide uses of coal for generation of electricity in power plants generate vast amounts of fly ash. The reuse of fly ash is rather limited and only a relatively small amount is used in construction works, the production of materials such as ceramics and concrete and in cement manufacturing.

Fly ash has been investigated as soil amendment for agricultural use due to its capacity to improve the physical and chemical properties of soils (Wong et al., 1997; Wong and Su, 1997; Petruzzelli, 1989). It has been shown that an increase in pH causes an increase in the fixation of metals by soils and therefore a decrease in the uptake by plant (Adriano et al., 1980, 1982; Chlopecka and Adriano, 1997; Tadesse et al., 1991). Hence, the alkaline properties of fly ash not only make it a good soil liming material for agriculture (Korcak, 1985; McCarty et al., 1994) but also help to eliminate the residual pathogens in the sludge compost (Wong and Lai, 1996). The utilization of fly ash for sludge conditioning and stabilization may substitute lime and minimize the costs of lime purchase and landfilling, and can comprise a feasible management option for the reuse of two solid materials produced in vast amounts worldwide. The examination of fly ash as an agent for sludge stabilization is limited.

Blending of commercial compost with small amounts of fly ash suggested an improvement of the availability of essential and non-essential plant food elements (e.g. Ca, Mg, and Na) from fly ash and has been proved in general as beneficial to plant production without causing any deleterious effects on plant growth and plant composition (Ghuman et al., 1994).

The objectives of this work were the examination of the stabilization of sewage sludge by mixing with fly ash, the comparison between lime and fly ash stabilized sludge and the investigation of the effect of operation parameters, such as mixing ratios and stabilization period on the properties of the mixtures.

2. MATERIALS AND METHODS

The work was conducted in two subsequent phases. In both phases the pilot studies took place in the landfill of Kozani in West Macedonia, Greece. The characteristics of each phase are described below Q

1st Phase

During this phase, mixtures of sewage sludge with lime and/or fly ash were prepared in pilot scale. The total weight of each mixture was 500 kg. The ratios of each mixture mixtures are shown in Table 4.

Each mixture was monitored for microbiological and physicochemical parameters during the stabilization period. Sampling was conducted at the beginning of the period (0 days), 1 and 2 weeks, 1 month, 2 months and 3 months of stabilization. Certain parameters such as pH, temperature and conductivity were measured on a daily basis. For the determination of all parameters standards methods were used (APHA, 1995).

Table 4. Characteristics of the mixtures and samples

Mixture	Sample number	Characteristics	Total Ca content in the mixture
1	0	Sludge only – control sample	0%
2	1	Sludge with lime	10%
3	2	Sludge with fly ash	10%
4	3	Sludge with fly ash	5%
5	4	Sludge with fly ash	15%

Three categories of microorganisms were measured Total Coliforms, E.coli, and Feacal coliforms as well as Salmonella sp. The physicochemical parameters that were monitored included:

1. Electrical conductivity
2. pH
3. Total and Volatile Solids
4. Water content
5. Nutrients (Total Kjeldahl, Nitrogen, Ammonia-Nitrogen, Total Phosphorus and Potassium).

2nd Phase

During this phase, sludge was mixed with lime and/or fly ash in the following ratios:
1. Sludge mixed with 10% lime based on dry weight
2. Sludge mixed with 30% of fly ash based on dry weight.
3. Moreover a control sample (raw sludge) was prepared.
The sampling and analysis of the samples followed the same protocol as in the 1st phase.

3. RESULTS AND DISCUSSION

1st Phase

During this phase the sludge obtained from the municipal wastewater treatment plant of Kozani was mixed with lime and fly ash in various ratios. The first parameter measured was pH, and the corresponding values as a function of stabilization time are shown in Figure 2.

In general, the mixtures upon the addition of alkaline agents exhibited high pH values, reaching up to values of 12. The highest pH values were observed for lime addition in sample 1. The increase of stabilization time resulted in the reduction of the pH values to 8. In the control sample, pH values of about 7 were observed up to the 40th day of stabilization time, while a significant decrease was observed at about the 60th day, with pH values of 4. This may be attributed to the reactions of anaerobic digestion that might have taken place.

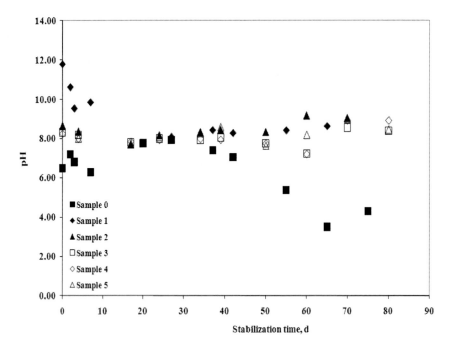

Figure 2. pH values of the mixtures as a function stabilization time.

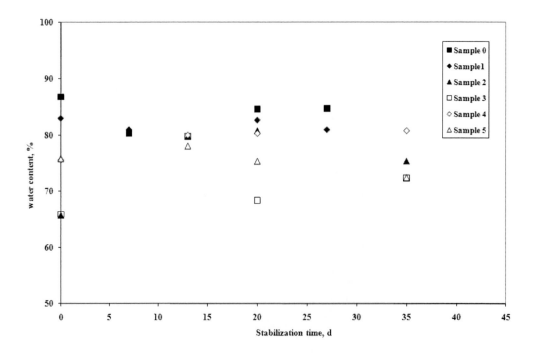

Figure 3. Water content of the sludge/alkaline mixtures as a function of stabilization time.

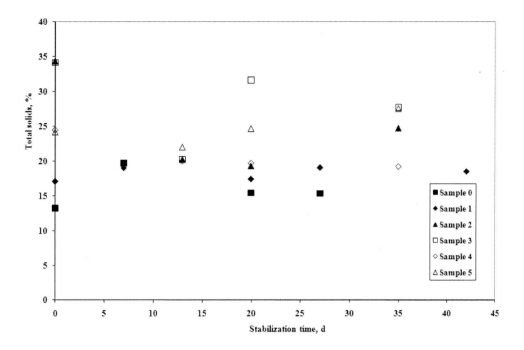

Figure 4. Total solids content of the sludge/alkaline mixtures as a function of stabilization time.

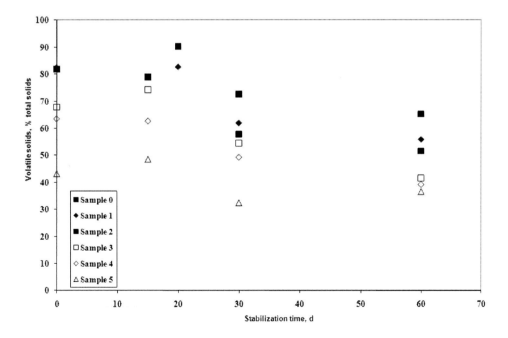

Figure 5. Total volatile solids contents of the sludge/alkaline mixtures as a function of stabilization time.

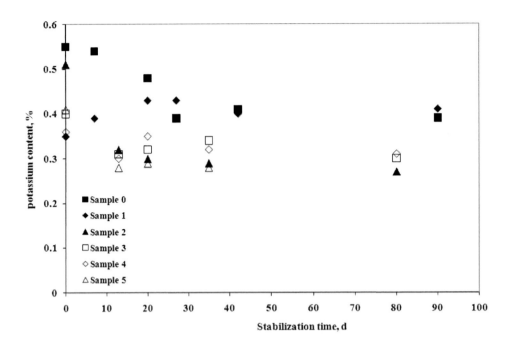

Figure 6. Potassium content of the sludge/alkaline mixtures as a function of stabilization time.

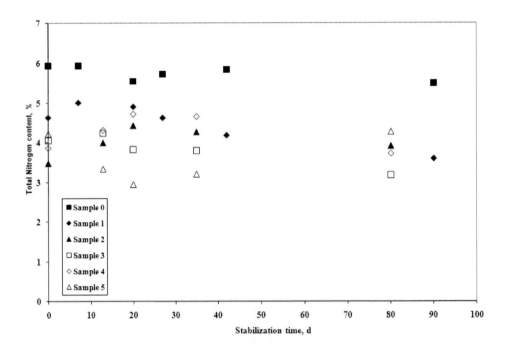

Figure 7. Total nitrogen content of the sludge/alkaline mixtures as a function of stabilization time.

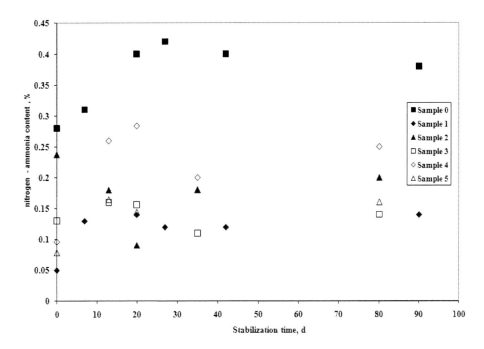

Figure 8. Nitrogen - ammonia content of the sludge/alkaline mixtures as a function of stabilization time.

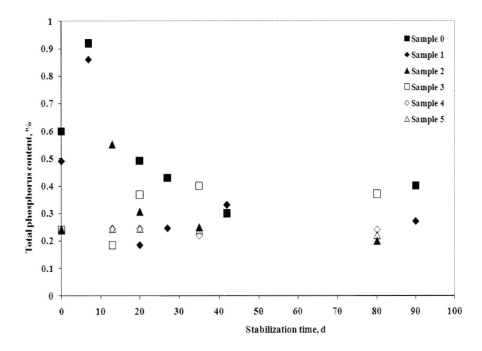

Figure 9.Total phosphorus content of the sludge/alkaline mixtures as a function of stabilization time.

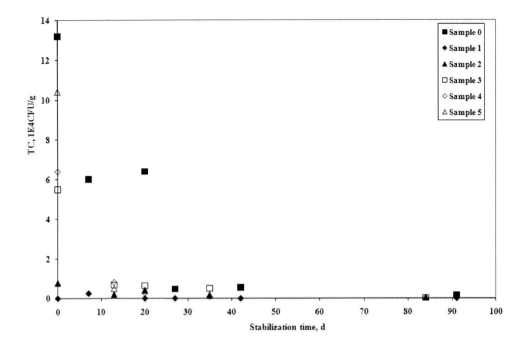

Figure 10.Total Coliforms content of the sludge/alkaline mixtures as a function of stabilization time.

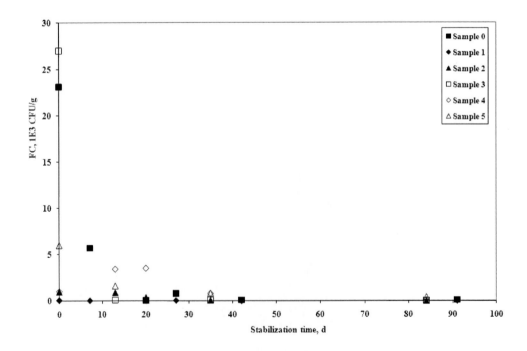

Figure 11. Feacal Coliforms content of the sludge/alkaline mixtures as a function of stabilization time.

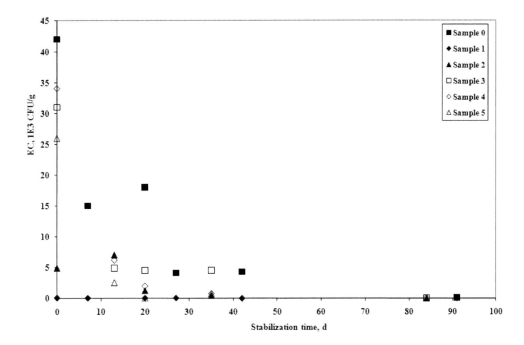

Figure 12. E. Coli content of the sludge/alkaline mixtures as a function of stabilization time.

The initial stage of such reactions is taking place through oxygen consuming bacteria, that break down the organic constituents to simpler form of organic acids; it is the presence of these organic acids which result to the decrease of the pH values.

The water content, total solids and volatile solids concentration as a function of stabilization time are shown in Figures 3, 4 and 5 respectively. The water content in the control sample was maintained over 80% during the whole stabilization time; however the addition of lime and fly ash resulted in the decrease of the water content with the increase of the stabilization time.

The samples containing fly ash above 10%, exhibited the lowest water content. The stabilization time had no effect on the water content in the mixtures. The total solid content in the control sample ranged between 13 and 20%, while in the samples with the addition of lime and fly ash the total solids content ranged between 17 and 34%. The highest total solids content was observed in the samples with the highest fly ash content.

Volatile solids content is a significant parameter for the stabilization process, as a low volatile solids content is related not only to the low microbial content but also minimizes the unpleasant smells during reuse. The decrease of volatile solids content with the increase of stabilization time is shown in Figure 5; as shown volatile solidsn percentages were ranging from 65 – 90% of total solids. The most significant reduction of volatile solids was observed in the mixtures with fly ash, where the volatile solids content was only 35% of the total solids content.

The profiles of potassium, nitrogen (total and ammonia - nitrogen) and total phosphorus in the mixtures are shown in Figures 6, 7, 8 and 9 respectively. In general, the control sample exhibited high content of nutrients, while the mixtures showed decreased contents due to the dilution of sludge with lime or sludge. Stabilization time seemed to have no effect on the potassium and total phosphorus content but variations were observed in nitrogen ammonia content. These variations may be attributed to several factors that enhance the ammonia removal, such as water content, pH and temperature.

One of the most important parameter in the stabilization of sludge is the efficient reduction of the microbial content. The microbial content in this study was assessed by the measurement of Total Coliforms, Feacal Coliforms and E. Coli, the corresponding results are shown in Figures 10, 11 and 12 respectively. It is observed that the control sample exhibited the highest microbial content, which was decreased with the stabilization time. The addition of lime and sludge decreased significantly the microbial loading; however the addition of lime achieved higher reduction of the microbial loading due to the higher pH of the mixtures respectively. The highest reduction was observed just after the mixing of the sludge with lime. A low reduction of the microbial loading was observed in the mixtures with the addition of fly ash; however at longer stabilization time significantly low microbial content was observed even in those mixtures. Moreover the addition of higher fly ash dosed was more efficient in the removal of microbial content.

2nd Phase

The pH value of the alkaline agents/sludge mixtures as a function of stabilization time in the 2nd phase is given in Figure 13. It is observed that the pH values of the mixture reached to values higher than 10. However, at extended stabilization time, pH values decreased to about 8. This may be attributed either to the interferences of the atmospheric CO_2 or to the depletion of the alkaline substances due to their interaction with sludge. In all cases, the pH of the mixtures was significantly higher than that of the control sample.

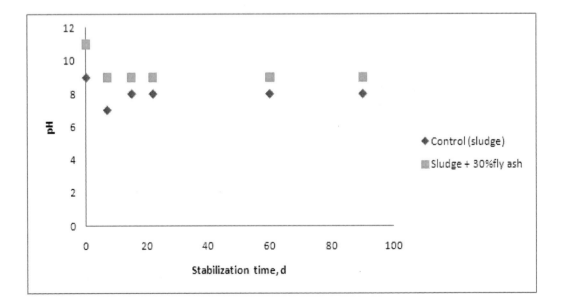

Figure 13. pH values of the sludge/alkaline mixtures as a function of stabilization time.

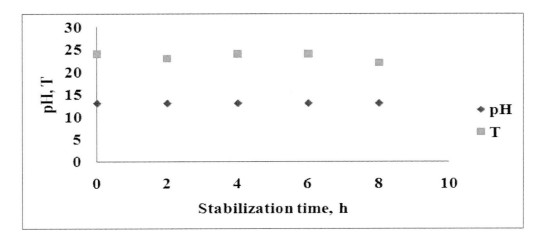

Figure 14. pH and temperature profiles of the mixture with limestone during the 8-hours stabilization period.

The pH and temperature profiles of the mixture with CaO are shown in Figure 14. The pH increased up to 13 just after the mixing; the stabilization time had negligible effect on pH. During the addition of the CaO, the temperature of the mixture increased due to the following hydrolysis reactions:

$$CaO + H_2O \rightarrow Ca(OH)_2 + 15 \text{ kcal}$$

However, during the subsequent short term stabilization period of eight hours, an additional increase of temperature was not observed.

The hydrolysis of CaO in short term experiments was certifies by the XRD analysis of the produced samples. The XRD plots of alkaline/sludge mixtures are given in Figures 15 – 19, together with the XRD plots of pure lime and limestone. As shown, the characteristic peaks of calcium hydroxide were not observed in the mixtures where CaO was added as the stabilization agent.

The short term stabilization time had negligible effect on the stability of the lime and sludge mixtures, since no change was observed in the position of the characteristic peaks (Figure 16). Similar results were observed for mixtures produced by addition of Ca(OH)2 and fly ash, as well as for extended stabilization periods (Figures 16 – 19).

As shown in Figures 16 to 19, the stabilization time had no effect on the type of chemical compounds that were formed during mixing, since the position of characteristic peaks remained constant in samples collected at different periods.

The changes in water content and the solids and volatile solids content of the mixture containing 30% fly ash with stabilization time are presented in Figures 20 and 21. As gobserved in Figure 20, water content in the control sample remained as high as 80% during the whole stabilization time, while inthe mixture produced by the addition of fly ash in sludge, the water content decreased with stabilization time up to 60%. In the mixture with CaO water content did not change with reaction time.

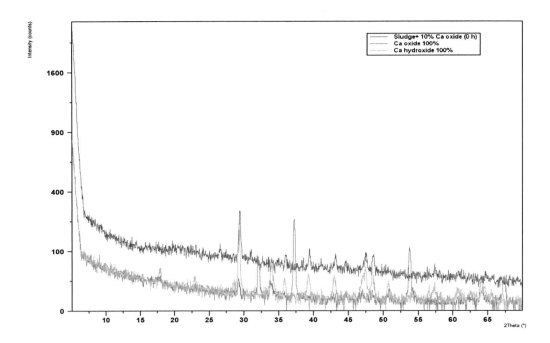

Figure 15. XRD analysis of the mixture with 10% CaO and sludge.

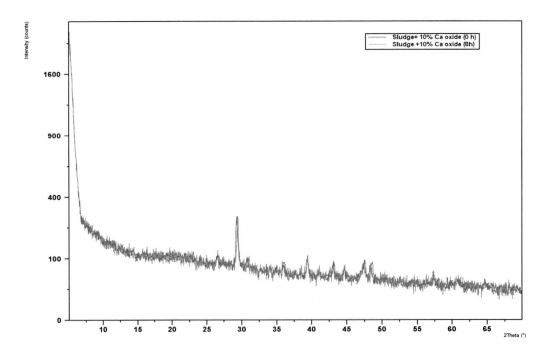

Figure 16. XRD plot of the mixture with 10% CaO at 0 and 8 h stabilization time.

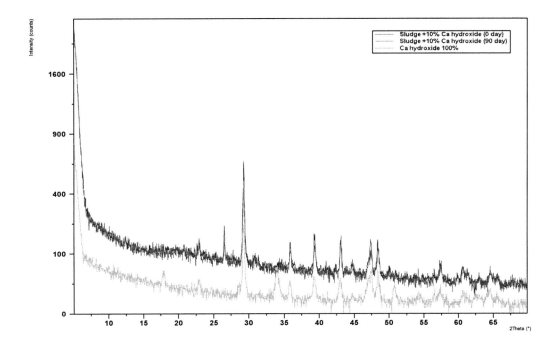

Figure 17. XRD plots of the mixture with 10% Ca(OH)2 at the beginning (0 d) and the end of the stabilization period (90 d).

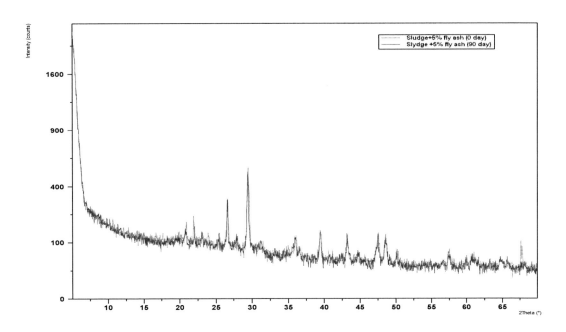

Figure 18. XRD plots of the mixture with 5% fly ash at the beginning (0 d) and the end of the stabilization period (90 d).

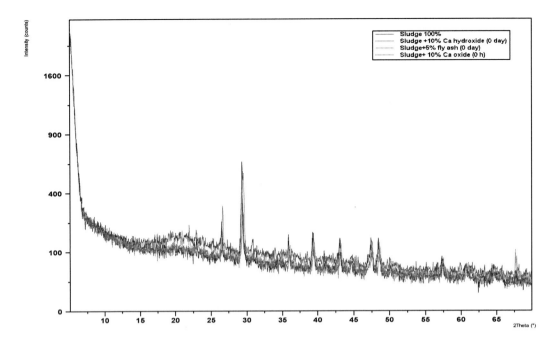

Figure 19. XRD plots of the control sample at the beginning (0 d) and the end of the stabilization period (90 d).

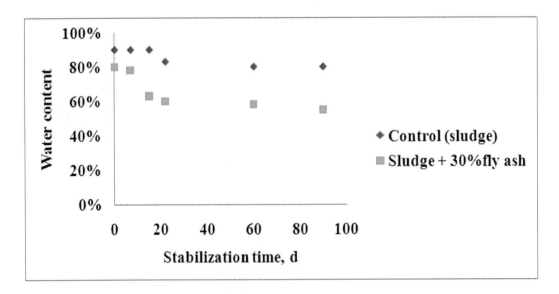

Figure 20. Water content as a function of stabilization time for the control sample and the mixture containing 30% fly ash.

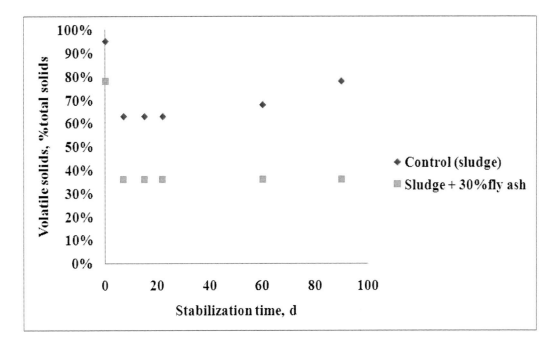

Figure 21. Volatile solids content as a function of stabilization time for the control sample and the mixture containing 30% fly ash.

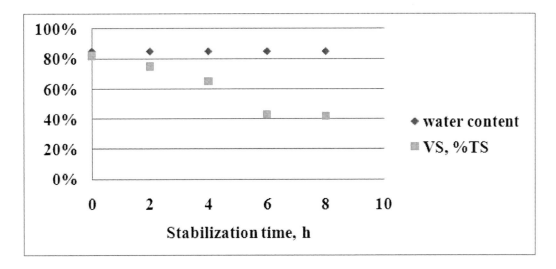

Figure 22. Water content as a function of stabilization time for the sludge/CaO mixture.

The volatile solids content is presented in Figure 21; as in the control sample, the initial solids content reached to 89% and was gradually decreased with stabilization time to 63%. In the mixture containing 30% fly ash, a significant reduction of the volatile solids content to about 40% was observed, in short stabilization time.

Similar results were obtained for the mixture of sludge with CaO (Figure 22). The volatile solids content decreased to 46%, while the water content was higher in this sample than in the mixtures with fly ash.

The nitrogen concentration as a function of stabilization time is shown in Figure 23. The control sample presented the highest nitrogen content up to 5.8%, while in the mixture of sludge with fly ash, the nitrogen content was about 3%. Stabilization time had a negligible effect on the nitrogen content.

The microbial content of raw sludge as a function of time is given in Table 5. In the mixtures with CaO and fly ash the microbial content decreased to non detection limits (below 9 CFU/g), in short times after the addition of the alkaline agents, and the low values were maintained for the whole stabilization period.

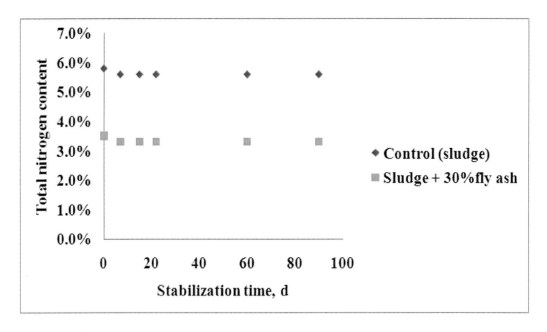

Figure 23. Nitrogen content as a function of stabilization time in the control sample and the mixture containing 30% fly ash.

Table 5. Microbial content in the control sample with stabilization time

Microorganism	0 day	14 days	60 days
Total Coliforms, $x10^4$	7.9	5.3	1.4
Feacal Coliforms, $x10^3$	35	17	8
E. Coli, $x10^3$	30	12	5.7

CONCLUSION

The stabilization potential of sewage sludge was examined in this work, by the addition of alkaline agents including fly ash, lime and limestone at concentrations reaching up to 30% on a dry weight basis. The produced mixtures were stabilized for a period of three months, and their properties were monitored for the determination of stabilization rate. The addition of fly ash at the highest dosage resulted in the production of a stabilized sludge product, due to the alkaline pH values of the mixture. This mixture presented a low volatile content, around 35% of the total solids, a low moisture content, about 63%, while pathogens were not observed from the beginning of the stabilization process. However, the addition of fly ash at lower dosages was not as effective as the highest dosage, resulting in mixtures with a lower pathogen removal rate, although at extended stabilization time negligible microbes were measured. Comparison of fly ash results to lime and limestone revealed that limestone and fly ash were the most efficient agents for sludge stabilization; however, the benefits of the latter over limestone, such as integrated use of a solid waste, availability and low cost, indicate that fly ash could become a potential agent for sludge stabilization. As a result, the proposed method could be used for the stabilization of sludge, taking into advantage that its implementation might have an additional benefit: the exploitation of two materials, sludge and fly ash, which are today treated as solid wastes and are produced in huge amounts. The produced mixtures could have several applications i.e. soil amendment, daily cover in sanitary landfills, restoration of closed coal mines etc.

ACKNOWLEDGMENTS

This work was financially supported by the technical company MESOGEOS S.A. The assistance of the personnel of DIADYMA and especially of A. Grigoropoulos and K. Tritskas, in the preparation of the experimental facilities, is greatly appreciated.

REFERENCES

Adriano, D.C., Page, A.L., Elseewi, A.A., Chang, A.C., Straughan, I.A., 1980. Utilization and disposal of fly ash and other coal residues in terrestrial ecosystems: a review. J. Environ. Qual. 9, 333–344.

APHA-AWWA-WPCF, 1995. In: Clesceri, L.S., Greenberg, A.E.,Russell R.R. (Eds.),StandardMethods for the Examination of Water and Wastewater. 18th ed., USA.

Chaney, R.L., 1983. Potential effects of waste constituents on the food chain. In: Parr, J.F., et al. (Eds.), Land Treatment of Hazardous Wastes. Noyes Data pp. 50–76.

Chlopecka, A., Adriano, D.C., 1997. Zinc uptake by plants on amended polluted soils. Soil Sci. Plant Nutr. 43, 1031–1036

Council of the European Communities. 1986. Council Directive 86/278/EEC of 12 June 1986 on the Protection of the Environment, and in Particular of the Soil, When Sewage Sludge is Used in Agriculture. Community Legislation in Force. Document 386L0278.

De Bertoldi, M., Vallini, G., Pera, A., 1983. The biology of composting: a review. Waste Manage. Res. 1, 157–176.

EPA (U.S. Environmental Protection Agency). 1995. A Guide to the Biosolids Risk Assessments for the EPA Part 503 Rule. EPA 832-B-93-005. Office of Wastewater Management, U.S. Environmental Protection Agency, Washington, DC. September1995.

EU, Disposal And Recycling Routes for Sewage Sludge, Scientific and Technical sub – component report, 23 October, 2001

European Union. 2000b. Working Document on Sludge, 3rd Draft. ENV.E.3/LM. European Union, Brussels. April 27, 2000. The European Union On-Line.

Gendebien, A., C. Carlton-Smith, M. Izzo, and J.E. Hall. 1999. U.K. Sewage Sludge Survey-National Presentation. R&D Technical Report P 165. Environmental Agency, Bristol,UK.

Ghuman, G.S., Menon, M.P., Chandra, K., James, J., Adriano, D.C., Sajwan, K.S., 1994. Uptake of multielements by corn from fly-ash compost amended soil. Water Air Soil Pollut. 72, 229–285.

Inbar, Y., Hadar, Y., Chen, Y., 1993. Waste management, recycling of cattle manure: the composting process and characterization of maturity. J. Environ. Qual. 22, 857–863.

Korcak, R.F., 1985. Effects of coal combustion wastes used as lime substitutes in nitration of apples in three soils. Plant Soil 85, 437–441.

Matthews, P., ed. 1996. Global Atlas of Wastewater Sludge and Biosolids Use and Disposal. Scientific and Technical Report No. 4. London: International Association on Water Quality. 197 pp.

McCarty, G.W., Siddaramappa, R., Wright, R.J., Codling, E.E., Gao, G., 1994. Evaluation of coal combustion byproducts as soil liming materials: their influence on soil pH and enzyme activities. Biol. Fertil. Soils 17, 167–172.

McGrath, S.P., A.C. Chang, A.L. Page, and E. Witter. 1994. Land application of sewage sludge: scientific perspectives of heavy metal loading limits in Europe and the United States. Environ. Rev. 2:108-118

Petruzzelli, G., 1989. Recycling wastes in agriculture: heavy metal bioavailability. Agric. Ecosys. Environ. 27, 493–503.

Rampton, S. 1998. Let them eat nutri-cake: Merriam-Webster thinks our "biosolids" don't stink. (how the word biosolid became a dictionary term). Harper's Magazine. November 1998.

Robert, L.J., Winkler, M., 1991. Sludge Parasites and Other Pathogens.Ellis Horwood, New York, NY.

Tadesse, W., Shuford, J.W., Taylor, R.W., Adriano, D.C., Sajwan, K.S., 1991. Comparative availability to wheat of metals from sewage-sludge and inorganic salts. Water Air Soil Pollut. 55, 397–408.

Wong, J.W.C., Lai, K.M., 1996. Effect of an artificial soil mix from coal fly ash and sewage sludge on soil microbial activity. Biol. Fertil. Soils 23, 420–424.

Wong, J.W.C., Su, D.C., 1997. The growth of Agropyron elongatum in an artificial soil mix from coal fly ash and sewage sludge. Bioresour. Technol. 59, 57–62.

In: Sewage Sludge Management
Editors: A. A. Zorpas and V. J. Inglezakis

ISBN: 978-1-61324-393-0
© 2012 Nova Science Publishers, Inc.

Chapter 11

OCEAN DUMPING: AN OLD AND KNOWN SEWAGE SLUDGE METHOD

Antonis A. Zorpas[*] *and Irene Voukkali*

Institute of Environmental Technology and Sustainable Development,
Paralimni, Cyprus

ABSTRACT

"Sludge: back into the ocean?"- An article by D. V. Feliciano in the October 1981 issue of the Journal of the Water Pollution Control Federation (U.S.) on the subject of sludge dumping at sea focus attention on a matter of considerable importance in the protection of the total environment. "Controversy over ocean dumping of sludge is once again under public scrutiny, courtesy of the US courts which have ruled that the ocean should be considered as a viable alternative disposal site when the action will not unreasonably degrade the marine environment" is Feliciano's statement of the situation. The choice of the word 'disposal', it was pointed out, was meant to signify dispersal of sludge at sea under controlled conditions without causing undesirable effects or ecological damage, as opposed to the harmful consequences that could follow unrestricted dumping of sludge at badly chosen sites. Present methods of disposal of high-bulk low-toxicity wastes, such as sewage sludge, into shallow marine environments are beginning to be questioned by many environmentalists. The alternative options generally considered all have environmental costs. A novel approach of discharging such wastes as slurries into the deep ocean at depths of 4,000 m close to the abyssal sea bed is considered. Such disposal would fulfil all the criteria required by the Oslo Convention before dumping at sea can be permitted, in that it isolates the waste from Man's ambit and offers no threat to present or foreseeable uses of the ocean. It also seems to fulfil the criteria of the World Conservation Strategy, so long as the oxygen levels in the deep waters of the ocean are not reduced excessively. The improvements in the quality of coastal seas would offset the doubling in the economic cost of disposal.

---

[*] Website: www.envitech.org Tel: +357-23743440, Fax: +357-23743441, P.O.Box 34073, 5309, Paralimni, Cyprus. Email: antoniszorpas@envitech.org, antoniszorpas@yahoo.com.

1. GENERAL REVIEW

The micronutrients in municipal sludge enhance ocean productivity in much the same way as they enhance plant productivity when applied to the land. The organic solids of municipal sludge can also directly enhance production of valuable biomass in the ocean, particularly benthic biomass. Studies of municipal sludge and wastewater discharge into marine waters throughout the world have demonstrated that, if the application rate of municipal sludge or wastewater was optimized by utilizing larger sites, beneficial enhancement of marine production could be achieved without the adverse effects associated with excessive application rates and overfertilization. Existing evidence indicates that, although not risk- or disadvantage-free, beneficial use of municipal sludges in the ocean is a viable management option which, in some instances, is likely to have fewer negative environmental side effects than other feasible beneficial use options.

The nutrient (nitrogen and phosphorus) value of municipal sludge has been used for many years to enhance the growth of plants in agricultural production, home gardens, and forests. In addition, municipal sludge has been used as a soil conditioner and a ground cover to accelerate reclamation of strip-mined and other land use areas. While the public has serious concerns about the use of municipal sludge for some of these applications, the concept that municipal sludge can provide benefits by enhancing the productivity of plants on the land is generally accepted. In contrast, the concept that the nutrient and nutritive value of municipal sludge could be beneficially utilized in the ocean under appropriate conditions is equally valid and has been largely ignored. Since the scientific community has ignored this potential beneficial use of municipal sludge in the ocean, the public and the policy-makers have an almost universal perception that municipal sludge which is applied to the land is beneficially recycled, while sewage sludge applied to the ocean is dangerous and damaging. The inconsistency in these two views is obvious.

Although both land and ocean disposal of sewage sludges have been intensively researched, the attitude difference toward these two management alternatives has resulted in different research hypotheses and approaches being applied to the two alternatives. Research into the land application of municipal sludges has been aimed at developing the knowledge necessary to optimize the practice for maximum benefit with minimum environmental damage and human health risk. Potential problems, when found, have led to recom-mendations for lower application rates, changes in composting methods, or other such improvements in management practice. In contrast, research into the ocean application of municipal sludge has been dominated by a search for environmental or human health effects, with the philosophy that any adverse effect, no matter how small, would be sufficient cause to end the practice.

It has become clear that there are minor environmental effects and human health risks inherent in the ocean application of sewage sludge. However, these are no more serious than the adverse effects of improper land application, and it is apparent that these minor adverse effects could be reduced or eliminated by optimizing the methods of ocean application of municipal sludge.

In this chapter, we intend to briefly review technical information which demonstrates that municipal sludge can enhance marine production when applied to the ocean environment. We will present discuss the conceptual approach that would be necessary in order to realize the maximum benefit from ocean application of sludge, while minimizing environmental damage and/or human health risk. It is hoped that this chapter will stimulate further study and research of this concept, since the potential for non-ocean beneficial use of sewage sludges from industrial societies is limited by the availability of suitable land and by air and groundwater contamination problems. In some regions, application of municipal sludge to the ocean may be the only beneficial use that is practicable.

2. CONCEPTUAL APPROACH AND SUPPORTIVE EVIDENCE

Municipal sludge essentially consists of a low concentration (generally 3-8%) of natural organic particles suspended in water. The natural organic particles in municipal sludge contain trace metals and trace concentrations of synthetic organics that are present in all foods, the human body, and human wastes. Also, most municipal sludges contain small additional quantities of trace metals and synthetic organics that are derived from other sources, including domestic cleaning products, street run-off, and industrial wastes. Municipal sludges also contain low concentrations of micronutrients, such as nitrogen and phosphorus compounds. The concentration of these micronutrients in municipal sludge is generally low since large proportions of the nitrogen and phosphorus entering sewage treatment plants are discharged with the treated effluents in soluble form and, therefore, are not incorporated into the municipal sludge.

Municipal sludges can be beneficially applied either to the land or to the ocean, if the micronutrient elements can be utilized to provide the growth requirements of plants in nutrient-limited environments. In the ocean, the natural particulate organic matter (POM) in municipal sludge additionally can provide food for appropriate consumer organisms. In obtaining all of these benefits, it is important that any offsetting adverse effects are minimized which may be caused by the toxic trace metal, synthetic organic or microbial contaminants in the municipal sludge. These considerations must be taken into account in developing a conceptual approach to beneficial uses of municipal sludge, either on land or in the ocean.

In the coastal marine environment, production valuable to Man takes place within two distinctly different strata: the euphotic zone of the water column, and the benthic community. Primary production takes place in the euphotic zone, where it is utilized by pelagic food chains, culminating in species which are beneficial as food resources to Man. In addition, a proportion of the organic matter created through primary production is transferred to the benthos where it provides food for other food chains which also lead to organisms which have food value to Man. Therefore, municipal sludge could produce beneficial effects if it was used to enhance primary and/or secondary production within the pelagic food chain or the benthic food chain. It must be borne in mind, however, that enhanced productivity in the ocean may not always lead to beneficial enhancement of the environment, since the health of many communities, particularly benthic communities, is controlled by the critical balance between respiratory use of oxygen and the resupply of this essential element through either photosynthesis or mixing.

The natural POM and micronutrients in municipal sludge are essentially-the same as the natural organic matter and nutrients found in treated sewage effluents. The differences between treated wastewater effluent discharge and sewage sludge application to the ocean are simply differences in the concentrations of the materials discharged into the marine environment, and differences in the technical means by which the materials are discharged. However, there have been many more studies of the effects of treated wastewater effluents than of the effects of sewage sludge on the ocean environment, and these studies have shown that treated sewage effluent can enhance marine productivity under appropriate conditions. Because of the similarity between the two sewage treatment byproducts, these effluent effects studies are relevant to demonstration that enhancement of ocean productivity through sewage sludge disposal is a viable concept.

There is considerable evidence that the application of micronutrient-containing sewage wastes to the ocean can lead to substantial increases in phytoplankton and zoo-plankton productivity within the water column (Pike & Gameson, 1970; Eppley et al., 1972). In this regard, Sin-dermann (1976) stated: "….. increases in marine productivity and abundance of fish stocks may well be related to the fertilizing effects of domestic sewage ….For example, the North Sea has been estimated …. to receive 256 tons of phosphorous and 1548 tons of nitrogen daily from sewage and run-off. Earlier studies of fish catches in the southern North Sea ... pointed out that high nitrates and phosphates outside the Thames estuary were derived from the sewers of London, and that the catch of fish per unit area was about double the corresponding catch for the rest of the North Sea, the English Channel, and the Kattegat-Skagerak region. The higher average catch was attributed to nutrients from London".

The productivity of pelagic food chains in coastal areas could be enhanced by the application of municipal sludge into surface layers of the ocean, since municipal sludge contains beneficial micronutrients, particularly nitrogen compounds, and since nitrogen is generally a limiting nutrient, at least for mid-shelf, mid-latitude marine productivity. In addition, it would appear obvious that the POM in municipal sludge is directly utilized by pelagic animals, resulting in enhanced secondary productivity. These two beneficial actions of municipal sludge and sewage effluents would manifest themselves in much the same manner in a specific ocean region. Therefore, the marine productivity enhancement, to which Sindermann (1976) for example refers, may be a function of both of these mechanisms. The doubling of fisheries production around the Thames estuary may be due to a combination of both nutrient-enhanced primary productivity and POM-enhanced secondary production.

Unfortunately, relatively little is documented concerning the availability and nutritive value of sewage POM to pelagic animals. Nevertheless, whatever the relative contributions to primary productivity from nutrient fertilization and to secondary productivity enhancement from POM, it is clearly possible for properly-applied, treated sewage effluent and municipal sludge to increase fisheries production in the ocean.

The micronutrient concentrations in municipal sludge are limited and much of the nitrogen and phosphorous contained in the municipal sludge is not available to phytoplankton, since it is bound to the POM in the sludge. Therefore, although municipal sludge clearly could enhance pelagic productivity under appropriate circumstances, it is probably not optional for this application since its nutrient value is low.

In most areas where treated sewage effluent or municipal sludges are discharged into the ocean, the water column is sufficiently deep that primary productivity is insignificant or does not occur at the bottom of the water column. Therefore, the benthos in these areas relies upon the transport of fixed organic carbon from the euphotic surface layers to the bottom. However, the POM contained in sewage-derived material can provide a food source for the benthos, which utilizes it in the same way as the naturally-produced, fixed organic carbon transported to the benthos from the surface layers. Depending upon its characteristics and location, the POM in municipal sludge can be utilized by benthic filter-feeders or deposit feeders.

Many municipal wastewater outfalls throughout the world are located on the continental shelf, and the effects of these discharges on the benthos within the dispersal area surrounding these pipeline discharges have been intensively studied. The effects of municipal sludge POM which is transported to the benthos have also been intensively studied in areas where municipal sludge is dumped or discharged. These studies of the benthic effects caused by municipal sludge and treated effluent discharges have been more intensive than studies of pelagic effects, since the benthos are less variable in their distribution and are sessile.

The addition of POM to the sediments through ocean outfalls is of particular interest, since the discharge location is fixed and, therefore, it is possible to observe the changes that take place with distance from the discharge point. A gradient of application rates of the suspended solids is thereby formed, with the highest application rates seen at the discharge point and locations extremely close to it, while locations remote from the discharge point represent areas where the application rate is very low.

Studies of the impact of treated sewage effluent, municipal sludge and other organically-rich materials on the sedimentary in fauna have been carried out in many areas, including the New York Bight (Pearce et al., 1976), the Saronikos Gulf (Sheppard, 1977), the Gulf of Napoule, France (Bellan, 1979), the Wadden Sea, Netherlands (Essink, 1978), the North Sea Island of Sylt (Otte, 1979), Vancouver, Canada (Otte, 1979), the Thames estuary (Shelton, 1971), the English Channel (Jenkinson, 1972), the Firth of Clyde (Mackay et al., 1972: Halcrow et al., 1973), the Firth of Forth, Scotland (McLusky et al., 1978; Read et al., 1982, 1983), Southern California (Bascom, 1977), the German Bight (Rachor, 1977), and the German North Sea Coast (Cas-pers, 1976). Each of these studies has obtained data which describe the distribution of benthic fauna in an area where organically-rich sediments are found at or near a point of discharge, or near a dumping site of an organically-rich material, usually municipal sludge.

In each of the above studies, the distributions that have been observed are essentially similar. In the sediments with the highest organic carbon content, the diversity of benthic in faunal populations is reduced, although the biomass is usually increased. In these organically-rich sediments, species that dominate are usually deposit-feeders, particularly polychaetes and bivalve molluscs. In an area ringing the central portion characterized by the highest organic carbon content in the sediments, there is always a region of lower organic carbon enrichment where diversity of species is normal, but the biomass is higher than areas further removed from the high organic sediments. In these transition zones, filter-feeding species and their predators dominate. The available evidence from these studies support a conclusion that the dominant factors controlling changes in ecology of the benthos are the increased sedimentation rate and the greater availability of organic detritus. Cumulatively, these factors lead to competitive advantages for organisms with different feeding strategies than those which dominate in slower-accumulating, lower organic sediments (e.g., Caspers, 1976). In most of these studies, higher concentrations of trace metals and synthetic organic compounds have been found in the organically-rich sediments, compared to the coarser-grained sediments found farther from the discharge. However, the available evidence suggests that toxic contaminant levels in these sediments are not sufficiently high to have caused the observed changes in benthic ecology.

In certain of these studies where organic loading rates were extremely high, the zone at the point of discharge exhibits both reduced species diversity and reduced biomass. It would appear, therefore, that the changes that take place in benthic fauna in an area to which municipal sludge POM is applied are systematically related to the application rate. As application rate increases, the first effect is a beneficial one, that of increased biomass without substantial change in the species composition (species diversity). At higher loading rates, the biomass continues to increase, but shifts in species composition may occur toward deposit-feeders which are, in some cases, thought to be less readily included in beneficial food chains (Boesch, 1982). At even higher application rates, both species diversity and biomass may be substantially reduced. These changes are entirely analogous to the effects that would be observed by increasing application rates of municipal sludge or composted municipal sludge to a terrestrial ecosystem. At low rates of application, beneficial enhancement of productivity is observed; some plant species would be inhibited at higher rates; and only resistant plant species would survive at extremely high continuous application rates.

If ocean application of municipal sludge is to be used for beneficial enhancement of the benthos, these studies indicate that the application rate must be optimized to maximize enhancement of biomass, while maintaining a diversity and species composition which ensures that the additional biomass is incorporated in useful food chains. It should be emphasized that this optimization process may not involve choosing an application rate which causes no changes in benthic species composition, since certain compositional shifts may be beneficial (i.e. larger total populations of commercially valuable species may be obtained even though the populations of certain commercial species are somewhat reduced).

Several well-documented studies of municipal sludge disposal at ocean sites in the United States and Europe are particularly relevant to an understanding of the potential for beneficial use of municipal sludge applied to the ocean. Caspers (1976) reports a detailed study of a municipal sludge dumpsite used by Hamburg, West Germany. This study covered a several-year period, beginning shortly after dumping commenced. Caspers reported that massive populations of the mollusc Abra alba, a suction-feeder, developed within the dumpsite region during the first several years of dumping. This population developed at the expense of filter-feeding bivalves, but other components of the benthic ecosystem were relatively unaffected. The *Abra alba* population was reported to be consumed by substantial populations of various species of flatfish which became established in the area within several years. As the dumping continued, the Abra alba grew extremely dense and, in certain areas, anaerobic sediments developed. In these sediments, species diversity was substantially reduced such that no macrofauna were detected and large populations of a polychaete had displaced other benthic populations. The sequence of events described by Caspers represents the response of the benthic community to increasing loading of municipal sludge that took place over a several-year period after dumping began. During the initial period, beneficial changes occurred in response to what can be considered low, long-term application rates, while the adverse effects, including reduced species diversity and development of large polychaete populations in anaerobic sediments, occurred only in response to much larger, long-term loadings or application rates. If the additional municipal sludge dumping occurring in this area had been moved to another location at the time when the maximum enhancement of useful biomass in the initial dumping area had been reached, the adverse effects observed from long-term, high application rates at the geographically-limited dump site would have been avoided. This study is a good example of the problems that are created by attempting to restrict the disposal of municipal sludges in the ocean to a very small area. Much the same situation, of course, would occur if the application of composted municipal sludge to the land were to be contained in a very small area. It is important to realize that in the region of the German Bight studied by Caspers, large areas of benthos exist that are very similar to the benthos at the studied disposal site. Therefore, as practiced on land, it would certainly have been possible to use dispersed dumping locations or to rotate locations from year-to-year, or on shorter time scales, to maintain an appropriate application rate at a given location.

Watling et al. (1974), in a study of a municipal sludge dumpsite off Delaware Bay performed a decade after dumping began, observed changes that were similar to those observed by Caspers in the German Bight. However, after a decade of dumping, the benthic populations observed by Watling closely resembled the populations observed by Caspers in the German Bight after only several years of dumping. At the Delaware site, Watling et al. reported that the benthos consisted of a diverse and abundant fauna with very large populations of Nucula proxima, a deposit-feeding bivalve. It seems likely that the extreme changes observed at the German Bight dumpsite were not observed at the Delaware site, because the effective application rate of sewage particulates to the benthos at the Delaware site was much smaller than that at the German Bight site. At the Delaware site, the long-term application rate was probably close to the optimum application rate which was needed to enhance productivity without the negative consequences associated with over-application rates.

Thompson & Psuty (1984) studied the annual growth rates of the commercially-valuable ocean quahog, Arctica islandica, at, and in the vicinity of, a municipal sludge dumpsite utilized by Philadelphia during the period 1973-1980. Growth rings observed in ocean quahog shells are analogous to the growth rings found in trees and. therefore, the quahog's rate of growth in any given year within its several-decades-long lifetime can be estimated from the width of the ring. It was found that those ocean quahogs which inhabited areas receiving municipal sludge particulates grew faster during the years that dumping took place than they had previously, and faster than quahogs from control areas outside the range of the dumping influence. This Philadelphia site is a highly dispersive site and relatively small amounts of municipal sludge were disposed there. Therefore, the application rate at this site was very low, and no adverse effects on the benthic faunal composition were observed. Instead, the only indication of an effect was one of a beneficial increase in the growth rate of the commercially-valuable ocean quahog.

Chen & Orlob (1972) studied the changes in benthic populations that occurred around an outfall off San Diego during a 2 year period before, and a 7 year period following, initiation of the discharge. During the third and fourth years of discharge, digested municipal sludge was discharged in addition to the treated effluents. The discharge resulted in increased biomass and species numbers, without significant changes in community composition, in the area surrounding the outfall (stations approx. 1 mile from the discharge point). This observed enhancement appeared to respond to the additional carbon loadings during the period of municipal sludge discharge, and appeared to diminish when the municipal sludge discharge was discontinued. Chen & Orlob were able to estimate the beneficial application rate, since they "tentatively concluded that an increase of more than 3 mg g^{-1} of BOD in the sediments is the threshold at which one may anticipate an adverse reaction of the benthic animal population. The environment can accept and beneficially use some organic enrichment up to the threshold".

3. STRATEGY FOR BENEFICIAL USE OF MUNICIPAL SLUDGE IN THE OCEAN

In order to understand the changes that are needed in municipal sludge management if the potential benefits of municipal sludge are to be utilized in the ocean, it is necessary to examine the characteristics of our current disposal strategies. At present, two methods of disposal of municipal sludge in the ocean are utilized: discharge through outfalls, and dumping from barges. Discharges through outfalls are inflexible and difficult to manage, since the discharge location is fixed, and the initial dispersion is generally low (approx. 100 times). However, the discharge depth may be selected to some degree. Ocean dumping from barges is considerably more flexible since the discharge site can be easily moved and the application rate can be controlled by rotating disposal among locations on appropriate time scales. Ocean dumping provides a high potential for dispersion with initial dilutions of approximately 10,000 times. However, ocean dumping has several limitations, including its higher cost compared to pipeline discharges, and the technological difficulty of discharging the municipal sludge into other than the upper euphoric zone of the ocean where the maximum benefits of the municipal sludge may not be realized. Nonetheless, at the present time, the greatest disadvantage of ocean dumping is the philosophy which drives the practice of ocean dumping. Although ocean dumping of sewage sludge is inherently a dispersive practice which relies upon assimilation of the material into the marine ecosystem to minimize adverse effects, current United States policy adheres to a containment philosophy, which restricts the disposal of municipal sludge to very small dumpsites in specific locations, in a misguided and naive attempt to restrict the adverse effects of sewage sludge to this small location. In fact, this containment philosophy results in the creation of adverse effects which need not occur, since this policy dictates that application rates in the chosen small site are extremely high.

Any strategy for ocean productivity enhancement must take into account the following principles:

1. Municipal sludge must be dispersed effectively to minimize potential impacts from excessive application rates.
2. The application rate must be controlled to maximize the productivity of valuable species.
3. As with all other municipal sludge management options, the concentrations of toxic trace metal and synthetic organics in the municipal sludges must be minimized through pre-treatment and other toxic control strategies.
4. Areas where beneficial application of sewage sludge can be practiced must be carefully selected. The existing information suggests that at least two types of application areas may be appropriate: mid-shelf areas characterized by sediments with low organic content and a relatively impoverished benthic fauna; and shallow areas with fine grained muddy sediments, such as those found in the German Bight.

4. COMPARATIVE ADVANTAGES OF BENEFICIAL USE OF SEWAGE SLUDGE ON LAND AND IN THE OCEAN

It is instructive to compare the advantages inherent in utilizing municipal sludge to enhance productivity on the land, which is an accepted practice, and in the ocean, which is at present not accepted. Application of municipal sludge to the land, usually in composted form, has several disadvantages, including the potential for: groundwater contamination; contamination of crops with trace metals and toxic organics; build-up of contaminant concentrations in the soil (particularly cadmium and copper) and, consequently, limited application rates and lifetime loadings; high cost; limited reliability of pathogen and odour control afforded by the existing composting technology; and adverse public reaction. In contrast, application of municipal sludge to the land has the advantage that the nutrient value of the municipal sludge is maximized by its use in a single-crop, high-yield system, and it is possible, although probably unrealistic, to envisage land application as a containment approach where the toxic contaminants are contained in a given location for future management.

In contrast to the use of municipal sludge for productivity enhancement on the land, ocean productivity enhancement would have several disadvantages including the inability to contain the toxic contaminants for future management, and the poor control at present of target organisms for enhancement, which would lead to a relatively low yield of harvestable food compared to land application. Of course, the first disadvantage is of no consequence if the concentrations of the toxic materials in municipal sludge are reduced to the lowest possible levels, and if the material is effectively dispersed and assimilated in the ocean. Productivity enhancement through sewage sludge disposal in the ocean has several major advantages. These advantages include: the availability of large areas for application of the increasing quantities of municipal sludge that are being generated by major coastal cities; the lack of long-term contaminant build-up in the sediments; the lack of biomagnification of metals through the marine food chain, as compared to land food chains; the reduced need for pathogen control since many pathogens are inactivated in sea water; the probable reduction in human health risk (seafood contamination vs. groundwater and crop contamination vs. air pollution); and the low cost of the management technique.

It is clear that the use of municipal sludges for ocean productivity enhancement is not without risk and potential disadvantages. However, the same statement is also true for the use of sewage sludges for land productivity enhancement and all other practicable municipal sludge management options. Perhaps the most compelling argument in favour of further research on the potential beneficial uses of municipal sludges in the ocean is the public perception regarding the disposal of municipal sludges. Although, at present, the public is not in favour of ocean disposal of materials since it considers the practice to be damaging to the ocean, the public, in most developed nations, is much more intensely opposed to the disposal of these same materials on the land or into the atmosphere, both of which are closer to their living environment. Therefore, the beneficial use of municipal sludge in the ocean, properly researched and managed, may be a more acceptable practice for long-term municipal sludge management than land application.

In summary, properly managed ocean disposal of municipal sludge would have several benefits, including the conservation of scarce land resources, the protection of groundwater, the minimization of atmospheric pollution, the minimization of human health risks, reduced treatment costs, and finally ocean productivity enhancement.

5. TOXICITY OF SEWAGE SLUDGE TO MARINE ORGANISMS

Sewage sludges generally consist of 1-5% dry solids, depending on factors such as treatment works capacity, thickening methods, and rainfall. Because sewers often collect effluent from agricultural and urban runoff, street drains, domestic sinks and toilets, hospitals and factories, almost anything can occur in sewage. In the UK, the four main disposal routes for sewage sludge are, tipping on land (22% and 25%), spreading on agricultural land (45% and 40%), incineration (4% and 10%) and disposal at sea (29% and 25%) (Collinge & Bruce, 1981 and Topping, 1987, respectively).

Guarino et al (1975) in the USA, and Collinge & Bruce (1981) in the UK, calculated that dumping at sea was the cheapest financial option. Disposal by this method has been practised since the last century from 1887 in the Thames estuary in south-east England (Norton et al, 1981), from 1888 off Garroch Head in south-west Scotland (MacKay, 1986), and from 1890 in Liverpool Bay in north-west England (Vivian, 1986). Countries of the European Community intend to stop disposal of sewage sludge at sea by dumping and via outfalls by the end of 1998 (Department of the Environment, 1990), and the USA intends to do so by January 1992 (Walker & Paul, 1989). The cessation of dumping depends on alternative and practicable means of disposal being found, and such alternatives have been found by most US cities (Walker & Paul, 1989). Countries outside the European community and USA may begin, continue to, or increase the amount of sewage sludge they dispose of at sea Sewage sludge is therefore likely to continue to come into contact with marine organisms for the fore-seeable future. While this chapter is largely concerned with the disposal of sewage sludge by dumping at sea, the toxicity data is equally relevant for other means of disposal into the marine environment (e g. outfalls).

6. CASE LABORATORY STUDIES ON SLUDGE TOXICITY

Lloyd & Tooby (1979) have pointed out that when the toxicant concentration has not been shown to be within 10% of the initial concentration or may not have remained constant during the test, then the LC_{50} (median lethal concentration for 50% of test organisms) should correctly be called $LC(I)_{50}$, as only the initial concentration of pollutant would be known. All sewage toxicity tests would fall into this category. It should be noted that significant toxic effects will occur at concentrations less than the LC_{50} value. Results of toxicity tests of sewage on marine organisms are listed in Table 1, where repeated tests showed different effective concentrations, the range of these is given. The concentrations are expressed as percent volume of sewage sludge in the test water. The solids content of these sludges varied, but the toxicological significance of such variation is unknown.

In Britain, Read (1977) found a 96 h LC_{50} of 19 8% sewage sludge at 20°C for one species of polychaete worm (Scolelepsis squamata (Muller)), but no mortality at up to 20% sludge for another (Malacoceros fuliginosus (Claparede)) for Edinburgh sewage sludge (Table 1) Costello & Gamble (1992) have found toxic effects on herring (Clupea harengus L) and cod (Gadus morhua L) embryos and larvae at 0 1% (i.e 1 part per 1000) and 1 0% sewage sludge from the Shieldhall treatment works in Glasgow. For this same sludge, respiration rate and clearance rate of mussels (Mytilus edulis L) were adversely affected at all concentrations studied, namely 0 021%,0 084% and 0 42% dilutions (Table 1). Franklin (1983) determined 24 and 96 h LC_{50} values of British sewage sludges for species of fish (Agonus cataphractus (L)), adult and larval shrimp (Crangon crangon (L)), mussel (M edulis), cockle (Cerastoderma edule (L)), and periwinkle (Littorina littorea (L)) (Table 1). She found the 96 h LC_{50} to generally fall between 0 1% and 1 0% dilution for 13 batches of sludge from 10 different treatment works in England. In some tests, the shrimp larvae were over 530 times more sensitive, with a 24 h LC_{50} of 0 0003% at 20°C (Table 1). However, Franklin (pers comm) believes the latter result to be anomalous. Furthermore, using a similar test design the lowest 24 h LC_{50} determined for shrimp larvae of the same species by Fretwall (1989) for Shieldhall sludge was 0 3% at 20°C (95% confidence limits, 0 36% and 0 42%), >2 0% at 15°C, and >5 0% at 10°C. Lack & Johnson (1985) found inhibition of growth of a hydroid (Laomedea flexuosa Alder) at concentrations of 1% and 2% sludge (Table 1), but water samples taken in the wake of the ship during sludge dumping off Plymouth (S W England) did not inhibit the hydroid growth. In tests with mussel larvae, Whitelaw & Andrews (1988) found a toxic threshold of 0 5% dilution for the soluble phase of sewage sludge, and did not observe any toxic effects of sediments taken from the Thames dump site.

In the United States, Miller et al (1987) found 96 h LC_{50} values of 1-8% for fish larvae (Cyprinodon variegatus, Menidia menida, M beryllina) and juveniles (M menida), 0 5-2 6% for mysid larvae (Mysidopsis bahia), and <0 02-0 36% for copepods {Eurytemora herdmani, Pseudocalanus minutus) for three New York and New Jersey sewage sludges (Table 1). In comparison, Chapman (1985) found a UK sludge to be marginally less toxic (Table 1) to another copepod (Acartia clausu Giesbrecht). From the results of regulatory toxicity testing of sewage sludges from 20 treatment facilities dumped in the New York Bight, 96 h LC_{50} averages of 3 2% and 1 3% were found for fish (n =138) (M menida) and mysids (n = 139) (M bahia) respectively, and a 96 h EC_{50} of 15 9% was obtained for a planktonic diatom (Skeletonema costatum) (Santoro & Fikslin, 1987). However, the minimum 96 h LC_{50} and EC_{50} found by Santoro & Fikslin (1987) was 30-times (for the fish), 260-times (for the mysids), and 20 times (for the diatom) less than the average toxic concentrations (Table 1). Concurrent with the MERL mesocosm study described below, Frithsen et al (1989) measured the fertilisation ability of sea-urchin sperm {Arbacia punctatula) which had been subjected to sewage sludge for 1 h, and generally found an EC_{50} of 1 0% sewage (range m Table 1). Anderson & Hunt (1988) reported a 'no observed effect concentration' of 0 56% for the germination of kelp zoospores (Macrocystis pyrifera Agardh).

Table 1. Summary of Minimum Effective Toxic Concentrations (Mortality, Hatching, Feeding, Growth, Respiration, Filtration, Fertilisation, Germination) and LC50. Averages of Sewage Sludge on Marine Organisms Effective Concentration is Expressed as Percentage Sludge in the Test Seawater Results of Tests on the Liquid Fraction Only are in Brackets

Organisms	Life stage	Duration of test[a]	Effect	Effective consentration range (%)	Ref[b]
Pisces					
Agonus cataphractus	Adult	96h	LC_{50}	0 28->1 0	2
Clupea harengus	Larvae	24h	Mortality	1 0	11
	Larvae	96h	Mortality	0 1-1 0	11
	Embryo	6d	Hatching	0 1-1 0	11
Cyprinodon variegatus	Larvae	96h	LC_{50}	5 0-7 5	6
Gadus morhua	Larvae	24h	Mortality	0 1-1 0	11
	Larvae	4d	Feeding	0 1-1 0	11
	Embryo	6&9 d	Hatching	1 0	11
	Juvenile	79d	Growth	0 01	12
Limanda limanda	Juvenile	85d	Growth	0 03	12
Menidia menidia	Adult	96h	LC_{50}	0 72-2 87	4
	Adult	96h	LC_{50}	(0 56-3 95)	4
	Juvenile	96h	LC_{50}	1 0-2 6	6
	Larvae	96h	LC_{50}	3 9-6 5	6
	Juvenile	96h	LC_{50}	0 11-1 65	7
Menidia beryllina	Lavrae	96h	LC_{50}	2 3-8 1	6
Crustacea					
Anostraca					
Artemia salina	Larvae	24h	LC_{50}	8 5-11 75	10
Copepoda					
Acartia clausu	Adult	48h	LC_{50}	0 1-2 0	3
Eurytemora herdmani	Adult	96h	LC_{50}	<0 02-0 36	6
Pseudocalanus minutus	Adult	96h	LC_{50}	0 11-0 19	6
Temora longicornis	Adult	24h	LC_{50}	0 1	3
Decapoda (Natania, shrimps)					
Crangon crangon	Adult	96h	LC_{50}	0 21->1 0	2
		24h	LC_{50}	0 16->10	2
	Adult	96h	LC_{50}	0 42-1 69	10
	Larvae	24h	LC_{50}	0 0003-0 1	2
	Larvae	24h	LC_{50}	0 36->5 0	10
Palaemonetes pugio	Adult	96h	LC_{50}	1 51-7 85	4
	Adult	96h	LC_{50}	(1 87-8 40)	4
Mysidacea					
Mysidopsis bahi	Adult	96h	LC_{50}	0 01-1 64	4

Organisms	Life stage	Duration of test[a]	Effect	Effective consentration range (%)	Ref[b]
	Adult	96h	LC$_{50}$	(0 36-2 80)	4
	Juvenile	96h	LC$_{50}$	0 51-2 6	6
	Juvenile	96h	LC$_{50}$	0 005-4 2	7
Mollusca					
Cerastoderma edule	Adult	96h	LC$_{50}$	0 54->1 0	2
Littorina littorea	Adult	96h	LC$_{50}$	>1 0	2
Mytilus edulis	Adult	96h	LC$_{50}$	>1 0	2
Mytilus edulis	Adult	31d	Respiration clearance rate condition	0 021-0 42	12
Echinodermata					
Arbacia punctatula	Sperm	1h	Fertilization	0 15-1 63	9
Polychaeta					
Malacoceros * fuliginosus	Adult	96h	LC$_{50}$	>20 0	1
Scolelepis squamata**	Adult	96h	LC$_{50}$	19 8	1
Hydrozoa					
Laeomedea flexuosa***	Adult	d	Growth	10-20	5
Algae					
Isocrysis galbana	Adult	48h	Loss from suspension	10	3
Skeletonema costatum	Adult	96h	EC$_{50}$	0 77-42 0	7
Macrocystis pyrifera	Zoospore	48h	germination	0 56-1 0	8
Mesocosm					
Phytoplankton	assemblage	4mo	Depressed bloom	1 0-10 0	9

[a] h = hours, d = days, mo = months.

[b] References 1 = Read (1977), 2 = Franklin (1983), 3 = Chapman (1985), 4 = Fava et al (1985), 5 = Lack & Johnson (1985), 6 = Miller et al (1987), 7 = Santoro & Fikslin (1987), 8 = Anderson & Hunt (1988), 9 = Frithsen et al (1989), 10 = Fretwell (1989), 11 = Costello & Gamble (1992), 12 = Costello, M J, unpublished (effective concentration is 24 h average of sewage sludge added once daily in flow-through aquaria).

Reported as *Scololepis fuliginosa, **Nerine cirratulus, ***Campanularia flexuosa.

7. MESOCOSM STUDIES

To date, some of the available mesocosm studies on the toxicity of sewage sludge in seawater have been at the land-based MERL (Marine Ecosystems Research Laboratory) mesocosms (Oviatt et al, 1986, 1987, Frithsen et al, 1989). These consisted of 5 5 m high and 1 8 m wide cylindrical tanks, containing 13 1 m^3 of seawater, and 0 4 m deep sediment. Water constantly flowed through these tanks and a plunger mixed the water Phytoplankton, zooplankton and benthic invertebrate populations were studied, and a wide range of nutrients and metals were routinely analyzed.

Early MERL mesocosm studies compared six nutrient-only additions, consisting of multiples of the nitrogen phosphorous silica ratio in sewage sludge, and three controls (Oviatt et al, 1986). For even the lowest nutrient treatments, effects on system production and respiration in the water column were detected. Subsequent work used three nutrient treatments and three sewage sludge treatments (Oviatt et al, 1987). No toxic effects on phytoplankton were reported and a reduction in benthic Crustacea was ascribed to hypoxia (Oviatt et al, 1987).

Subsequently, the toxicity of three duplicated sludge treatments, representing 0 1%, 1 0% and 10% concentrations was studied (Frithsen et al, 1989). The 0 1% and 1 0% treatments, and duplicate controls had nutrients added to approximate nutrient levels in the 10% treatment. At a 10% addition of sewage sludge the system became hypoxic Phytoplankton blooms were depressed in both 10% and 1% treatments despite an excess of nutrients, good transparency, and no changes in zooplankton abundance (Frithsen et al, 1989). Treatments generally stimulated production, reflected in greater abundance of zooplankton and benthic Crustacea and Polychaeta. Effluent toxicity, as measured by a sea-urchin sperm cell test, was not correlated with that of the mesocosms. However, this comparison was complicated by the phytoplankton blooms (in controls and treatments) which may themselves have had toxic effects (Frithsen et al, 1989).

8. DISPERSION AND DILUTION AT FIELD SITES

Disposal sites may be selected because of their ability to disperse or accumulate waste. In dispersal sites (e g Firth of Forth, Thames estuary, Dublin Bay), tracing the distribution and fate of sludge components is limited by their rapid dilution. MacKay (1986) argues that an advantage of accumulating sites (e g Garroch Head, New York Bight) is that the dispersal of persistent contaminants (e g PCBs) is localised, and hence their short and long term effects more easily detectable. However, anaerobic conditions are more likely to occur at accumulating disposal sites, and hence biodegradation of sludge (by bacteria and invertebrates) and mineralisation would be slower, and release of metals from sediments greater (Grunseich & Duedall, 1978, Nedwell & Lawson, 1990).

Sludge is largely dispersed from dump sites by tidal water currents, but summer thermoclines and winter winds may also be important (Eagle et al, 1979a, Murray et al, 1980b, Lewis & Riddle, 1989). The occurrence of an accumulation of sewage sludge at dump sites is difficult to quantify due to the differing rates of decay, dispersion, and biological processing of different sludge components being confounded by continuous inputs of freshly dumped material. However, at the Firth of Forth in south-east Scotland dumping is alternated between St Abb's Head and Bell Rock sites every six months, and this intermission allows sufficient time for almost all traces of sludge to be lost from the sites (Moore, 1981, Topping, 1987, Moore & Davies, 1987). Components of sewage, such as faecal bacteria, tomato seeds, mammalian sterols (e g coprostanol), spores of the bacterium Clostridium welchu (perfringens), and ratios of carbon and nitrogen isotopes, may be used to estimate the dispersion of dumped sludge on the seabed if the area is uncontaminated by other sources (e g sewage outfalls, estuarine inputs). (Shelton, 1971, Rau et al 1981, O'Connor et al, 1985, Vivian, 1986, Chapman et al, 1986, Venkatesan & Santiago, 1989, Rees et al, 1990, MAFF, 1990a, 1991b). In the Thames estuary, Talbot et al (1982) and Whitelaw & Andrews (1988) tracked sewage sludge labelled with radioactive silver for eight hours, and detected it in the bottom sediments six and four months later respectively.

Dilution varies according to the length and speed of discharge vessel, rate of discharge, and local hydrographic conditions (Lewis & Riddle, 1989). Spot dumping from a stationary vessel results in a concentrated discharge plume (Duedall et al, 1977), but discharge is normally from a moving vessel and below the waterline in its turbulent wake.

The immediate dilution of sludge may be estimated by its properties (e g salinity, pH, oxygen demand) and from a wide range of its components. Most estimates of sewage sludge dilution and dispersion have used parameters associated with the particulate fraction of the sludge, and concentrated on the settlement of the primary plume. Recent field studies by Clark (1989) tracked both the surface and primary plumes, and used measures of both particulate matter and soluble components as tracers.

Measurements of suspended solids by filtered or dried water samples, coulter counter, light transmission or acoustic tracking have been used to track particulate matter in the water column (e g Rohatgi & Chen, 1976, Eagle et al, 1978, Chapman et al, 1986, Jenkins & Gibbs, 1988, Clark, 1989), and remote sensing has been used to track surface plumes (e g Johnson et al, 1977). Generally, the sludge solids are diluted to less than 1% within 10 min (Eagle et al, 1978; Murray et al, 1980b, Norton et al, 1981, Jenkins & Gibbs, 1988, Whitelaw & Andrews, 1988, Clark, 1989), and to between 0 1% and 0 01% in 1 h (Chapman et al, 1986, Whitelaw & Andrews, 1988, Clark, 1989). The primary plume, comprising the bulk (but not all) of the suspended solids, generally settles to the seabed within about 4 hours of discharge (Eagle et al, 1978, 1979a, Murray et al, 1980b, Norton et al, 1984a) but may take up to 8 hours at the Thames dump site (Talbot et al, 1982, Whitelaw & Andrews, 1988).

Dissolved components, such as soluble metal complexes, phosphate and ammonia, have received rather limited attention. The dilution of the surface plume was estimated to be only 0 7% (1 140) in 30 min from measurements of ammonia at the New York Bight dump site (Duedall et al, 1977) Chapman et al (1986) found the surface plume lasted over 6 hours at the Tyne dump site (N E England), and reckoned it would survive over several tidal cycles. At the Liverpool Bay dump site, concentrations of metal associated with suspended matter are elevated on days when dumping occurs, and return to background levels within three days (Norton et al, 1984b). Thus substantial differences may be recorded in dilution rates depending on both the hydrodynamics at the site and which parameter is used as an indicator of sludge concentration.

9. FIELD STUDIES

In relation to the effects of dumping at sea, most research has consisted of environmental surveys of dump sites (e g Shelton, 1971, MacKay et al, 1972, Eagle et al, 1978, 1979a, 1979b; Murray et al, 1980a, 1980b, Norton et al, 1981, Murray & Norton, 1982; Bucke et al, 1983, Norton et al, 1984a, MacKay, 1986, Pearson & Blackstock, 1986, Moore & Davies, 1987, McVicar et al, 1988, Whitelaw & Andrews, 1988, Department of the Marine, 1989a, Rees et al, 1990, MAFF, 1990a, 1991a, 1991b). Factors studied include water quality, sediment characteristics, metal content of sediments and biota, distribution of invertebrates, fish diseases, and occurrence of faecal bacteria.

The emphasis of field surveys has been to determine effects on the seabed and its fauna (Norton & Rolfe, 1978). In most sewage sludge dump sites some degree of enrichment of the seabed occurs Increased levels of organic carbon and trace metals have been recorded in sediments at the dump sites at Garroch Head, S W Scotland (MacKay et al, 1972, MacKay, 1986, Topping, 1987, Clark, 1989, MAFF, 1991a), Lyme Bay, S W England (Eagle et al, 1978), Thames estuary, S E England (Norton et al, 1981, Talbot et al, 1982, Whitelaw & Andrews, 1988), and off the River Humber, N E England (Murray et al, 1980a). Increased metal levels have also been recorded at the dump site in Liverpool Bay (Norton et al, 1948a, MAFF, 1991a), and off the Humber estuary (MAFF, 1991b). At the Firth of Forth, two dump sites are used alternately at six monthly intervals, and while sediment organic carbon and trace metals may increase during the dumping period (Moore, 1981), there is a return to background levels before dumping resumes (Moore & Davies, 1987).

10. EFFECT ON BENTHIC FAUNA

Reductions in the number of species of benthic invertebrates have been recorded for the Lyme Bay (Eagle et al, 1978) and Thames dump sites (Norton et al, 1981, Talbot et al, 1982), and at sewage outfalls (e g Anger, 1975a,b, Aschan & Skullerud, 1990). Enrichment of the seabed is associated with high numbers of individuals at the Tyne site (MAFF, 1990a), and with both high numbers of species and individuals at the Dublin Bay dump site (Department of the Marine, 1989b). The change in the benthos to a fauna dominated by certain polychaete worms (e g Notomastus latericeus Sars, Polydora ciliata (Johnston), Cirratulus cirratus (Muller), Capitella capitata (Fabncus)), as occurs at the Thames (Norton et al, 1981, Talbot et al, 1982) and Garroch Head (MacKay et al, 1972, MacKay, 1986, Topping, 1987, Clark, 1989) sites, is indicative of the organic enrichment of the area (Dauer & Conner, 1980, Eleftheriou et al, 1981).

The rate of recovery of a dump site will depend on the physicochemical and biological characteristics of the site. While no impact on the benthic fauna of the Firth of Forth dump sites was detectable after four years (Moore & Davies, 1987), the benthos of the old (pre-1974) Garroch Head dump site was still recovering after 11 years (Clark et al, 1990).

At some sites, where sewage sludge effects can be detected at the macrobenthic community level, it may be difficult to detect toxic effects at the individual level as actual presence on a disposal ground may be because the individual organism is tolerant of (or have acclimated to) sewage sludge toxicity. However, Blackstock et al (1986) found enzyme activity levels in the polychaete Glycera alba (Muller) to correlate well with ecological measures of sewage effects.

Pollutants may have sublethal effects and reduce the ability of organisms to resist other stress, such as disease, parasitism or extremes of weather. At the New York Bight dump site and in laboratory aquaria, Young & Pearce (1975) associated shell disease of crabs (Cancer irroratus) and lobsters (Homarus americanus) with sewage sludge contaminated sediments Cross (1988) found the incidence of fin erosion and epidermal tumours in fish {Microstomus pacificus (Lockington)) was correlated with contaminant concentrations in sediments, effluent, fish tissues, and distance from a sewage outfall in California. However, while exposure to sewage sludge for 12 weeks significantly decreased growth, it did not affect the health (including body condition, blood parameters, disease, parasites) of a flatfish, the dab (L limanda), and roundfish, the cod (G morhua) (Costello, M J , unpublished). In dab, sewage sludge has also been shown to decrease protein growth by increasing protein turnover (Houlihan et al, in press), and while some immunological effects have been observed their biological significance at the individual level are uncertain (Secombes et al, 1991, 1992). Furthermore, disease surveys have been unable to link diseases and pollution in the open sea (Bucke et al, 1983, McVicar et al, 1988, Dethlefsen, 1988, Whitelaw & Andrews, 1988). Cellular and physiological bioassays on mussels placed at and near the Plymouth (southwest England) dump site also proved inconclusive due to insufficient environmental information, especially on sewage dispersion and natural conditions (e g food supply) which may have influenced mussel health (Lack & Johnson, 1985). However, mussels suspended in cages in the path of the sludge plume at the Thames dump site did exhibit greater sublethal stress ('scope-for-growth' bioassay) than mussels placed nearby (Whitelaw & Andrews, 1988).

11. EFFECTS ON WATER QUALITY

Within minutes of discharge, and depending on water stratification, the sludge separates into three phases, a surface slick, a low salinity surface plume, and a primary plume containing the bulk of the suspended solids (Jenkins & Gibbs, 1988, Clark, 1989).

Grunseich & Duedall (1978) found decomposition of sludge faster in aerobic than anaerobic conditions, and took about 12 weeks in the laboratory Oxygen depletion may occur at sites where sludge accumulates Dissolved oxygen levels less than 1 mg l^{-1} have occurred at the 12-mile New York Bight dump site (Young & Pearce, 1975). At UK dump sites, oxygen levels generally remain high (Topping, 1987, Nedwell & Lawson, 1990).

Total ammonia is comprised of non-ionised and ionised forms, whose relative proportions depend on pH and salinity. The former is considered the toxic component, and safe levels of 0 01-0 02 mg l^{-1} non-ionised ammonia have been proposed (Chin & Chen, 1987, Seager et al, 1988). High ammonia levels were recorded at sewage sludge dump sites by Duedall et al (1975, 1977) (up to 17 5 mg l^{-1} total ammonia, equal to 7 0 mg l^{-1} non-ionised), Draxler (1979) (up to 3 6 mg l^{-1} total ammonia) and Norton et al (1981) (up to 0 81 mg l^{-1} total ammonia). At the Garroch Head dump site in Scotland, Clarke (1989) found total ammonia concentrations from 1 14 mg l^{-1} 1 h after dumping to decrease to <0 22 mg l^{-1} after

2 h. In the New York Bight dump ground (an accumulating site), Duedall et al (1975) found ammonia generally increased from the surface to bottom waters in the range 0 014 - 0 126 mg l^{-1} non-ionised ammonia, and concentrations of non-ionised ammonia of about 0 07 mg l^{-1} remained after 12 5 h at 21 m depth. In laboratory tests, Grunseich & Duedall (1978) found ammonia continued to be produced for over 12 weeks in aerobic conditions during the decomposition of sewage sludge. However, ammonia production decreases as the sludge decomposes, for example, ammonia reached control levels within six days with a 0 3% dilution of sewage sludge in aerated seawater at 20°C (Fretwell, 1989).

Most of the trace metals are associated with suspended solids (Norton et al 1984b), and 80-90% of these are on fine particles of less than 10 μm (Chapman, 1986). Cadmium and nickel (Rohatgi & Chen, 1976, Norton et al, 1981, 1984b), and ammonia and phosphate (Draxler, 1979), may remain in solution for longer than less soluble metals and nutrients Hence, the fate of sewage sludge is determined by the differing physical and chemical characteristics (e g sedimentation rate, solubility) of its components, and their uptake by biota.

12. BIOACCUMULATION AND PERSISTENCE

Two reports assessed bioaccumulation of metals from sewage sludge in animals. Although unfed, the test animals may have eaten some of the sludge. Over 10 days, exposure to different sludges of up to 1% concentration showed no significant accumulation of cadmium, zinc, or copper in any of the species of fish (A cataphractus), adult and larval shrimp (C crangon), mussel (M edulis), cockle (C edule), and periwinkle (L littorea) studied by Franklin (1983). Over 60 days, one sludge at 0 1% concentration resulted in significant accumulation of zinc, and some accumulation of lead and copper in fish. Two sludges at 0 03% and 0 1% concentrations resulted in significant accumulation of zinc and copper in shrimp, and copper and lead in mussels Cadmium did not accumulate, and possible mercury accumulation may have been due to contamination of the seawater (Franklin, 1983). Maciorowski et al (1985) subjected fish (M menidia), hard clams (Mercenaria mercenaria (L)), and grass shrimp (Palaemonetes pugio), to dilutions of 0 033-0 0042% sewage sludge for 10 days. They found no bioaccumulation of cadmium, mercury, PCBs, DDT, or petroleum hydrocarbons, but stressed that such tests should only be considered to be part of the assessment of a wastes bioaccumulation potential.

Franklin (1983) noted that if her bioaccumulation values were considered alone, they would suggest that more mercury could be dumped at sea (as part of the sludge), than had already been shown to be desirable from field studies Mercury concentrations in fish muscle had decreased following a reduction in mercury content of sewage dumped at the Thames site (Preston & Portman, 1981, Murray & Norton, 1982, Franklin, 1983).

Chapman et al (1988) examined the ability of radioactive silver in sewage sludge to accumulate in Crustacea through their diet. They found assimilation and accumulation of the silver by suspension feeding barnacles (Elminius modestus Darwin), and mysids (Schistomysis spiritus (Norman)), and subsequent assimilation and accumulation by shrimp (C crangon) feeding on the treated mysids. As suspension feeding, scavenging and predatory

Crustacea are an important prey of most marine fish, it seems likely that substances from the sewage could travel further in the food chain. Indeed, Chapman (1985) previously found evidence of sludge ingestion by the copepod Temora longicornis (Muller) Subsequent production of faecal pellets by planktonic copepods may aid settlement and biological cycling of the sludge (Chapman, 1985). Chapman (1985) pointed out that an unknown proportion of fine particulates are ingested by zooplankton, while soluble components, such as ammonia, may be assimilated by phytoplankton and bacteria in the water column.

Faecal coliforms, which may survive 2-3 weeks in seawater, occur in sediments of at least the Thames, Bristol and Liverpool sites (Murray et al, 1980b, Norton et al, 1981, 1984a, Whitelaw & Andrews, 1988) Pearson & Blackstock (1986) recorded faecal bacteria in fish guts from the Garroch Head dump site Dab (L limanda) collected at the St Abb's dump site in south-east Scotland in 1989 had been feeding on sweetcorn, presumably derived from dumped sewage sludge (Stagg, R M, pers comm). Acoustic surveys of the passage of sludge through the water column at the Garroch Head dump site repeatedly recorded groups of fish associated with the primary sludge plume (Clark, 1989). It is therefore clear that marine macrofauana, as well as zooplankton, ingest components of sewage sludge.

Elevated levels of metals, organochlorines and polychlorinated biphenyls (PCBs), at least partly derived from sewage sludge, have been recorded in the tissues of finfish (Scomber scombrus L, L limanda, G morhua, Pleuronectes platessa L , Solea solea (L), Merlangius merlangus (L)) and crustaceans (Crangon sp , Cancer pagurus L , Pagurus bernhardus (L)) in Liverpool Bay (Murray & Norton, 1982, Norton et al, 1984a), organochlorines and mercury for finfish (Platichtys flesus, Clupea harengus, G morhua, P platessa, L limanda, M merlangus) in the Thames estuary (Murray & Norton, 1982), and metals in fish (Cheilodactylus fuscus) caught off the coast of Sydney, Australia (McLean et al, 1991). However, river and estuarine inputs can be a greater input of such contaminants to coastal waters than sludge dumping (e g Norton et al, 1984b). While MacKay et al (1972) recorded elevated metals in molluscs (Buccinium undatum L) and crustaceans (Crangon allmani Kinahan, Pandulus montagui Leach) from Garroch Head, subsequent surveys (Pearson & Blackstock, 1986) found no evidence of bioaccumulation of metals. The survey by Murray & Norton (1982) suggested that contaminant levels m biota were generally declining in Liverpool Bay and the Thames estuary None of the recorded contaminant levels were considered to have represented a hazard to public health (Murray & Norton, 1982, MacKay, 1986).

Considerable variation occurs in sludge toxicity between treatment works and days collected (Franklin, 1983, Miller et al, 1987, Frithsen et al, 1989), but subsequent storage conditions also influence toxicity in a temperature dependent manner (Miller et al, 1987, Frithsen et al, 1989). Toxicity may thus last longer at winter than summer temperatures (Frithsen et al, 1989). Using the microtox bacterial bioassay, Miller et al (1987) found the toxicity of sludge supernatant to increase by c 57% (from an average sludge EC_{50} of 5 4% to 3 1%) over 24 h. However, in laboratory tests with shrimp larvae, whole sludge (initial concentration 0 3%) toxicity decreased to control levels over six days at 20°C (Fretwell, 1989). Furthermore, in mesocosms there was no apparent build-up of toxicity over 4 months at 1% and 0 1% additions of sewage sludge (Frithsen et al, 1989).

13. CAUSE OD TOXICITY

Some laboratory toxicity tests have emphasised the importance of ammonia (Read, 1977, Fava et al, 1985, Costello & Gamble, 1992). Ammonia is highly soluble, toxic, and present in both liquid and whole fractions of sewage sludge. That both fractions had similar toxicity to fish and shrimp suggested to Fava et al (1985) that ammonia was the dominant toxicant, as anticipated by Read (1977). However, the liquid fraction was generally less toxic than whole sludge for a mysid (Fava et al, 1985) (Table 1) Fava et al (1985) thus suggested that particulate matter may have interfered with feeding, respiration and orientation in the mysid. The entrapment of planktonic copepods and fish larvae by sludge as been observed by Chapman (1985) and Costello & Gamble (1992) respectively Santoro & Fikslin (1987) found a significant relationship between solids (both total and suspended) and toxicity to fish and phytoplankton (see Table 1 for species tested). However, whether the toxicity related to physical interferences with the test organisms or interactions between particulates and toxic compounds, or both, is unclear (Santoro & Fikslin, 1987).

In mesocosm studies (Frithsen et al, 1989), toxicity could not be related to carbon, nutrients, chlorine, cadmium, chromium, copper, iron, nickel or zinc. However, Santoro & Fikslin (1987) correlated metal concentrations in sewage sludge with toxicity to fish, mysids and diatoms. They also found that sludges with more industrial waste were the most toxic. In experimental studies of the toxicity of sewage sludge to marine fish, concentrations of seven metals (cadmium, chromium, copper, lead, mercury, nickel, zinc) and 29 organic contaminants (17 organochlorine insecticides, 8 polychlorinated biphenyls, 4 synthetic pyrethroids) were below their individual toxic levels at dilutions of 0 1% (Costello & Gamble, 1992) and in seawater and aquarium sediments (Costello, M J, unpublished). The toxicity of sewage sludge at levels at which its components are below their individual toxic threshold would indicate an additive or synergistic toxic effect, but other undetermined toxicants may also be involved.

In field studies, changes in benthic macroinvertebrate communities appear to be caused by nutrient enrichment. The magnitude of the effects of such organic pollution may conceal other toxic effects.

14. DEEP-SEA BIODIVERSITY AND THE IMPACTS OF OCEAN DUMPING

The human imagination can inspire great scientific discovery. Such was the case when Captain Nemo and the crew of the Nautilus marveled at the abundance of animal life in the deep sea. It was novelist Jules Verne who came up with this idea for his submarine adventure Twenty Thousand Leagues Under the Sea published in 1870. Sixty years later, William Beebe's Half Mile Down described his dive sphere descents to the deep sea and provided glimpses of a strange new world of flashing lights and sea monsters. However, disproving the myth that the deep sea was an inhospitable desert took practically a whole century before an astonishing number of real animals from the seabed were unearthed by scientists Bob Hessler and Howard Sanders from the Woods Hole Oceanographic Institution (WHOI).

More recent studies have given scientists plenty of room to argue that the ocean bottom supports as diverse a habitat as any community on Earth. But between our limitless imaginations about the natural world and public recognition for its real value, can the resource be protected from exploitation, let alone be understood. The discovery that the deep sea may be every bit as biodiverse as a tropical rainforest comes at a time when pressure is mounting to use every available square foot of land for development. A major problem of the 21st century will be deciding what to do with the vast increases in waste that a growing population -forecast to double from five billion to 10 billion in the next century- will produce. The oceans (which cover seven-tenths of the planet's surface) are likely to become an option for waste management in the future.

At present, ocean dumping is predominantly banned by international law. The motivation for banning ocean dumping was sparked by the shutdown of beaches due to contaminated wastes from sewage-derived micro-organisms, the closing of shellfish beds due to metal contamination, and infection of fish by lesion-causing parasites. Coastal oceans continually enriched by nutrients in waste products that run off the land suffer from eutrophication resulting in an increase in toxic algal blooms and decreased oxygen levels, both of which can kill fish populations.

With more than 80 percent of the ocean at depths of more than 3000 m, the deep-sea floor may seem safe from the man-made disturbances that threaten terrestrial and coastal ocean environments. And yet most environmental litter from both natural and artificial waste (such as sewage sludge, mining tailings, fly ash from power stations, dredged spoils from harbors and estuaries, dangerous man-made organic compounds used for pesticides, weapons, and industrial uses, as well as packaged goods) makes its way to the sea floor over time.

The vast and remote deep-sea floor could make it appear like an attractive alternative for dumping. To determine the impact of waste disposal on bottom-living animals, the National Undersea Research Program (NURP) has supported numerous projects in the oceans and Great Lakes. Of particular concern to researchers are the effects of dumping on living resources and deep-sea biodiversity, as well as the transmission of contaminants back to the human population. In the most detailed study ever done related to the impacts of ocean dumping, NURP-funded scientists documented the impact of 42 million tons of wet sewage sludge dumped 2500 m (8000 ft) off the Mid-Atlantic coast between 1986 and 1992. One of the most significant environmental impacts detected at the "106-mile dumpsite," named for its location 106 nautical miles southeast of New York Harbor, was the restructuring of a community of deep sea organisms.

Two momentous developments laid the foundation for observations made at the 106-mile dumpsite. The first development was the invention of the box corer, a stainless steel trap that takes relatively undisturbed bites out of the seafloor, enabling biologists to count the number of species in each core and compare them to cores collected elsewhere in the deep ocean. The second development was an ecological survey conducted for the U.S. Minerals Management Service in the mid-1980s by Fred Grassle, a benthic ecologist (formerly of WHOI), who now serves as director of the Institute of Marine and Coastal Sciences, and Nancy Maciolek of Battelle Ocean Sciences.

In a series of 233 cores taken for the survey along a 176-kilometer track off the coast of New Jersey and Delaware during a two-year period, Grassle and Maciolek found an incredible diversity of animals, most of which were unknown. They picked out 798 species, 171 families, and 14 phyla at around 2100 m (6720 ft) (a sampling that revealed much richer life at those depths than earlier samples had hinted). They reserved their count to the tremendous diversity of tiny invertebrate mud dwellers too big to slip through their sieves.

As they sampled sites to the north and south, the number of new species they found had doubled without reaching a plateau, suggesting that species diversity was much richer than ever imagined. If a generalization were to be drawn for how many new species could be found every square kilometer of the sea floor beneath more than 1000 meters of water (excluding the abyssal depths thought to have less species diversity) the researchers came up with a ballpark form of 10 million species. "This sampling revealed that the deep-sea may, in fact, rival tropical rainforests in terms of the numbers of species present," Grassle said. "Thus the deep sea may physically resemble a desert, but in terms of species composition it is more like a tropical rainforest." One aspect of the deep-sea biodiversity study was not apparent until sewage sludge dumping began at the 106-mile dumpsite around the same time period. A shallower site in the New York Bight Apex had shown unacceptably high pathogen levels and signs of fish disease, which led to its closure. As an alternative, roughly eight million tons of sludge a year began to be dumped at the 106-mile deepwater site on the continental rise adjacent to the New York Bight starting in 1986. Researchers Grassle and Maciolek found themselves with a baseline of information on deep-sea organisms right around the dumpsite, which other scientists could use for comparison in determining whether damage might be caused to deep-living communities.

During the course of the next six years, NURP sponsored studies to determine the fate and effects of the sewage sludge at the sea floor. All of the evidence indicated that the sludge material dumped by barges did reach the ocean bottom slightly west of the area where it was discharged, and that it had significant effects on the metabolism, diet, and composition of organisms that lived there. There was a presence of sludge in sediments at the dumpsite, and the level of silver was 20 times higher at the site relative to an unaffected reference area. This was confirmed by chemist Michael Bothner of the U.S. Geological Survey. The submersible Alvin used by Bothner and his colleagues helped them collect the silver samples in sediment cores to make the determination. They were also able to observe how contaminants introduced to the sediments from dumping penetrated to a depth of 5 cm below the sea floor as organisms living in the sediments burrowed through them. However, during a 10-month period of sampling, researchers observed seven occasions where the currents were strong enough to resuspend the contaminated sediments. During the same period, chemist Hideshige Takada of Tokyo University and Bothner reported elevated levels of linear alkylbenzenes (LABs), widely used as surfactants in synthetic detergents, and coprostanols, a fecal marker of animals, at the dumpsite. The increased flux of sludge caused measurable changes in the benthic ecology near the dumpsite. Dr. Cindy Lee Van Dover (a biological oceanographer), Grassle, and colleagues observed a tenfold increase in the abundance of urchins, starfish and sea cucumbers at the dumpsite, and the ingestion of sludge-derived organic matter by sea urchins. By entering into the benthic food web, researchers believe that it is likely that long-term disposal programs would result in the restructuring of the benthic community favoring species that can exploit the organic material available in sewage sludge. Two species of polychaete worms, normally not abundant in the area, also increased at the dumpsite. Sludge disposal at the 106-mile dump site was curtailed in July 1992. "This provided additional opportunities to examine the long-term dispersal and effects of waste material in the deep-sea environment," Bothner said. Bothner found that silver levels in sediment samples had begun to decline after the dumping stopped. In subsequent studies by Bothner, and chemical oceanographers Elizabeth Lamoureux and Bruce Brownawell of the State University of New York at Stony Brook, it was found that while LABs had tapered off, elevated levels of organic contaminants including PCBs and PAHs in surface sediments around the dumpsite could still be detected. Van Dover found that the density of benthic communities at the dumpsite was decreasing and ingestion of sewage-derived organic matter was also subsiding at the dumpsite

While the effects of sludge dumping appeared to be abating in the vicinity of the dumpsite, an additional chapter to the story of the 106-mile site still remains to be written. Levels of silver appeared to be on the increase 50 nautical miles south of the dumpsite, as did the densities of sediment-dwelling organisms. This suggests that the recovery of the dumpsite had led to changes in other habitats as resuspended materials were transported to the south of the dumpsite, according to a 1993 study by Grassle, Paul Snelgrove, associate chair in fisheries conservation at Memorial University of Newfoundland, and Rosemarie Petrecca, a senior marine scientist at Rutgers University. Many questions remain to be answered about the potential short- and long-term effects of toxic compounds found to accumulate in deep water sediments from ocean dumping. Are toxins diluted to acceptable concentrations? Do heavy metals present as serious an environmental problem to deep-sea benthic communities as they do in shallow seas?

Two arguments in favor of deep-ocean dumping are that the material dilutes during sinking and is stable on the sea floor. The present body of research suggests that dilution does not completely abate the effects of dumping, nor does the waste sit still once it gets to the bottom. By establishing a long-term observatory at the 106-mile dumpsite, Grassle hopes future research will be directed towards better understanding the interactions of deep-sea organisms with their natural environment, and towards monitoring the potential effects of pollution necessary for the future wise management of ocean resources. "Almost nothing is known about the tolerances of deep-sea organisms to the gradual build-up of anthropogenic chemicals, and there is a potential for changes to be widespread if they do occur," Grassle said. "Measurement of pollutants, descriptions of deep-sea communities from many parts of the ocean, and *in situ* toxicity studies are urgently needed." Real aquanauts have replaced the imaginary crew of the Nautilus descending by submarine into the abyss. Life persists in the absence of sunlight, and the aquanauts find a remarkable diversity of creatures including sea cucumbers with their porcupine bristles, the brilliant yellow and red star shaped crinoids, branched gorgonians, crabs, shrimp, and rat tail fish, and examples of the millions of mud-dwellers, most of which have never been viewed by the human eye. Along the journey, they also find the results of hundreds of years of waste disposal at sea including bottles, cans, pottery, cocktail sticks off the great luxury liners, and big chunks of clinker from coal-fired steam engines. But the more the aquanauts discover, the less they understand. The deep sea—the largest, most diverse environment on Earth—still awaits us as we enter the 21st century. (Collie and Russo, 2010)

CONCLUSION

At present, ocean dumping is predominantly banned by international law. The motivation for banning ocean dumping was sparked by the shutdown of beaches due to contaminated wastes from sewage-derived micro-organisms, the closing of shellfish beds due to metal contamination, and infection of fish by lesion-causing parasites. Coastal oceans continually enriched by nutrients in waste products that run off the land suffer from eutrophication resulting in an increase in toxic algal blooms and decreased oxygen levels, both of which can kill fish populations.

REFERENCES

Anderson, B S & Hunt, J W (1988) Bioassay methods for evaluating the toxicity of heavy metals, biocides, and sewage sludge using microscopic stages of giant kelp Macrocystis pyrifera (Agardh) a preliminary report Mar Envir Res, 26, 113-34.

Anger, K (1975a) On the influence of sewage pollution on inshore benthic communities in the south of Kiel Bay Part 1 Qualitative studies on indicator species and communities Herentutkimuslait Julk /Havsforskningsinst Skr , 239, 116-22.

Anger, K (1975b) On the influence of sewage pollution on inshore benthic communities in the south of Kiel Bay Part 2 Quantitative studies on community structure Helgolandeer wiss Meeresunters, 27, 408-38.

Aschan, M M & Skullerud, A M (1990) Effects of changes in sewage pollution on soft-bottom macroinvertebrate communities in the inner Oslofjord, Norway Sarsia, 75, 169-90.

Bascom, W. (1977). Marine Pollution in Southern California. Mar. Pollut. Bull., 8,221-225.

Bellan, G. L. (1979). An attempted pollution abatement in the Gulf of La Napoule (Cannes, France). Mar. Pollut. Bull., 20. 163-166.

Blackstock, J, Johannessen, P J & Pearson, T H (1986) Use of a sensitive indicator species in the assessment of biological effects of sewage disposal in fjords near Bergen, Norway Mar Biol, 93, 315-22.

Boesch, D. F. (1982). Ecosystem consequences of alterations of benthic communitystructure and function in the New York Bight region. In Ecological Stress and the New York Bight: Science and Management (G. F. Mayer, ed.), pp. 543-568. Estuarine Research Federation. Columbia, South Carolina.

Bucke, D, Norton, M G & Rolfe, M S (1983) The field assessment of effects of dumping wastes at sea II Epidermal lesions and abnormalities of fish in the outer Thames Estuary Fishing Research Technical Report, MAFF Directorate of Fisheries Research, Lowestoft, 71, 16 pp.

Caspers, H. (1976). Ecological effects of sewage sludge on benthic fauna off the German North Sea coast. Prog. Water Technol, 9,951 -956.

Champalbert, G & Macquart-Moulin, C (1970) Les peracarides de l'hyponeuston nocturne du Golfe de Marseille Cah Biol Mar,11, 1-29.

Chapman, D V (1985) Preliminary observations on the interaction between plankton and sewage dumped at sea Marine Environmental Quality Committee, Int Co Explor Sea CM 1985/E 26, 11 pp.

Chapman, D V (1986) The distribution of metals in sewage sludge and their fate after dumping at sea Sci Total Environ ,48, 1-11.

Chapman, D V, Byrne, C D, Law, R J & Waldock, M J (1986) A comparison of methods for measuring the dispersion in the water column of sewage sludge dumped at sea EP

Consultants, Edinburgh, pp 131-3 2nd International Conference of Environmental Contamination Proc CEP.

Chapman, D V, White, S Rainbow, P S & Taylor, M (1988) Interactions between marine crustaceans and digested sewage sludge Mar Poll Bull, 19, 115-19.

Chen, C. W. & Orlob, G. T. (1972). The accumulation and significance of sludge near San Diego outfall. J. Water Pollut. Contr. Fed., 44. 1362-1371.

Chin, T-S & Chen, J-C (1987) Acute toxicity of ammonia to larvae of the Tiger Prawn, Penaeus monodo, Aquaculture, 66, 247-53.

Clark, G K (1989) The fate of trace metals at the Garroch Head sewage sludge dump site in the Firth of Clyde, Scotland Napier Polytechnic of Edinburgh, Edinburgh, 370 pp.

Clark, G K, Moore, D C & Davies, I M (1990) Recovery of a sewage sludge dumping ground Int Co Explor Sea, Marine Environmental Quality Committee CM 1990/E 27, 21 pp.

Collinge, V E & Bruce, A M (1981) Sewage sludge disposal a strategic review and assessment of research needs Technical Rep TR 166, Water Research Centre (UK).

Costello, M J & Gamble, J C (1992) Effects of sewage sludge on embryos and larvae of marine fish Mar Envir Res, 33, 49-74.

Cripe, G M, Hansen, D J, Macauley, S F & Forester, J (1986) Effects of diet quantity on sheepshead minnows (Cyprinodon variegatus) during early life-stage exposures to chlorpyrifos In Aquatic Toxicology and Environmental Fate Ninth volume ASTM STP 921, ed T M Poston & R Purdy American Society for Testing and Materials, Philadelphia, 450-60.

Cross, J N (1988) Fin erosion and epidermal tumours in demersal fish from Southern California In Oceanic Processes in Marine Pollution, Vol 5 Urban wastes in coastal marine environments, ed D A Wolfe & T P O'Connor R E Krieger Publ Co, Florida, 57-64.

Dauer, D M & Conner, W G (1980) Effects of moderate sewage input on benthic polychaete populations Estuar Mar Set, 10, 335-46.

Deep-Sea Biodiversity and the Impacts of Ocean Dumping by Marcia Collie and Julie Russo http://www.oar.noaa.gov/spotlite/archive/spot_oceandumping.html

Department of the Environment (1990) Third International Conference on the Protection of the North Sea UK Guidance note on the ministerial declaration Department of the Environment (UK).

Department of the Marine (1989a) Remots® reconnaissance survey of Dublin sewage sludge and dredge spoil disposal grounds July 1988 Department of the Marine, Dublin, 32 pp.

Department of the Marine (1989b) Environmental impact assessment sewage sludge and dredge spoil dumping in Dublin Bay, 1971-1988 Department of the Marine, Dublin, 11 pp.

Dethlefsen, V (1988) Ten years fish disease studies of the Federal Research Board Fisheries Hamburg International Council for Exploration of the sea (ICES) C M 1988/E 23.

Draxler, A F J (1979) Transient effects of ocean waste water sludge dumping J Wat Poll Contr Fed, 51, 741-8

Duedall, I W, O'Conners, H B, Oakley, S A & Stanford, H M (1977) Short-term water column perturbation caused by wastewater sludge dumping in the New York Bight apex J Wat Poll Contr Fed, 10, 2074-80

Duedall, I W, Bowman, M J & O'Connors, H B Jr (1975) Sewage sludge and ammonium concentrations in the New York Bight apex Estuar Coast Mar Sci, 3, 457-63.

Eagle, R A, Hardiman, P A , Norton, M G & Nunny, R S (1978) The field assessment of effects of dumping wastes at sea 3 A survey of the sewage sludge disposal area in Lyme Bay Fish Res Tech Rep, MAFF Directorate of Fisheries Research, Lowestoft, 49, 22 pp

Eagle, R A, Hardiman P A, Norton, M G & Nunny, R S (1979a) The field assessment of effects of dumping wastes at sea 4 A survey of the sewage sludge disposal area off Plymouth Fish Res Tech Rep, MAFF Directorate of Fisheries Research, Lowestoft, 50, 24 pp.

Eagle, R A, Hardiman, P A, Norton, M G. Nunny, R S & Rolfe, M S (1979b) The field assessment of effects of dumping wastes at sea 5 The disposal of solid wastes off the north-east coast of England Fishing Research Technical Report, MAFF Directorate of Fisheries Research, Lowestoft, 51, 34 pp.

Eleftheriou, A, Moore, D C, Basford, D J & Robertson, M R (1981) Experimental studies on the effects of sewage sludges on a benthic community International Council for Exploration of the Sea (ICES) CM 1981/E 43.

Eppley. R. W., Carlucci, A. F, Holm-Hansen. O., Giefer, D., McCarthy, J. J., & Williams. P. M. (1972). Evidence for eutrophication in the sea near southern California coastal sewage outfalls. Coli. Co-op. Oceanic Fish. Invest. Rep., 16, 74.

Essink. K. (1978). The effects of pollution by organic waste on macro-fauna in the eastern Dutch Wadden Sea. Pub. Neth. Inst. Sea. Res., 1, 1-135.

Fava, J A, Gift, J J, Maciorowski, A F, McCulloch, W L & Reisinger II, H J (1985) Comparative toxicity of whole and liquid phase sewage sludges to marine organisms In Aquatic toxicology and hazard assessment seventh symposium ASTM STP 854, ed R D Cardwell, R Purdy & R C Banner American Society for Testing and Materials, Philadelphia, 229-52.

Franklin, F L (1980) Assessing the toxicity of industrial wastes, with particular reference to variations in sensitivity of test animals Fishing Research Technical Report, MAFF Directorate of Fisheries Research, Lowestoft, 61, 10 pp.

Franklin, F L (1983) Laboratory tests as a basis for the control of sewage sludge dumping at sea Mar Poll Bull, 14, 217-23.

Fretwell, K (1989) The effects of sewage sludge on selected marine Crustacea M Sc thesis, Napier Polytechnic of Edinburgh, Scotland.

Frithsen, J B , Nacci, D, Oviatt, C, Strobel, C J & Walsh, R (1989) Using single-species and whole ecosystem tests to characterize the toxicity of a sewage treatment plant effluent In Aquatic Toxicology and Environmental Fate eleventh vol, ASTM STP 1007, ed G W Suter II and M A Lewis American Society for Testing and Materials, Philadelphia, 231-50.

Gibbs, R J (1982) Particle dynamics of sewage sludge dumping in the ocean Oceans, 82, 1058-62.

Guarino, C F, Nelson, M D, Townsend, S A , Wilson, T E & Ballotti, E F (1975) Land and sea solids management alternatives in Philadelphia J Wat Poll Contr Fed, 47, 2551-64

Grunseich, G S & Duedall, I W (1978) The decomposition of sewage sludge in seawater Water Res, 12, 535-45.

Halcrow, W., Mackay, D. W. & Thornton. I. (1973). The distribution of trace metals and fauna in the Firth of Clyde in relation to the disposal of sewage sludge, J. Mar. Biol. Assoc. U. K., 53, 721-739.

Hardy, J, Kiesser, S, Antrim, L, Stubm, A, Kocan, R & Strand, J (1978a) The sea-surface microlayer of Puget Sound Part I Toxic effects on fish eggs and larvae Mar Environ Res, 23, 227-49.

Hardy, J T, Crecelius E A , Antrim L D, Broadhurst, V L , Apts, C W, Gurtisen, J M & Fortman, T J (1987b The sea-surface microlayer of Puget Sound Part II Concentrations of contaminants and relation to toxicity Mar Environ Res, 23, 251-71

Hislop, J R G (1979) Preliminary observations on the near surface fish fauna of the northern North Sea in late autumn J Fish Biol, 15, 697-704.

Houlihan, D, Costello, M J, Secombes, C J, Stagg, R & Brechin, J Effects of sewage sludge exposure on growth, feeding and protein synthesis of dab, Limanda limanda Mar Environ Res (in press).

Jenkinson. I. R. (1972). Sludge dumping and benthic communities. Mar. Pollut. Bull., 3,102-105.

Jenkins, W A & Gibbs, R J (1988) Settling and dispersion of ocean dumped sewage sludge Mar Poll Bull, 19, 120-4.

Johnson, R W , Duedall, I W, Glasgow, R M, Proni J R & Nelsen, T A (1977) Quantitative mapping of suspended solids in wastewater sludge plumes in the New York Bight apex J Wat Poll Contr Fed, 10, 2063-73.

Kocan, R M, Westernhagen, H von, Landolt, M L & Furstenberg, G (1987) Toxicity of sea-surface microlayer effects of hexane extract on Baltic herring (Clupea harengus) and atlantic cod (Gadus morhua) embryos Mar Environ Res, 23, 291-305

Lack, T J, & Johnson, D (1985) Assessment of the biological effects of sewage sludge at a licensed site off Plymouth Mar Poll Bull, 16, 147-52.

Lewis, R E & Riddle, A M (1989) Sea disposal modelling studies of waste field dilution Mar Poll Bull, 20, 124-9.

Lloyd, R & Tooby, T E (1979) New terminology required for short-term static fish bioassays LQ I50 Bull Environ Contam Toxicol, 22, 1-3.

Maciorowski, A F, McCulloch, W L & Fava, J A (1985) An approach to sewage sludge bioaccumulation potential tests In Aquatic toxicology and hazard assessment, ed R D Cardwell, R Purdy, & R C Banner Americ Soc Test Mat, Philadelphia, 253-68.

MacKay, D W (1986) Sludge dumping in the Firth of Clyde—a contaminant site Mar Poll Bull, 17, 91-5.

MacKay, D W, Halcrow, W & Thornton, I (1972) Sludge dumping in the Firth of Clyde Mar Poll Bull, 3, 7-10.

McLusky, D. S., Elliott, M., & Warnes, J. (1978). The impact of pollution on the intertidal fauna of the estuarine Firth of Forth. In Physiology and Behaviour of Marine Organisms (D. S. McLusky & A. J. Berry, eds.), pp. 203-210. Pergamon Press, New York.

Ministry of Agriculture, Fisheries and Food (MAFF) (1989) First report of the Marine Pollution Monitoring Management Group's Co-ordinating Group on Monitoring of Sewage-Sludge Disposal Sites Aquat Environ Monit Rep, MAFF Directorate of Fisheries Research, Lowestoft, 20, 64 pp.

Ministry of Agriculture, Fisheries and Food (MAFF) (1990a) Monitoring and surveillance of non-radioactive contaminants in the aquatic environment, 1984-1987 Aquat Environ Monit Rep , MAFF Directorate of Fisheries Research, Lowestoft, 22, 60 pp.

Ministry of Agriculture, Fisheries and Food (MAFF) (1990b) Utility of experimental measures of biological effects for monitoring marine sewage-sludge disposal sites Aquat Environ Monit Rep, MAFF Directorate of Fisheries Research, Lowestoft, 24, 45 pp.

Ministry of Agriculture, Fisheries and Food (MAFF) (1991a) Second report of the Marine Pollution Monitoring Management Group's Co-ordinating Group on monitoring of sewage-sludge disposal sites Aquat Environ Monit Rep, MAFF Directorate of Fisheries Research, Lowestoft, 25, 39 pp.

Ministry of Agriculture, Fisheries and Food (MAFF) (1991b) Monitoring and surveillance of non-radioactive contaminants in the aquatic environment, and activities regulating the disposal of wastes at sea, 1988-1989 Aquat Environ Monit Rep , MAFF Directorate of Fisheries Research, Lowestoft, 26, 90 pp.

Mclntyre, A D & Johnston, R (1975) Effects of nutrient enrichment from sewage in the sea In Discharge of sewage from sea outfalls, ed A L H Gameson Pergammon Press, Oxford.

McLean, C, Miskiewicz, A G & Roberts, E A (1991) Effect of three primary treatment sewage outfalls on metal concentrations in the fish Cheilodactylus fuscus collected along the coast of Sydney, Australia Mar Poll Bull, 22, 134-40.

McVicar, A H, Bruno, D W & Fraser, C O (1988) Fish diseases in the North Sea in relation to sewage sludge dumping Mar Poll Bull, 19, 169-73.

Miller, D C, Marcy, M, Berry, W, Deacutis, C, Lussier, S, Kuhn, A, Herber, M, Schimmel, S C & Jackim, E (1987) The acute toxicity of sewage sludge to marine fish, mysids, and copepods In Oceanic processes in marine pollution, vol 5 Urban wastes in coastal marine environments, ed O A Wolfe & T P O'Connor Robert E Krieger Publ Co, Malabar, Florida, 103-13.

Moore, D C, (1981) Studies on environmental effects of sewage sludge dumping off the Firth of Forth International Council for Exploration of the Sea (ICES) CM 1981/E42.

Moore, D C & Davies, I M (1987) Monitoring the effects of the disposal at sea of Lothian Region sewage sludge Proc R Soc Edinb, 93B, 467-77.

Murray, A J & Norton, M G (1982) The field assessment of effects of dumping wastes at sea 10 Analysis of chemical residues in fish and shellfish from selected coastal regions around England and Wales Fishing Research Technical Report, MAFF Directorate of Fisheries Research, Lowestoft, 69, 42 pp.

Murray, L A, Norton, M G, Nunny, R S & Rolfe, M S (1980a) The field assessment of effects of dumping wastes at sea 6 The disposal of sewage sludge and industrial waste off the River Humber Fishing Research Technical Report, MAFF Directorate of Fisheries Research, Lowestoft, 55, 35 pp.

Murray, L A, Norton, M G, Nunny, R S & Rolfe, M S (1980b) The field assessment of effects of dumping wastes at sea 7 Sewage sludge and industrial waste disposal in the Bristol Channel Fishing Research Technical Report, MAFF Directorate of Fisheries Research, Lowestoft, 59, 40 pp.

Nedwell, D B & Lawson, P A (1990) Degradation of digested sewage sludge in marine sediment—water model systems, and fate of metals Mar Poll Bull, 21, 87-91

Norton, M G & Rolfe, M S (1978) The field assessment of effects of dumping at sea I An introduction Fishing Research Technical Report, MAFF Directorate of Fisheries Research, Lowestoft, 45, 9 pp

Norton, M G, Eagle, R A, Nunny, R S, Rolfe, M S, Hardiman, P A & Hampson, B L (1981) The field assessment of effects of dumping wastes at sea 8 Sewage sludge dumping in the outer Thames estuary Fishing Research Technical Report, MAFF Directorate of Fisheries Research, Lowestoft, 62, 62 pp.

Norton, M G, Franklin, A, Rowlatt, S M, Nunny, R S & Rolfe, M S (1984a) The field assessment of effects of dumping wastes at sea 12 The disposal of sewage sludge, industrial wastes and dredged spoils in Liverpool Bay Fishing Research Technical Report, MAFF Directorate of Fisheries Research, Lowestoft, 76, 50 pp.

Norton, M G, Jones, P G W, Franklin, A & Rowlatt, S M (1984b) Water quality studies around the sewage sludge dumping site in Liverpool Bay Estuar Coastal Shelf Sci, 19, 53-67.

O'Connor, T P, Walker, H A, Paul, J F & Bierman, V J Jr (1985) A strategy for monitoring of contaminant distributions resulting from proposed sewage sludge disposal at the 106-mile ocean disposal site Mar Envir Res, 16, 127-50.

Otte, G. (1979). Investigations on the effects of domestic sewage on the benthic ecosystem of marine intertidal flats. Helgol. Wiss. Meeresunters., 32, 73-148.

Oviatt, C A, Keller, A A, Sampou, P A & Beatty, L L (1986) Patterns of productivity during eutrophication a mesocosm experiment Mar Ecol Prog Ser, 28, 69-80.

Oviatt, C A, Quinn, J G, Maughan, J T, Ellis, J T, Sullivan, B K, Gearing, J N, Gearing, P J, Hunt, C D, Samou, P A & Latimer, J S (1987) Fate and effects of sewage sludge in the coastal marine environment a mesocosm experiment Mar Ecol Prog Ser, 41, 187-203

Paper first presented at OCEANS 84, Annual Meeting of the Marine Technology Society, Washington D.C., September 1984.

Pearce, J. B., Caracciolo, J. V., Halsey, M. B. & Rodgers, L. H (1976). Temporal and spatial distributions of benthic macroinvertebrates in the New York Bight. In Middle Atlantic Continental Shelf and the New York Bight, Am. Soc. Limno. Oceanogr. Special Symp., Vol. 2. (M. G. Gross, ed). pp. 33-403. Allen Press. Lawrence, Kansas.

Pearson, T H & Blackstock, J (1986) Garroch Head sludge disposal ground survey Final report on the monitoring survey carried out on 21st – 29th May 1985 Unpubl Rep, Scottish Marine Biological Association, Oban.

Pike, E. B. & Gameson, A. L. H. (1970). Effects of marine sewage disposal. Water Pollut. Control, 69, 355.

Preston, A & Portman, J E (1981) Critical path analysis applied to the control of mercury inputs to the United Kingdom coastal waters Envir Pollut (Series B), 2, 4512-64

Rachor, E. (1977). Faunenverarmung in einem Schlikgebict in der Nahe Helgolands. Helgo. Wiss. Meeresunters., 30, 633-651.

Rau, G H, Sweeney, R E, Kaplan, I R , Mearns, A J & Young, D R (1981) Difference in animal 13C, 15N, and D abundance between a polluted and an unpolluted coastal site likely indicators of sewage uptake by a marine food web Estuar Coast Shelf Sci, 13, 701-7

Read, P A (1977) An investigation of the influence of water quality on the intertidal macrobenthos of selected areas of the south shore of the outer Firth of Forth Ph D thesis, Napier College of Commerce and Technology, Edinburgh.

Read, P. A., Anderson, K. J., Matthews, J. E., Watson, P. G., Halliday. M. C. & Shiells, G. M. (1982). Water quality in the Firth of Forth. Mar. Pollut. Bull., 13, 421-425.

Read, P. A., Anderson, K.J., Matthews, J. E., Watson, P. G. Halliday, M. C. & Shiells, G.M. (1983). Effects of pollution on the benthos of the Firth of Forth. Mar. Pollut. Bull., 14, 12-16.

Rees, H L, Moore, D C, Pearson, T H, Elliott, M , Service, M , Pomfret, J & Johnson, D (1990) Procedures for the monitoring of marine benthic communities at UK sewage sludge disposal sites Scot Fish Infor Pamp No 18, 79 pp.

Rohatgi, N K & Chen, K V (1976) Fate of metals in wastewater discharge to ocean J Environ Eng Div, Am Soc Civ Eng, 102, 675-85.

Santoro, E D & Fikslin, T J (1987) Chemical and toxicological characteristics of sewage sludge ocean dumped in the New York Bight Mar Poll Bull, 18, 394-9.

Seager, J, Wolff, E W & Cooper, V A (1988) Proposed environmental quality standards for list II substances in water—Ammonia Water Research Centre, TR 260, 61 pp.

Secombes, C J, Fletcher, T C, O' Flynn, J A, Costello, M J, Stagg, R & Houlihan, D (1991) Immunocompetence as a measure of the biological effects of sewage sludge pollution m fish Comp Biochem Physiol, 100C, 133-6.

Secombes, C J, Fletcher, T C, White, A, Costello, M J, Stagg, R & Houlihan, D (1992) Effects of sewage sludge on immune responses in the dab, Limanda limanda (L) Aquatic Toxicology, 23, 217-30.

Shelton, R G J (1971) Sludge dumping in the Thames estuary Mar Poll Bull, 2, 24-7.

Sheppard, C. R. C. (1977). Effects of Athens pollution outfalls on marine fauna of the Saronikos Gulf. Int. J. Environ. Stud., 11, 39-44.

Sindermann, C. J. (1976). Effects of coastal pollution on fish and fisheries-with particular reference to the Middle Atlantic Bight. Amer. Soc. Limnol. Oceanogr. Spec. Symp., 2, 281-301.

Suter, G W II & Rosen, A E (1988) Comparative toxicology for risk assessment of marine fishes and crustaceans Environ Sci Techno), 22, 548-56.

Swartz, R C, Schults, D W, Ditsworth, G R & DeBen, W A (1984) Toxicity of sewage sludge to Rhepoxynius abronius, a marine benthic amphipod Arch Environ Comtam Toxicol, 13, 207-16.

Talbot, J W, Harvey, B R, Eagle, R A & Rolfe, M S (1982) The field assessment of effects of dumping wastes at sea 9 Dispersal and effects on benthos of sewage sludge dumped in the Thames estuary Fishing Research Technical Report, MAFF Directorate of Fisheries Research, Lowestoft, 63, 42 pp.

Thompson, I. & Psutsy, N. P. (1984). Growth of the ocean quahog, Arctica islandica, in the area of the Philadelphia dumpsites. Final Report NAI-RAD00023, Office of Marine Pollution Assessment, National Oceanic and Atmospheric Administration.

Topping, G (1987) Sewage sludge dumping in Scottish waters current practices and future outlook Publ Health Engineer, 14, 49-51.

Tully, O & O'Ceidigh, P (1987α) Investigations of the plankton of the west coast of Ireland— VIII The neustonic phase and vertical migratory behaviour of benthic Peracandea in Galway Bay Proc R Ir Acad, 87B, 43-64.

Tully, O & O'Ceidigh, P (1987b) The seasonal and diel distribution of lobster larvae (Homarus gammarus (Linnaeus)) in the neuston of Galway Bay J Cons Int Explor Mer, 44, 5-9.

Venkatesan, M I & Santiago, C A (1989) Sterols in ocean sediments novel tracers to examine habitats of cetaceans, pinnipeds, penguins and humans Mar Biol, 102, 431-7.

Vivian, C M G (1986) Tracers of sewage sludge in the marine environment a review Sci Total Envir, 53, 5-40

Young, J S & Pearce, J B (1975) Shell disease in crabs and lobsters from New York Bight Mar Poll Bull, 6, 101-5.

Walker, H A & Paul, J F (1989) Ocean dumping of sewage sludge Maritimes, 33, 15-17.

Walker, H A, Paul, J F & Bierman, V J Jr (1987) Methods for waste load allocation of municipal sewage sludge at the 106-mile ocean disposal site Environ Tox Chem, 6, 1-15

Watling, L., Leatham, W., Kinner, P., Wethe, C. & Maurer, D. (1974). Evaluation of sewage sludge dumping off Delaware Bay. Mar. Pollut Bull., 5, 39-42.

Westernhagen, H von, Landolt, M, Kocan, R, Furstenberg, G, Janssen, D & Kremling, K (1987) Toxicity of sea-surface microlayer effects on herring and turbot embryos Mar Environ Res, 23, 273-90.

Whitelaw, K. & Andrews, M J (1988) The effects of sewage sludge disposal to sea the Outer Thames estuary, U K Wat Sci Tech , 20, 183-91.

Williams, A B & Bynum, K H (1972) A ten-year study of meroplankton in North Carolina estuaries amphipods Chesapeake Sci, 13, 175-92.

In: Sewage Sludge Management
Editors: A. A. Zorpas and V. J. Inglezakis

ISBN: 978-1-61324-393-0
© 2012 Nova Science Publishers, Inc.

Chapter 12

POTENTIAL USE OF SEWAGE SLUDGE AS A CONSTRUCTION MATERIAL

*Demetris Nicolaides**

Department of Civil Engineering, Frederick University, Nicosia, Cyprus

ABSTRACT

The common practices of sludge treatment used until recently are seen as not being sufficient any more, mainly due to severe limitations set by national legislations. Therefore, alternative practices, which will be more environmental friendly, financially viable and practical must be adopted for the treatment of the specific material. Such disposal problems can be drastically reduced, if sludge can be recycled and reused as a value-added constituent for the development of construction materials. It is the aim of this chapter to thoroughly review research that focuses on the potential use of sewage sludge as a construction material.

1. Introduction

Sewage sludge is an inevitable by-product of the treatment process of wastewater, which is rapidly increasing due to the fast industrial development, population growth, the fact that landfill sites are becoming increasingly difficult to come by and the more stringent environmental regulations. The continuously increasing generation of dry sewage sludge is regarded as one of the most challenging problems of environmental control. Common practices of sludge disposal and/or treatment used until recently have either been abandoned (e.g. ocean dumping), or are being looked upon less favourably, especially when viewed through the prism of sustainable development and practices, (e.g. landfilling), or have been connected to peripheral problems (e.g. spreading of sludge on reclaimed land, which has been connected to possible contamination of sub-surface waters). Overall, existing practices for the

---

* Email: d.nicolaides@frederick.ac.cy.

disposal of wastewater sludge are seen as not being sufficient any more, a view corroborated by several, and often forbidding limitations set by national legislations. Therefore, alternative practices, which will be more environmental friendly, financially viable and practical must be adopted for the treatment of the specific material. Such disposal problems can be drastically reduced, if sludge can be recycled and reused as a value-added constituent for the development of construction materials.

Despite the fact that the possible incorporation of sludge as a constituent product of materials used in the construction industry has been initially received with certain reluctance and scepticism, extensive research has been carried out to examine the inclusion of sludge in several products. Specifically, research has been conducted for the use of sewage sludge in brick-making, as a lightweight aggregate material, as a cementitious binder used in concrete mixes, as a ceramic-making material, or as a constituent in asphaltic paving mixes. Further, work has been carried out for the use of sewage sludge in more specialised applications, such as heat insulating material, or in geotechnics, for the stabilisation treatment of soft subgrade soils. In most of these cases sludge has been used either in the form of ash (product of sludge incineration), or in the form of dry sludge, following stabilisation and solidification treatment processes.

This chapter critically explores recent research advances that focus on the potential use of sewage sludge as a construction material. The discussion will be divided into three main sections: 1) Sewage Sludge as Lightweight Aggregate Material, 2) Sewage Sludge Ash as Brick and Tile Making Material, and 3) Sewage Sludge Ash as a Value-Added Product in the Construction Industry.

2. SEWAGE SLUDGE AS LIGHTWEIGHT AGGREGATE MATERIAL

The production of artificial lightweight aggregates from wastewater sludge has been investigated by several researchers, from as far back as 1984. Bhatty and Reid (1989a) presented a study which described the procedure of the manufacture of lightweight aggregates from a sludge ash, which has been fired at elevated temperatures. The sludge ash has been pelletised or slabbed, and subsequently fired until it sintered and expanded to form the lightweight products. The pellets and slabs were subjected to a series of tests (i.e. size, grading, bulk density, absorption and compressive strength), in order to result to the product characterisation and the investigation of their suitability as aggregates for the manufacture of moderate strength concrete. The compressive strengths of the produced concrete mixes were recorded and compared with those of control concrete specimens prepared from commercial expanded clay aggregates, tested under identical conditions. The moderate strength concretes produced from the pellets have better strength characteristics than those made of slabs and other commercial aggregates, mainly because of their spherical shape, uniformity of size and low moisture absorptions. The aforementioned factors contributed to improved workability and optimum packing of the artificial aggregates during casting, resulting in more compact specimens. On the other hand, the slab and clay aggregates were of angular shapes with wide size distribution. They also had large moisture absorption and they were weaker than the pellets; therefore they produced weaker concrete mixes. Bhatty and Reid (1989b) also investigated the potential for using sludge ash as fine aggregates in mortar, setting the resulting compressive strength of concrete as a success criterion. The authors completed a comprehensive test program, during which the strength characteristics of mortars prepared from sludge ash with the addition of fly ashes and varying amounts of cement and lime were determined. The experimental results indicated that although it was possible to produce some reasonable mortar strengths by incorporating major amounts sludge ash, the mortars remained much weaker than the required level of strength for normal mortars, even after prolonged curing times (e.g. 28 days). The authors attributed the received poor strength to the finely dispersed and porous nature of the ash which, despite its high silica, alumina and lime contents, lacked pozzolanic activity. The same authors (1989c) developed a range of artificial aggregates from incinerated sewage sludge ash. The ash was either pelletised or slabbed and subsequently fired until it sintered and expanded to form lightweight products which were later graded for use as aggregates in concrete. The results from the tests conducted on size grading, bulk density, absorption and fracture strength, and the incorporation of these materials in concrete mixes revealed that both the pellets and slabs were potentially suitable as lightweight aggregates in concrete (Bhatty and Reid, 1989a).

Tay and Show (1992) presented a review paper on the utilisation of municipal wastewater sludge as building and construction materials. Amongst others, they highlight the work of several researchers on the production of artificial lightweight aggregates from sludge. Specifically, they state the work done by Kato and Takesue (1984), who claimed that artificial lightweight fine aggregates could be produced from pulverised sludge ash. The sludge ash which was collected after incineration was pulverised and mixed with water and waste liquor from an alcohol plant. The mixture was pelletised and fired at 1050 ^{0}C. The resulting residues were of lower specific gravity and slightly lower strength than the conventional artificial fine lightweight aggregates. Tay and Show (1992) also report the works of Tay and Yip (1989, 1990), who digested and dewatered sludge incinerated at 1050 ^{0}C, which was subsequently used in the production of lightweight coarse and fine aggregates. It was claimed that the low thermal conductivity and high fire resistance of the artificial sludge ash aggregates may render them suitable for use in thermal insulation and fire protection of concrete. Moreover, the 28-day compressive strength of the lightweight concrete was comparable to those using other commercially available lightweight aggregates. The results of the fire resistance test revealed an identical behaviour between ordinary and sludge ash concrete for the first six minutes of burning, whereas thereafter the temperature of the reinforcement bar in sludge ash concrete was substantially lower than the corresponding in ordinary concrete. Tay and Show (1991) concluded that incinerated clay-blended sludge was a potential material for the production of lightweight aggregate concrete for structural use. During this study, digested and dewatered sludge, collected from a sewage treatment plant, was mixed with clay and subsequently fired at 1050 ^{0}C - 1080 ^{0}C. The clay composition in the mixture was ranging from 10% to 40%. The ash produced was crushed and graded to the required aggregate sizes. All concrete mixtures could easily achieve the recommended 28-day compressive strength of 15 N/mm^2. Similar conclusions were drawn by Elkins et al. (1985), who also supported that mixtures of sludge and clay could be used to produce lightweight building aggregates.

More recent studies, investigated the potential manufacture of lightweight aggregates from sewage sludge. Cheeseman and Virdi (2005) studied the properties and microstructure of lightweight aggregate produced from sintered sewage sludge ash. The ash produced by fluidised bed incineration of sewage sludge. As the authors report, the sewage sludge ash was used "as received", without any particle size reduction by milling. The ash appeared to be a homogeneous fine powder, which did not contain any significant lumps or agglomerates. However, through the characterisation experiments of the ash, it was revealed that it contained relatively high concentrations of a range of metals of potential environmental concern including Zn, Cu, Pb, Mn, Cr, Ni and Cd. The sewage sludge ash was thoroughly mixed with different amounts (1-16 wt%) of clay, in order to aid pellet formation. A 1 wt% addition of an organic binder was also used to replace the clay binder in some of the mixes. A controlled amount of water was used until the mix consistency allowed formation of approximately spherical, 8-10 mm diameter pellets. The pellets were then sintered by passing them through a rotary tube furnace at temperatures ranging between 1020 ^{0}C and 1080 ^{0}C. The properties (i.e. density, water absorption and compressive strength) of the produced pellets were similar or superior to those of commercially available lightweight aggregate. The effect of increasing the clay addition had negligible effect on sintered properties. In addition, lightweight pellets could be formed either using a clay, or 1 wt% addition of an organic binder. Conclusively, the study claims that the production of low density, low water absorption, and relatively high strength sintered pellets is feasible using simple processing methods involving pelletising and relatively low temperature sintering. However, the authors suggested that additional work was necessary in order to fully characterise sintered sewage sludge ash pellets, reduce the sintering temperatures and investigate effects of variations in chemical composition of these ashes. Chiou et al. (2006) conducted a study for which sewage sludge ash with similar characteristics to expansive clay was used as the principal material and sewage sludge as the admixture to sinter lightweight aggregates. The experimental results showed that both sewage sludge and sewage sludge ash could be sintered to produce synthetic aggregates individually or mixed. However, it was suggested that the amount of sewage sludge added should be less than 20%, due to the significant difference in specific gravity between sewage sludge and sewage sludge ash, resulting to difficulty of the mixed powder to form spherical particles in a rotational pelletiser. The authors reported that when only sewage sludge was used, porous and loose aggregates were produced due to the higher organic matter content, whereas when only sewage sludge ash was used more energy was consumed because of the lower organic matter content. Therefore, for energy savings the authors suggested that a mixture of sewage sludge ash would be more suitable to be sintered as a normal weight aggregate. A mixture with less sewage sludge (0-10%) was good to produce a low/medium density aggregate, and a mixture with more sewage sludge (20%-30%) was adequate to sinter a lower density aggregate. Finally, it was reported that the sintering temperature affected the properties of sewage sludge ash lightweight aggregate more than the retention period did. In a similar study, conducted by Mun (2007), the development and testing of lightweight aggregate using sewage sludge for non-structural concrete was investigated. Five types of sintered lightweight aggregate manufactured with various mass ratios of clay to sewage sludge. The physical properties were compared to those of a commercial lightweight aggregate for non-structural concrete, in order to recommend the optimum sewage sludge content. Specifically, the aggregates were tested for density, water absorption, abrasion loss, crushing value, impact value and heavy metal leaching. Based on the experimental results, it was concluded that an experimentally manufactured lightweight aggregate with a recommendable sewage sludge content of 75% had similar or even superior physical properties comparing to the corresponding values of the commercial product. Specifically, the water absorption of the manufactured lightweight aggregate was about 50% or less than that of the commercial lightweight aggregate. Moreover, no toxic heavy metals were detected from any of the manufactured aggregates with the highest or lowest sewage sludge contents, concluding to the statement that the artificial lightweight aggregate could be regarded as environment-conscious material. Finally, concrete specimens were made with four types of lightweight aggregate and the specimens were tested for compressive and flexural strength, density, water absorption, and thermal conductivity. Due to a higher mixing ratio of sewage sludge the compressive and flexural strengths of concrete gradually decrease. Nevertheless, the compressive strength was consistently above 15 MPa and the flexural strength consistently above 3 MPa. It was concluded that non-structural lightweight concrete with a density range of 1400-1500 kg/m^3 and thermal conductivity range of 0.59-0.73 W/mK could be manufactured. Considering that thermal conductivity of ordinary concretes using crushed stones is about 1.50-1.60 W/mK, when concrete is made of lightweight aggregate using sewage sludge, the insulation effect becomes twice as strong as that of ordinary concrete. Wang et al. (2009) used dry sewage sludge as the principal material, which was blended with coal ash in order to produce lightweight aggregate. The results indicated that the sintered products produced using pure dry sewage sludge were porous and loose (due to the formation of large pores during sintering) and had low compressive strength due to high organic matter content. The addition of coal ash increased the content of SiO_2 and Al_2O_3, thus decreasing the pore size of the sintered products and, concurrently, increasing the compressive strength. Furthermore, the sintering temperature and the proportion of coal ash were found to be the primary factors affecting the properties of the sintered products, and the addition of 18-25% of coal ash coupled with sintering at 1100 ^{0}C for 30 minutes produced the highest quality lightweight aggregates.

Another work presented by Wang et al. (2005) investigated the use of sewage sludge ash for the production of lightweight, foamed materials. The authors examined how the mixing proportions affected the foaming behaviour, including the lightweight characteristics and pore structure of the materials produced. Sewage sludge ash was mixed with cement and aluminium powder, which acted as a foaming agent. A series of specimens were produced for the study of the engineering properties, the microstructure analysis and the thermal conductivity analysis. When the cement amount was more than 20%, the compressive strength of sewage sludge ash foamed materials was greater than 8.52 MPa, whereas if the cement amount was only 10%, the fresh paste still had obvious foaming reaction. However, when cured in lime-saturated water, the specimens became crushed and deformed. Therefore, the minimum cement amount was decided according to the compressive strength requirement, not the alkalinity. The suggested optimum cement amount was 15-30% of the total solids weight, that is, the sewage sludge ash amount could be about 70-85% of the total. Furthermore, the thermal conductivity of sewage sludge ash foamed material produced was between 0.084 and 0.102 W/mK, meeting the requirements as a heat insulating material. In addition, sewage sludge ash could be used as a lightweight filler material. Similar work has been conducted by Chen et al. (2006), where sewage sludge ash was used as the main material to produce a foamed lightweight material. During the study, a number of firing tests were conducted at different temperatures, in order to investigate the effect of temperature on the engineering properties and microstructure of the foamed materials. The bulk density became greater after fired at 1093 ^{0}C. This was attributed to the significant volume shrinkage and denser structure resulted at this high temperature. It was also found that after being fired 1093 ^{0}C, the water absorption of the material decreased. It was concluded that the more sewage sludge ash used, the more significant was the water absorption decreased.

Furthermore, the compressive strength of the foamed lightweight material after fired at 1093 ^{0}C was higher than before firing. It was found that specimens with 60%, 70% and 80% of sewage sludge ash had an increase of their compressive strengths, which it was proportional to the amount of sewage sludge ash used. In contrast, the compressive strength was found to be inversely proportional to water/solid ratio, whether before or after fired. Finally, after fired at 1093 ^{0}C for 4 hours, the sewage sludge ash foamed lightweight material completed the enlarging stage of liquid phase sintering, and thus decreased total pore volume by 30% than before fired.

The effect of SiO_2-Al_2O_3-flux ratio change on the bloating characteristics of lightweight aggregate material produced from recycled sewage sludge was investigated by Tsai et al. (2006). The study investigated the characteristics of lightweight aggregates sintered from sewage sludge ash by modifying the proportion of the main components (SiO_2-Al_2O_3-flux). The results of the study showed that amorphous SiO_2 lowered the melting point and increased foaming, Al_2O_3 raised the compressive strength, whereas fly ash lowered the sintering temperature required. It was therefore concluded that the co-treatment of sludge ash with cullet powder (amorphous SiO_2) was a favourable option for the production of lightweight aggregates from sludge ash.

A number of researchers focused on the stabilisation and solidification of sewage sludge. Specifically, Valls and Vazquez (2000) investigated the formation of a binding and stabilising matrix by combining sludge and cement and also sludge, cement and coal fly ash. The authors studied the effects of various percentages of waste and binder and the behaviour and evolution of sludge in the pastes formed with cement, and with cement-coal fly ash. The parameters of study were the setting speed, the hydration products formed and the chemical stability, as this resulted by the pH values attained and the concentrations of micro-organisms and organic compounds in the systems formed. It was found that higher amounts of sludge in the mixture, resulted to a higher delay of the beginning and end of setting procedure. The hydration speed was improved substantially with the addition of an accelerate additive ($CaCl_2$) in all of the mixes. All of the mixes tested attained setting with time, whereas the setting speed was improved in mixes with Portland cement I 45/A (medium-high strength), in comparison with that of the Portland cement I 35/A (medium strength). The setting time was also increased proportionally to amount of coal fly ash added as a cement substitute. The hydrates formed in all of the cement pastes with sludge were the same as those formed in pastes without it. The hydrates of the Portland cement formed guaranteed a high alkalinity of the system and no variation of pH was observed between cement pastes with and without addition of coal fly ash. The alkalinity of the medium favoured the neutralisation and decomposition of the microorganisms and organic compounds contained in sewage sludge, guaranteeing the stability and solidification of the system.

Donatello et al. (2009) investigated the effect of milling and acid washing on the pozzolanic activity of incinerator sewage sludge ash, which contains significant levels of phosphates. Phosphate can be extracted from the incinerator sewage sludge ash by acid washing, as reported by Oliver and Carey (1976). It was found that the pozzolanic activity of incinerator sewage sludge ash improved after milling, as this confirmed by both the Frattini and the Strength Activity Index (SAI) test. Moreover, if the incinerator sewage sludge ash was washed to recover phosphate, it was likely to produce an acid insoluble material with little or no pozzolanic activity. However, as received the incinerator sewage sludge ash could be milled to produce a useful cement-replacement material. Finally, the sulphuric acid washing decreased the pozzolanic activity of all materials according to both the Frattini and the SAI test.

The potential use of sludge ash in construction products was reported by Tay (1986). Specifically, the author suggested the use of sludge ash as filler in concrete. It was resulted that there was no significant difference in segregation for concrete samples with various percentage of pulverised sludge ash used to replace the cement in the concrete mixes. In general, the workability was observed to be improved with the higher percentages of sludge ash, as this was indicated by the results of the slump and compaction factor tests, whereas the setting times remained unaffected. The compressive strength for the concrete cubes with various percentages of pulverised sludge ash showed a general decrease in strength as the amount of sludge increased. The same author (Tay, 1989) reported that pulverised sludge ash had insignificant effect on the segregation, bulk density, shrinkage, and water absorption of concrete cubes, whereas an increase of the workability with an increase of pulverised sludge ash content was observed. Furthermore, the compressive strengths of concretes with 10% of pulverised sludge ash were similar to the control samples, whereas contents beyond 10% resulted in a decrease of the compressive strength, which was proportional to the sludge content. It was, therefore, concluded that pulverised sludge ash could be used as a partial replacement of cement in concrete. In addition, the study revealed that using reclaimed wastewater as mixing water for concrete did not affect the properties of concrete, such as segregation, shrinkage, water absorption, bulk density, and setting times, while the use of reclaimed wastewater had a favourable effect on the strength development at earlier ages and an insignificant effect for ages of three months and beyond. Tay and Show (1994) investigated the potential for using wastewater sludge in combination with limestone to produce cementitious building materials. Experimental results showed that under controlled incineration, it was possible to produce the so-called "Bio-cement" from wastewater sludge that would satisfy the strength requirements for masonry cement. The compressive strength test results of the blended cements indicated that up to 30% of the Portland cement could be replaced by the "Bio-cement" without deteriorating the strength. In addition, blended cements with up to 10% replacement level showed slightly higher strength than the control strength and a higher rate of strength development at early ages. However, a considerable strength reduction observed at replacement levels beyond 30%.

Valls et al. (2004) examined the possibility of adding dry sludge from a sewage treatment plant to plain concrete, in combination with Portland cement. During the experimental work specimens containing different percentages of dry sludge were prepared, aiming to the study of both the physical (density, porosity, and absorption capacity) and the mechanical (compressive strength, flexural strength, and elastic modulus) properties over time. Four concrete mixtures containing different percentages of dry sludge were studied (reference concrete or 0% sludge, 2.5% sludge, 5% sludge and 10% sludge). The authors set as a maximum sludge percentage the 10%, as previous work in the literature claimed that for sludge content of more than 10% in the cement mix, the final setting time was extremely high (Yague et al., 2001). Furthermore, as the authors report, the introduction of a waste material, such as sludge, into a matrix with cement causes changes in the normal behaviour of cement, mainly due to the organic material which acts as a setting retardant and also due to some heavy metals which interfere with the hardening reactions. The added dry sludge did not contain any clay material, which could have caused stability problems in the concrete. The presence of sludge reduced the mechanical strengths of concrete. The density of the concrete decreased with an increasing sludge content, while the porosity and the absorption coefficient increased with increasing sludge content. Increase of the curing time had a beneficial effect on all properties (i.e. increase in strengths, increase of density and decrease of porosity and absorption coefficient). The experimental results confirmed that up to 10% of treatment plant sludge can be added to concrete for use in very specific applications (e.g. road bases and sub-bases, and as a filling material).

The combination of different types of waste sludge ash (including sewage sludge ash and water purification sludge ash) with other raw materials, such as limestone and ferrate, was investigated for the production of eco-cement clinkers. Specifically, Lin et al. (2005) prepared three blends of eco-cements (A, B and C) with different proportions of raw materials and they studied the hydration characteristics of the products. The blend ratios (wt.%) of the raw materials (sewage sludge ash – water purification sludge ash – ferrate – limestone) was 4.69% - 13.03% - 1.88% - 80.40% for eco-cement A, 8.98% - 9.34% - 2.08% - 79.60% for eco cement B and 4.24% - 12.45% - 1.85% - 81.45% for eco-cement C. The authors studied the setting time, the compressive strength, and the distribution of pores at various ages of the materials. The burning process of the raw materials was including incineration, drying and pulverisation of sewage and water purification sludge materials. The produced ashes were mixed with ferrate and limestone, and burned at 1400 ^{0}C for 6 hours. The resultant clinkers were cooled, mixed with plaster and pulverised to form the eco-cement clinkers. An ordinary Portland cement paste was used as the control mix. The major components of the Portland cement C_3S, C_2S, C_3A and C_4AF were all found in both the eco-cement A clinker and eco-cement C clinker. The 28-day compressive strength of eco-cement C paste was higher than the corresponding value of ordinary Portland cement paste due to the large quantity of limestone used, which enhanced the formation of C_3S. Eco-cement A paste had an initial setting time and final setting time similar to ordinary Portland cement paste, whereas Eco-cement B paste had an observed delay in the setting time. Finally, it was found that, with increasing curing ages, the gel pores (<0.01 μm) increased and the total porosity and capillary pores (>0.01 μm) decreased, which was attributed to the hydration that filled the pores.

Malliou et al. (2007) investigated the possibility of using sewage sludge from urban wastewater treatment plants as an additive in order to develop new construction materials. Several mixtures of sludge-cement-calcium chloride and calcium hydroxide were prepared. Calcium chloride and calcium hydroxide were used as accelerating additives. The specimens were tested in terms of setting time, compressive strength and toxicity characteristics. The mortars with the lower amount of sludge solids content gave the best results in compressive strength. Moreover, the samples containing calcium chloride (3%) improved the compressive strength. The inclusion of sewage sludge in the cement paste increased the setting time and therefore, it was recommended the addition of an accelerating additive, such as calcium chloride. The setting time was reduced as the percentage of calcium chloride added increased, however, the decrease in setting time was not satisfactory and further work was suggested to be done. A sample containing High Alumina cement showed a significant reduction in the setting time. Tests have shown high percentages of retention of heavy metals in the hydrated cement phases. Therefore, the authors suggested that if sewage sludge is to be used directly in mortar samples, then their durability should be concerned. Cyr at al. (2007) studied the physical, chemical and mineralogical characteristics of sewage sludge ash and evaluated its possible use in cement-based materials. The experimental results revealed that sewage sludge ash was composed of irregular grains with high specific surface area, leading to a significantly high water demand. The sewage sludge ash induced short delays of cement hydration. Although the oxides responsible for the pozzolanicity of materials (i.e. SiO_2 and Al_2O_3) were found in limited contents in a crystallised form, the authors suspected a slight pozzolanic activity of sewage sludge ash, as they observed a long-term positive effect on the compressive strength of mixtures containing sewage sludge ash. Nevertheless, the compressive strength of mixes containing 25% and 50% of sewage sludge ash were always lower than those of reference mortars. The main heavy metals found were zinc, chromium and copper, which occurred in concentrations greater than 2 g/kg of residue. The authors suggest that heavy metals of sewage sludge ash probably affected the hydration of cement and therefore the setting time and the compressive strength development of the mortars at early ages. They also considered the environmental impact of sewage sludge ash used in mortars, in terms of their leaching behaviour. It was shown that the leaching behaviour of mortars containing sewage sludge ash was of the same order of magnitude as the reference mortar without residue.

Sales and Souza (2009) went a step further, as they evaluated the possibility of recycling the water treatment sludge in construction and demolition waste recycling plants. The authors investigated the axial compressive strength and water absorption of concretes and mortars produced with the exclusive and joint addition of these two types of waste. The production of concretes and mortars with the joint addition of water treatment sludge for partial replacement of natural fine aggregates and recycled concrete rubble aggregates proved to be a viable recycling alternative from the standpoint of axial compressive strength, modulus of elasticity, water absorption and indirect tensile strength. Concretes and mortars under study were produced for three construction applications, namely medium strength structural concrete, underlayment concrete, and blocklaying mortar. It was concluded that sludge could be applied as a regulator of consistency and plasticity and, in suitable quantities, could even increase the compressive strength of concretes and mortars. Moreover, in all the additions of sludge studied, there was a decrease in the percentage of water absorption. The results showed that there wasn't notable difference between the modulus of elasticity of concrete produced with coarse aggregates of recycled construction rubble and the modulus of elasticity of concrete produced with coarse aggregates of recycled construction rubble and partial substitution of sand for sludge (5% in mass). Similar results (i.e. without significant difference) obtained for the tensile strength, as this was determined by the Brazilian method. The eco-efficiency of this recycling was also evaluated by determining the concentration of aluminum in the leached extract resulting from the solubilisation of the recycled products. The results indicated aluminum concentrations that exceeded water potability standards, without, however, rendering the leachate from this concrete harmful.

Chen et al. (2006) studied the sintering effect on cement bonded sewage sludge ash, which was used as the main material to produce a sewage sludge ash foamed lightweight material. The experimental results revealed that when the sintering temperature was lower than 600 ^{0}C, the engineering properties and microstructure of the lightweight material were mainly affected by dewatering and hydrates decomposition, whereas when the temperature was over 600 ^{0}C, the results were mainly affected by sintering effects. It was also concluded that at high temperatures, the more sewage sludge ash was used, less crack formation resulted, and after fired at 1093 ^{0}C for 4 hours, the compressive strength was substantially improved and the total pore volume decreased.

3. Sewage Sludge Ash as Brick and Tile Making Material

Extended studies have been conducted, investigating the potential use of dried sludge as a brick-making material, with very promising results (Alleman and Berman, 1984; Tay, 1986; Trauner, 1993; Okuno and Takahashi, 1997; Wiebusch and Seyfried, 1997). Most of the studies agree that the ash proportion and the firing temperature are the two key factors determining the quality of the bricks. Most recent studies by Lin and Weng (2001) investigated also the possible combination of incinerated sewage sludge ash and clay for the manufacture of bricks. In order to obtain the suitable manufacturing conditions, they investigated the proportion of sludge ash in the brick and the firing temperature that might affect the qualities of bricks. The dewatered sewage sludge sample was collected from the belt press dewatering machine of a municipal water treatment plant. Upon collection, the sludge sample was dried in an oven at 103 ^{0}C for 1 day. The sludge was incinerated in a combustion chamber at 800 ^{0}C to remove the organic substance and then was used as a clay substitute without further treatment. As the authors report, the recommended proportion of sludge ash in brick is 20 to 40%, with 13 to 15% optimum moisture content prepared in the molded mixtures and fired at 1000 ^{0}C to produce a good quality brick. Increase of the firing temperature and decrease of the amount of ash in the bricks resulted in a decrease of water absorption. Moreover, the results of Atterberg limits tests of molded ash-clay mixtures indicated that both plastic index and dry shrinkage decrease with an increasing amount of ash in the mixture. In addition, the compressive strength of the bricks made from both clay and sludge ash indicate that the strength (and the overall quality) is greatly dependent on the amount of ash in the brick and the firing temperature. This is also supported by the work published by Weng et al. (2003), where, however, is referred that the recommended proportion of sludge ash in brick is 10%, with 24% optimum moisture content prepared in the molded mixtures and fired at 880-960 ^{0}C to produce a good quality brick. It was also concluded that the incineration process made metals (chromium and zinc) less leachable. Other leached metals from either dried sludge or clay were of insignificant concern. A report compiled by Petavratzi (2007) presented a case study investigating the technical and economic feasibility of using incinerated sewage sludge ash in facing bricks. The investigators concluded that the use of incinerator sewage sludge ash as a filler for bricks is technically feasible, despite of the high water absorption values recorded during the experimental work, and the fact that the addition of the sludge ash does not provide any desirable aesthetic result. It is claimed in the report that the use of alternative materials as filler substitutes are considered by the brick sector only when suitable primary sources are not found in close proximity or if the cost of secondary materials in combination with their properties is advantageous. It was concluded that although the use of incinerated sewage sludge ash could be applicable to brickworks, it was anticipated that it will be utilised only if profit made by substituting primary sand was enough to cover the additional water and energy consumption.

Others investigated the application of sewage sludge to manufacture tiles as construction materials (Lin et al., 2005; Jordan et al., 2006; Lin et al., 2008; Chen and Lin, 2009; Montero et al., 2009), or for the preparation of multi-application ceramic products (Merino et al., 2006). Chen and Lin (2009) replaced various portions of potter's clay and porcelain clay with incinerated sewage sludge ash, in order to manufacture tile specimens. Nano-SiO$_2$ particles were added in the clay-sludge ash materials, in order to investigate their effect as strengthening additives. Mechanical tests were performed to measure shrinkage, water absorption, abrasion and bending strength. The experimental results showed that firing could reduce water absorption of both porcelain and potter's clay based tiles, with the kiln temperature effect to be more prominent in porcelain tiles. Furthermore, the addition of nano-SiO_2 had a positive effect on the shrinkage and abrasion resistance of both types of tiles. The bending strength was also improved with the addition of nano-SiO_2 in both tile specimens, with the improvement to be depended of the percentage of incinerated sewage sludge ash replacement and the sintering kiln temperatures. The authors concluded that the nano-SiO_2 additive has positive influences on the improvement of properties for both tile specimens, however they suggest that care should be taken to optimise both clay-sludge ash ratio and kiln temperature. Lin et al. (2007) also investigated the effects of nano-SiO_2 on tiles manufactured with clay and incinerated sewage sludge ash, obtaining similar results. Montero et al., (2009) investigated the possible manufacture of ceramic tile bodies, by making use of sewage sludge and marble residues, in substitution of clay raw materials. The experimental work revealed that despite of the substantial environmental and economic benefits deriving by the use of these residues as additives in the manufacture of traditional ceramics (i.e. recycling of wastes, immobilisation of heavy metals, and reduction in raw materials cost), their presence have a negative influence on the properties of these products. Specifically, it was observed an increase of water absorption and a decrease of the bending strength. Therefore, it was suggested that the amount of added wastes should be controlled in such a way that the ceramic product adequately meets the relevant standards. Other kind of additives, such as montmorillonite, illitic clay or powder flat glass were found to increase compressive strength and generally improve the overall characteristics of ceramic products including sewage sludge ashes, even at lower treatment temperatures. In contrast, improvements were not possible when kaolin added, due to its refractory character (Merino et al., 2006).

4. SEWAGE SLUDGE ASH AS A VALUE-ADDED PRODUCT IN THE CONSTRUCTION INDUSTRY

A few researchers have also been involved in studies investigating the possible use of sewage sludge in applications of special character, aiming to reuse sewage sludge as a value-added product. Al Sayed et al. (1995) presented the experimental results on the utilisation of sewage sludge ash as a replacement for mineral filler in asphalting paving mixes in hot regions. The Marshall Test method was used to evaluate the suitability of the waste material as filler in asphaltic concrete wearing courses. It was concluded that all the mixes satisfied the national standard specifications regarding stability, flow, voids in the mix, voids in the aggregates, voids filled with bitumen and Marshall stability and flow at elevated temperatures of 70 and 80 ^{0}C, suggesting the technical suitability of the waste as a filler in hot environments. However, the authors suggested that in order to replace the limestone filler, which is generated at no extra cost as a dust from aggregate crushing operations, by the sludge ash, was not justified if it was intended solely for this purpose, due to the cost of crushing and grinding the sludge in order to be suitable as filler. They added that if the crushing and grinding of the sludge waste is done as part of a bigger operation, then it will be economical to use the sludge as a mineral filler in asphaltic mixes.

Sewage sludge ash found ground of application in geotechnics also, as it has been successfully used to enhance the properties of soft soils (Lin et al., 2007; Chen and Lin, 2009). Chen and Lin (2009) mixed incinerated sewage sludge ash with cement for use as a stabiliser to improve the strength of soft, cohesive, subgrade soil. The soil samples mixed with sludge ash and cement admixture tested in terms of Atterberg limits, compaction, California bearing ratio (CBR), unconfined compressive strength, and triaxial compression. The sludge ash and cement admixture effectively improved the basic properties of soft subgrade soil, by decreasing the Plasticity Index. Moreover, soil swelling was clearly improved as more sludge ash and cement admixture added, contributing positively to the volumetric stabilisation of soft subgrade soil. Finally, the unconfined compressive strength and the CBR test results substantially improved with admixture amount.

Finally, several legislative and environmental issues regarding the use of municipal sewage sludge and ash from coal, co-firing as construction material are explicitly analysed by Cenni at al. (2001), who introduced the European legislation pertaining to the use of ash for construction purposes, and presented the environmental issues relating to its use. Specifically, it was concluded that excluding a priori the use of ash from co-firing as suitable additive for construction material could cause an unnecessary burden to the environment, since in absence of reutilisation ash might have to be disposed of in landfill. Also, allowing the use of ash from co-firing requires the modification of the standards to include limitations on all elements and compounds that are deleterious to the quality of construction materials.

CONCLUSION

Selected studies that focused on the potential uses of sewage sludge as a construction material were reviewed in this chapter. As it was amply indicated, sewage sludge can be used as a value-added constituent in a wide range of construction materials and applications, such as bricks, lightweight aggregates, cementitious binder of concrete mixes, ceramic materials, or as a constituent in asphaltic paving mixes. Further, sewage sludge ash was shown to be a successful admixture for soil improvement and stabilization, thus establishing itself in the field of geotechnical engineering, as well. In most of these cases, sludge has been used either in the form of ash (product of sludge incineration), or in the form of stabilized and solidified dry sludge. However, the common understanding remains that, for sewage sludge to be accepted as a true value-added product in construction, the modification of existing legislation and practice standards to incorporate specifications as to the chemical composition of these wastes is required.

Considering all the possible technical, environmental and economical benefits, the induction of sewage sludge in the ranks of construction materials appears as a promising and reliable means for the final disposal of this, otherwise, wasted, and difficult to handle, material. However, for sludge to become a real competitor to conventional, cheap, raw materials in the construction industry, further research is still needed towards overcoming technical obstacles and particularly in reducing the relatively high cost of sludge treatment.

Moreover, if wastewater sludge is to be truly exonerated, as a construction material, especially with regards to cost, studies that would indicate the overall economic, societal and environmental benefits remain to be seen.

REFERENCES

Al Sayed, M.H., Madany, I.M. and Buali, R.M., 1995. Use of sewage sludge ash in asphaltic paving mixes in hot regions. Construction and Building Materials, Vol. 9, No. 1, 19-23.

Alleman, J.E. and Berman, N.A., 1984. Constructive sludge management: Biobrick. Journal of Environmental Engineering, ASCE, 110 (2), 301-311.

Bhatty, J.I. and Reid, K.J., 1989a. Moderate strength concrete from lightweight sludge ash aggregates. The International Journal of Cement Composites and Lightweight Concrete, Vol. 11, No. 3, 179-187.

Bhatty, J.I. and Reid, K.J., 1989b. Compressive strength of municipal sludge ash mortars. ACI Materials Journal, 86, 394-400.

Bhatty, J.I. and Reid, K.J., 1989c. Lightweight aggregates from incinerated sludge ash. Waste Management & Research, Vol. 7, No. 1, 363-376.

Cenni, R., Janisch, B., Spliethoff, H. and Hein, K.R.G., 2001. Legislative and environmental issues on the use of ash from coal and municipal sewage sludge co-firing as construction material. Waste Management, 21, 17-31.

Cheeseman, C.R. and Virdi, G.S., 2005. Properties and microstructure of lightweight aggregate produced from sintered sewage sludge. Resources, Conservation and Recycling, 45, 18-30.

Chen, C.H., Chiou, I..J. and Wang, K.S., 2006. Sintering effect on cement bonded sewage sludge ash. Cement and Concrete Composites, 28, 26-32.

Chen, L. and Lin, D.F., 2009. Applications of sewage sludge ash and nano-SiO2 to manufacture tile as construction material. Construction and Building Materials, 23, 3312-3320.

Chen, L. and Lin, D.F., 2009. Stabilization treatment of soft subgrade soil by sewage sludge ash and cement. Journal of Hazardous Materials, 162, 321-327.

Chiou, I..J., Wang, K.S., Chen, C.H. and Lin, Y.T., 2006. Lightweight aggregate made from sewage sludge and incinerated ash. Waste Management, 26, 1453-1461.

Cyr, M., Coutand, M. and Clastres, P., 2007. Technological and environmental behaviour of sewage sludge ash (SSA) in cement-based materials. Cement and Concrete Research, 37, 1278-1289.

Donatello, S., Freeman-Pask, A., Tyrer, M. and Cheeseman, C.R., 2009. Effect of milling and acid washing on the pozzolanic activity of incinerator sewage sludge ash. Cement and Concrete Composites, In Press.

Elkins, B.V., Wilson, G.E., and Gersberg, R.M., 1985. Complete reclamation of wastewater and sludge. Water Science and Technology, 17, 1453-1454.

Jordan, M.M., Almendro, M.B., Romero, M. and Rincon, J.M., 2006. Application of sewage sludge in the manufacturing of ceramic tile bodies. Applied Clay Sciences, 30, 219-224.

Lin, D.F. and Weng, C.H., 2001. Use of sewage sludge ash as brick material. Journal of Environmental Engineering, ASCE, Vol. 127, No. 10.

Lin, D.F., Chang, W.C., Yuan, C. and Luo, H.L., 2008. Production and characterization of glazed tiles containing incinerated sewage sludge. Waste Management, 28 (3), 502-508.

Lin, D.F., Lin, K.L., Hung, M.J. and Luo, H.L., 2007. Sludge ash/hydrated lime on the geotechnical properties of soft soil. Journal of Hazardous Materials, 145 (1-2), 58-64.

Lin, D.F., Luo, H.L. and Sheen, Y.N., 2005. Glazed tiles manufactured from incinerated sewage sludge ash and clay. Journal of Air Waste Management Association, 55 (2), 163-172.

Lin, D.F., Luo, H.L. and Zhang, S.W., 2007. Effects of nano-SiO2 on tiles manufactured with clay and incinerated sewage sludge ash. Journal Materials Civil Engineering, ASCE, 19 (10), 801-808.

Lin, K.L., Chiang, K.Y. and Lin, C.Y., 2005. Hydration characteristics of waste sludge ash that is reused in eco-cement clinkers. Cement and Concrete Research, 35, 1074-1081.

Malliou, O., Katsioti, M., Georgiades, A. and Katsiri, A., 2007. Properties of stabilised/solidified admixtures of cement and sewage sludge. Cement and Concrete Composites, 29, 55-61.

Merino, I., Arevalo, L.F. and Romero, F., 2006. Preparation and characterization of ceramic products by thermal treatment of sewage sludge ashes mixed with different additives. Waste Management, 27, 1829-1844.

Montero, M.A., Jordan, M.M., Hernandez-Crespo, M.S. and Sanfeliu, T., 2009. The use of sewage sludge and marble residues in the manufacture of ceramic tile bodies. Applied Clay Science, doi: 10.1016/j.clay.2009.10.013.

Mun, K.J., 2007. Development and tests of lightweight aggregate using sewage sludge for nonstructural concrete. Construction and Building Materials, 21, 1583-1588.

Okuno, N. and Takahashi, S., 1997. Full scale application of manufacturing bricks from sewage. Water Science and Technology, 36 (11), 243-250.

Oliver, B.G. and Carey, J.H., 1976. Acid solubilisation of sewage sludge and ash constituents for possible recovery. Water Resources, 10, 1077-1081.

Petavratzi, E., 2007. Case Study: Incinerated sewage sludge ash in facing bricks. Characterisation of Mineral Wastes, Resources and Processing technologies – Integrated waste management for the production of construction material, WRT 177 / WR0115.

Sales, A. and de Souza, F.R., 2009. Concretes and mortars recycled with water treatment sludge and construction and demolition rubble. Construction and building materials, 23, 2362-2370.

Tay, J.H. and Show, K.Y., 1991. Clay-blended sludge as lightweight aggregate concrete material. Journal Environmental Engineering, American Society of Civil Engineers, 117, 834-844.

Tay, J.H. and Show, K.Y., 1992. Utilisation of municipal wastewater sludge as building and construction materials. Resources, Conservation and Recycling, 6, 191-204.

Tay, J.H. and Show, K.Y., 1994. Municipal wastewater sludge as cementitious and blended cement materials. Cement and Concrete Composites, 16, 39-48.

Tay, J.H., 1986. Potential use of sludge ash as construction material. Resources and Conservation, 13, 53-58.

Tay, J.H., 1989. Reclamation of wastewater and sludge for concrete making. Resources, Conservation and Recycling, 2, 211-227.

Trauner, E.J., 1993. Sludge ash bricks fired to above and below ash vitrifying temperature. Journal of Environmental Engineering, ASCE, 119 (3), 506-519.

Tsai, C.C., Wang, K.S. and Chiou, I..J., 2006. Effect of SiO2-Al2O3-flux ratio change on the bloating characteristics of lightweight aggregate material produced from recycled sewage sludge. Journal of Hazardous Materials, B134, 87-93.

Valls, S. and Vazquez, E., 2000. Stabilisation and solidification of sewage sludges with Portland cement. Cement and Concrete Research, 30, 1671-1678.

Valls, S., Yague, A., Vazquez, E. and Mariscal, C., 2004. Physical and mechanical properties of concrete with added dry sludge from a sewage treatment plant. Cement and Concrete Research, 34, 2203-2208.

Wang, K.S., Chiou, I..J., Chen, C.H. and Wang, D., 2005. Lightweight properties and pore structure of foamed material made from sewage sludge ash. Construction and Building Materials, 19, 627-633.

Wang, X., Jin, Y., Wang, Z. and Nie, Y., 2009. Development of lightweight aggregate from dry sewage sludge and coal ash. Waste Management, 29, 1330-1335.

Weng, C.H., Lin, D.F. and Chiang, P.C., 2003. Utilization of sludge as brick materials. Advances in Environmental Research, 7, 679-685.

Wiebusch, B. and Seyfried, C.F., 1997. Utilisation of sewage sludge ashes in the brick and tile industry. Water Science and Technology, 36 (11), 251-258.

Yague, A., Valls, S., Vazquez, E. and Kuchinow, V., 2001. Study of the hydration of cement pastes and dry sewage sludge. Proceedings of the International Symposium organised by the Concrete Technology Unit, University of Dundee, Scotland, UK Thomas Telford, 253-262.

In: Sewage Sludge Management
Editors: A. A. Zorpas and V. J. Inglezakis

ISBN: 978-1-61324-393-0
© 2012 Nova Science Publishers, Inc.

Chapter 13

PROBLEMS RELATED TO THE MANAGEMENT OF SEWAGE SLUDGE

Vassilis J. Inglezakis[*]

SC European Focus Consulting srl, Banatului 16, Bacau, Romania

ABSTRACT

Sewage sludge management is a serious and difficult to problem worldwide. Traditionally, sludge is landfilled, incinerated or landspread, however, is true that the major part is still landfilled, a practice that is far from being a sustainable management practice. The problems associated to health effects concern the untreated sludge as well as its disposal and treatment methods and sewage sludge became an issue of strong conflict within the scientific society as well as between governmental policies and citizens organizations. In the same time, sludge is rich in nutrients such as nitrogen and phosphorous and contains valuable organic matter, useful when soils are depleted or subject to erosion. The development of sustainable sludge management depends largely on the possibilities to improve the quality of the sludge itself and increase confidence in sludge quality which highlights the need for prevention of pollution of the wastewater at source. In the present Chapter the three main management options are presented, namely, agricultural land application, incineration and landfilling.

1. INTRODUCTION

Sludge originates from the process of treatment of wastewater and the quantity of sludge produced in a plant is approximately 1% of the quantity of treated wastewater. Sludge is usually treated before disposal or recycling in order to reduce its water content and toxicity, a process which is difficult and costly as it takes 40 to 50% of the total wastewater treatment costs [4]. Several treatment processes exist, such as thickening, dewatering, stabilisation and

---

[*] Website: www.efcon-group.ro, inglezakis@efcon-group.ro.

Table 1. Sewage sludge production and quantities recycled to agriculture in the 27 EU Member States [9]

Country	Year	Production (tons DS)	% used in agriculture
Austria	2005	266100	18
Denmark	2006	140021	35
France	2007	1125000	70
Germany	2007	2056486	29
Greece	2006	125977	≈0
Italy	2006	1070080	18
Netherlands	2003	550000	≈0
Spain	2006	1064972	65
UK	2006	1544919	68
Cyprus	2006	7586	41
Poland	2006	523674	17
Romania	2006	137145	≈0

disinfection and thermal drying. Following treatment, the sludge is basically recycled in agriculture (landspreading), incinerated or landfilled. Less developed routes exist as use in silviculture, land reclamation, and other developing combustion technologies as pyrolysis and gasification [5].

Traditionally, sludge is landfilled, incinerated or landspread, however, is true that the major part is still landfilled, a practice that is far from being a sustainable management practice. Apart from the problem of landfill overload and thus decreased life span, biodegradable waste, as sewage sludge is, decomposes in landfills following a long ecological cycle, producing landfill gas and leachates, both harmful for the environment and human health. In the same time, major part of the nutrients contained in the waste remains in the landfill practically unavailable for plant growth. The last 30 years wastewater and sewage sludge management is regulated in the developed countries and only in European Union more than 50 000 wastewater treatment plants are operating and sludge production is more than 10 million tons annually [16, 17]. In Europe, dry weight per capita production of sewage sludge resulting from primary, secondary and even tertiary treatment is in average 90 g/person/day.

During the last decades there has been a major change in the ways sludge is disposed, at least in Europe and USA. Since the decade of '80s and mid-'90s municipal sludge was primarily disposed at seawaters or landfilled and a smaller part was used as a fertilizer on agricultural land or incinerated. Since mid-'90s the sea disposal of sewage sludge is banned and gradually sludge deposits in landfills were or are phased out and the agricultural use has become the principal disposal method for sewage sludge. The latest trends in the field of sludge management, i.e. combustion, wet oxidation, pyrolysis, gasification and co-combustion of sewage sludge with other materials for further use as energy source, have generated significant scientific interest [16]. From economical point of view, landspreading of solid and semi-solid sludge entail on average the lowest cost from an overall economic point of view [5].

Table 2. Cost of sewage sludge utilization (2002 prices)

Use	Cost (€/ton DS)
Landspreading of solid and semi-solid sludge	110- 160
Landfilling, mono-incineration and co-incineration with other wastes	260-350
Landspreading of composted sludge, use of sludge in land reclamation, and use of sludge in silviculture	210-250

Due to the physical-chemical processes involved in the wastewater treatment, the sludge tends to concentrate potentially toxic elements as heavy metals, poorly biodegradable trace organic compounds and potentially pathogenic organisms. In the same time, sludge is rich in nutrients such as nitrogen and phosphorous and contains valuable organic matter, useful when soils are depleted or subject to erosion [1]. The problems associated to health effects concern the untreated sludge as well as its disposal and treatment methods and sewage sludge became an issue of strong conflict within the scientific society as well as between governmental policies and citizens organizations. Scientific results, official surveys and publications and NGO's opinions are frequently controversial causing confusion to citizens and policy makers. Concerning the final disposal on land, there are studies concluded that the practice is not dangerous while others showed that it could be harmful to human health [18, 20, 24].

The health and environmental effects of sewage sludge are largely centered around its ability to harbor bacteria, viruses, fungi and intestinal worms. Many of the pathogens contained within sewage sludge like salmonella, shigella and e-coli and parasites have the ability to cause diseases that can debilitate or even kill humans. Sewage sludge also contains large amounts of heavy metals as Cu, Cr and Zn, which are potentially toxic elements, and above certain concentrations, may interfere with or inhibit the actions of cellular enzymes. The chemical form and corresponding bioavailability of potentially toxic elements to plants, fungi, micro-organisms and animals are also important. Organic substances is another potential threat to the environment and human health. Sewage sludge potentially contains thousands of organic compounds derived from industrial, domestic, atmospheric and natural dietary sources. It has been said that over 60000 toxic substances and chemical compounds can be found in sewage sludge and each year 700 - 1000 new chemicals are being developed [21]. Finally, there are studies indicating the existence of radionuclides in sewage sludge samples [19].

2. LANDSPREADING AND USES IN AGRICULTURE

Landspreading of sludge or sludge-derived material partially replaces the use of conventional fertilisers, since it contains organic matter and compounds of agricultural value like nitrogen and phosphorus. In EU about 10 million tones of sewage sludge (dry basis) produced in 2006 of which 39% was used in agriculture [9].

Table 3. Summary of typical characteristics of sewage sludge land application practices in USA[1] [8]

Characteristics	Agricultural land application	Forest land application	Land applications at reclamation sites	Application to public contact sites, lawns and home gardens
Application rates	Range[2]: 2-70 t/ha/yr Typical: 10 t/ha/yr	Range[2]: 10-220 t/ha/yr Typical: 18 t/ha/yr	Range[3]: 7-450 t/ha/yr Typical: 112 t/ha/yr	Range[2]: - Typical: 18 t/ha/yr
Application technology	Annually, between harvesting and planting	Annually or 3- to 5- year intervals	Usually a one-time application	Varies depending on end use

[1] All values are approximate and they refer to usual rates and situations.
[2]Depends on type of crops, soil, turf, tree species, sewage sludge characteristics, etc.
[3] Usually a one-time application.

It is well known that the main route of entry of environmental contaminants into the human food chain is by uptake into the edible parts of crop plants and thus, the accumulation of potentially toxic elements in sewage sludge is an important aspect of sludge quality, which should be considered in terms of the long-term sustainable use sludge on land [2]. The constraints on the land application of sludge vary according to the treatment methods and the crops which are produced subsequent to the sludge application.

Of the major problems in the utilization of sewage sludge is the presence of heavy metals and studies have shown that more than 70% of the metals contained in urban wastewater is transferred in the sludge. Scientific evidence shows that they accumulate in the upper layers of the soil, due to binding to the different existing organic or mineral particles. Leaching to groundwater appears to be a negligible phenomenon while runoff may play a significant role in metal transfer [5]. Heavy metals are concentrated in the roots and vegetative parts of plants and are less present in the generative parts. Uptake of metals by animals occurs through contaminated plant consumption or soil ingestion. Several studies describe several consequences of metal provision to the soil on the micro-organisms population and biodiversity.

Table 4. Medium composition (mg/kg dm) for heavy metals in mineral fertilizers and urban sludge [10]

Metal	Mineral fertilizer	Urban sludge
As	1.8	-
Cd	1.6	5.3
Cr	63	80
Cu	-	334
Ni	20	39
Pb	7	133
Hg	0.1	2.7

Table 5. Comparison of potentially toxic element concentrations (mg/kg dm) in sewage sludge applied to agricultural land in Germany and the United Kingdom in 1996 [10]

Metal	Germany (average)	UK (weighted average)	UK (median)
Zn	776	792	559
Cu	305	568	373
Ni	24	57	20
Cd	1.45	3.3	1.6
Pb	57	221	99
Cr	40	157	24
Hg	1.35	2.4	1.5

A problem of particular significance is the sludge organic chemicals content. Production of artificial organic chemicals has been increased dramatically and in the United States, the rate of growth has been exponential in '90s. In the UK, the number of chemicals available could be increasing by about 1000 per annum. It is obvious, therefore, that a proportion of these compounds will get into waste water treatment works during their production, transport, use and disposal [15]. Organic compounds, once applied to the land, they are distributed throughout all soil media and undergo several retention and transport processes and they gradually transformed in other intermediary compounds during their mineralization. Leaching of organic pollutants to ground water appears to be insignificant but cannot be neglected in some cases while runoff may play an important role in the transfer of organic compounds [5].

In contrast to heavy metals, for organic pollutants there is no evidence for soil-crop transfer and thus, it can be stated that most organic pollutants are not taken up by plants [1, 5, 10, 11]. This is because many organic compounds exhibit some solubility and rapid degradation and volatilization processes in soil. The problem is concentrated on more persistent compounds which usually have very low solubility and are strongly adsorbed by the soil matrix in non-bioavailable forms. For persistent organic pollutants the persistence varies between different groups and specific compounds within each group, increasing generally in the order: PCBs>CBs>PAHs>PCDD/Fs [10]. However, a risk of contamination of the food chain exists when spreading sludge directly onto crops, especially on plants which are to be consumed raw or semi-cooked [5]. Soil and sludge ingestion on land used for grazing is the main route for animal contamination and in contrast to heavy metals, accumulation of bioaccumulative compounds such as PCDD/Fs, PCBs or PAHs may occur in meat and milk. However, there has been no demonstrable relationship between sludge application and transfer of organics to animal tissues or milk [1].

Table 6. Overview on the behavior of some organic substances in soils [7]

Substance	Mammalian and human toxicity (acute)	Water solubility	Persistence
LAS	Medium	High – enhances mobility of other pollutants	Medium
DEHP	Low – suspected estrogenic effect	Low	Medium
PCBs	Medium – tumour promoting	Low	High
PCDD/Fs	High - carcinogenic	Low	High

Finally, pathogens are mainly present on the soil surface or at shallow depths where sludge is applied and transfer to groundwater is only assumed to occur in some particular cases. Surface water contamination is more likely to occur when runoff water transports pathogens which are bound to soil particles. Survival of pathogens on plants is shorter than in soil, due to the effects of desiccation and sunlight. Transmission to grazing domestic and farm animals takes place via ingestion of contaminated feed and soil and humans can mainly be affected by consuming raw or semi-cooked contaminated vegetables or meat [5].

The constraints apply particularly to conventionally treated sludges where there is a greater risk of pathogens being present. In general, the advanced treatments reduce the pathogen content to insignificant levels while conventional to certain limit levels. Advance treatments are windrow composting, aerated pile and in vessel composting, thermal drying, thermophilic digestion (aerobic or anaerobic), heat treatment followed by digestion and treatment with lime (CaO). Other forms of sludge treatment, such as mesophilic anaerobic digestion, use of slaked lime, and storage in various forms, will not produce a sludge that can be considered to be free of pathogens. Sludges treated to the suggested standards for advanced treatment will not add to the pathogen burden of the environment and present no risk to human, animal or plant health [6].

The European Union Directive 86/278/EEC and the Part 503 of Section 40 of the Code of Federal Regulations (CFR) in USA prohibit the use of sludge in agriculture if the heavy metals concentrations exceed specific limit values (Table 8).

Table 7. Suggested constraints on land use if treated sludge [6]

Crop	Advanced treatment	Conventional treatment
Pasture	Yes	Injection and 3-week no-grazing
Forage	Yes	3-weeks no-harvest
Arable	Yes	Injection or plough-in
Vegetables in ground contact	Yes	10-month no-harvest
Fruit and vegetables eaten raw (salads)	Yes	30-month no-harvest
Fruit trees, vineyards	Yes	Injection and 10-month no-access
Parks and urban open spaces	Well stabilized and odorless	No
Land reclamation	Yes	10-month no-access

Table 8. Maximum permissible limits for land application in EU and USA (mg/kg dw)

Metal	86/278/EEC (range)	USA (Part 503)
Zn	2500-4000	7500
Cu	1000-1750	4300
Ni	300-400	420
Cd	20-40	85
Pb	750-1200	840
Cr	-	3000
Hg	16-25	57
Mn	-	-

Sewage Sludge Directive prohibits the use of untreated sludge on agricultural land unless it is injected or incorporated into the soil. However, in order to provide protection against potential health risks from residual pathogens:

- Sludge must not be applied to soil in which fruit and vegetable crops are growing or grow
- Grazing animals must not be allowed access to grassland or forage land less than three weeks after the application of sludge

The Directive sets out requirements for the keeping of detailed records of the quantities of sludge produced, the quantities used in agriculture, the composition and properties of the sludge, the type of treatment and the sites where the sludge is used. Limit values for concentrations of heavy metals in sewage sludge intended for agricultural use and in sludge-treated soils are set in Directive Annexes. Directive 86/278/ EEC was adopted over 20 years ago and the European Commission is currently assessing whether the current Directive should be reviewed – and if so, the extent of this review and the changes proposed included the following:

- Revision of current limit values for heavy metals;
- Introduction of limit values for organic pollutants;
- Introduction of pathogen concentration limits; and
- Introduction of a quality assurance system

3. INCINERATION

Incineration is a combustion reaction, which in turn is the sequence of exothermic chemical reactions between a fuel and an oxidant accompanied by the production of heat and conversion of chemical species. Incineration temperatures are $850\text{-}1450°C$ and the reaction takes place under surplus of oxygen. Basically, waste incineration is the oxidation of the combustible materials contained in the waste the objective being the treatment of waste so as to reduce its volume and hazard, whilst capturing or destroying potentially harmful substances. Incineration processes can also provide a means to enable recovery of the energy, mineral and/or chemical content from waste. There are different types of processes, namely mono-incineration when sludge is incinerated in dedicated incineration plants, incineration with other wastes and co-incineration when sludge is used as fuel in energy or material production [5]. The incineration technology is viable for the thermal treatment of high quantities of solid waste (more than 50-100000 tn/year).

Incineration of sewage sludge takes place in rotary kilns, multiple hearth, or fluidized bed incinerators. Co-combustion in grate-firing systems, coal combustion plants and industrial processes is also applied. Raw as well as oxidises and digested sludges have a natural water content of between 92% and 99%, the sludge water being well combined with the solid particles [10]. Water reduction means energy demand and sludge water content is an important operation problem. Therefore, sludge needs needs dewatering, using energy intensive processes such as centrifugation or thermal dehydration. Centrifugation requires less

Table 9. Typical heating values for several types of sewage sludge [16] and other fuels

Fuel	Heating value (MJ/kg dw)
Hard coal	30
Waste engine oils	29
Tyres	26
Raw sludge	25.5
Plastic waste	22
Activated sludge	21
Biological filter sludge	19.5
Raw chemically precipitated primary sludge	16
Wood	14
Anaerobically digested primary sludge	11
Paper	10
Municipal waste	9.3
Brown coal	9

energy but surplus heat from incineration that can be used for thermal dehydration. Net treatment cost, including revenue from energy generation, is about 230-300 €/t for small plants (50000 t/a) and decreases to 80-100 €/t for big plants (300000 t/a). The high costs involved along with the residues management issues turns incineration a difficult option for small municipalities. Detailed information on specific best available techniques (BAT) for sewage sludge incineration are provided elsewhere [14].

Incineration generates emissions into the [5]:

- air (particles, acid gases, greenhouse gases, heavy metals, volatile organic compounds, etc.),
- soil (disposal of ashes and flue gas treatment residues to landfill, atmospheric deposition of air emissions) and
- water (flue gas treatment wet processes)

Although fumes are filtered before being released into the atmosphere many different kind of chemicals is possible to be released, the most toxic being dioxin. Concerning the ashes, the final disposal is either in landfills or in building material. Residues are about 30% of the waste input mass and they contain high concentrations of heavy metals and may be regarded as hazardous waste especially if the sludge is incinerated along with municipal waste [10, 16].

The relevant European Union legislative act is the Directive 2000/76/EC on Incineration of Waste. The aim of the WI Directive is to prevent or to reduce as far as possible negative effects on the environment caused by the incineration and co-incineration of waste.

The WI Directive makes a distinction between:

- incineration plants, which are dedicated to the thermal treatment of waste and may or may not recover heat generated by combustion and
- co-incineration plants, such as cement or lime kilns, steel plants or power plants whose main purpose is energy generation or the production of material products

Is important to mention that in European Union, according to the Waste Framework Directive (2008/98/EC), incineration of waste is considered to be a disposal operation unless is energy-efficient and may be considered a recovery operation. In particular, this is true, when the waste is used principally as a fuel or other means to generate energy and the energy efficiency is equal to or above:

— 0.60 for installations in operation and permitted in accordance with applicable Community legislation before 1 January 2009
— 0.65 for installations permitted after 31 December 2008

The amount of sludge being incinerated in Denmark has already reached the percentage of 24% of the sludge produced, 20% in France, 15% in Belgium, 14% in Germany while in USA and Japan the percentage has increased to 25% and 55%, respectively [16]. According to the latest available data, in EU, in total 1444 incineration and co-incineration plants have been reported – 595 plants dedicated waste incineration and 849 co-incineration. The majority of incineration plants have been indicated for France (155), followed by the UK (85) and Germany (70) while the majority of co-incineration plants has been reported by Italy (487). The dedicated incinerator types are: municipal waste incinerator (MWI), hazardous waste incinerator (HWI), clinical waste incinerator (CWI), sewage sludge incinerator (SSI) and other type of waste incinerators (427 of 595 reported). According to 2007 data, sewage sludge incinerator are 57 with a capacity of 1.66 Mt (dw)/yr. Concerning co-incineration, 4% of industrial and 4.3% of municipal sewage sludge are used as feed while the sectors of the most incinerators in use are the energy industries (combustion plants) and cement industry (24.9% and 16.8%, respectively) [13, 14].

Incinerators

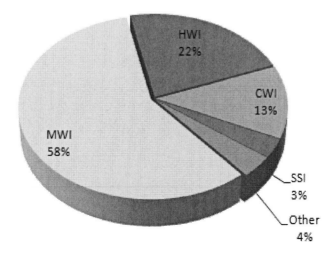

Figure 1. (Continued).

Waste Input Capacity

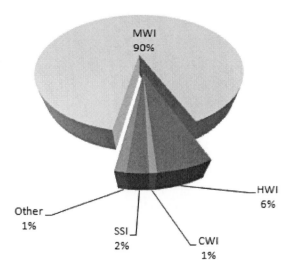

Figure 1. Number of (dedicated) Incinerators and Waste input Capacity (2007) [13].

The advantages of incineration are:

- Volume (approx. 90%) and weight (approx. 75%) reduced
- Waste reduction is immediate and destruction in seconds
- Incineration can be done at generation site
- Air discharges can be controlled and ash residue is usually inert
- Small disposal area required
- Cost can be offset by heat recovery/ sale of energy

The disadvantages of incineration are:

- High capital cost
- Skilled operators are required (particularly for boiler operations)
- Some materials are noncombustible and others require supplemental fuel
- Air contaminant potential
- Volume of gas from incineration is 10 times as great as other thermochemical conversion processes, greater cost for gas cleanup/pollution control
- Public disapproval

4. LANDFILLING

As is the common practice in many occasions, after digesting for an extended period, the digested sludge is disposed of in municipal landfills along with other solid waste. Sludge landfilling methods generally involve mixing the concentrated sludge with other municipal solid waste. However, stricter waste disposal legislation and higher landfill taxes are forcing the waste industry to look for more efficient disposal strategies. Furthermore, landfill operators are also increasingly reluctant to accept sludge material in the slurry state because of construction difficulties and the potential for instability of the landfill slopes [25]. Landfill disposal of organic wastes, such as sludge, is not a sustainable option due environmental pollution by landfill gas and leachate emissions, landfill capacity and lifespan reduction and loss of organic matter and nutrients that decomposes in a landfill and is not available for plant growth. In particular, co-disposal of municipal solid waste and sewage sludge has a significant effect on the generation and quality of leachate due to the increase of the moisture input by sludge [26].

These motivations, among others, have contributed to the adoption of the European Union Landfill Directive 1999/31/EC that introduced targets for the reduction of biodegradable municipal waste to landfill as follows:

- reduction to 75% of total biodegradable municipal waste (weight) produced in 1995 by 2006;
- reduction to 50% by 2009;
- reduction to 35% by 2016.

There are two possibilities in terms of sludge landfilling: mono-deposits, where only sludge is disposed of and most commonly, mixed-deposits when the landfill is also used for municipal wastes. Landfill operation generates emissions into [5, 10]:

- the air, mainly greenhouse gases like methane and carbon dioxide, reduced when biogases are collected and burnt. Landfill gas, if not captured, contributes considerably to the greenhouse effect as is mainly composed of methane, which is twenty times more powerful than carbon dioxide in terms of climate change effects
- the soil, leachates rich in several pollutants, like heavy metals and
- water at dumpsites, where various compounds such as heavy metals, organic compounds and micro-organisms in leachate are found their way in underground waters

The co-disposal of dewatered sewage sludge and municipal solid waste may cause significant impacts (e.g. slope instability and leachability) on landfill operations and the subsurface environment [26]. The sludge must be adequately dried for more the efficient handling and trafficability by machine plant, optimum compaction, and an adequate safety level against slope instability. After drying, sludge is a soil-like material and its behavior in the landfill can be modeled using soil mechanics theory [25]. Once placed in landfills, sludge can be viewed as a geotechnical material, and it should be compared to non-consolidated cohesive soils with high organic content [26]. Finally, is important to note that the landfill design must take into account the effects of digestion on the engineering properties of the sludge when determining the long-term safety level against instability of the landfill slopes [25].

REFERENCES

[1]   European Commission, DG Environment, http://ec.europa.eu/environment/waste
       /sludge/index.htm
[2]   Pollutants in urban waste water and sewage sludge, European Commission, DG
       Environment, Luxembourg: Office for Official Publications of the European
       Communities, 2001
[3]   European Commission, DG Environment, http://ec.europa.eu/environment/water/water
       -urbanwaste/legislation/directive_en.htm
[4]   Izrail S. Turovskiy, P. K. Mathai, Wastewater sludge processing, John Wiley & Sons,
       Inc., USA, 2006
[5]   Disposal and Recycling Routes for Sewage Sludge, Synthesis report, DG Environment
       B/2, European Commission, 2002
[6]   Evaluation of sludge treatments for pathogen reduction, Luxembourg: Office for
       Official Publications of the European Communities, 2001
[7]   H. Langenkamp, P. Part, W. Erhardt, A. Prüeß, Organic contaminants in sewage sludge
       for agricultural use, European Commission, Joint Research Centre, Institute for
       Environment and Sustainability, Soil and Waste Unit, 2001
[8]   Process Design Manual, Land Application of Sewage Sludge and Domestic Septage,
       United States Environmental Protection Agency, 1995
[9]   Environmental, economic and social impacts of the use of sewage sludge on land,
       Consultation Report on Options and Impacts, Report by RPA, Milieu Ltd and WRc for
       the European Commission, DG Environment under Study Contract DG
       ENV.G.4/ETU/2008/0076r, 2009
[10]   Workshop on problems around sludge, Jointly organised by: the Directorate-General
       for the Environment, the Joint Research Centre of the European Commission, Edited
       by Heinrich Langenkamp, Luca Marmo, Proceedings, 18-19 November 1999 - Stresa
       (NO) Italy, 1999
[11]   Researching the Sludge Directive - A conference on sewage sludge, Brussels on 30-31
       October 2001, Jointly organised by DG Environment and UKWIR, 2001
[12]   Howard R. Rogers, Sources, behaviour and fate of organic contaminants during sewage
       treatment and in sewage sludges, The Science of the Total Environment, 185, 3-26,
       1996
[13]   Knut Sander, Christian Tebert, Stephanie Schilling, Dirk Jepsen, Assessment of the
       application and possible development of community legislation for the control of waste
       incineration and co-incineration, Ökopol GmbH, European Commission Service
       Contract No.070501/2006/446211/MAR/C4, Final Report, November 2007
[14]   Reference Document on the Best Available Techniques (BREF) for Waste Incineration,
       European Commission, 2006
[15]   Min-Jian Wang and Kevin C. Jones, Behaviour and fate of chlorobenzenes (cbs)
       introduced into soil-plant systems by sewage sludge application: a review,
       Chemosphere, Vol. 28, No. 7, pp. 1325-1360, 1994
[16]   D. Fytili, A. Zabaniotou, Utilization of sewage sludge in EU application of old and new
       methods—A review, Renewable and Sustainable Energy Reviews, 12, 116–140 (2008)

[17] Eva Eriksson, Nina Christensen, Jens Ejbye Schmidt, Anna Ledin, Potential priority pollutants in sewage sludge, Desalination 226, 371–388 (2008)

[18] Lewis, D. L., D. K. Gattie, M. E. Novak, S. Sanchez, and C. Pumphrey. 2002. Interactions of Pathogens and Irritant Chemicals in Land-Applied Sewage Sludges (biosolids). BMC Public Health. 2:11.

[19] The presence of radionuclides in sewage sludge and their effect on human health, Environmental Radiation Program, WDOH/320-013,Washington State Department of Health, December, 1997

[20] Dorn CR, Reddy CS, Lamphere DN, Gaeuman JV, Lanese R., Municipal sewage sludge application on Ohio farms: health effects, Environ Res. 1985 Dec;38(2):332-59.

[21] Stauber J and Rampton S, 1995. Toxic Sludge is Good for you! Lies Damn Lies and the Public Relations Industry. Centre for Media and Democracy, http://www.prwatch .org /cmd/index.html

[22] Report on waste incineration and possible contamination of the food supply with dioxins, Food Safety Authority of Ireland, 2003

[23] Draft discussion document for the ad hoc meeting on biowastes and sludges 15-16 January 2004, Brussels, 18 December 2003

[24] Recycling of Biosolids to Land, Water UK, March 2006

[25] Brendan C. O'Kelly, Sewage sludge to landfill: some pertinent engineering properties, Journal of the Air & Waste Management Association, June 2005

[26] Irene M. C. Lo, Wenwen Zhou and Ken K. H. Ng, Environmental and Landfill Operational Aspects of Co-disposal of Dewatered Sewage Sludge and Municipal Solid Waste, Submitted to HKIE Environmental Division, 2003 Peter HK Chan Award for Best Environmental Paper (BEPA), 2003

In: Sewage Sludge Management
Editors: A. A. Zorpas and V. J. Inglezakis

ISBN: 978-1-61324-393-0
© 2012 Nova Science Publishers, Inc.

Chapter 14

IMPACT ASSESSMENT FROM SEWAGE SLUDGE

Antonis A. Zorpas, Vasilis Inglezakis and Irene Voukalli*
Institute of Environmental Technology and Sustainable Development,
Paralimni, Cyprus

ABSTRACT

Sludge is formed during wastewater treatment. Wastewater may contain certain undesirable components, including organic, inorganic and toxic substances, as well as pathogenic or disease-causing micro-organisms. Sewage sludge is being generated in ever increasing amounts and the dilemma centers on how to find an environmentally acceptable way to utilize sewage. The treatment and disposal of sewage sludge is an expensive and environmentally sensitive problem. It is also a growing problem world-wide since sludge production will continue to increase as new sewage treatment works are built and environmental quality standards become more stringent. With some traditional disposal routes coming under pressure, and others such as sea disposal having been phased out, the challenge facing sludge managers is to find cost-effective and innovative solutions whilst responding to environmental, regulatory and public pressures. Recycling and use of wastes are the preferred options for sustainable development, rather than incineration or landfilling, but with sewage sludge this is not straight forward because of perceptions over contaminants, pathogens and its faecal origin, particularly by the food retailers. If future quality standards for sludge and the receiving environment are made too stringent, the agricultural outlet may become untenable for the water utilities resulting in sludge being disposed of by other means that offer the utilities greater operational and financial security, but which may be less acceptable in the long-term.

1. INTRODUCTION

The disposal of sludge always requires very positive and careful management but the ease, or difficulty, with which disposal is actually achieved, and the associated costs depend

---

* Website: www.envitech.org Tel: +357-23743440, Fax: +357-23743441, P.O.Box 34073, 5309, Paralimni, Cyprus.
E-mail:antoniszorpas@envitech.org, antoniszorpas@yahoo.com.

very much on circumstances. Local and national geographical, agronomic, economic and stakeholder perception factors have considerable influence. The general trend in recent years in most developed countries has been for the disposal of sludge to become more, rather than less difficult, and this has been exacerbated by a considerable increase in EU and national environmental legislation. The management of sewage sludge in an economically and environmentally acceptable way is a matter of increasing importance. The stricter regulations on sludge disposal in both industrialised and emerging countries are going through significant, prompt and continuous changes.

At this moment, the European legislation applicable to sludge is evolving rapidly [1]. In particular, a new Directive on sludge has been at the stage of its third draft since April 2000 [2], a new Directive on incineration was issued in December 2000 [3] and another document regarding biological treatment of biowaste is being drawn up [4]. Historically, most of the sludge generated has been directed to incineration, landfilling or disposed in the sea1. Only a small portion has been reused in agriculture, mainly because of the apprehension that the application of treated sludge to agricultural land may cause the transfer of pathogens, viruses, heavy metals and organochlorine residues to the crops to humans. In this sense, anaerobic digestion cannot provide a 100% reduction of pathogens and viruses, while a thermal drying technology, consisting on gasification and combustion with recovery of energy, can effectively attain that [5]. Furthermore, a study from the late seventies regarding new technologies reported the advantages of sludge pyrolysis over sludge incineration or landfill, such as less air and no fuel consumption or less transport requirements [6]. Life Cycle Assessment (LCA) is a tool for evaluating the environmental performance not only of goods, but also of processes or services (collectively termed products). ISO 14040 defines LCA as a compilation and evaluation of the inputs, outputs and the potential environmental impacts of a system throughout its Life Cycle [7]. This practice is obtained by means of a systematic, four-step procedure: goal and scope definition, inventory analysis, impact assessment and interpretation. Since its early stages, the first studies to look at life cycle aspects of products date from the late sixties and early seventies, and focused on issues such as energy efficiency, the consumption of raw materials and, to some extent, waste disposal [8], LCA has proven to be a valuable tool to verify and analyse environmental considerations of goods and service systems that need to be part of the decision-making process towards sustainability [9]. The application of LCA to sludge management has turned out to be quite extensive. Bridle and Skrypski Mantele [5] reported qualitative conclusions on sludge management by means of four criteria based on a sustainable life-cycle approach. Hwang and Hanaki [10] exclusively quantified the amount of CO_2 produced from the construction, operation and dismantlement of a sewage sludge treatment system, in order to estimate its global warming potential. Suh and Rousseaux [11] carried out a comparison of five alternative scenarios, but the only results from their relative environmental profile (with no absolute numbers) were presented. A more detailed analysis was performed by Houillon and Jolliet [12] inside the framework of the Ecosludge project. Six scenarios were compared but, as in this first paper of the study only energy and emissions contributing to global warming were quantified, no final conclusions on their global environmental impact could be obtained. Finally, Lundin et al. [13] have gone further and performed an environmental and economic analysis of four options, including agricultural application, co-incineration with waste and two recent technological approaches for phosphorus recovery (Bio-Con and Cambi-KREPPO). While the former two proved to have environmental or economical restrictions, the latter two turned out to be a promising solution for sewage sludge treatment. The general aim of this work is to complete the analysis performed at a wastewater treatment plant (now referred to as WWTP) [14], where sludge post-treatment was identified as one of the main contributors to the environmental performance of the global system. To achieve this target, several options will be compared with the actual post-treatment of the sludge generated in the wastewater plant mentioned considering a broad number of impact categories, which will provide a comprehensive knowledge of the environmental performance of the process under study.

1.1. Formation of Sewage Sludge

There are currently over 50000 wastewater treatment works (WWTPs) operating in the European Union yielding a total of about 7.9 million tonnes of dry solids (tds) in year 2000. The amount of sludge will continue to increase as the Urban Waste Water Treatment Directive continues to be implemented, reaching at least 8.3 million tds/y by 2006.

Sludge is formed during wastewater treatment. Wastewater is a combination of the liquid- or water-carried wastes removed from residential, institutional, commercial and industrial establishments, together with ground water, surface water and storm water, as may be present [15]. Wastewater may contain certain undesirable components, including organic, inorganic and toxic substances, as well as pathogenic or disease-causing micro-organisms. Table 1 shows a typical composition of untreated municipal wastewater. The main groups of the organic solids in wastewater are proteins, carbohydrates, fats and oils. Protein contains about 16% nitrogen, and together with urea are the major sources of nitrogen in wastewater [16]. In its untreated form, wastewater can not be disposed of for several reasons. First, the biological decomposition of the organic materials in wastewater consumes oxygen and thus reduces the quantity available in the receiving waters for the aquatic life. The decomposition also produces large quantities of malodorous gases. Secondly, the numerous pathogenic or disease-causing micro-organisms in untreated wastewater are health hazards to human beings. Third, its toxic compounds, especially heavy metals, can be dangerous to both plants and animals, and finally the presence of phosphates and nitrogen may lead to uncontrolled growth of aquatic plants [15]. It is therefore necessary to reduce the organic components, nitrogen and phosphorus, toxic compounds, as well as to destroy the pathogenic or disease-causing micro-organisms from wastewater before its disposal.

Table 1. Typical domestic sewage characteristics, in mg/l (source: [16])

Weak		Medium	Strong
Total suspended solids	100	200	350
Volatile suspended solids	75	135	210
BOD$_5$	100	200	400
COD	175	300	600

Table 1. (Continued)

Weak		Medium	Strong
Ammonia-N	5	10	20
Organic-N	8	20	40
PO_4-P	7	10	20

There are regulations governing the quality of the effluent discharged from Sewage Treatment Works (STW). The regulations set limits for the concentrations of BOD5, COD, total suspended solids, phosphorus, nitrogen, mercury and phenol, among others, in the discharge from STW. In the European Union (EU), the quality of effluent is currently regulated through the Urban Wastewater Treatment Directive (91/271/EEC) [17] of May 1991. The directive also requires that all communities with above 2000 population equivalents install adequate collection, treatment and disposal systems for the wastewater generated. Table 2 gives some values of the national standards of some countries for illustration. A comparison of the limits in Table 2 with the concentrations in Table 1 shows that a significant reduction of the various components must be affected in the STW to meet the limits required. It is important to note that even the Newly Industrialized Countries (NIC) are already moving towards stricter limits for pollutants in their effluent discharges. To achieve the set limits for effluent discharges, a modern wastewater treatment plant may employ three stages of wastewater treatment, i.e. mechanical, biological, and an additional (third) stage for the elimination of nitrogen and phosphorous. The mechanical stage provides a primary clarification. The incoming untreated wastewater passes through racks and screens where coarse solids are removed, and then into grid chambers, where sand, gravel, cinders and other heavy solid materials are separated. Thereafter, in sedimentation tanks, readily settleable solids and floating materials are removed. In the mechanical stage, 50–70% of the suspended solids and 25–40% of the BOD5 can be removed [15]. The materials removed form primary sludge, having 3–5 wt% solids which consist of 30% inorganic and 70% organic matter [19]. In the biological treatment phase, the coagulation and removal of the non-settleable colloidal solids and the stabilization of the organic matter is accomplished biologically using a variety of micro-organisms, principally bacteria [15]. The micro-organisms convert the colloidal and dissolved carbonaceous organic matter into various gases and cell tissues. The cell tissues have a specific gravity slightly higher than that of water and can therefore be removed through sedimentation. The sludge removed at this stage is normally referred to as secondary sludge. In the third treatment stage, the elimination of phosphorous and nitrogen takes place. Nitrogen elimination may be achieved by a two-stage process: nitrification and denitrification. Ammonia in the wastewater is first oxidized to nitrate (nitrification) and thereafter the nitrate is converted to free nitrogen (denitrification) before it is discharged [20]. Denitrification is an important part of the nitrogen cycle, since through the release of the nitrogen to the atmosphere; the cycle is completed [21]. Phosphorous compounds can be eliminated through chemical precipitation using additives followed by the sedimentation of the sludge formed, or through biological treatment, where phosphorous compounds are incorporated in the cell tissues which are thereafter removed through sedimentation [22]. Nitrogen and phosphorus sludges may be formed at this stage.

Table 2. Comparison of national standards for effluents in mg/l (sources: [17, 18])

	Hong Kong	Japan	Malaysia	Singapore	Taiwan	France	NL	Portugal	EU [17]
BOD$_5$	20–40	160	20–50	100	100	30	20	40	25
COD	80–85					90		150	125
SS	30–40					30	30	60	35
N	20–50					10	10–15	15	10–15
Hg	0.001	0.05	0.005–0.05	0.001–0.05	0.005				
Phenols	0.1	5	0.001–1	◆ 5	5				

SS—suspended solids, NL—Netherlands and EU—European Union.

1.2. Processing of Sewage Sludge

Sludge is by far the largest in volume amongst the byproducts of wastewater treatments, and its processing and disposal is perhaps one of the most complex environmental problems facing the engineer in this field. This is because the sludge resulting from the wastewater treatment operations and processes is usually in the form of a very dilute suspension, which typically contains from 0.25 to 12% solids, depending on the operation and process used. Apart from that, sludge is composed largely of the substances responsible for the offensive, pathogenic and toxic characteristics of the untreated wastewater [15]. Sludge processing is therefore intended to reduce smell, reduce the quantity of the organic solids, eliminate disease causing bacteria, improve the dewatering characteristics of sludge, and reduce the water content so that the end product can be treated further or disposed of with less handling problems and environmental consequences. These can be achieved through sludge stabilization, conditioning and dewatering.

1.2.1. Stabilization

There are three methods of stabilization normally practiced in wastewater treatment plants. These are digestion, lime stabilization and heat treatment. Digestion involves partial conversion of the organic matter into a gas and a stabilized residue with the help of bacteria either in the absence of (anaerobic) or presence of oxygen (aerobic), during which disease-causing bacteria are also destroyed. Digestion is used widely for municipal sludge, but may not be applied to industrial sludge which may contain a high content of indigestible substances and/or chemicals likely to destroy the micro organisms [23]. Lime stabilization and conditioning involves the addition of lime to untreated sludge in sufficient quantity to raise the pH to 12 or higher, an environment that is not conducive to the survival of micro-organisms. Consequently, the sludge will not putrefy, create odours or pose a health hazard as long as the pH is maintained at this level [15]. Heat treatment also stabilizes and conditions sludge. It involves two-stage heating of sludge, at 30–75 and 75–190°C, and at a pressure of 26 bar for a short time [24]. This coagulates the solids, breaks down the gel structure and reduces the water affinity of the sludge solids. As a result, the sludge is sterilized, practically deodorized and is dewatered readily on vacuum filters or filter presses to 35–40% d.m. without additives [24]. The methods of sludge stabilization and conditioning chosen are important since they determine the quantity of sludge to be further treated. They also change the physical and chemical characteristics of the sludge, and these may have significant effects on sludge incineration. Stabilization through digestion, e.g. leads to a decrease in the quantity of sludge, since part of the organic substance is released as CH4 and CO2. This, however, reduces its calorific value from around 17.5 MJ/kg d.m. for raw sludge to about 10.5 MJ/ kg d.m. for digested sludge [25], so it can be argued that incorporating digestion process would be undesirable if the sludge is to be incinerated. But, Albrink et al. [16] have shown that digestion of sludge prior to incineration has several advantages which outweigh the effect of reduced calorific value. The sludge volume is reduced by 50% and its dewatering properties increased by two–three points. Furthermore, in the event of a breakdown of the incinerator, digested sludge can be handled easier than raw sludge. The digestors could also provide short term storage of sludge. Lime stabilization leads to an increase in the quantity of sludge and lowers its calorific value. For sludge disposal through land filling, the addition of lime would increase the disposal cost. Thermal stabilization and conditioning cause insignificant changes in the quantity of sludge, but experience from large scale application shows that due to the return of the dissolved organic substance, heavy metals and NH_3 are returned back to the raw wastewater, the quality of effluent decreases [24]. In the European Union, 50% of the sewage treatment plants have anaerobic digestion, 18% aerobic digestion and 4% lime stabilization, whereas 24% of the plants undertake no sludge stabilization [17]. Digestion is therefore the most popular sludge stabilization method within the EU. Significant reduction in sludge quantity is achieved through digestion, which is in line with European Waste Management Policy of sustainable development having the following hierarchy of waste management priority: avoidance, minimization, recycling, incineration with energy recovery and landfilling [27].

1.3 Types of Sewage Sludge

As a result of treatment, six main types of sewage sludge product can be identified:

Liquid sewage sludge - This is raw liquid sewage sludge produced by sedimentation of screened sewage and contains between 2% and 7% dry solids. Up to 75% of the solid fraction is organic matter. Stabilisation of the sewage sludge through mineralization of the organic matter via anaerobic digestion typically results in a 40% reduction in organic matter. This leads to a dry solid content of between 2% and 5%.

Untreated sewage cake - Sewage is screened to remove litter and foreign objects but will not have been stabilised. The sewage sludge cake is formed by dewatering of liquid sewage sludge and has a consistency similar to soil. Cake presents a decreased risk of diffuse pollution in comparison to liquid sewage sludge and has a slow release of nutrient content. Microbiologically, untreated cake is comparable to farmyard manures and slurry.

Conventionally treated sewage sludge - In this case the sewage has been subject to defined treatment processes, such as digestion and standards to ensure a 99% reduction in microbiological content.

Enhanced treated sewage sludge - Sewage treated by a process capable of virtually eliminating any pathogens which maybe present in the untreated sewage sludge. Enhanced treated sewage sludge will be free from salmonellae and will have been treated so as to ensure a 99.9999% reduction in pathogens. This is an energy intensive process. Techniques include drying of the sewage sludge to form granules (98% dry solids), lime pasteurisation or digestion followed by heat treatment.

Composted sewage sludge - In this case the sewage sludge is composted to produce an odourless, friable, soil-like material. Any type of sewage sludge can be used but dewatered undigested sewage sludge composts most readily. Cereal straw or another bulking agent is added to the sewage sludge to enhanced aeration, a process often facilitated by mechanical mixing or enforced aeration. The product is then allowed to mature for a period of time. The availability of nitrogen within composted sewage sludge is usually low but this form is relatively rich in phosphorus and provides a good source of stabilised organic matter.

Lime treated sewage sludge - This results from undigested sewage sludge cake being mixed with lime and waste lime materials in controlled conditions to produce a friable and easily handed product. The resulting high pH drives off ammonia reducing the nitrogen content and the availability of phosphorus maybe reduced. If waste lime materials have been added the product will have improves potassium levels.

2. REDUCING HEALTH HAZARDS FROM LAND APPLICATION OF SEWAGE SLUDGE

Sewage sludge contains a great variety of human pathogenic organisms which originate directly from the excreta of man or animals, or which have multiplied in the wastewater during transport to the treatment plant. Use of sludges in agriculture may therefore create risks to the health of man or animals. These health risks may be rather direct as through the consumption of contaminated crops, or indirect as through the contamination of food animals which may become healthy carriers of pathogenic agents. The routes of contamination of animals in their turn may be direct through contaminated feeds, or indirect through a general increase in the contamination of the environment with disease causing agents. Proper knowledge of the organisms concerned, of their life cycles and epidemiologal features is necessary to understand and manage the risks involved. The EEC Concerted Action 68 [28] has provided scientists throughout Europe a framework to meet and discuss these problems, but more importantly to exchange information within a multidisciplinary group.

Risk management of pathogenic organisms in sewage sludge involves basically three different options, which are:

-Restrictions on sludge use
-Restrictions on land use after sludge application
-Decontamination of sludge before application

It is important to realize that all pathogenic organisms can be effectively eliminated from sludge's by appropriate disinfection methodologies but at relatively high costs. Hence, the central question in relation to pathogens in sludge is to find the most effective combination of the above mentioned general principles which leads to an acceptable risk at lowest costs. It is implied that in order to find these solutions, microbiologists and epidemiologists should work closely together with engineers and sludge producers, and this undoubtedly is one area where the COST 68 [28] project has stimulated better communication.

Bacteria, Viruses and Parasites

There are a number of relevant issues which need to be addressed which build what is otherwise an only partially helpful listing of pathogens which have at some stage been found in sewage sludge.

A wide range of micro-organisms can be found in untreated sewage sludge but the majorities are not known to be pathogenic to man. Although treatment processes limit pathogen levels in the treated sewage sludge, there have been reports in the literature of water contamination which may result in human exposure and increase risks of infection.

A study from Sweden [29] surveyed the presence of bacterial pathogens (Salmonella, Listeria, Campylobacter and E coli) in eight Swedish sewage treatment plants employing four different treatment methods Both raw (64 samples) and treated (69 samples) sewage sludge samples were tested with Salmonella, the most common organism found in both raw (67% positive) and treated sewage sludge (55% positive). In a later study from the same Swedish group [30] associations were shown between the presence of Salmonella in sewage sludge and cases of human salmonellosis, the authors further concluding that Salmonella species isolated in sewage treatment plants were of human origin and survived the treatment process.

Levels of faecal coliforms in soil treated with sewage sludge decrease with time [31]. In this study from the Yangtze River basin, levels of these organisms in sewage treated land was similar to control soils at 56 days, with a further reduction to extremely low levels at 3 months. If air-dried sewage sludge was applied the levels were 50 times higher than land treated with dewatered sewage sludge. The authors suggested that a period of waiting of two months before human access would reduce the risk of infection.

Drug resistant organisms are of high concern at the present time and sewage will inevitably contain such organisms [32] carried out a study to determine whether the frequency of faecal coliforms (including drug resistance patterns) in a river used for recreational purposes was influenced by its proximity to a farmland treated with sewage sludge. The authors did not state whether the farmland was used for arable or livestock purposes. Three sites in the area were tested for 40 days, results showing that treated sewage sludge was positive for faecal coliforms and for drug resistant bacteria (species not defined) found in the surface waters, particularly in the site which ran off to the river. They found levels of faecal coliforms ranging between 108 and 385 cfu/100ml and for ampicillin resistant bacteria between 110 and 780 cfu/ml. The site which contributed most to runoff had the highest levels of both faecal coliforms and resistant bacteria. It is not possible to separate out whether the source of these bacteria was from the sewage or potentially from farm animals using the land under observation. A further study of water from waste water treatment plants revealed relatively high levels of resistant E.coli [33] at 103 cfu/ml resistant E.coli, of which 17% had a six-fold resistance to antibiotics. While it is difficult to quantify accurately what these two studies mean in terms of increased risk to human health, the fact that as many as 100/ml of E.coli survived treatment indicates that if some were pathogenic strains then an infective dose could be reached by ingestion of small volumes. Ten to one hundred E.coli 0157 are thought to have caused infection in known outbreaks.

Listeria is an important human pathogen although infection rates in the UK are low. Its existence in sewage sludge has been demonstrated in an Italian study [34] , testing a large number of samples from five different types of sewage sludge, where the highest frequency of the pathogen was detected in activated sewage sludge whereas the lowest frequency was observed in conventionally treated sewage sludge.

A wide range of viruses are found in sewage sludge (Table 3) of which the most relevant are the hepatitis viruses and the polio virus.

Barbier et al (1990) [35] in a study on the parasitic hazard from sewage sludge applied to land, reported that although a three week no grazing period is stipulated after application of biosolids to the land, levels of viable Taenia saginata (T.saginata) eggs remain high (20m/ha). So at an application rate of 9 tonnes of sewage as dry solids the number of T.saginata eggs applied will be over 250million, of which around 20 m will be viable with resulting increased risk of infection to cattle and humans.

**Table 3. Pathogens demonstrated to be present
in at least some samples of sewage sludge**

Bacteria	Protozoa
Salmonella spp.	Cryptosporidium spp.
Shigella spp.	Entamoeba histolytica
Tersubua spp.	Giardia lamblia
Vibrio cholerae	Balantidium coli
Campylobacter jejuni	Toxopasma gondii
Escherichia coli (pathogenic strains)	Sarcocystis
Clostridium perfringens	
Legionella pneumophila	**Helminths**
Listeria monocytogenes	Ascaris lumbricoides
Staphylococus aureus	Ascaris sum
Leptospira icterohaemorrhagiae	Trichuris trichiura
Mycobacterium spp.	Toxocara canis
Bacillus anthracis	Tenia solium
Leptospira spp.	Necator americanus
	Toxocara cati
Viruses	Ancylostoma duodenale
Poliovirus	Echinococcus gramulosus
Coxsackievirus	Diphyllobothrium latum
Echovirus	Hymenolepis nana
Hepatitis A virus	
Rotavirus	**Fungi**
Human caliciviruses	Aspergillus spp.
Reovirus	Aspergillus fumigatus
New enterovirus	Geotrichium candidum
Coronavirus	Trichophton spp.
Norwalk-like calicivirus	Epidermophyton spp.
Small round viruses	
Parvovirus	**Yeast**
Adeno-associated viruses	Candida albicans
Hepatitis B, C and E virus	Candida krusi
Astroviruses	Candida tropicalis
Adenoviruses	Candida guillermondii
	Cryptococcus neoformans
	Trichosporon

3. REDUCTION OF PATHOGENIC MICRO-ORGANISMS IN SLUDGE TREATMENT

Before sludge is applied to land, some form of stabilization is usually applied. There is a great variety in types of stabilization within the EC. Bruce and Fisher [28] distinguished between biological and chemical stabilization and identified only one type of stabilization that is well established and widely used in all countries: heated, anaerobic, mesophilic single-stage digestion. Seven other methods of stabilization were identified which were well established but much more common in some countries than in others. These included unheated anaerobic or aerobic digestion, stacking of sludge cakes, composting in windrows, drums or aerated vessels and lime addition. Eleven other processes were identified which were in limited use or only in the developmental stage. Taking into account this variety of methods and greatly differing local conditions, particularly climatic aspects, one must conclude that there is no way of indicating the general effect of stabilization on pathogenic micro-organisms. Working Party 3 has already indicated in the final report of the COST 68 bis phase that in general stabilisation processes can be grouped into three categories in relation to their sanitizing effect and has suggested applications appropriate to the achieved degree of pathogen reduction (Table 4). This approach, which has also been adopted in USA guidelines on land applications of sewage sludge appears to be realistic and cost-effective. Unfortunately, little of this knowledge can be found in the EC guidelines which simply require sludge to be "stabilized" before use in agriculture.

Examination of the EC guidelines will also show that several key elements of the above table have been incorporated in the final text as such or in a modified form. The no-grazing period in the EC guidelines is 6 weeks rather than 6 months and it is left to the individual

Table 4. Classification of sewage sludge stabilization processes in relation to pathogen reduction and land application (Final report COST project 68 Bis)

Type of treatment
Pathogen reduction
Application

Stabilized liquid, dewatered or naturally dried: composting I
Especially vegetative bacteria; more resistant forms such as cysts, eggs and spores depending on time-temperature effects
Arable land during vegetation rest periods, vineyards and orchards; pasture and forage land after harvest and until the end of the year; grassland with no grazing period (6 months safe under all conditions); not on crops used uncooked for human consumption
Lime conditioning; low dose irradiation
Seduction except for parasites and spores, vegetative bacteria destroyed
Less restrictions on period of application than above
Pasteurization; (quick) lime treatment; composting II; high dose irradiation; aerobic thermophilic stabilisation; high temperature treatment
Sanitized   No restrictions

Notes: refer to original table for further details such as time-temperature combinations or radiation doses.

countries to extend the period, e.g. if the local climate requires this. Hence, although micro-organisms are not mentioned explicitly in the guideline, some articles have been prepared particularly to reduce risks of infection.

The above table 3 also includes some processes which do not stabilize the sludge but have specifically been designed for disinfection of sewage sludge, such as pasteurization and irradiation. Their obvious disadvantage is that they do not stabilize the sludge, although their application may have favourable effects on the suitability of sludges for other processes like dewatering. The advantage of using these processes is that they can accurately be designed and controlled to obtain a desired degree of pathogen reduction, and they can be monitored by continuous physical methods, leading to a high degree of reliability. An enquiry into the use of "sanitizing" processes in EC member states again demonstrated the extreme variability encountered in this field [36].

The characterisation of the microbicidal effects of different sludge treatment processes has been an area of permanent attention in WP3, and much information has been exchanged. However, there are still several unresolved questions in this domain. One problem is related to the inherent difficulties of directly working with the pathogens of concern. Indigenous levels in sludges are often too low to be quantified by available methods and safety considerations preclude seeding experiments on large scales. It is often difficult to assess the viability and infectivity of organisms encountered in treated sludges, especially of parasites. Isolation and characterisation often require specialized techniques which are not routinely used by water authorities or even research institutes because they are labour-intensive and costly. Widespread application of pathogen assays are also hampered by the low priority that is assigned to them by policy makers and managers as compared to man-made chemical pollution of sludges. This problem certainly is not restricted to the field of sewage sludge hygiene, but is a general imbalance in many areas of environmental protection. To overcome the problems with direct assays for pathogens, investigators have attempted to use model organisms instead. Principles and methods from drinking water and food microbiology were adopted, depending on local traditions and preferences.

4. EPIDEMIOLOGY OF SLUDGE-TRANSMITTED PATHOGENS

The priority microbial pollutants of sewage sludge have been defined as follows: Salmonella spp., Taenia saginata, Ascaris spp., and for certain applications human viruses. More information is needed on protozoan parasites such as Cryptosporidium and Sarcocystis. The agents involved are either transmitted directly between humans (viruses or Ascaris), or may involve food animals (zoonoses, such as Salmonella spp., Taenia saginata, Cryptosporidium, and Sarcocystis). All agents have complex infectious cycles which may involve many routes of transmission of which sewage sludge is only one. The effect of sludge application on health can therefore not be assessed accurately and will strongly depend on local conditions. One major drawback is that the incidence of disease caused by the above mentioned organisms is only poorly known. Salmonella may serve as an example in this respect. The official system of notification to the Medical Health Inspectorate leads to an incidence of ca. 200 cases annually in the Netherlands. Food Inspection Services register an annual number of 10-100 cases related to the consumption of contaminated food. Some major outbreaks like the 1984 "Eurotop" explosion in Maastricht may include up to 1000 cases. The real incidence of salmonellosis in the Netherlands is considerably higher as has been learned from an active surveillance study (A.M.M. Hoogenboom, unpublished results). General practitioners in an urban and a rural area, who served a population of 100,000 registered all cases of acute gastroenteritis on a continuous basis from 1987 onwards. An average incidence of 15 cases/1000 persons/year was found. This figure corresponds well to those obtained in other sentinel studies. All patients were asked to complete a questionnaire and to submit faecal specimens for laboratory examination. Between 4 and 7% of samples yielded Salmonella, leading to a total estimate of 12,000 cases of salmonellosis which are seen by general practitioners in the Netherlands annually. As it is estimated that only 5% of all who suffer from acute gastro-enteritis consult their physician, the total number of cases of salmonellosis in the Netherlands may well be 240,000 per year instead of the officially reported few hundreds! The study also demonstrated that the majority of cases of salmonellosis is sporadic rather than outbreak-associated, and vehicles of infection arising from outbreak investigations may not be the most significant. Detailed case-control studies on non-outbreak related salmonellosis will be necessary to establish the major sources of infection, although it may safely be assumed that animal food products, in particular poultry and associated products will turn out to be the major source of infection. Even if this knowledge is available, it must be realized that land application of sewage sludge is only an indirect determinant in the epidemiology of human salmonellosis, leading to increased pollution of the environment in general and possibly to direct infection of animals grazing on treated land.

The fight against human salmonellosis requires a multifactorial approach and those producing or applying sewage sludge should be prepared to contribute their share by reducing Salmonella-levels in sludges as far as possible and applying the sludge in an approved manner. This approach should take into account local factors on land availability, climate, density of animal-rearing units etc. In this view the EEC Directive is a minimum approach, applicable to all countries under all circumstances and should not be seen as sufficient in most situations.

5. THE CHEMICAL CONTAMINATION OF SEWAGE SLUDGE

Chemical contaminants arise from two main sources: industrial and domestic effluents; others also arise from urban runoff. Assessing the relative proportions which arise from these sources is difficult particularly in the UK where information concerning industrial discharges to sewers is currently not in the public domain. It is likely that the industrial sources outweigh the domestic [37]. Very little detailed work on this problem has been done and although such research would be interesting the real solution lies in preventing the contaminants reaching the sewage works. The levels of contaminants can be reduced most effectively by reduction at source [38, 39]. Removal of contaminants once they have reached the sewage sludge is neither economically feasible [40] nor likely to be practically possible.

5.1 Chemicals

A large number of chemicals have been identified in sewage sludge and the recognised number of constituents is almost certainly a considerable underestimate. Concentrations of known organic chemicals found in sewage sludge are summarized in Table 5 and a wider list of chemicals found in sewage and their likely original sources but without specific concentrations is shown in Table 6. Pharmaceuticals found in sewage are summarized in Table 7 [41] [42] [43] .

Organic Chemicals

The sources of these chemicals are varied but we can find no guide as to what extent concentrations vary within sewage sludge according to the contribution from different sources. It is clear, however that, where measured, the ranges are wide.

It is also unclear as to how content will vary over time when additional applications come from different sources or what the contribution to run off concentrations might be from other sources such as industry or rain (although we expect these latter contributions to be small). However, it should be noted that it is rare within the environmental medicine literature for ranges of exposure concentrations to be available for any specific exposure, making quantitative risk assessment difficult. We are not aware of data which demonstrate the effects of known treatment processes on levels of metals or chemicals in sewage sludge although some processes will dilute or concentrate levels. Equally, some forms of treatment will in all probability segregate certain molecules into the sewage sludge or another component arising from the treatment process. Rogers (1996) called for a better understanding of the effects of sewage sludge treatment on the behavior and fate of organic substances. He stated that:

> "During treatment it is likely that many organic compounds, particularly hydrophobic/lipophilic compounds are sorted onto sewage sludge where they consequently occur in much higher concentrations than in the sewage from which the sewage sludge was derived."

...and...

> "Other compounds may be unaffected by sewage treatment and remain in the aqueous effluent, are completely degraded or mineralised during treatment or partially degraded to produce breakdown products."

For some substances clearance is low following treatment (e.g. 80-99% of PAHs appearing in sewage sludge rather than the aqueous effluent). (Rogers, 1996) He concluded that a combination of both oxidative and reductive steps was likely to be the best approach to clearing organic compounds from sewage. Some generalisations can be made about likely fate of these molecules, however. Some molecules (e.g. PCBs) are resistant to oxidative approaches, those with highly branched hydrocarbon chains are generally less amenable to bio-degradation while reductive dechlorination is needed for toxic substances such as polychlorinated pesticides. Without quantitative data, however, it is not possible to define the effect of treatment on reduced or increased risk on health to those exposed to sewage sludge.

Table 5. Typical concentrations (mg/kg) of selected organics found in sewage sludge (44). * It should be noted that many of these pesticides are banned in the UK and current levels in sewage sludge are thus likely to be lower

Compound	Concentration	Compound	Concentration (mg/kg)
Pesticides		**PCB**	<1
Aldrin	0.08	**Halogenated aliphatics**	
Dieldrin	0.1	Chloroform	1
DDT/ DDE/ DDD	0.3	Carbon tetrachloride	0.1
2,4-D	5	Tetrachloroethylene	0.4
Hepatochlor	0.07	Trichloroethylene	1
Lindane	0.1	Vinyl chloride	6
Malathion	0.05		
Monocyclic aromatics		**PCDD/F**	<0.01
Benzene	0.3	Phenols	
Toluene	0.03	Chlorophenol	3
Xylene	0.03	Pentachlorophenol	5
Ethylbenzene	6	Phenol	5
PAH		**Phthalate esters**	
Pyrene	5	Diethylhexylphthalate	10^2
Fluoranthrene	9	**Surfactants**	
Benzo-[a]-pyrene	3	LAS	10^4
Benzo-[b]-pyrene	8	Nonylphenol	10^3

Table 6. Organic contaminants in sewage sludge

Organic chemical group	Chemicals	Use
Adsorbable organic halogen compounds (AOX)	Trihalomethanes	Paper industry : Bleaching of paper
Nonylphenoles (NP) and nonyl phenol ethoxylates (NPE)		Cleaning products, cosmetics, emulsification of paints and pesticides
Linear alkyl benzene sulfonic acids (LAS) and sulfonates		Anionic surfactants in cleaners and detergents
Phthalates	Benzene dicarboxylic acids, phthalic acid	Plastic industry, inks, glues
Polycyclic aromatic hydrocarbons (PAH)	Napthalenes. polychloronapthalene	Not produced intentionally for particular application except naphthalene-mothproofing, dyestuffs, synthesis of phthalate esters
Polychlorinated biphenyls (PCB)		PVC, artificial rubbers, printing inks, wax, polishes
Polychlorinated dibenzodioxins and dibenzofurans (PCDD/F)	TeCDDs, PeCDD, HxCDD, HpCDD	Production of chlorophenols, chlorophenoxy herbicides, pulp bleaching
Organotins	Tributyltin oxide	Stabilisers in PVCs, transformer oils, polyurethane foams, epoxy resins, silicone rubbers, anti foulants as biocides, agrochemicals
Musk ketones and musk xyxlenes		Perfumes, cosmetics, soaps

Organic chemical group	Chemicals	Use
Adsorbable organic halogen compounds (AOX)	Trihalomethanes	Paper industry : Bleaching of paper
Chlorobenzenes (CB) and pesticides	Chlorobenzenes, Lindane (g-HCH), Aldrin, Dieldrin, Endrin	Paint removers, engine cleaners, coolants, glass tempering, grain fungicide, pesticides
Chlorophenols and Chlorophenoxy acids	4-Chlorophenol, 2-Chlorophenol, 2-Chloro-6-methyl-phenol, 4-Chloro-2-methyl-phenoxy acetic acid (MCPA), 2,4-Dichlorophenoxy acetic acid (2,4-D)	Herbicides
Oranophosphorus compounds	Diazinon, malathion, Reofos 95 (mixture of isopropylphosphate esters)	Flame retardants, textile industries, insecticides
Nitrosamines and Nitoraromatics	Dimethylnitrosamine (DMN), N-Nitrosomorpholine (NMOR), Nitrotoluene	Dyestuffs and urethane products, in batteries
Lipids	Phospholipids (phosphatidyl choline, phosphatidyl ethanolamine)	
Acrylamide monomers	Polyacrylamide	Coagulant during wastewater treatment

Stevens et al. (2003) carried out a survey of conventionally treated sewage sludge from 14 UK wastewater plants to obtain data on certain classes of persistent organic compounds and to assess whether UK sewage sludge was likely to comply with the sewage sludge limits for PCBs and PAHs suggested by the European Union. Their results for the some of the groups listed in Table 4 are as follows:

- PAHs: The "EU total PAH" concentrations ranged from 18-50 mg/kg dw. Concentrations in all the sewage sludge samples exceeded the proposed EU limit of 6mg/kg. However the authors state that the values are consistent with data sets in the UK and in other EU countries.
- PCBs: Total PCB concentrations were 110-440 µg/kg dw, below the EU limit.
- PCNs (polychlorinated naphthalenes). These were used in the early 1900s for wood, paper and textile impregnation but were replaced by PCBs. There are no EU limit values for these compounds. The survey showed a mean value 83 µg/kg dw, much lower than an archived UK sewage sludge sample (concentration 250 µg/kg dw).
- Organochlorine pesticides: Out of a wide range of organochlorine pesticides analysed in the samples, two gave values above the detection limit and were present in all the samples 1) HCB hexachlorobenzene: 6.4-260 µg/kg dw and 2) ρ,ρ-DDE (breakdown product of HCB): 6.0-28 µg/kg dw.
- Synthetic musks: The most abundant compounds in this group were Galaxolide (HHCB) and Tonalide (AHTN) which reflects common usage (they share > 95% of the market for polycyclic musks). The data on musks from the survey were comparable to those from other EU countries.

- Polychlorinated n-Alkanes (PCAs): The total concentrations for short and medium chained PCAs ranged between 7-200 mg/kg and 37-9700 mg/kg respectively. These results were consistent with earlier data from England and Wales. The authors added that these concentrations were higher relative to the other groups studied and this may be a consequence of widespread use (plasticizers, flame retardants, paint additives).

Persistence of PAHs has been studied in soils amended by varying doses of sewage sludge containing PAHs (45), showing that degradation of PAHs was dependent both on the degree of sewage sludge application and the type of plants grown.

5.2 Pharmaceuticals

A huge range of drugs, both prescribed and over-the-counter medications are taken daily in the UK and those which are excreted unchanged will appear in sewage sludge along with the break down products of those which the body metabolizes. While a number of these have been identified in the past to be present in sewage (Table 5) there will be others which simply haven't been looked for. In terms of risk, there are no data for levels of these drugs in sewage to be able to estimate such risks.

5.3 Odour

Odour is the main complaint of populations non-occupationally exposed to sewage sludge. There is a large literature on odour in a range of settings including approaches to understanding mechanisms which will not be covered here. However, some important threads can be identified which are relevant to the sewage sludge issue. Schiffmann & Williams (2005) have suggested three mechanisms by which ambient odours may produce health effects:

Table 7. Pharmaceutical products found in sewage

Anti-inflammatory drugs and analgesics	Other products
Aspirin	Amitryptilline
Sulphasalazine	Caffeine
Codeine phosphate	Chlorhexidine
Dextropropoxyphene	Chlorhexidine
Ibuprofen	Ephedrine
Meprobamate	Menthol
Naproxen	Methyldopa
Paracetamol	Phenylpropanolamine
	Theobromine
Antibiotics	Theophylline
Ampicillin	Tolbutamide
Erythromycin	Oestrogens/progestagens
Metronidazole	Prozac
Sulphamethoxazole	
Tetracycline	

1. Exposure to odorants at levels that also cause irritation and/or toxicological effects. Irritation may be the cause of these health effects whereas the odour may act as an exposure marker.
2. Health symptoms from odorants at non-irritant concentrations can be as a result of genetically coded or learned aversions.
3. Symptoms observed may also be due to co-pollutant exposures (e.g. in this context endotoxin) that is part of the odorant mixture.

It is also generally accepted that unpleasant odours act as warning signs or as indicators of potential risk to health but not as direct triggers to ill health.

Measurement of odour has been reviewed by Gostelow et al. (2001) who stressed the importance of distinguishing between odorants and odours. They describe an odorant as "the compound responsible for imparting an odour", whereas an odour is "the perceived effect of the odorant as detected and interpreted by the olfactory system".

Measuring odour is not simple, but can be addressed through two separate approaches:

Sensory measurements - which are subjective and rely on perception Analytical measurements - for the odorants themselves.

The nose is a very sensitive organ being able to detect very low concentrations of some aromatic compounds. The fact that a smell can be detected does not necessarily mean that the cause of that smell is adverse in terms of inducing an organic response in an individual. But where there is a perception that the source of the smell is potentially toxic, the fact that a smell is detected is often taken by the public as proof that the cause of the odour is truly toxic. Separating out these perceived effects from genuine effects can be impossible.

With respect to sewage sludge, four main groups of compounds contribute to the characteristic odours:

1. Sulphurous compounds
2. Nitrogenous compounds
3. Acids and aldehydes
4. Ketones.

These are broad groups for which it is difficult to create group-specific odour thresholds at a population level. Nevertheless, by understanding the dispersion of odour with respect to distance from a source some idea of likely health impact may be determined. Dispersion modelling techniques can be used to predict odour concentrations, although such models are based on the assumption that odour concentrations measured by olfactometry are additive and this may not be true due to the nature of odourant mixtures.

Repeated exposure to odours often leads to tolerance and loss of recognition of the odour. In some cases this can be potentially dangerous as in occupational exposure to hydrogen sulphide where loss of appreciation of the presence of the gas can lead to increasing exposure. In parallel, workers in smelly occupations regard this tolerance effect as a benefit as they continue to work in what others would regard an impossible environment, providing a continuing income.

There are a number of examples of odours from point sources (usually manufacturing processes such as from paper pulp mills or sugar beet processing plants) which give rise to impaired quality of life and reported ill health. In many cases the effects relate to direct effects such as nausea while other effects such as fatigue feature in some studies. This literature is wide but the overall impression given is that odour can cause ill health through these sort of impacts although the mechanisms may in many cases be exaggerated through psychologically driven responses. However, in some situations, cases of clinical depression have been recorded. There is a tendency to down play the effects of odour on an exposed community but this is done at cost, as there is no doubt that such pressures on quality of life can spread to wider impacts on society through behavioural disruption and knock on effects arising from an apparent disregard for the population's concerns.

5.4 Heavy Metals

Historically sludge and/or trace metal application limits were developed on the basis of phytotoxicity but more recently these were lowered to take into account the possibility of accumulation in the food chain [39]. High levels of trace metals applied to land with sludge can lead to increased concentrations in vegetation [46]. Following concerns about the contamination of sewage sludge discussed in European fora, an EC Directive has been adopted which details acceptable levels of trace metals in sludge and in the land to which it is to be applied [47]. While this is a necessary step in the right direction it does not address the real problem of limiting the contamination of sludge.

Cadmium presents a significant risk to human health [46, 48] and accumulates in many plants such as soyabeans [49], wheat [50], corn [51] and vegetable crops [52]. However, recently (1970s), sludge containing high levels of cadmium as a result of discharges from a lightbulb factory, was being applied to soil in the US [53]. Attempts to correct this have involved a long term programme of soil replacement, liming and ploughing. However, it is unlikely that all soils contaminated by this source will ever be identified and remediation carried out [53].

A more insidious problem may be the effect that elevated trace metal concentrations have on soil microbial populations [48]. In extensive experiments carried out on the same site for over 40 years workers have found that the nitrogen fixing activity of the bacterium Rhizobia, is seriously affected by moderate levels of trace metals in the soil. The root nodules of clover grown in soil with a high trace metal content were found to be abnormally small and contain a strain of Rhizobia which tolerate high levels of trace metals but are not effective at fixing nitrogen [54]. Other soil microbes are also affected by high trace metal concentrations in the soil [48].

Awareness of problems with trace metal contamination has resulted in a considerable reduction of levels in sewage sludge and a concomitant lowering of the regulated acceptable levels [39, 55]. This fall in trace metals is laudable but could be considered to be too late in view of the areas of land which are presently contaminated. Trace metals applied to soils have long residence times with off take in crops and leaching (unless the soil is very acidic) being very low [56]. Around 40% of applied trace metal was recoverable after 40 years (20 years after the last sludge application) but two thirds of the 'losses' were accounted for by lateral transport [56]. The levels of trace metals which do affect microbial action are very close to those permissable under the recently adopted EC guidelines [47, 57] and considering the long residence times levels of trace metals reaching soils must be minimised and industrial wastes excluded from domestic sewage as a matter of urgency [48].

The uncontrolled growth of the human population and industrial development in the last century have caused tremendous changes in the Earth's environment. The most difficult environmental problems we face as we cross into the new millennium are land degradation and the loss of irreplaceable, non-renewable natural resources, including the natural environment with its consequent loss of biodiversity. On the other hand, pollution, ozone depletion and global warming changes can be reversed by the maintenance of a sustainable human population [58]. In soil development the main types of metal inputs to the soil are atmospheric deposition and rock weathering. Higher quantities of certain heavy metals are also released into the environment by anthropogenic activities, for instance, by industrial processes, fossil fuel combustion, manufacturing, and the disposal of domestic and industrial refuse and waste material, [59, 60]. Sewage sludge from municipal wastewater treatment plants contains organic compounds, macronutrients, essential and non-essential trace elements, organic micropollutants, microorganisms and eggs of parasitic organisms. The disposal alternatives for it are landfill, incineration, composting or application to the land. The high content of organic matter and substantial N and P concentrations suggest its use, in the most cost-effective manner, as a fertiliser in agriculture or as a regenerator for soil. Such application may cause several environmental pollution hazards, e.g., odour, water pollution, veterinary hazards and contamination by heavy metals. Heavy metal contamination in soils and groundwater may represents a longterm threat for the environment and man, [61, 62, 63]. For evaluation of the heavy metal burden in soil, sediment or sewage sludge, total metal concentrations are determined. Additional information for assessment of the mobility, bioavailability and related eco toxicity of heavy metals in the environment is provided from speciation studies, [64, 65, 66]. An extensive amount of literature exists on the issue of so-called operational speciation. In this approach different chemical extractants are applied in single or sequential extraction procedures to solid environmental samples. Many of the sequential extraction procedures [63, 67] are based on the five-step method developed by Tessier et al. [68] or its modifications, [65, 69]. Also widely used is the BCR three-step sequential extraction procedure, [62, 67, 70, 71, 72]. From the results of such partitioning studies the distribution of certain heavy metals between easily and sparingly soluble fractions is estimated [62, 63, 65, 67, 68, 69,70 , 71, 72]. Nevertheless, some caution should be applied in interpreting the results due to the nonselectivity of the reagents and the redistribution of the metals between the phases during the multistage process [67, 71, 72]. Various methods have been developed for the remediation of heavy metal contaminated sites, [73, 74, 75, 76, 77]. Liming has been used traditionally for abiotic remediation to increase the alkalinity and hence reduce the mobility of heavy metals. However, an increase of soil pH can also provoke negative effects, such as reduced manganese and phosphate bioavailability, changes in soil biocenoses, increased nitrate concentrations in drainage water or mobilisation of copper and lead due to the decay of humus [74, 78]. Sewage sludges containing pathogenic microorganisms and helminth eggs have been treated with lime for post-stabilisation prior to disposal or reuse in the environment. An increase in Cu mobility caused by lime treatment of sewage sludge has been reported. In this work a representative dewatered sewage sludge sample was collected at the Domz˘ale–Kamnik Wastewater Treatment Plant, the largest presently operating in Slovenia. It receives discharges from both residential areas and industry. Analyses of the total concentrations of Cd, Cr, Cu, Pb, Fe, Ni and Zn in the sewage sludge generated at this plant were performed in our laboratory over the period from June 1997 to June 1999. The results indicated that relatively high concentrations of heavy metals, which vary mainly according to the activities of local metal works, were contained in the sludge [50]. Over the time span investigated the potentially pathogenic bacterium Salmonella sp. was occasionally detected in the sludge. Prior to its final disposal sludge was amended with quicklime (CaO) for disinfection and sawdust for post-stabilisation. The amount of quicklime added to the sludge or sludge mixed with sawdust was selected according to the criterion of raising the pH above 12.

Heavy Metal Content in Sludge Identification of Regulatory Authority Affected by Changes to the Directive

Most regulatory authority has set more stringent standards than those in the current Directive.. The costs of changing the legislation are not expected to be significant in comparison with the costs that may arise from changes in disposal [79].

In practice however, information on the quality of sludge seems to indicate that the quality of sludge may be better that the national limits given in Table 8. There is limited information however on the percentile distribution of metal in sludge by regulatory authority. Thus, the information presented in Table 9 is based on country averages and has been used for estimating the impacts (this information was provided to the consultants by the Commission services). Although the quality of the sludge seems to be better than those given under the proposed new limits, it can not be stated that all sludge arising within these are compliant with the new limits. Indeed the first consultation revealed that the content can vary significantly, so these figures need to be read with caution. (In addition, the data do not cover all Member States).

Table 8. Proposed limit values on Potentially Toxic Elements (PTE) in sewage sludge

PTE	CEC 2003 (mg/kg)
Cd	10
Cr	1000
Cu	1000
Hg	10
Ni	300
Pb	750
Zn	2500

Table 9. Countries with national limits less stringent than those proposed under Option: introduce certain more stringent standards, especially for heavy metals, standards for some organics and pathogens, and more stringent requirements on the application, sampling and monitoring of sludge; e.i. setting limits on Maximum level of heavy metals (mg per kg of dry substance) - in grey

PTE	Cd	Cr	Cu	Hg	Ni	Pb	Zn
New limits	10	1000	1000	10	300	750	2500
Bulgaria	30	500	1600	16	350	800	3000
PTE	Cd	Cr	Cu	Hg	Ni	Pb	Zn
Cyprus	20-40	-	1000-1750	16-25	300-400	750-1200	2500-4000
Denmark	0.8	100	1000	0.8	30	120	4000
Estonia	15	1200	800	16	400	900	2900
France (4)	10	1000	1000	10	200	800	3000
Germany (1)	10	900	800	8	200	900	2500
Greece	20-40	500	1000-1750	16-25	300-400	750-1200	2500-4000
Hungary	10	1000/1(3)	1000	10	200	750	2500
Ireland	20		1000	16	300	750	2500
Italy	20		1000	10	300	750	2500
Lithuania	-	-	-	-	-	-	-
Luxembourg	20-40	1000-1750	1000-1750	16-25	300-400	750-1200	2500-4000
Portugal	20	1000	1000	16	300	750	2500
Spain	20-40	1000-1750	1000-1750	16-25	300-400	750-1200	2500-4000
Czech Republic	5	200	500	4	100	200	2500

Table 10. Quality of sewage sludge (on dry solids) recycled to agriculture (2006) against new Option: introduce certain more stringent standards, especially for heavy metals, standards for some organics and pathogens, and more stringent requirements on the application, sampling and monitoring of sludge

Parameter	Cadmium	Chromium	Copper	Mercury	Nickel	Lead	Zinc
New limits for the above option	10	1000	1000	10	300	750	2500
BE –Flanders	1	20	72	0.2	11	93	337
BE-Walloon	1.5	54	167	1	25	79	688
Bulgaria	1.6	20	136	1.2	13	55	465
New limits for the above option	10	1000	1000	10	300	750	2500
Parameter	Cadmium	Chromium	Copper	Mercury	Nickel	Lead	Zinc
Germany	1	37	300	0.4	25	37	713
Spain	2.1	72	252	0.8	30	68	744
Finland	0.6	18	244	0.4	30	8.9	332
France	1.3	43	272	1.1	21	50	598
Italy	1.3	86	283	1.4	66	101	879
Portugal	<0.4	20	12	<1	15	27	341
Sweden	0.9	26	349	0.6	15	24	481
UK	1.3	61	295	1.2	30	112	574
Cyprus	6.9	37	180	3.1	21	23	1188
Czech Republic	1.5	53	173	1.7	29	40	809
Estonia	2.8	14	127	0.6	19	41	783
Hungary	1.4	57	185	1.7	26	36	824
Lithuania	1.3	34	204	0.5	25	21	534
Latvia	3.6	105	356	4.2	47	114	1232
Portugal	4	127	153	4.6	32	51	996
Slovenia	0.7	37	190	0.8	29	29	410
Slovakia	2.5	73	221	2.7	26	57	1235

Impacts from the component – Costs and Benefits

The following costs (Table 11) are calculated on the basis of the costs of the alternative disposal options. It is important to note that owing to the nature of the unit costs, such costs include both environmental and human health costs in addition to financial costs. The

Table 11. Costs from New Limits of PTE in sludge: Option: introduce certain more stringent standards, especially for heavy metals, standards for some organics and pathogens, and more stringent requirements on the application, sampling and monitoring of sludge (EAC, €2009)

MS	Costs from switch to mono-incineration	Costs from switch to co-incineration	Costs from switch to landfill	TOTALS
France	980000	513000	110000	1602000
Greece	158000	53000	43000	254000
Ireland	1381000	446000	391000	2217000
Italy	1770000	1166000	492000	3428000
Luxembourg	111000	91000	16000	217000
Portugal	1789000	723000	391000	2903000
Spain	3185000	2120000	878000	6182000
United Kingdom	10527000	-	-	10527000
EU15	19900000	5111000	2320000	27331000
Bulgaria	-	2000	2000	4000
Cyprus	-	47000	39000	86000
Estonia	-	17000	13000	30000
Hungary	-	494000	379000	872000
Latvia	-	114000	89000	203000
Lithuania	-	10000	7,000	17000
Poland	-	1364000	1042000	2406000
Romania	-	-	-	-
Slovakia	-	456000	350000	805000
Slovenia	-	1000	-	1000
EU-new	-	**2504000**	**1920000**	**4424000**
EU-TOTAL	**19900000**	**7614000**	**4241000**	**31755000**

environmental costs, on the basis of the degree of quantification possible to date however, represent around 10% of the total costs (although in the case of incineration, the externality are closer to the 10% value of the total quantifiable costs).

Trace Organics

Although the problem with trace metals has been acknowledged, the problems with contamination by trace organic materials have not been addressed in the same comprehensive way. In a survey carried out in Canada of 144 priority organic contaminants analysed 106 were found to be present in sewage or sewage sludge [80] and the authors noted that there were a further 80 organic compounds on the Canadian Environmental Monitoring Priority Pollutant. List which were not determined. A review of the literature by Rogers [81] also noted a very wide range of organic contaminants in sewage.

At present, few organic contaminants are routinely monitored in UK sludges. The EC has a list of 129 dangerous substances but the procedure to initiate the Directives is very slow [81]. Moreover, the inherent flaw in any procedure to list prohibited chemicals is that contaminants have to be seen to cause damage before action is taken. A more prudent philosophy, given the range of chemicals available and the rate at which new ones are developed, would be to adopt the precautionary approach and prevent contamination at source.

The treatment that sewage requires is dependent on its end use. However, in general, it must be stabilised by a digestion or composting procedure which will remove some of the organic contaminants added by industrial or domestic users. However a considerable amount of work has shown that this is not the case for all contaminants [82,83]. An example of this is chlorinated organic compounds which in many cases persist on land for considerable lengths of time (more than 10 years) [82].

Polychlorinated biphenyls (PCBs) in the environment have received considerable attention because of their toxicity, persistence and tendency to bioaccumulate. During sewage treatment chlorinated organics accumulate in sludge with accumulation being greater with higher degrees of chlorination [83]. Work by Fairbanks et al. [84] found that after 240 days between 8 and 33% of PCBs added to soil with sludge had been removed. However, only 1-11% of this loss was due to degradation the rest being lost by volatilisation. Thus, although calculated 95% disappearance times ranged between 6.1 and 26.3 years, breakdown rates would be very much slower. Results from another study indicated no loss of PCBs, within the error of analysis (about 20%), from soil over a period of a year [85].

A review by Jacobs et al. [82] discusses differing levels of trace organic uptake by plants depending on a variety of factors such as soil, plant and compound type. A recent study showed little uptake of PCBs from sludge amended soil into lettuce and fescue but did observe some uptake into the skins of carrots [86]. However, the author's conclusion that this is of minimal concern because the contamination would be removed by normal culinary practices' is unacceptable.

Limits on Organics: Identification of Regulatory Authorities Affected by Changes to the Directive

Currently, some regulatory authorities have limits on organics although this is not the general norm. Some countries such as UK, USA and Canada have not set any limit on organic contaminants (OCs) in sludge suggesting that concentrations present are not hazardous to human health, the environment or soil quality. However, other countries have set limits for some OC groups. For example, Germany has set limits for PCBs and dioxins but not PAHs. France has limits for PAHs and PCBs but not dioxins. Denmark has set limits for a range of OCs including linear alkyl sulphonates, nonylphenol and nonylphenol ethoxylates and the phthalate, di(ethylhexyl)phthalate (DEHP). The following Table shows the different limits on organics based on previous consultation [79].

Table 12. Existing legislative limits on organics

	Polycyclic aromatic hydrocarbon (PAH) mg/kg DS	Polychlorinated biphenyls (PCB) mg/kg DS
Option 2	6	0.8
Austria		
Lower Austria	-	0.2 c)
Upper Austria		0.2 c)
Vorarlberg		0.2 c)
Carinthia	6	1
Denmark (2002)	3a)	

France	Fluoranthene: 4 Benzo(b)fluoranthene: 2.5 Benzo(a)pyrene: 1.5	0.8c)
Germany (BMU 2002)		0.2 d)
Germany (BMU 2007) e)	Benzo(a)pyrene: 1	0.1 d)
Sweden	3a)	0.4b)
Hungary	10	1
Czech Republic	-	0.6

Notes:
a)sum of 9 congeners.
b)sum of 7 congeners: PCB 28, 52, 101, 118, 138, 153, 180.
c)sum of 6 congeners:PCB28,52,101,138,153,180.
d)Per congener.
e)Proposed new limits in Germany (BMU 2007).

6. THE BIOLOGICAL CONTAMINATION OF SEWAGE SLUDGE

The contamination of sewage with pathogens has long been recognized widely as a potential health problem. Reductions of certain pathogens can be achieved 'at source' by disease prevention schemes [87] but generally contamination by pathogenic organisms cannot be avoided and treatment is required to produce a microbiologically acceptable sludge for the proposed use. The traditional indicator of pathogenic contamination has been coliform bacteria, however it has been shown that in media such as seawater these are not always appropriate as indicators as their decay does not mirror that of other pathogens [87].

Work has been carried out on the survival of a wide range of pathogenic organisms in sludges. This has shown that the application of untreated sludge may result in risks to the health of humans, animals and crops due to a variety of pathogens [88, 89, 90, 91, 92]. Numerous viruses are found in sewage sludge after traditional cold digestion treatments [93]. However, the health hazard from viruses can be minimized by appropriate waste water treatment [90]. The most important factors in the destruction of pathogens are thermal treatment, loss of moisture through evaporation, microbial antagonism, exposure to high pH and irradiation [90].

Common methods of eradicating pathogens generally involve heat treatment such as thermophilic digestion, composting and other technologies discussed at this conference. However, problems can occur if temperature regimes are not adhered to during processing as microbes may persist at temperatures a few degrees lower than recommended microbes [94]. In addition, it is essential that there is total separation within the plant of the treated sludge from incoming wastes and sources of contamination [95, 96].

Seventeen respondents to the first consultation specifically mentioned or discussed pathogens in sludge. Most of these either inferred or specifically described the evidence that there have been no adverse health effects on humans, animals or plants whilst using sludge for agriculture treated and recycled in accordance with the Sludge Directive requirements. Five of the respondents specifically described a desire for pathogen controls to be based on

Table 13. Standards for maximum concentrations of pathogens in sewage sludge

	Salmonella	Other pathogens
Denmark a)	No occurrence	Faecal streptococci:< 100/g
France a)	8 MPN/10 g DS	Enterovirus: 3 MPCN/10 g of DS Helminths eggs: 3/10 g of DS
Finland (539/2006)	Not detected in 25 g	Escherichia coli <1000 cfu
Italy	1000 MPN/g DS	
Luxembourg	-	Enterobacteria: 100/g no eggs of worm likely to be contagious
Hungary	-	Faecal coli and faecal streptococci decrease below 10% of original number
Poland	Sludge cannot be used in agriculture if it contains salmonella	

different standards for different purposes, and possibly even with requirements adjusted by location as well, whilst three respondents would prefer consistent or harmonized controls.

None of the respondents made any specific recommendations other than by referring to existing quality limits or more stringent recycling controls used in some Member States either as regulatory controls or as codes of practice.

Currently, only a few regulatory authorities [79] are known to have limits on pathogens, shown in Table 13. The 2002 CBA concluded that pollution prevention for pathogens by reducing at source was not feasible. However, local controls which specify indicator pathogen limits in the sludge have been implemented in several of the EU15 countries, driven by stakeholder demands. Sludge producers have installed new treatment processes that achieve more reliable and greater levels of pathogen destruction during treatment. Countries without equivalent systems to conventional standard however are using anaerobic digestion or aerobic digestion but this may not reliably achieve the standards.

7. THE PRACTICAL ASPECTS OF SEWAGE SLUDGE USE

The last broad area of concern is the practical aspects of sewage sludge transport to where it can be utilized and its acceptability. These points are often raised especially when considering sludge from urban areas.

Transport

Concern about carbon dioxide emission is often raised in discussions about transport but decisions are generally based on economics. Analyses of environmental impacts must include the implications of all elements of the equation. For example in a discussion of the greenhouse gas emissions from different sewage sludge treatment and disposal routes, Frost and Bruce [97] conclude that dumping at sea involves the least generation of greenhouse gases whilst application to agricultural land ranks middle and landfill dumping last, generating the greatest amount of greenhouse gases. However, no consideration is given to the saving in greenhouse gas generation from reduced fertilizer production and transport which ought to have been included in a comprehensive survey.

That there is a market for properly treated sewage sludge is demonstrated by the importation of 20,000 tonnes of pasteurized filter cake from Switzerland to Swansea in Wales which was sold for agricultural use and land reclamation schemes [98]. Whilst not wanting inefficient long distance hauling of sludge it is very important to utilize the benefits of sludge. The volume and weight of sludge can be reduced by various digestion and dewatering processes. In the US sludge has been dried before transport and then rewetted to enable application. In addition, at present the only transport commonly discussed is by road, but other means need to be investigated, such as sea (possibly in old dump ships), by rail and pipeline.

Agriculture

At present 40% of the UK sludge arising and 34% of EC sludge is used on agricultural land [98]. It is in this use that chemical and biological contamination, transport and acceptability are most commonly highlighted as limiting factors. In the UK sewage sludge is supplied free to farmers and the authorities regard the farmers as doing them a favour by providing an outlet for the sludge. Thus sludge is being viewed as a disposal problem, rather than as a resource, though it is unlikely that farmers would use untreated sludge if asked to pay for it. Hence, sludge should be treated and converted to a useful product rather than simply dumped on land in the hope that nothing will go wrong. Perhaps the sludge is given free because the authorities feel that they are dealing with a substance of suspect quality which they are glad to 'get rid' of. This could lead to the authorities being nervous of increases in the quantity of sludge applied to land.

The product is valuable to farmers as a good, cheap fertilizer and soil conditioner. The value of the nitrogen and phosphorous content has been estimated to be £5.8 million and could contribute 2% of the total farming fertilizer requirement [98] which, with animal wastes, could make a real contribution to these needs in the UK. Sludge acts as a slow release fertilizer which reduces land runoff of fertilizers [99, 100] and with the return of interest in sustainable and organic farming there must be a role for these wastes [101, 102].

Agricultural land is not available for sludge application throughout the year due to crop cycles and the weather. Nonetheless, as part of an integrated scheme it can play an important part in any programme of sewage sludge utilization. The key to the use of sludge in agriculture lies in the knowledge that the sludge is of guaranteed chemical and pathological quality. This guarantee can only be supplied if there is no industrial input to the domestic sewage system and appropriate treatment is undertaken by the authorities.

Composting

Composting sludge reduces biological contamination of sewage sludge and makes it acceptable for application to land. This treatment has the advantage of being able to produce a variety of different media, increasing the scope of markets available [103]. It may be used to prepare agricultural fertilizer and soil conditioner, materials for use in land reclamation, forestry, and for the horticultural markets [104, 105, 106]. However, this application, as with all the others, requires sludge free from chemical contamination.

Composting has been used in the United States since the 1970s and is now of growing interest there and in Europe [106, 107]. Large composting facilities have recently been built to serve major population centres, the largest being in Philadelphia which has a maximum capacity of 367,400 dry kg per day [104]. Composting can also utilize other waste materials such as straw, urban leaves and grass cuttings, and the organic component of domestic refuse [106, 108]. In the UK one technique has been developed which can use untreated liquid sewage and straw to produce compost [109] which may be widely used in the future when straw burning is banned.

The limitation on the use of composting is generally considered to be the size of the market however work suggests that for a satisfactory product there is a readily available market [104, 106, 109]. It is also reported that there are few problems with public perception once the product has been seen and used. Since composting already requires some handling during processing it is amenable to manipulation to create the required product Work is being carried out to try to produce an acceptable sewage based peat substitute [110]. At present in the UK damage is caused by peat extraction in large quantities from bogs a considerable distance from the population centres where it is required.

Forestry

In the UK the potential for sewage sludge as a forest fertilizer has been estimated to be between 6 and 11 % depending on the transport distance accepted [101]. It is already being used extensively in some US states and is being investigated in the Netherlands and France [112, 113, 114].

It has much to offer forestry because it fertilizes and conditions the often poor soil used in forestry and requires less frequent application than conventional fertilization systems by inducing natural nutrient cycling [115, 116]. Significant improvements in growth have been observed following application of sludge and scheduling application according to season is less restrictive than agriculture [113, 117]. Forestry offers a long term outlet which can be used for the bulk of the year.

However the key to any land application programme including forestry is high quality sludge [113]. Pollution problems of water runoff and the build up of contaminants in wildlife needs to be guarded against [118].

Land Reclamation

A wide variety of land types, including derelict mined land and colliery spoil, may be reclaimed to permanent vegetation by the addition of sludge [119, 120, 121]. Large quantities of sludge can be applied (ensuring a long term outlet) and experiments confirm that sludge is superior to more traditional methods at enabling a self-sustaining vegetation cover to be established [119, 122].

Transport is less likely to be a barrier to this use because many sites are situated around industrialized centres thus enabling short hauling distances [122]. Although reclaimed land is seldom used for arable pasture it is not acceptable to use contaminated sludge as this may again lead to build up in wildlife, grazing farm animals and ground water.

Novel Uses

A variety of other options have been discussed; two of these, thermal conversion and use in building materials, seem promising [123, 124]. There has been some interest in the use of sludge as a feedstuff for a variety of animals and fish [124]. However, in the light of the recent food scares in the UK it would seem very unlikely that this would now be acceptable to the public (supposing it ever would have been).

Thermal conversion of sewage sludge to fuel oil and char has been researched in Canada where a full-scale demonstration plant is being built [125]. Sludge disposal costs by this route are estimated to be considerably less (around half) than existing methods. In the UK an analysis comparing low temperature conversion and incineration found that at 1985 fuel costs thermal conversion would be 40% more economic than incineration [123]. However, no extensive environmental analysis appears to have been carried out and this must be a prerequisite of any implementation programme. Only uncontaminated sludges should be used as the resulting fuels or wastes would contain any contaminants which had been present in the original sludge and any which were synthesized during the processing.

The use of sludge in building materials has received considerable attention in the US where sludge has been used successfully to make 'Biobricks' [124]. It has been estimated that 15% of municipal sludge production in the US could be incorporated in bricks [123].

8. THE DEVELOPMENT OF A SLUDGE USE STRATEGY

The essential prerequisite for any sludge use programme is that it is not contaminated with industrial wastes or domestic products which may lead to adverse effects or a build up of contaminants in the environment. Industrial pollutants can be removed by disconnecting their effluents from the domestic sewage system. Industries will then have to treat their own effluents thus forcing them to accept responsibility for their wastes.

It is very difficult to control domestic pollutants once they are available on the open market. Certainly, public education should be able to prevent some materials reaching the sewage system and the European Commission 'eco-labelling' scheme may help. The working paper suggests that products should only be considered after a full 'cradle to grave' assessment of the environmental [126]. This should involve studies of the effects on sewage and sludge quality.

Of course, wide ranging questions should be raised about products that damage the environment. Using the precautionary approach, producers should prove that their products do no harm to the environment. Industry should develop clean production which ensures that production does not cause pollution and that the product is not polluting either during or at the end of its life.

Sewage sludge must be treated to remove all pathogens which will enable the general public and the water workers to have confidence in its safety in handling and use. A variety of treatment methods are available which can be used to develop a wide spectrum of products for different applications.

In developing any sludge use scheme all aspects of the environmental equation must be considered even those which do not have a direct economic influence but have an environmental benefit. The US Environmental Protection Agency has emphasized the importance of beneficial reuse of sewage sludge and assessments by certain states have come to the conclusion that this outweighs certain economic arguments [127]. In The UK House of

Commons Environment Committee Third report it is stressed that 'disposal routes which convert sludge into useful products seem more desirable and their viability should be researched and reviewed thoroughly' [128].

The key to the whole problem must lie in the water authorities being willing to develop a wide range of sludge outlets and not expecting a single strategy to fulfil all their requirements. The present thinking is exemplified in the discussion by North West Water of their options for sewage sludge disposal [129]. They discuss only agricultural land disposal, landfill and incineration as alternatives to sea disposal. The other options discussed in this paper, such as forestry and composting, are grouped in a separate section where it is noted that none have the 'capability to provide substantial and continuous outlet capacity for the quantities of sludge' and that 'it would be imprudent to place early reliance on an untried method' [129].

More creative thinking is essential if we are to realize the benefits of sewage sludge and establish a long term environmentally secure programme. Strategies cannot be entirely dependent on one disposal method but must be able to incorporate a range of uses to enable the practical difficulties to be overcome.

CONCLUSION

Sewage sludge is a resource which if treated properly can be of use in a variety of different products. When developing a strategy for using sewage sludge effectively, it is essential that assessments are not made solely on economic grounds but also recognize the environmental factors. The various options for using sewage sludge must be fully investigated in the decision making process. In a world of diminishing resources it is vital that we make maximum use of those resources available to us.

REFERENCES

[1]   Spinosa L (2001): Evolution of sewage sludge regulations in Europe. Wat Sci Tech 44(10) 1–8
[2]   European Commission (2000): Working document on sludge 3rd draft: <http://europa.eu.int/comm/environment/ waste/ sludge/workingdoc3.htm>
[3]   European Commission (2000): Directive 2000/76/EC of the European Parliament and of the Council of 4 December 2000 on the incineration of waste. Official Journal of the European Communities, L 332/91-111
[4]   European Commission (2001): Working Document on Biological Treatment of Biowaste (2nd Draft), Brussels
[5]   Bridle T, Skrypski-Mantele S (2000): Assessment of sludge reuse options: a life cycle approach. Wat Sci Tech 41 (8) 131–135
[6]   Galandak J, Racstain M (1979): Design considerations for pyrolysis of municipal sludge. J WPCF 5 (2) 370–377
[7]   International Standard Organisation (2000): ISO 14000. Environmental Management. ISO Standards Collection on CDROM, Genève

[8] Jensen AA, Hoffman L, Møller BT, Schmidt A, Christiansen K, Elkington J, van Dijk F (1997): LCA – A guide to approaches, experiences and information sources. Environmental Issues Series, report n° 6 from the European Environmental Agency, Copenhagen

[9] United Nations Environment Program (2003): Evaluation of Environmental Impacts in Life Cycle Assessment. Report from the Division of Technology, Industry and Economics (DTIE), Paris

[10] Hwang Y, Hanaki K (2000): The generation of CO2 in sewage sludge treatment systems: life cycle assessment. Wat Sci Tech 41 (8) 107–113

[11] Suh YJ, Rousseaux P (2002): An LCA of alternative wastewater sludge treatment scenarios. Res Conserv & Rec 35, 191–200

[12] Houillon G, Jolliet O (2004): Life cycle assessment for the treatment of wastewater urban sludge: energy and global warming analysis. J Cleaner Prod 13 (3) 287–299

[13] Lundin M, Olofsson M, Pettersson GJ, Zetterlund H (2004): Environmental and economic assessment of sewage sludge handling options. Res Conserv & Rec 41, 255–278

[14] Hospido A, Moreira MT, Fernández-Couto MP, Feijoo G (2004): Environmental Performance of a Municipal Wastewater Treatment Plant. Int J LCA 9 (4) 261–271

[15] Chobanoglous G. Wastewater engineering, treatment, disposal and reuse. New Delhi: Tata McGraw-Hill, 1987.

[16] McGhee TJ. Water supply and sewerage. New York: McGraw-Hill, 1991.

[17] Hall JE, Dalimier F. Waste management—sewage sludge: survey of sludge production, treatment, quality and disposal in the EC. EC Reference No: B4-3040/014156/92, Report No: 3646, 1994.

[18] Holmes PR. Effluent control, the Hong Kong approach. J CIWEM 1996;10:199–204.

[19] Mudrack K, Kunst S. Biologie der Abwasserreinigung (Biology of Wastewater Treatment). Gustav-Fischer, 1988.

[20] Mathias B. Basiswissen Umwelttechnik [Basic knowledge of environmental technology], 2nd edn. Wu¨rzburg: Vogel Buchverlag, 1994.

[21] Mathys RG. Nitrifikation/Denitrifikation-Neue Aspekte und Verfahren [Nitrification and denitrification—new aspects and processes]. gwa 2/94, 1994:89–103.

[22] So¨lter K, Peschen N, Ro¨hricht K, Gehrke R. Simultanfa¨llung it Kalk im großtechnischen Betrieb-Ergebnisse, Kosten, Vorteile [Simultaneous precipitation using lime in largescale operation—results, costs and advantages]. In: Awt-Abwassertechnik, vol. 1, 1994:19–23.

[23] Zubiller C, Hahn N. Bedingungen der Entsorgung von Kla¨rschla¨mmen aus der Behandlung industrieller Abwa¨sser [Requirements for disposal of sewage sludges from industrial wastewater treatment]. In: Entsorgungs-Praxis Spezial, No. 3, 19:18–20.

[24] Reimann DO. Thermische schlammkonditionierung [Thermal conditioning of sludge]. In: Kla¨rschlammentsorgung. Du¨sseldorf: VDI-Bildungswerk, 1989:12–26.

[25] Reimann DO. Heizwert und schwermetall aus kla¨rschlamm [Heating values and heavy metals from sludge]. In: Kla¨rschlammentsorgung, Beiheft No. 28 to journal Mu¨ll and Abfall, pp. 16–19.

[26] Albrink B, Hanssen H, Nicolai V. Hamburger Kla¨rschlammentwa¨sserungs- und -trocknungsanlage in Betrieb [Operation of Hamburg's sewage sludge dewatering and

drying plant]. In: Abwassertechnik, Abfalltechnik und Recycling, 43rd ed., vol. 10. 1992:11–13.

[27]   Davis RD. The impact of EU and UK environmental pressures on the future of sludge treatment and disposal. J CIWEM 1996;10:65–69.

[28]   Bruce, A.M. and Fisher, W.J. Sludge stabilization - methods and measurement. In: Sewage sludge stabilization and disinfection (A.M. Bruce, ed.), pp. 23-47. Chicester: Ellis Horwood, 1984.

[29]   Sahlstrom L, Aspan A, Bagge E, Danielsson-Tham ML, and Albihn A, 2004 : Bacterial pathogen incidences in sewage sludge from Swedish sewage treatment platns. Water research 38:46-52

[30]   Sahlstrom L, De Jong B, and Aspan A, 2006 : Salmonella isolated in sewage sludge traced back to human cases of salmonellosis. Letter in applied microbiology 43:46-52

[31]   Sun Y.H , Luo Y.M, Wu L.H, Li Z.G, Song J and Christie P, 2006 . Survival of faecal coliforms and hygiene risks in soils treated with municipal sewage sludges. Environ. Geochem. Health 28:97-101

[32]   Selvaratnam S, and Kunberger J.D, 2004: Increased frequency of drug-resistant bacteria and fecal coliforms in an Indiana Creek adjacent to farmland amended with treatedsewage sludge. Can. J. Microbiol.50:653-6

[33]   Reinthaler F.F, Posch,J., Feierl, G., Wust,G, Martin G, Haas D, Ruckenbauer G, Mascher F, and Marth E, 2003: Antibiotic resistance of E.coli in sewage and sewage sludge. Water Res, 37:1685-90

[34]   De Luca,G., Zanetti,F., Fateh-Mofhadm,P., and Stampi,S., 1998: Occurrence of Listeria monocytogenes in sewage sludge. Zentralbl. Hyg. Umweltmed., 201:269-77

[35]   Barbier, D., Perrine,D., Duhamel,C., Doublet,R., and Georges,P., 1990 Parasitic hazard with sewage sludge applied to land. Appl Environ Microbiol, 56:1420-2

[36]   Havelaar, A.H. and BRUCE, A.M. Disinfection of sewage sludge. An enquiry among member countries of the EEC and some other countries. In: Disinfection of sewage sludge: technical, economic and microbiological aspects (A.M. Bruce, A.H. Havelaar and P. L'Hermite, eds), pp.245-252. Dordrecht, D. Reidel, 1982.

[37]   Hathaway, S.W. (1980). Sources of Toxic Compounds in Household Wastewater. United States Environmental Protection Agency Report No. EPA-600/2-80-128. Municipal Environmental Research Laboratory, Cincinnati, Ohio, pp 83.

[38]   Moller, U. (1986). Overview Of Sewage Sludge Treatment Processes: An Outlook. In "Processing And Use Of Organic Sludge And Liquid Agricultural Wastes" (Ed L'Hermite, P.) Commission of European Communities 4th International Symposium, D. Rendel Publishing Co., Dordrecht, ρ 2-24.

[39]   Tjell, J.C. (1986). Trace Metal Regulations For Sludge Utilization In Agriculture: A Critical Review. In "Processing And Use Of Organic Sludge And Liquid Agricultural Wastes" (Ed L'Hermite, P.) Commission of European Communities 4th International Symposium, D. Rendel Publishing Co., Dordrecht, ρ 348-361.

[40]   Bastian, R.K. (1986). Overview On Sludge Utilization. In The Forest Alternative For Treatment And Utilisation Of Municipal And Industrial Wastes' (Eds Cole, D.W., Henry, C.L. and Nutter, W.L.) University of Washington Press, ρ 7-25

[41]   Langerkamp H, and Part P , 2001. : Organic contaminants in sewage sludge for agricultural use. European Commision

[42] Rogers H.R, 1996 : Sources , behaviour and fate of organic contaminants during sewage treatment and in sewage sludges. Sci Total Environ, 185:3-26

[43] Harrison E.Z, Oakes S.R, Hysell M, and Hay A, 2006 : Organic chemicals in sewage sludges . Sci Total Environ. (in press)

[44] Cameron KC, Di H.J and McLaren R.G 1997: Is soil an appropriate dumping ground for our wastes? Austral J Soil Res 35:995-1035

[45] Oleszczuk P, 2006 : Persistence of polycyclic aromatic hydrocarbons (PAHs) in sewage sludge-amended soil. Chemosphere 65:1616-26

[46] Chang, A.C., Hinesly, T.D., Bates, T.E., Doner, H.E., Dowdy, R.H. And Ryan. J.A. (1987). Effects Of Long-term Sludge Application On Accumulation Of Trace Elements By Crops. In "Land Application of Sludge" (Eds Page, A.L., Logan, T.J. and Ryan, J.A.) Lewis Publishers Inc. ρ 53-66.

[47] Ec (1986). Council Directive (86/278/Eec) On The Protection Of The Environment, And In Particular Soil, When Sewage Sludge Is Used In Agriculture. Official Journal Of The European Communities, No L 181/6 -181/12.

[48] Giller, K. And Mcgrath, S. (1989). Muck, Metals And Microbes. New Scientist, 4 November 1989, P 31-32.

[49] Heckman, J.R., Angle, J.S. And Chaney, R.L. (1987). Residual Effects Of Sewage Sludge On Soyabean: I. Accumulation Of Heavy Metals. J. Environ. Qual. 16,113-117.

[50] Lubben, S. And Sauerbeck, D. (1989). The Uptake Of Heavy Metals By Spring Wheat And Their Distribution In Different Plant Parts. Presented To 'Alternative Uses For Sewage Sludge' Held At York, September 1989.

[51] Rappaport, B.D., Martens, D.C., Reneau Jnr, R.B. And Simpson, T.W. (1988). Metal Availability In Sludge-Amended Soils With Elevated Metal Levels. J. Environ. Qual. 17, 42-47.

[52] Keefer, R.F., Singh, R.N. And Horvath, D.V. (1986). Chemical Composition Of Vegetables Grown On An Agricultural Soil Amended With Sewage Sludges. J. Environ. Qual. 15,146-152.

[53] Lagan, T.J. And Cassler, D.E. (1989). Correcting Widespread Cadmium Contamination. Wat. Environ. Technol. 1,312-315.

[54] Giller, K.E., Mcgrath. S.P. And Hirsch, P.R. (1989). Absence Of Nitrogen Fixation In Clover Grown On Soil Subject To Long-Term Contamination With Heavy Metals Is Due To Survival Of Only Ineffective Rhizobium. Soil Biol. Biochem. 21,841-848.

[55] Candinas, T., Gupta, S.K., Zaugg, W., Besson, J.M. And Lischer, P. (1989). Changes In Heavy Metal Contents Of Swiss Sewage Sludge And Future Prospectives. Presented To 'Alternative Uses For Sewage Sludge' Held At York, September 1989.

[56] Mcgrath, S.P. And Lane, P.W. (1989). An Explanation For The Apparent Losses Of Metals In A Long-Term Field Experiment With Sewage Sludge. Environ. Pollut. 60,235-256.

[57] Mcgrath, S.P., Brookes, P.C. And Giller, K.E. (1988). Effects Of Potentially Toxic Metals In Soil Derived From Past Applications Of Sewage Sludge On Nitrogen Fixation By Trifolium Repens L. Soil Biol. Biochem. 20,415-424.

[58] K. Christen, Environ. Sci. Technol., 2000, 34(5), 123A.

[59] S. M. Ross, Toxic metals in soil-plant system, Wiley, Chichester, 1994

[60] O. Chadwick, L. A. Derry, P. M. Vitousek, B. J. Huebert and L. O. Hedin, Nature, 1999, 397, 491.

[61]. P. Baveye, M. B. McBridge, D. Bouldin, T. D. Hinesly, M. S. A. Dahdoh and M. F. Abdel-Sabour, Sci. Total Environ., 1999, 227, 13.

[62]. J. Sˇcˇancˇar, R. Milacˇicˇ, M. Strazˇar and O. Burica, Sci. Total Environ., 2000, 250, 9.

[63]. R. Zufiaurre, A. Olivar, P. Chamorro, C. Nerin and A. Callizo, Analyst, 1998, 123, 255.

[64]. J. Sˇcˇancˇar, R. Milacˇicˇ and M. Horvat, Water, Air, Soil Pollut., 2000, 118, 87.

[65]. R. Milacˇicˇ and J. Sˇ tupar, Environ. Sci. Technol., 1995, 29, 506.

[66]. P. Planquart, G. Bonin, A. Prone and C. Masiani, Sci. Total Environ., 1999, 241, 161.

[67]. B. Perez-Cid, I. Lavilla and C. Bendicho, Analyst, 1996, 121, 1479.

[68]. A. Tessier, P. G. C. Cambel and M. Bisson, Anal. Chem., 1979, 51, 844.

[69]. R. H. C. Emmerson, J.W. Birkett, M. Scrimshaw and J. N. Lester, Sci. Total Environ., 2000, 254, 75.

[70]. G. Rauret, J. F. Lopez-Sanchez, A. Sahiquillo, R. Rubio, C. Davidson, A. Ure and Ph. Quevauviller, J. Environ. Monit., 1999, 1, 57.

[71]. M. D. Ho and G. J. Evans, Environ. Sci. Technol., 2000, 34, 1030.

[72]. J. L. Gomez Ariza, I. Giraldez, D. Sanchez-Rodasand and E. Morales, Sci. Total Environ., 2000, 246, 271.

[73]. A. J. Francis and C. J. Dodge, Environ. Sci. Technol., 1998, 32, 3993.

[74]. B. Lothenbach, G. Furrer and R. Schulin, Environ. Sci. Technol., 1997, 31, 1452.

[75]. C. M. Lytle, F. W. Lytle, N. Yang, J. H. Qian, D. Hansen, A. Zayed and N. Terry, Environ. Sci. Technol., 1998, 32, 3087.

[76]. A. Davis, L. E. Eary and S. Helgen, Environ. Sci. Technol., 1999, 33, 2626.

[77]. S. Yoshizaki and T. Tomida, Environ. Sci. Technol., 2000, 34, 1572.

[78]. L. J. Evans, G. A. Spiers and G. Zhao, Int. J. Environ. Anal. Chem., 1995, 59, 291.

[79] Environmental, economic and social impacts of the use of sewage sludge on land Final Report Part II: Report on Options and Impact 2009, http://ec.europa.eu/environment

[80] Ontario Ministry Of The Environment (1988). Thirty Seven Municipal Water Pollution Control Plants, Pilot Monitoring Study. Volume I, Interim Report, Queen's Printer For Ontario.

[81] Rogers. H.R. (1987). Organic Contaminants In Sewage Sludge (Ec 9322 Sld): Occurrence And Fate Of Synthetic Organic Compounds In Sewage And Sewage Sludge -A Review. Wrc Environment, Medmenham Laboratory, Henley Road, .Medmenham, Bucks Sl7 2hd.

[82] Jacobs, L.W., O'connor, G.A., Overcash, Ma. And Rygiewicz, P. (1987). Effects Of Trace Organics In Sewage Sludges On Soil-Plant Systems And Assessing Their Risk To Humans. In 'Land Application Of Sludge' (Eds Page, A.L., Logan, T.J. And Ryan, J.A.) Lewis Publishers Inc P 101-143.

[83] Kirk, P.W.W., And Lester, J.N. (1988). The Behaviour Of Chlorinated Organics During Activated Sludge Treatment And Anaerobic Digestion. Wat Sci. Tech. 20,353-359.

[84] Fairbanks, B.C., O'connor, G.A. And Smith, S.E. (1987). Mineralsiation And Volatilisation Of Polychlorinated Biphenyls In Sludge-Amended Soils. J. Environ. Qual. 16, 18-25.

[85]   Marcomini, A., Capel, P.D., Lichtensteiger, Th., Brunner, P.H. And Giger, W. (1989). Behaviour Of Aromatic Surfactants And Pcbs In Sludge-Treated Soil And Landfills. J. Environ. Qual. 18,523-528.

[86]   O'connor, G.A., Kiehl, D., Eiceman, G.A. And Ryan, J.A. (1990). Plant Uptake Of Sludge-Borne Pcbs. J. Environ. Qual. 19,113-118.

[87]   Poyry, T., Stenvdc, M. And Hovi, T. (1988). Viruses In Sewage Waters During And After A Poliomyelitis Outbreak And Subsequent Nationwide Oral Poliovirus Vaccination Campaign In Finland. App. Environ. Microbiol. 54,371-374.

[88]   Wheeler, D. (1990). The Pollution Of Beaches By Viruses. In: House Of Commons Committee Enquiry Into The Pollution Of Beaches. Appendices To The Minutes Of Evidence 11,24-39. Hmso, London.

[89]   Watkins, J. And Sleath, K.P. (1981). Isolation And Enumeration Of Listeria Monocytogenes From Sewage, Sewage Sludge And River Water. J. App. Bacteriol. 50,1-9.

[90]   Hurst, C.J. (1989). Fate Of Viruses During Wastewater Sludge Treatment Procedures. Crc Critical Reviews In Environmental Control. 18,317-343.

[91]   Tharaldsen, J. And Helle, O. (1989). Survival Of Parasite Eggs In Livestock Slurry Utilised For Compost Heat Acta Agric. Scand. 39,381-387.

[92]   Spaull, A.M., Mccormack, D.M. And Pike, E.B. (1989). Effects Of Various Sewage Sludge Treatment Processes On The Survival Of Potato Cyst-Nematodes (Globodera Spp.) And The Implications For Disposal. Wat. Sci. Tech. 21,909-916.

[93]   Nash1da, M., Shimohara, E., Sugishima, S. And Kaneko, M. (1989). Viruses Isolated From Activated Sludges. 1984-1986. Wat. Sci. Tech. 21,279-282.

[94]   Pike, E.B., Carrington, E.G. And Harman, S.A. (1988). Destruction Of Salmonellas, Enteroviruses And Ova Of Parasites In Wastewater Sludge By Pasteurisation And Anaerobic Digestion. Wat, Sci. Tech. 20,337-343.

[95]   Fujioka, R.S., Hirano, W.M. And Loh, P.C. (1988). Disinfection, Recontamination And Regrowth Of Indicator Bacteria In A Wastewater Sludge Heat Treatment System. Wat Sci. Tech. 20,329-335.

[96]   Burge, W.D., Enkiri, N.K. And Hussong, D. (1987). Salmonella Regrowth In Compost As Influenced By Substrate. Microb. Ecol. 14,243-253.

[97]   Frost, R.C. And Bruce, A.M. (1989). Energy From Sludge. Presented To 'Alternative Uses For Sewage Sludge' Held At York, September 1989.

[98]   House Of Commons (1990). Toxic Waste Disposal In Wales. Minutes Of Evidence Taken Before The Welsh Affairs Committee, Session 1989-1990, 6 December 1989. Hmso, London.

[99]   Davis, R.D. (1989). Agricultural Utilisation Of Sewage Sludge: A Review. J. Iwem, 3,351-355.

[100] Wrc Soil Injection Of Sewage Sludge, A Manual Of Good Practice (2nd Edition). Ref No. Fr 0008. Foundation For Water Research, Water Research Centre, Medmenham, Bucks. Ppl5.

[101] Reganold, J.P., Papendick, R.I. And Parr, J.F. (1990). Sustainable Agriculture. Sci. Am. June, 72-78.

[102] Stanhill, G. (1990). The Comparative Productivity Of Organic Agriculture. Agric. Ecosystems Environ. 30,1-26.

[103] Border, D., Coombes, C. And Shellens, M. (1988). Composting Straw With Untreated Liquid Sludge. Biocycle. July, 54-55.

[104] Kuchenrither, R.D., Diemer, D.M., Martin, W.J. And Senske, F.J. (1987). Composting's Emerging Role In Sludge Management: A National Perspective. J. Wpcf, 59,125-131.

[105] Lopez-Real, J.M., Witter, E., Midmer, F.N. And Hewett, B.A.O. (1989). Evaluation Of Composted Sewage Sludge/Straw Mixture For Horticultural Utilisation. Wat. Sci. Tech. 21,889-897.

[106] Goldstein, N. (1988). Steady Growth For Sludge Composting. Biocycle, November-December, 27-32.

[107] Martel, J.L. (1989). Economics And Marketing Of Urban Sludge Composts In The Eec. Presented To 'Alternative Uses For Sewage Sludge' Held At York. September 1989.

[108] Pereira, J.T., Stentlford, E.I. And Smith, D.V. (1986). Survival Of Faecal Indicator Micro-Organisms In Refuse/Sludge Composting Using The Aerated Static Pile System. Waste Management And Res. 4,397-406.

[109] Matthews, P.J. And Border, D. (1989). Compost: A Sewage Sludge Resource For The Future. Presented To 'Alternative Uses For Sewage Sludge' Held At York, September 1989.

[110] Bragg, N. (1990). Ministry Of Agriculture Fisheries And Food, Personnel Communication.

[111] Taylor, C.M.A. And Moffat, A.J. (1989). The Potential For Utilising Sewage Sludge In Forestry In Great Britain. Presented To 'Alternative Uses For Sewage Sludge' Held At York, September 1989.

[112] Lavergne, G. (1989). The Evaluation Of Dewatered Sludge From The Marseille Treatment Plant For Use In Forestry. Presented To 'Alternative Uses For Sewage Sludge' Held At York, September 1989.

[113] Nichols, C.G. (1989). Us Forestry Uses Of Municipal Sewage Sludge. Presented To 'Alternative Uses For Sewage Sludge' Held At York, September 1989.

[114] Olesen, S.E. And Mark, H.S. (1989). Long-Term Effects Of Sewage Sludge Application In Conifer Plantation On Sandy Soils. Presented To 'Alternative Uses For Sewage Sludge' Held At York, September 1989.

[115] Moffat, A.J. And Bird, D. (1989). The Potential For Using Sewage Sludge In Forestry In England And Wales, Forestry, 62, 1-17.

[116] Bayes, C.D., Taylor, C.M.A. And Moffat, A.J. (1989). Sewage Sludge Utilisation In Forestry: The Uk Research Programme. Presented To 'Alternative Uses For Sewage Sludge' Held At York, September 1989.

[117] Burd, R.S. (1986). Forest Land Applications Of Sludge And Wastewater. In The Forest Alternative For Treatment And Utilisation Of Municipal And Industrial Wastes' (Eds Cole, D.W., Henry, C.L. And Nutter, W.L.) University Of Washington Press, P 3-6.

[118] Haufler, J.B. And West, S.D. (1986). Wildlife Responses To Forest Application Of Sewage Sludge. In The Forest Alternative For Treatment And Utilisation Of Municipal And Industrial Wastes' (Eds Cole, D.W., Henry, C.L. And Nutter, Wx.) University Of Washington Press, P .110-116.

[119] Metcalf, B. And Lavin, J.C. (1989). Consolidated Sewage Sludge As Soil Substitutes In Colliery Spoil Reclamation. Presented To 'Alternative Uses For Sewage Sludge' Held At York, September 1989.

[120] Sopper, W.E. And Seaker, E.M. (1983). 'A Guide For Revegetation Of Mined Land In Eastern United States Using Municipal Sludge', School Of Forest Resources And Institute For Research On Land And Water Resources, The Pennsylvania State University, March 1983.

[121] Werner, W. And Scherer, H.W. (1989). Experiences Of The Usage Of Heavy Amounts Of Sewage Sludge For Reclaiming Opencast Mining Areas And Amelioration Of Very Steep And Stony Vineyards. Presented To 'Alternative Uses For Sewage Sludge' Held At York, September 1989.

[122] Byrom, K. And Bradshaw, A.D. (1989). The Potential Value Of Sewage Sludge In Land Reclamation. Presented To 'Alternative Uses For Sewage Sludge' Held At York, September 1989.

[123] Frost, R.C. And Campbell, H.W. (1986). In 'Processing And Use Of Organic Sludge And Liquid Agricultural Wastes' (Ed L'hermite, P.) Commission Of European Communities 4th International Symposium, D.Reidel Publ. Co., Dordrecht. P. 94-109.

[124] Webber, M.D. (1989). Resource Recovery Through Unconventional Uses Of Sludge. Presented To Alternative Uses For Sewage Sludge' Held At York, September 1989.

[125] Campbell, H.W. And Bridle, T.R. (1989). Conversion Of Sludge To Oil: A Novel Approach To Sludge Management. Wat. Sci. Tech. 21, 1467-1475.

[126] Ends Report. (1990). Brussels Reveals Eco-Labelling Plans. Ends Report 182, 23.

[127] Sludge Management Study Blue Plains Wastewater Treatment Plant -Washington D.C. Final Environmental Impact Statement, Usepa March, 1990.

[128] House Of Commons. (1987). Third Report From The Environment Committee. Session 1986-7. Pollution Of Rivers And Estuaries. Volume 1.

[129] Ministry Of Agriculture Fisheries And Food (1990). Report On The Disposal Of Waste At Sea 1986 And 1987. Hmso, London.

In: Sewage Sludge Management
Editors: A. A. Zorpas and V. J. Inglezakis

ISBN: 978-1-61324-393-0
© 2012 Nova Science Publishers, Inc.

Chapter 15

LEGISLATION ON SEWAGE SLUDGE

Vassilis J. Inglezakis[*]
SC European Focus Consulting srl, Banatului 16, Bacau, Romania

ABSTRACT

The present Chapter presents the basic legislative aspects on the sewage sludge management in European Union and USA. Following the trends in environmental legislation the last 30 years in both continents, sewage sludge management is regulated by a number of legislative tools. Although the positive steps, there is the need for continuous improvement and more intense enforcement. The major gaps identified are the need for limits setting for organic substances and regulations on recycling and final disposal.

1. EUROPEAN UNION LEGISLATION

Water policy in the European Union is regulated by the Water Framework Directive (2000/60/EC), which targets the long-term progressive reduction of contaminant discharges to the aquatic environment in urban wastewater [2]. Several Directives have an influence on sludge management but the ones which have the strongest impact on sludge management are Directives 91/271/EEC on urban waste water treatment and 86/278/EEC on the use of sludge in agriculture. Sewage sludge is defined in Article 2(a) of the Sewage Sludge Directive 86/278/EEC2 as *"(i) residual sludge from sewage plants treating domestic or urban waste waters and from other sewage plants treating waste waters of a composition similar to domestic and urban waste waters; (ii) residual sludge from septic tanks and other similar installations for the treatment of sewage; (iii) residual sludge from sewage plants other than those referred to in (i) and (ii)"*.

---

[*] Website: www.efcon-group.ro, inglezakis@efcon-group.ro.

The Council Directive 91/271/EEC adopted in 1991, concerns urban waste water treatment and aims the protection of the water environment from the adverse effects of discharges of urban waste water and from certain industrial discharges [3]. Sewage sludge is also a product of wastewater treatment and the Urban Waste Water Treatment Directive encourages the use of sludge whenever appropriate. In particular, the Article 14 of the Urban Waste Water Treatment Directive states that "*Sludge arising from waste water treatment shall be re-used whenever appropriate. Disposal routes shall minimise the adverse effects on the environment.*" The requirement for wastewater treatment before discharge and the crucial requirement for termination of sludge discharge into surface waters after 31 December 1998 had serious and direct effects on sewage sludge management. The progressive implementation of the said Directive in all Member States increased the quantities of sewage sludge requiring disposal and the effective implementation of Sewage Sludge Directive 86/278/EEC was a necessity. The Directive seeks to encourage the use of sewage sludge in agriculture and to regulate its use in such a way as to prevent harmful effects on soil, vegetation, animals and man [1].

Furthermore, the EC Directive 99/31/EC on the Landfill of Waste also impacts on the disposal of sewage sludge particularly with the stringent new standards relating to landfill of biodegradable waste. Biodegradable waste is defined in Article 2(m) of the Landfill Directive as "*waste that is capable of undergoing anaerobic or aerobic decomposition, such as food and garden waste, and paper and paperboard*". The Directive is setting mandatory targets for the reduction of biodegradable waste to landfill and thus there are certain conditions that limit the disposal of sludge to landfills. The targets and deadlines for reduction of biodegradable waste to landfill are as follows:

- reduction to 75% (by weight) of total biodegradable municipal waste produced in 1995 by
- 2006;
- reduction to 50% by 2009;
- reduction to 35% by 2016.

These requirement s coupled with the increasing restrictions on the use of sewage sludge on land and the potential unsustainability of landfill, have led to incineration as a possible disposal route. As is well known, sewage sludge contains many pollutants that could be emitted during incineration. The Directive 2000/76/EC on the Incineration of Waste, regulates waste (including sewage sludge) incineration technology. In particular, the Directive lays down emission limit values for selected heavy metals and chemical compounds (e.g. NO_x, SO_x, HCl, particulates, heavy metals and dioxins). The limit values are set in order to prevent and limit as far as practicable negative effects on the environment and the resulting risks to human health.

The application of the relevant EU legislation as described above has driven the management practices the last 25 years. For example, in England and Wales the disposal by dumping at sea, which previously accounted for about one quarter of production, was banned in 1998. Recycling to farmland became the main disposal route (73% in 2005, up from 59%

in 2000), followed by incineration (18% down from 22%). Disposal to landfill has significantly decreased in recent years (6% down to 0.5%) and use in land restoration/reclamation has increased (6% from 0%).

In the majority of Member States, the specific regulations which have been introduced covering the disposal and recycling of sludge mainly concern the use of sludge in agriculture, while the disposal of sludge is addressed by general legislation on landfill and incineration of waste. In the same way, these regulations mostly focus on urban sludge, as landspreading of industrial sludge is covered in the majority of countries by the regulations on the use of waste on land or on waste management.

In the beginning of the 1980, some countries established limit values for some selected heavy metals in sewage sludge to be used for agricultural applications [9]. In accordance with EU Directive 86/278/EEC, national legislation which has been established in the Member States prohibits the use of sludge in agriculture if the heavy metals concentrations exceed specific limit values (Table 1). The limit values for heavy metals in sludge defined in national regulations are presented in Table 1. In most cases, these limit values have been set significantly below the requirements of Directive 86/278/EEC. From the same table it is evident that the legislative limits as well as the typical concentration of heavy metals vary considerable from country to country (Tables 2 and 3).

Table 1. Maximum permissible limits for land application, EU, US and National Legislation (mg/kg dw) [8]

Metal	86/278/EEC (range)	Sweden	Netherlands	France	USA (Part 503)
Zn	2500-4000	800	300	3000	7500
Cu	1000-1750	600	75	1000	4300
Ni	300-400	50	30	200	420
Cd	20-40	2	1.25	20	85
Pb	750-1200	100	100	800	840
Cr	-	100	75	1000	3000
Hg	16-25	2.5	0.75	10	57
Mn	-	-	-	-	-

Table 2. Differences between the EU MS on the maximum permissible limits for land application [5]

Much more stringent	Denmark, Finland, Sweden, Netherlands
More stringent	Austria, Belgium, France, Germany
Similar	Greece, Ireland, Italy, Luxembourg, Portugal, Spain, United Kingdom, Estonia, Latvia

Table 3. Average metal content in sludge from EU and US (mg/kg dw)

Metal	EU	US
Zn	1222	1740
Cu	337	850
Ni	37	82
Cd	2.8	16
Pb	124	500
Cr	141	890
Hg	2.2	5
Mn	-	260

Sewage Sludge Directive prohibits the use of untreated sludge on agricultural land unless it is injected or incorporated into the soil. Treated sludge is defined as having undergone "*biological, chemical or heat treatment, long-term storage or any other appropriate process so as significantly to reduce its fermentability and the health hazards resulting from its use*". However, in order to provide protection against potential health risks from residual pathogens:

- Sludge must not be applied to soil in which fruit and vegetable crops are growing or grow
- Grazing animals must not be allowed access to grassland or forage land less than three weeks after the application of sludge

The Directive sets out requirements for the keeping of detailed records of the quantities of sludge produced, the quantities used in agriculture, the composition and properties of the sludge, the type of treatment and the sites where the sludge is used. Limit values for concentrations of heavy metals in sewage sludge intended for agricultural use and in sludge-treated soils are set in Directive Annexes (Table 4) [1].

Table 4. Annexes IA, IB and IC of Directive 86/278/EEC

Metal	Limit values for concentrations of heavy metals in soil (mg/kg dm)[1]	Limit values for heavy metal concentrations in sludge for use in agriculture (mg/kg dm)	Limit values of heavy metals which may be added annually to agricultural land, based on a 10 year average (kg/ha/y)
Cadmium	1-3	20-40	0.15
Copper	50-140	1000-1750	12
Mercury	1-1.5	16-25	0.1
Nickel	30-75	300-400	3
Lead	50-300	750-1200	15
Zinc	150-300	2500-4000	30

[1] For $6 < pH < 7$.

Directive 86/278/ EEC was adopted over 20 years ago and the European Commission is currently assessing whether the current Directive should be reviewed – and if so, the extent of this review [1].

The experience gained so far in the EU has shown three main weaknesses of this Directive, which are [2, 3]:

- covers only urban sludge but does not consider other non-hazardous sludges (e.g. paper sludges or textile sludges) that may have the similar negative and positive implications as urban sludge when they are spread on land;
- regulates the spreading of sewage sludge to agricultural land only but does not provide for any measure as regards other types of land use as for example non-agricultural land (e.g. tree plantations, green areas, landscaping purposes etc) which may have potential adverse impacts on human health and on wildlife and biodiversity;
- it is not conservative enough in taking into account the effects of long term accumulation of heavy metals to the topsoil

The changes proposed included the following:

- Revision of current limit values for heavy metals;
- Introduction of limit values for organic pollutants;
- Introduction of pathogen concentration limits; and
- Introduction of a quality assurance system

Some of the proposals are presented in Table 5 [7].

**Table 5. Proposed limit values on potentially toxic elements (PTE)
in sewage sludge and in soil (mg/kg dw)**

Metal	Sludge	Soil		
		5<pH<6	6<pH<7	pH>7
Cd	10	0.5	1	1.5
Cr	1000	50	75	100
Cu	1000	30	50	100
Hg	10	0.1	0.5	1
Ni	300	30	50	70
Pb	750	70	70	100
Zn	2500	100	150	200

Concerning organics, the following are considered of primary importance for EU as limits are to be set in the revision of Sewage Sludge Directive [6]:

- AOX, the so-called "sum of halogenated organic compounds"
- linear alkylbenzene sulphonates (LAS)
- di(2-ethylhexyl)phthalate (DEHP)
- NPE (nonylphenole and nonylphenole ethoxylates with 1 or 2 ethoxy groups)
- polynuclear aromatic hydrocarbons (PAHs)
- polychlorinated biphenyls (PCBs)
- polychlorinated dibenzo-p-dioxins and -furans (PCDD/Fs).

**Table 6. Limit values for concentrations (mg/kg dm) of organic compounds
in sludge of different countries and as suggested in the 3rd draft
of the "*Working paper on sludge*" for EU [6]**

	AOX	DEHP	LAS	NP/NPE	PAH	PCB	PCDD/F ng TEq/kg dm
EU	500	100	2600	50	6[1]	0.8[2]	100
Denmark	-	50	1300	10	3[1]	-	-
Sweden	-	-	-	50	3[3]	0.4[4]	-
Lower Austria	500	-	-	-	-	0.2[5]	100
Germany	500	-	-	-	-	0.2[5]	100

[1] Sum of acenapthene, phenanthrene, fluorine, fluoranthene, pyrene, benzo(b+j+k) fluoranthne, benzo(a)pyrene, benzo(ghi)perylene, indeno(1,2,3-c,d)pyrene.
[2] Sum of 6 congeners PCB 28, 52, 101, 138, 153, 180.
[3] Sum of 6 compounds.
[4] Sum of 7 congeners.
[5] Each of the six congeners PCB 28, 52, 101, 138, 153, 180.

Table 7. Standards for maximum concentrations of pathogens in sewage sludge [7]

Country	Salmonella	Other pathogens
Poland	Mo occurrence	Faecal streptococci: < 100/g
France	8 MPN/10g DM	Enterovirus: 3 MPCN/10g of DM
		Helminths eggs: 3/10g of DM
Finland	Not detected in 25 g	Escherichia coli < 1000 cfu
Italy	1000 MPN/g DM	-
Luxembourg	-	Enterobacteria: 100/g no eggs of worm likely to be contagious
Hungary	-	Faecal coli and faecal streptococci decrease below 10% of original number
Poland	Sludge cannot be used in agriculture if contains salmonella	-

The EU working document contains limit values for concentrations of seven groups of compounds: 11 PAHs; LAS; DEHP; NPE; AOX; and 7 PCBs as well as PCDD/F i.e. chlorinated dioxins and furans. Finally, in the Water Framework Directive a list of 33 priority substances of importance for the water phase that are primarily based on monitoring data in European surface waters are established [9]. Some European countries have been implementing their own limit values for several XOCs (xenobiotic organic compounds) in sludge [9]. For example, the Danish Statutory Order 1650 regulates the presence and levels of four groups of XOCs: 11 PAHs; LAS; DEHP; and NPE. The French sludge regulation includes limit values in sludge for 3 PAHs and a summary of 7 PCB congeners.

Finally, concerning pathogens, a moderate proposed change to the current Directive is to introduce standards for pathogens in line with the conventional treatment as given in the Commission Communication in 2003 [7]. Conventional treatment means any sludge treatment capable of achieving a reduction in Escherichia coli to less than 5×10^5 colony forming units per gram (wet weight) of treated sludge. Currently, only a few MS are known to have limits on pathogens, shown in Table 7.

2. LEGISLATION IN USA

The protection of the environment in US is regulated by the Section 40 of the Code of Federal Regulations (CFR). Sewage sludge regulations are included in Chapter I-*Environmental Protection Agency*, Subchapter O-*Sewage Sludge* which includes Parts 501 and 503 [11]:

- State sludge management program regulations
- Standards for the Use or Disposal of Sewage Sludge

In particular, Part 503 establishes numeric limits, management practices, and operational standards to protect public health and the environment from adverse effects of chemical and microbiological pollutants in sewage sludge [10]. Relevant is Section 405(d) of the Clean Water Act (Subchapter IV— *Permits And Licenses*, Sec. 405. [33 U.S.C. 1345] *Disposal of Sewage Sludge*) which requires the Environmental Protection Agency to review existing sewage sludge regulations at least every two years in order to identify additional toxic pollutants that may be present in sewage sludge and, if appropriate, to promulgate regulations for those pollutants consistent with the requirements set forth in the CWA [10, 12]. Is important to mention that in US, the term biosolids is used for sewage sludge, defined as *"treated residuals from wastewater treatment"* or in more words *"the nutrient-rich organic materials resulting from the treatment of sewage sludge"* [13, 14].

The Part 503 of CFR sets national standards for use or disposal of sewage sludge. Regulatory options include: land application, landfilling or surface disposal and incineration. States may adopt additional or more stringent requirements for the land application of sewage sludge [10]. The regulation is divided into five subparts: (1) general provisions, (2) land application, (3) surface disposal, (4) pathogens and vector attraction reduction, and (5) incineration. The subparts are described in the following sections [4, 11].

The first of three parameters that must be assessed to determine the overall quality of biosolids is the level of pollutants and in order to allow land application of biosolids of variable quality the law provides four sets of pollutant limits: ceiling concentration limits, concentration limits, cumulative loading rates and annual loading rates (see Table 8) [4, 11].

Table 8. Land application pollutant limits

Pollutant	Ceiling concentration limits[a,b] (mg/kg)	Pollutant concentration limits[a,c,d] (mg/g)	Cumulative pollutant loading rates (kg/ha)	Annual pollutant loading rates (kg/ha/365-day period)
Arsenic	75	41	41	2.0
Cadmium	85	39	39	1.9
Copper	4300	1500	1500	75
Lead	840	300	300	15
Mercury	57	17	17	0.85
Molybdenum	75	-	-	-
Nickel	420	420	420	21
Selenium	100	36	100	5
Zinc	7500	2800	2800	140
Applies to:	All biosolids that are land applied	Bulk biosolids and biosolids sold or given away in a bag or other container	Bulk biosolids	Biosolids sold or given away in a bag or other container

[a] Dry weight basis.
[b] Absolute values.
[c] Monthly average.
[d] Exceptional quality biosolds.

The second parameter in determining biosolids quality is the presence or absence of pathogens. The third parameter of biosolids quality is the attractiveness of biosolids to vectors. Vectors are animals and insects, such as rodents, flies, and birds, that might be attracted to biosolids and therefore could transmit pathogenic organisms to humans or to domestic animals or livestock [4].

The US regulation on the use of sewage sludge in agriculture does not establish numerical pollutant limits of any organic pollutants, because at least one of the following criteria applied for the organics considered [6]:

- the pollutant is banned for use, has restricted use or is not manufactured for use in the US;
- the pollutant is detected infrequently in sludge and is present in 5% of sludge samples;
- the limit for an organic pollutant derived from the 503 exposure assessment is greater than the 99th percentile concentration in sludge.

REFERENCES

[1] European Commission website, Environment/Waste webpage: http://ec.europa.eu /environment /waste/sludge/index.htm

[2] Pollutants in urban waste water and sewage sludge, European Commission, DG Environment, Luxembourg: Office for Official Publications of the European Communities, 2001

[3] European Commission website, Environment/Water webpage: http://ec.europa.eu/environment/water/water-urbanwaste/legislation/directive_en.htm

[4] I.S. Turovskiy, P. K. Mathai, Wastewater sludge processing, John Wiley & Sons, Inc., USA, 2006

[5] Disposal and Recycling Routes for Sewage Sludge, Synthesis report, DG Environment B/2, European Commission, 2002

[6] H. Langenkamp, P. Part, W. Erhardt, A. Prüeß, Organic contaminants in sewage sludge for agricultural use, European Commission, Joint Research Centre, Institute for Environment and Sustainability, Soil and Waste Unit, 2001

[7] Environmental, economic and social impacts of the use of sewage sludge on land, Consultation Report on Options and Impacts, Report by RPA, Milieu Ltd and WRc for the European Commission, DG Environment under Study Contract DG ENV.G.4/ETU/2008/0076r, 2009

[8] M.A. Stylianou, V.J. Inglezakis, K.G. Moustakas, M.D. Loizidou, Improvement of the quality of sewage sludge compost by adding natural clinoptilolite, Desalination, 224, 240–249 (2008)

[9] Eva Eriksson, Nina Christensen, Jens Ejbye Schmidt, Anna Ledin, Potential priority pollutants in sewage sludge, Desalination 226, 371–388 (2008)

[10] Targeted National Sewage Sludge Survey Overview Report, EPA-822-R-08-014, U.S. Environmental Protection Agency, 2009

[11] US Electronic Code of Federal Regulations, http://ecfr.gpoaccess.gov/

[12] Federal Water Pollution Control Act, Title 33—Navigation And Navigable Waters, Chapter 26—Water Pollution Prevention And Control, As Amended Through Pub.L. 110-288, July 29, 2008

[13] US Environmental Protection Agency (US EPA), Sewage Sludge (Biosolids), http://water.epa.gov/polwaste/wastewater/treatment/biosolids/index.cfm

[14] A Plain English Guide to the EPA Part 503 Biosolids Rule, US EPA, Office of Wastewater Management, Washington DC, US, 1994

In: Sewage Sludge Management
Editors: A. A. Zorpas and V. J. Inglezakis
ISBN: 978-1-61324-393-0
© 2012 Nova Science Publishers, Inc.

Chapter 16

APPROACHES TO THE IMPLICATIONS OF THE EU DIRECTIVE ON SLUDGE: ANALYTICAL METHODOLOGIES, CONCENTRATION LEVELS AND OCCURRENCE OF ORGANIC POLLUTANTS IN DIFFERENT TYPES OF SEWAGE SLUDGE

Juan Luis Santos, Julia Martín,
Irene Aparicio and Esteban Alonso[*]
Department of Analytical Chemistry, University of Seville,
C/ Virgen de África, Seville, Spain

ABSTRACT

In April 2000, the European Commission issued the 3rd draft of a working document on sludge from wastewater treatment plants in order to lay the basis of a new European Union Directive regarding the regulation of sludge application to soils. One of the most controversial aspects of the document was the proposal to fix concentration limits for certain organic compounds in sludge. Organic compounds that have to be monitored according to the 3rd draft of the future Directive are the sum of halogenated organic compounds (AOX), di(2-ethylhexyl)phthalate (DEHP), linear alkybenzene sulphonates (LAS), nonylphenol and nonylphenol ethoxylates with 1 or 2 ethoxy groups (NPE), the sum of nine polycyclic aromatic hydrocarbons (PAH), the sum of seven polychlorinated biphenyl congeners (PCB) and polychlorinated dibenzodioxins/dibenzofuranes (PCDD/F). In this chapter, the monitoring of these organic compounds in sludge samples is reported. Mean concentrations were estimated after sampling in several Andalusian wastewater treatment plants (South of Spain). The overall efficacies of sludge treatment technologies in the removal of these compounds were assessed. Nonylphenolic

---

[*] E-mail: ealonso@us.es.

compounds, di-(2-ethylhexyl) phthalate and linear alkylbenzene sulphonates were, in that order, the compounds most often found at concentration limits above those set in the European Union Directive draft.

1. INTRODUCTION

The increase of the number of urban wastewater treatment plants (WWTP) in Europe, fomented by the Directive 91/271/CEE [1], has resulted in a significant increase of the amount of sludge generated. For instance, more than one million tons of sludge was produced in Spain in 2003 [2]. Therefore, in addition to choosing effective wastewater treatment systems, it is essential to design waste removal systems with the utmost care, to ensure non-pollutant and economical removal of sludge, or its recovery in a rational manner. In the European Community, the most widely used solutions are: controlled transport to dumps, application to farmlands, incineration, expulsion into the sea, application to impoverished soils and recovery of energy or mineral products.

In Spain, there is a widespread application of sludge to farmlands (64 %), and this fact is even more pronounced in Andalusia (South of Spain) (84 %) [3] due to the combination of several factors as the high production of sludge in Andalusia, the decline of soil quality from the loss of organic fraction, erosive processes and other environmental problems. The use of sludge in agriculture is included in the paradigm of reutilization, which implies an economic value for this by-product of wastewater treatment.

Specific legislation is required to establish the recommended quality of sludge for each destination. To the date, the most restrictive legislation related to sludge from wastewater treatment plants is the European Directive 86/278/CEE [4] that is focused in the protection of the environment, in particular the soils, when residual sludge is used in agriculture.

The European Directive 86/278/CEE limits the total amount of some heavy metals, in relation to soil pH, and proposes recommendations for the control and dosage of sludge application, permitting national and regional limits depending on the specific problems of sludge generated in each case. However, nowadays these restrictions are considered to be insufficient, since a global evaluation of the application process of sludge to crop lands should also take into account factors such as human health, crop yields, animal health, and the quality of underground and surface waters, air quality, soil fertility and natural ecosystems.

In this context, the European Environmental Department produced in April 2000 the third draft of a working paper on sludge from wastewater treatment plants to lay the basis for a new European Directive for sewage sludge application to soil [5].

In this European Union (EU) Directive draft more restrict concentration limit values are fixed for heavy metals and concentration limit values for some organic compounds are included for the first time.

One of the most controversial aspects of the document is the question of which organic contaminants should be monitored and which limit values for these compounds in sludge should be set for land application of sludge. There are no unanimous criteria about the organic pollutant groups that must be controlled in sewage sludge (Table 1), nor which compounds in each organic pollutant group should be monitored, neither about the limiting concentrations that would condition the applicability of sludge to crop soils. This lack of unanimity is corroborated by most of the recent scientific studies in which determinations of different families of organic compounds in sludge [6, 7], or the same families but different compounds in each family [8, 9] are described. Hence, there is no consensus between the legislative texts, the Directive draft and the scientific literature about the organic compounds that should be regulated and the recommended limit concentrations.

This lack of common criteria can be explained by: (1) an incomplete knowledge about the global dynamics of these organic pollutants in sludge, in the treatment processes of urban wastewaters and in the subsequent application of sludge, especially, to agricultural soils and (2) the need to adopt national or local actions tailored to the specific problems of sludge generated in each region. Hence, the countries that have adopted national legislations (Denmark, Sweden, Austria and Germany) to limit the concentrations of some of these compounds, have also been the most prolific producers of scientific-technical works in their geographical region.

Similarly, there is no complete agreement either about the analytical methodologies to be employed to determine the organic compounds included in the EU Directive draft. Although the European Commission is making a considerable effort, through the European Committee for Standardisation, by developing environmental legislation on sludge from wastewater treatment plants for the EU, there is no analytical methodology yet, validated and widely accepted in the literature, to determine the majority of these compounds in the sludge matrix. This fact is reflected by the difficulty to find approved reference materials for all the classes of compounds found in sludge from wastewater treatment plants. Hence, the Institute for Reference Materials and Measurements (IRMM) of the European Union has approved only reference materials for PAH (BCR-088) and PCDD/F (BCR-677) in sludge, in contrast, for heavy metals included in the working paper there are already four reference materials approved for different types of sewage sludge. Neither the National Institute of Standards and Technology (NIST) of the United States has reference materials for organic compounds in sludge, only has reference materials for heavy metals (SRM-2781 and SRM-2782).

In this chapter, the implications of the first time inclusion of concentration limits for some organic compounds proposed by the European Directive for the agricultural use of sludge from wastewater treatment plants on the management of sludge generated in the Autonomous Community of Andalusia (South of Spain) is evaluated. To achieve this aim it was necessary to determine the mean concentration levels of these compounds in different types of sludge from Andalusian wastewater treatment plants and to assess the global efficacy of the degradation of these compounds with the sludge treatment technologies most used currently in Andalusia.

Table 1. Concentration limits of several groups of organic contaminants in sewage sludge in different EU countries

	AOX (mg/kg dm)	DEHP (mg/kg dm)	LAS (mg/kg dm)	NPE (mg/kg dm)	PAH (mg/kg dm)	PCB (mg/kg dm)	PCDD/F (ng ITEQ/kg dm)	Reference
Future EU Directive	500	100	2600	50	6[1]	0.8[2]	100	[5]
Denmark	-	50	1300	10	3[1]	-	-	[38]
France					1.5-5[3]	0.8	-	[39]

Table 1. (Continued)

Austria	500	-	-	-	-	0.2[4]	100	[40]

	AOX (mg/kg dm)	DEHP (mg/kg dm)	LAS (mg/kg dm)	NPE (mg/kg dm)	PAH (mg/kg dm)	PCB (mg/kg dm)	PCDD/F (ng ITEQ/kg dm)	Reference
Germany	500	-	-	-	-	0.2[4]	100	[41]

dm (dry matter).

[1] Sum of acenapthene, phenanthrene, fluorene, fluoranthene, pyrene, benzo(b+j+k)fluoranthene, benzo(a)pyrene, benzo(ghi)perylene, indeno(1, 2, 3-c,d)pyrene.

[2] Sum of PCB congeners 28, 52, 101, 138,153, 180.

[3] Fluoranthene, benzo(b)fluoranthene, benzo(k)fluoranthene, benzo(ghi)perylene, benzo(a)pyrene or indeno(1,2,3-cd)pyrene.

[4] Each of the PCB congeners 28, 52, 101, 138, 153, 180.

2. Sample Collection And Analytical Metodology

2.1 Sewage Sludge Treatment Plants and Monitoring Program

Sludge samples were collected from 20 WWTPs located in Andalusia region (South of Spain) (Figure 1). In order to obtain information about the occurrence of organic compounds in different sludge treatment technologies, WWTPs with different sludge stabilization technologies were studied. Sludge treatments applied in each plant were anaerobic or aerobic digestion of sludge (in anaerobic treatment plants (AnTPs) and aerobic treatment plants (AeTPs), respectively), anaerobic wastewater stabilization ponds (AnWSPs), dehydration (in a dehydration treatment plant (DTP)), and composting of anaerobically-digested and dehydrated sludge (in composting plants (CPs)). Types of sludge monitored were primary, secondary and anaerobically-digested and dehydrated sludge (three of them from AnTPs), mixed sludge (mixture of primary and secondary sludge) and aerobically-digested and dehydrated sludge (both of them from AeTPs), lagoon sludge from AnWSPs, dehydrated sludge from DTP and compost from CPs.

Sludge was sampled following the recommendations of the United States Environmental Protection Agency [10], with a frequency adapted to the characteristics of the processes and technologies used in each WWTP. Samples from AnTPs, AeTPs, DTP and CPs were collected monthly during one-year period from March 2005 to April 2006 according to the diagrams in Figure 2. Samples from AnWSPs were collected twice in a year. Primary, secondary, mixed, digested and dehydrated sludge grab samples were collected from each sampling site. Combined compost samples were obtained by mixing the aliquots collected from at least five sampling sites at approximately 30 cm deep in the compost battery. Combined samples from AnWSPs were obtained by mixing the aliquots collected from four sampling points located at the bottom of each AnWSP.

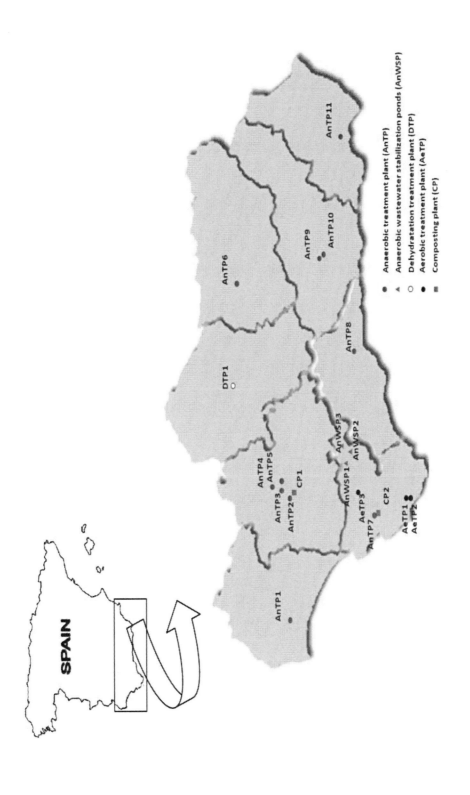

Figure 1: Location of the studied treatment plants.

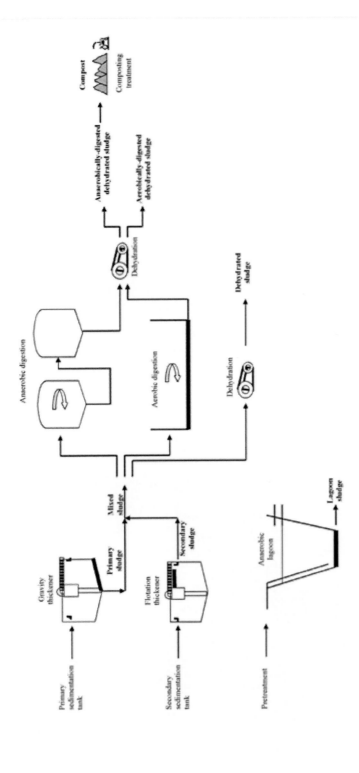

Figure 2. Scheme of the sludge treatments in the studied treatment plants and localization of the sampling sites.

2.2 Analytical Techniques Used for the Determination of the Concentration Levels of the Organic Compounds

Except for AOX and PCDD/F, for which standard methods are available from the United States Environment Protection Agency (USEPA) and the European Committee for Standardization (CEN), no analytical methodology has been validated and widely accepted to determine the presence in sludge matrix of the organic compounds included in the future EU Directive.

Several analytical methods have been reported in the scientific literature for the determination of organic pollutants in sewage sludge [11-16]. Traditional methods such as soxhlet extraction have been predominantly used for the extraction of organic compounds from sludge samples [11-14, 17-24]. However, although soxhlet extraction is recognized as efficient and robust, is tedious, time-consuming and use large volumes of toxic, expensive and inflammable solvents. Actually, several extraction methods such as sonication-assited extraction [13, 17, 21, 22, 25-27], pressurized liquid extraction [28-30] and microwave assisted extraction [31, 32] have been reported for the extraction of organic compounds from sludge samples.

High performance liquid chromatography equipped with ultraviolet [14, 16, 21, 31] and fluorescence detectors [13, 16, 21, 32] has been applied to LAS, NPE, DEHP and PAH determination. However, chromatography coupled to mass spectrometry detectors is being increasingly applied to the determination of these compounds in sludge samples, mainly to compounds which usually are present at µg or even ng per kilogram levels [15, 17-19, 24, 25, 33].

Concentration levels showed in this chapter were obtained by normalized analytical methods [34, 35], in the case of AOX and PCDD/F, and by two optimized and validated analytical techniques in the case of DEHP, NPE, PCB, LAS and PAH [15, 16]. These methodologies, not only possess the main analytical properties desirable in analytical methods but also, present optimum complementary characteristics for the routine analytical control of organic compounds included in the future EU Directive as reasonable analysis time and cost and the use of low toxicity solvents.

2.3 AOX Determination

AOX were determined in sludge samples according to CEN standard method [34]. The method was based on the simultaneous adsorption of water soluble organic compounds on activated carbon by shaking a mixture of sludge and active carbon with acidified nitrate solution, combustion of the loaded mixture in an oxygen stream, absorption of the hydrogen halides produced and determination of the halide ions by microcoulometry. Limits of detection (LODs), limits of quantification (LOQs) and recoveries of each compound are summarized in Table 2.

Table 2. Limits of detection (LOD), limits of quantification (LOQ) and recoveries of AOX, DEHP, NPE and LAS in the analyzed sludge samples

Group	Compound	LOD (mg/kg dm)	LOQ (mg/kg dm)	Recovery (%)
AOX	AOX	3.00	5.00	60.0
DEHP	DEHP	0.02	0.07	105
NPEs	NP	0.19	0.63	77.9
	NP1EO	0.75	2.50	88.6
	NP2EO	0.42	1.40	61.4
LAS	C10	8.90	42.8	97.2
	C11	46.7	69.5	87.5
	C12	4.40	42.0	85.9
	C13	23.4	77.3	81.9

2.4 DEHP, NPE and PCB Determination

For DEHP, NPE and PCB extraction, dried and lyophilized sludge was extracted with 5 mL of hexane by sonication-assisted extraction after shaking. The liquid phase was separated by centrifugation and the extraction procedure was repeated with the residue. The combined extracts were evaporated to dryness and diluted with hexane as described in a previously reported method [15].

Simultaneous determination was performed on an Agilent 6890N gas chromatograph coupled to an Agilent 5973N mass spectrometer [15]. LODs, LOQs and recoveries of DEHP and NPE are shown in Table 2. LODs, LOQs and recoveries of PCB are shown in Table 3.

Table 3. Limits of detection (LOD), limits of quantification (LOQ) and recoveries of PCB, PAH, PCDD and PCDF in the analyzed sludge samples

	Compound	LOD (µg/kg dm)	LOQ (µg/kg dm)	Recovery (%)
PCB	PCB 28	2	6	92
	PCB 52	3	5	80
	PCB 101	3	10	86
	PCB 118	2	7	92
	PCB 138	2	6	108
	PCB 153	2	7	105
	PCB 180	3	11	56
PAH	Naphthalene	29	92	85
	Acenaphthylene	5.0	16	96
	Acenaphthene + Fluorene	0.4	1.0	96
	Phenanthrene	0.5	1.0	72
	Anthracene	1.0	2.0	87
	Fluoranthene	4.0	12	109
	Pyrene	78	259	105
	Benzo[a]anthracene	6.0	21	98

	Compound	LOD (μg/kg dm)	LOQ (μg/kg dm)	Recovery (%)
	Chrysene	0.2	1.0	93
	Benzo[b]fluoranthene	1.0	2.0	108
	Benzo[k]fluoranthene	2.0	8.0	84
	Benzo[a]pyrene	24	80	100
	Dibenzo[a,h]anthracene	11	36	84
	Benzo[ghi]perylene	45	150	76
	Indeno[1,2,3-cd]pyrene	29	92	97
PCDD	2,3,7,8-TCDD	0.002	0.004	75
	1,2,3,7,8-PeCDD	0.002	0.004	80
	1,2,3,4,7,8-HxCDD	0.002	0.005	77
	1,2,3,7,8,9-HxCDD	0.002	0.005	85
	1,2,3,4,6,7,8-HpCDD	0.003	0.005	67
	OCDD	0.002	0.004	70
PCDF	2,3,7,8-TCDF	0.002	0.004	72
	1,2,3,7,8-PeCDF	0.002	0.005	81
	2,3,4,7,8-PeCDF	0.002	0.005	85
	1,2,3,6,7,8-HxCDF	0.002	0.004	83
	1,2,3,7,8,9-HxCDF	0.002	0.003	86
	1,2,3,4,7,8-HxCDF	0.001	0.004	83
	2,3,4,6,7,8-HxCDF	0.002	0.004	80
	1,2,3,4,6,7,8-HpCDF	0.002	0.004	80
	1,2,3,4,7,8,9-HpCDF	0.001	0.003	86
	OCDF	0.003	0.005	68

2.5 LAS and PAH Determination

LAS and PAH were simultaneously extracted from sewage sludge with 50 mL of methanol by sonication-assisted extraction according to Santos et al. (2007) [16]. The sample extract was filtered and subjected to clean-up by solid phase extraction. Fractions of 1 mL and 5 mL of the cleaned extract were separately evaporated to dryness by a gentle nitrogen stream and the residues were dissolved in 1 mL of water or 0.5 mL of acetonitrile, respectively. LAS were determined in the fraction dissolved in water and PAH were determined in the fraction dissolved in acetonitrile. Analytical determinations were carried out in an Agilent 1100 series high performance liquid chromatograph (HPLC) with diode array (DAD) and fluorescence (Fl) detectors according to Santos et al. (2007) [16]. LODs, LOQs and recoveries of LAS are summarized in Table 2. LODs, LOQs and recoveries of PAH are summarized in Table 3.

2.6 PCDD/F Determination

PCDD and PCDF were determined according to US EPA method 8290 [35]. Method was based on soxhlet extraction of lyophilized and sieved sludge sample, clean-up by solid phase extraction and determination by gas chromatography coupled to mass spectrometry detection (GC/MS). LODs, LOQs and recoveries of PCDD and PCDF are summarized in Table 3.

3. Occurrence and Fate of Organic Pollutants in Sludge Stabilization Processes

Concentration levels of the analyzed organic compounds are represented as boxplots (Figures 3 and 4). Each box shows the lower quartile ($\leq 25\%$), median ($\leq 50\%$), upper quartile ($\leq 75\%$) and mean concentrations. Maximum and minimum values are represented by the whiskers.

From the results obtained in this chapter and the statistical treatment of these data by box diagrams, the following data will be given: (1) the concentration ranges obtained as the sum of the compounds specified in the working paper and the percentages of samples that exceed limits set in this paper for different types of sludge; (2) the time course of these concentrations expressed as relative standard deviation for each sludge treatment technology; (3) the spatial evolution, when relevant, for each of the sludge treatment stages; and (4) the influence of sludge treatment technologies on the degradation of the compounds. Finally, some overall considerations will be made concerning all the organic compounds analysed.

3.1 AOX

AOX was present in a concentration range between 185 mg/kg dry matter (dm) and 274 mg/kg dm for anaerobically-digested and dehydrated sludge, composted sludge, dehydrated sludge, aerobically-digested and dehydrated sludge and sludge from anaerobic stabilization ponds (Figure 4). Therefore, AOX content in all the samples analysed (20) was far below the limit of 500 mg/kg dm proposed in the EU Directive draft. The highest concentrations were found in anaerobically-digested and dehydrated sludge.

3.2 DEHP

The concentration range for anaerobically-digested and dehydrated sludge, composted sludge, aerobically-digested and dehydrated sludge, and sludge from anaerobic stabilization ponds were 13-713 mg/kg dm, 24-616 mg/kg dm, 19-150 mg/kg dm, and 25-144 mg/kg dm, respectively (Figure 3).

Around 42 % of the anaerobically-digested and dehydrated sludge samples (88 samples) exceeded the limit established in the working paper (100 mg/kg dm). Around 36 %, 17 % and 33 % of composted sludge samples (11 samples), aerobically-digested and dehydrated sludge (18 samples) and sludge from anaerobic stabilization ponds (6 samples), respectively, exceeded the limit of 100 mg/kg dm.

The variability of DEHP concentration with time, estimated as the relative standard deviation, was ranged between 50 % for anaerobically-digested and dehydrated sludge and 132 % for composted sludge.

Anaerobically-digested and dehydrated sludge presented higher concentration levels of DEHP than composted, primary and secondary ones from anaerobic treatments, in that order. In the case of aerobic treatments, aerobically-digested and dehydrated sludge presented higher concentrations of DEHP than mixed ones.

Regarding to the differences of concentrations in relation to the technologies, anaerobically-digested and dehydrated sludge presented higher concentrations than aerobically-digested and dehydrated sludge.

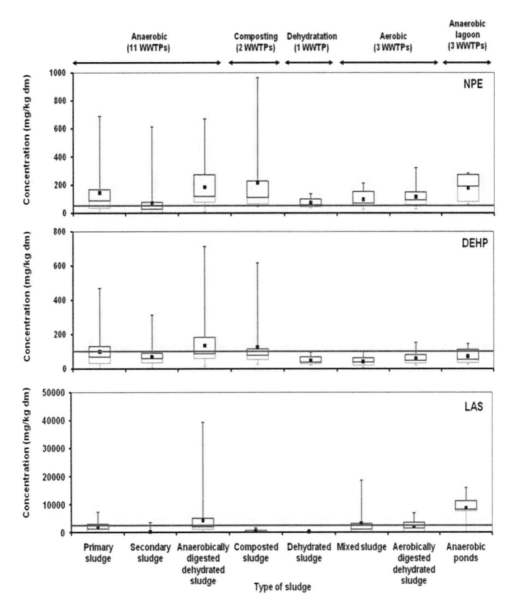

Figure 3. Box and whisker plots of NPE, DEHP and LAS concentrations in the analyzed sewage sludge samples. Horizontal lines show concentration limits fixed in the directive draft for land application of sludge [5].

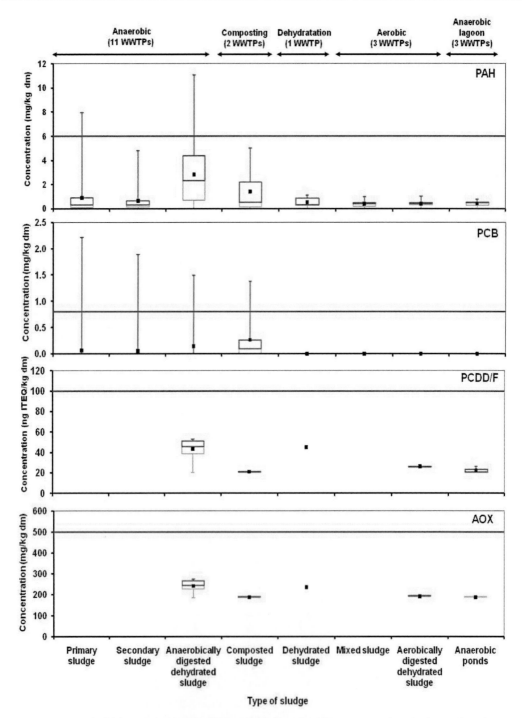

Figure 4. Box and whisker plots of PAH, PCB, PCDD/F and AOX concentrations in the analyzed sewage sludge samples. Horizontal lines show concentration limits fixed in the directive draft for land application of sludge [5].

Regarding to the differences of concentrations in relation to the technologies, anaerobically-digested and dehydrated sludge presented higher concentrations than aerobically-digested and dehydrated sludge.

3.3 LAS

LAS were found at high concentrations in the analyzed samples (Figure 5). C10 homologue was present at the lowest concentrations in the sludge samples analyzed while C11, C12 and C13 were present at higher and similar concentration levels. The higher concentrations were found in mixed sludge (up to 6683 mg/kg dm for LAS C12), anaerobically-digested and dehydrated sludge (up to 14539 mg/kg dm for LAS C12), and sludge from stabilization ponds (up to 5948 mg/kg for LAS C13). The concentration ranges of LAS (sum of C10, C11, C12 and C13) for anaerobically-digested and dehydrated sludge, composted sludge, aerobically-digested and dehydrated sludge, and sludge from anaerobic stabilization ponds were 22-39340 mg/kg dm, 44-1766 mg/kg dm, 6-7041 mg/kg dm, y 7763-16015 mg/kg dm respectively (Figure 3).

Around 40 %, 33 % and 83 % of the anaerobically-digested and dehydrated sludge samples (88 samples), aerobically (18 samples) and anaerobic stabilization ponds (6 samples) respectively exceeded the limit of 2600 mg/kg dm established in the working paper. None of composted sludge samples exceeded this limit (11 samples).

The variations of LAS concentrations with time, measured as relative standard deviations, ranged from 30% for sludge from tanks to 142% for anaerobically-digested and dehydrated sludge samples.

In the anaerobic sludge treatments, digested sludge presented higher concentrations of LAS than primary, secondary and composted sludge, in that order. In the aerobic processes, digested sludge had a higher concentration level than mixed ones.

3.4 NPE

NP was present at the highest concentration levels in the analyzed sludge samples followed by NP1EO and NP2EO, in this order (Figure 5). This distribution could be due to the fact that NP1EO and NP2EO are degradation products of nonylphenol polyethoxylates which end up to be degraded to NP under aerobic and anaerobic conditions by the loss of ethoxy groups [36]. This fact, together with the loss of weight achieved during sludge dehydration process, explains the high concentration levels of NP in anaerobically-digested dehydrated sludge compared to the concentration levels in primary and secondary sludge samples and the decrease of the concentrations of NP1EO and NP2EO from primary and secondary sludge to anaerobically-digested dehydrated sludge. The ranges of concentration recorded for NPE were 8-669 mg/kg dm, 44-962 mg/kg dm, 27-319 mg/kg dm and 61-282 mg/kg dm for anaerobically-digested and dehydrated sludge, composted sludge, aerobically-digested and dehydrated sludge and sludge from anaerobic stabilization ponds, respectively (Figure 3).

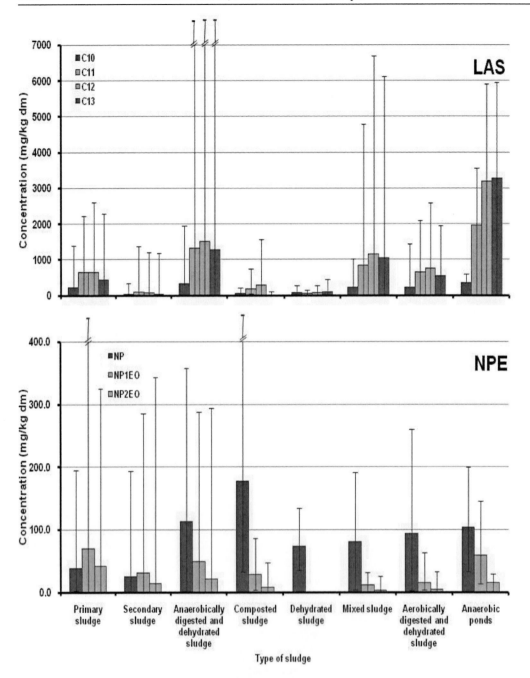

Figure 5. Concentration levels of LAS homologues and NPE compounds in the analyzed sewage sludge samples.

All the types of sludge analysed presented concentrations higher than the 75% of the limit set in the EU Directive draft (50 mg/kg dm): 78 % for aerobically-digested and dehydrated sludge (18 samples), 93 % for anaerobically-digested and dehydrated sludge (88 samples), 91 % for composted sludge (11 samples) and 100 % for sludge from anaerobic stabilization ponds (6 samples).

The variations in NPE concentrations with time ranged from 33% for anaerobically digested sludge and 133% for composted sludge, measured as relative standard deviations.

The anaerobically-digested and dehydrated sludge presented higher concentrations of NPE than composted, primary and secondary ones in that order, for WWTPs with anaerobic treatment. For aerobic WWTP, aerobically-digested and dehydrated sludge presented higher concentrations than mixed ones.

3.5 PAH

Some of the PAH included in the 3rd Draft of the future sludge Directive, it is fluoranthene, benzo[a]pyrene, benzo[b]fluoranthene, benzo[g,h,i]perylene, indeno[1,2,3-cd]pyrene, and others as acenaphthylene and dibenzo[a,h]anthracene were not detected in sludge samples (Figure 6). The sum of PAH acenapthene, phenanthrene, fluorene, fluoranthene, pyrene, benzo(b)fluoranthene, benzo(k)fluoranthene, benzo(a)pyrene, benzo(ghi)perylene and indeno(1, 2, 3-c,d)pyrene) was in the concentration ranges: 0.09-11.1 mg/kg dm, 0.14-5.0 mg/kg dm, 0.005-1.0 mg/kg dm and 0.22-0.78 mg/kg dm for anaerobically-digested and dehydrated sludge, composted sludge, aerobically-digested and dehydrated sludge and sludge from anaerobic stabilization ponds, respectively (Figure 4). Only 9 % of anaerobically-digested and dehydrated sludge samples (88) exceeded the limit set in the EU directive draft (6 mg/kg dm).

The temporal variations of PAH concentration, measured as relative standard deviations, was ranged from 1 % for sludge from stabilization ponds to 160 % for anaerobically-digested and dehydrated sludge.

Evolution of PAH concentrations alongside sludge treatment steps was similar to those observed in the case of DEHP and NPE: concentration levels in sludge anaerobically-digested and dehydrated sludge were higher than in composted sludge, primary sludge and secondary sludge in the case of anaerobic treatments and concentration levels in aerobically-digested and dehydrated sludge were higher than in mixed sludge in the case of aerobic treatments.

Finally, WWTPs with anaerobic sludge treatments presented higher concentrations of PAH than the other ones. This could be explained by the fact that the anaerobic treatments studied correspond to WWTP from large urban nuclei with larger percentages of these types of compounds.

3.6 PCB

PCB congeners 28, 52, 101 and 118 were not detected in any of the analyzed sludge samples. PCB congeners 138, 153 and 180 were found in sludge samples at concentration levels lower than 0.2 mg/kg dm (Figure 6). Detectable concentrations of PCB were recorded in digested and composted sludge (Figure 4), with maximum values of 1.49 mg/kg dm and 1.38 mg/kg dm, respectively. Between 7 % and 9 % of these sludge samples (of a total of 99) exceeded the limit set in the EU Directive draft for PCB (0.8 mg/kg dm).

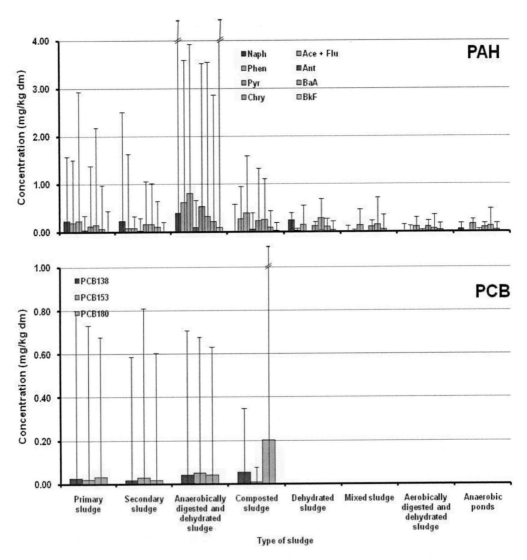

Figure 6. Concentration levels of PAH compounds and PCB congeners in the analyzed sewage sludge samples (Naph, naphthalene; Phen, phenanthrene; Pyr, pyrene; Chry, chrysene; Ace, acenaphthene; Flu, fluorene; Ant, anthracene; BaA, Benzo(a)anthracene; BkA, benzo(k)fluoranthene).

3.7 PCDD/F

Concentrations of PCDD/F ranged from 20 ng ITEQ/kg dm (International Toxicity Equivalent per kilogram) to 53 ng ITEQ/kg dm for anaerobically-digested and dehydrated sludge, composted sludge, dehydrated sludge, aerobically-digested and dehydrated sludge, and sludge from anaerobic stabilization ponds (Figure 4), respectively. Therefore, the concentrations of PCDD/F in all the samples analyzed (20) were far below the limit of 100 ng ITEQ/kg dm proposed in the EU directive draft. The highest concentrations, as in the case of AOX, were recorded in anaerobically-digested and dehydrated sludge.

To summarize the above:

NPE, DEHP and LAS are, in this order, the organic pollutants most frequently found above the concentration limits set in the EU directive draft for the group of samples analyzed. The values found for these pollutants are comparable to those described by Gómez-Rico et al. (2007) in Valencia (NPE: 190-3500 mg/kg dm; DEHP: 8-350 mg/kg dm and LAS: 130-32000 mg/kg dm) [37], Abad et al. (2005) in Catalonia (NPE: 14.3-3150 mg/kg dm) [24] and Jones and Northcott (2000) in the United Kingdom (LAS: 100-500 mg/kg dm in anaerobically-digested sludge and 5000-15000 mg/kg dm in aerobically-digested sludge [38]. Between 7 and 9 % of anaerobically-digested and dehydrated sludge samples and composted sludge exceeded the limit set for PCB. Around 9 % of composted sludge samples exceeded the limit for PAH (a percentage far below that the ones obtained by Jones and Northcott in the United Kingdom [38], where concentrations of PAH between 18 and 50 mg/kg dm are reported. None of the samples exceeded the limits set for AOX and PCDD/F.

Three variables have an important influence on the concentrations of these majority pollutants in the different types of sludge: the original concentration in wastewater, their predominant aerobic degradation and the concentration effect in the digestion processes due to the loss of weight. Hence, in general, in anaerobic treatment plants the highest concentrations of organic pollutants are reached in digested sludge, followed by composted sludge, primary sludge and, finally, secondary sludge. In aerobic treatment plants, higher concentrations in digested than in mixed sludge are detected. The concentration enrichment observed in digested sludge seems to be caused by the loss of weight of sludge by degradation of the organic matter in the digestion process: consequently, the pollutant ends to be concentrated. The lower enrichment of aerobically-digested and dehydrated sludge is due to the fact that during the digestion process, together with the loss of weight, a greater degradation of the organic pollutant occurs. The greater the degradation of the pollutant is, the smaller concentration increase in the digested sludge takes place. Composting causes a reduction (not always sufficient) in the concentration of organic compounds.

4. IMPLICATIONS OF THE FUTURE EU DIRECTIVE

Due to the presence of organic compounds in sewage sludge, the implementation of the future EU Directive would imply changes in the current sludge management and control strategies such as control of industrial wastes, management of WWTPs, technologies in sludge treatments or the search of alternative destinations of sludge.

4.1 Control of Industrial Wastes

The low concentrations of heavy metals and some classes of organic compounds (AOX, PAH, PCB and PCDD/F) could indicate a low incidence of industrial wastes in the WWTPs studied, either due to the progress made in the recent years in relation to the prevention, control and regulation of these wastes, or due to the absence of large industry in most of the population nuclei. In contrast, wastes containing detergents (NPE), surfactants (LAS) and plastifying agents are all essentially domestic wastes.

Hence, to prevent pollution at source by organic compounds, it is necessary to control dumping from smaller industries (which tend to be numerous in larger towns) in which some of these organic pollutants can be identified (launderettes, garages, car wash tunnels and others).

4.2 Management of WWTPs

Reduction of the concentrations of some of the compounds studied in the final product by management of the WWTPs, while maintaining their sludge treatment technologies (anaerobic digestion, composting, aerobic digestion and anaerobic processes), can be achieved by: (1) controlling the mixing ratio of primary and secondary sludge, which has lower concentrations of organic compounds, and, (2) selecting optimum residence times for sludge treatment to favour the degradation of these compounds.

4.3 Technologies in Sludge Treatment

On the whole, aerobic digestion seems to be more effective than anaerobic digestion in the degradation of the predominant organic compounds. In any case, both treatments have been shown to be ineffective maintaining the concentrations of some organic compounds below the limits set in the EU directive draft (among other reasons, as mentioned previously, this is due to the loss of weight that takes place in both digestion treatments). Composting, however, could prove to be a useful technique to degrade these compounds provided that the technological management of this technique is improved. Anaerobic tanks, however, have been shown to be ineffective to maintain concentrations of NPE, LAS and DEHP below the limits set.

Taking all of this into account, composting is, therefore, proposed as the most suitable and economic technology to guarantee the safety of applying sludge to cultivated soils in relation to the organic compounds present and other sludge quality parameters as presence of pathogens and biodegradable organic material. However, bearing in mind the concentrations of NPE and DEHP found in this type of samples, it will be necessary to introduce new technologies to: (1) optimise variables of residence time, oxygen supply and mixtures with other wastes, all key factors in the degradation and reduction of these organic compounds; and (2) to develop a more industrialised control of compost production to guarantee a uniform quality of the resulting product. For sludge from smaller treatment plants, such as those from anaerobic tanks with high concentrations of LAS and NPE, the best solution would be to send them to larger plants to be treated together with other kind of sludge.

New technologies based on processes of advanced oxidation have proven to be effective removing these organic compounds, although the feasibility of applying them to sewage treatment systems has not been sufficiently demonstrated.

4.4 Alternative Destinations for Sludge

The entry into force of a new directive, apparently dependent on the drafting of a macro-directive on the prevention of soil pollution, which sets concentration limits for organic compounds in sludge applied to soils as proposed in the working paper, could require to divert the sludge produced in Andalusia to different treatment plants (as occurs in Northern European countries), depending on the percentage of samples having concentrations of NPE, DEHP and LAS above the accepted limits. In spite of requiring transport to the dump (limiting factor) and incineration (costly), the recovery of mineral products is one of the most widely accepted alternatives. This process vitrifies sludge in a combustion chamber, which stabilises it and reduces its volume, as well as producing a material that can be used in construction: concrete, brick, glass, pottery, crystallised slag and others.

In any case, it is unlikely that any of these destinations replaces, in a significant degree, the application of sludge to soils (crop land or others) in Andalusia for several reasons: (1) the large production of sludge and, (2) the quality decline of agricultural soils by loss of the organic fraction, (3) erosive processes and (4) other environmental problems. Moreover, the application of sludge from wastewater treatment plants can be classed in the paradigm of reuse, which implies assigning a value to this by-product of wastewater treatment plants.

On the other hand, a special characteristic of organic pollutants, in contrast to heavy metals, is their biodegradation potential. Although variable, many of these compounds have a slow but significant biodegradation potential. Therefore, as mentioned previously, intensification of the biological treatment processes of sludge (such as composting) would increase the removal yields of these compounds. It must be also taking into account that this biodegradation can occur after spreading sludge over soils. Climate and soil characteristics where the sludge is to be applied will be, therefore, essential variables to predict the degradation of organic pollutants. In Andalusia, the low organic matter content of the soils, their good aeration, the long periods of sunshine and the high summer temperatures could favour the rapid degradation of these compounds.

These findings should encourage the reduction of these organic compounds in sludge for application to soils in the whole European Union, and to permit national or local actions adapted to the specific problems of sludge encountered in each case.

CONCLUSION

NPE, DEHP and LAS are, in this order, the organic pollutants most frequently found in the samples analyzed at concentrations above the concentration limits set in the EU Directive draft. PCB and PAH were present at concentration levels higher than the limit fixed by the future EU Directive in the 9 % of composted sludge samples. The most contaminated sludge samples were anaerobically-digested and dehydrated sludge. These results show the need of several technological changes in sludge management to reduce the concentration of these organic pollutants to levels that allow land application of sewage sludge according to the future European Directive.

ACKNOWLEDGMENTS

This work has been financed by the Secretaría General para la Prevención de la Contaminación y el Cambio Climático del Ministerio de Medio Ambiente y Medio Rural y Marino (Projects n° 072/2004/3, 4.1-008/2005/2-B, 004/2006/1-4.1) and by the Agencia Andaluza del Agua (with FEDER co-financing), de la Consejería de Medio Ambiente de la Junta de Andalucía (Project n° 2007/1481).

REFERENCES

[1]   Council Directive 91/271/EEC of 21 May 1991 concerning urban waste water treatment. The Council of the European Communities.

[2]   Registro Nacional de Lodos, Ministerio de Agricultura, Pesca y Alimentación, 2003.

[3]   Ministerio de Medio Ambiente. II Plan nacional de lodos de depuradoras de aguas residuales-EDAR IIPNLD (2007-2015). Plan nacional integrado de residuos (PNIR).Madrid, Spain.

[4]   Commission Directive 98/15/EEC of 27 February 1998 amending Council Directive 91/271/EEC with respect to certain requirements established in Annex I. The Commission of the European Communities.

[5]   Working Document on Sludge, Third Draft, 27 April 2000. European Union, Brussels, Belgium.

[6]   Nie, Y., Qiang, Z., Zhang, H., & Adams, C. (2009). Determination of endocrine-disrupting chemicals in the liquid and solid phases of activated sludge by solid phase extraction and gas chromatography–mass spectrometry. Journal of Chromatography A,1216 (42), 7071-7080.

[7]   García-Galán, M. J., Díaz-Cruz, M. S., & Barceló, D. (2009). Determination of triazines and their metabolites in environmental samples using molecularly imprinted polymer extraction, pressurized liquid extraction and LC–tandem mass spectrometry. Journal of Hydrology. In Press.

[8]   Busetti, F., Heitz, A., Cuomo, M., Badoer, S., & Traverso, P. (2006). Determination of sixteen polycyclic aromatic hydrocarbons in aqueous and solid samples from an Italian wastewater treatment plant. Journal of Chromatography A, 1102 (1-2), 104-115.

[9]   Núñez, L., Turiel, E., & Tadeo, J. L. (2007). Determination of nonylphenol and nonylphenol ethoxylates in environmental solid samples by ultrasonic-assisted extraction and high performance liquid chromatography-fluorescence detection. Journal of chromatography A, 1146 (2), 157-163.

[10]   USEPA. (1989). POTW Sludge Sampling And Analysis Guidance Document. EPA 833-B-89-100. Available from http://www.epa.gov/npdes/pubs/owm012.pdf. (11th November 2009)

[11]   Bagó, B., Martín, Y., Mejía, G., Broto-Puig, F., Díaz-Ferrero, J., Agut, M., & Comellas, L. (2005). Di-(2-ethylhexyl)phthalate in sewage sludge and post-treated sludge: Quantitative determination by HRGC-MS and mass spectral characterization. Chemosphere, 59 (8), 1191-1195.

[12]   Sablayrolles, C., Montréjaud-Vignoles, M., Benanou, D., Patria, L., & Treilhou, M. (2005). Development and validation of methods for the trace determination of phthalates in sludge and vegetables. Journal of Chromatography A, 1072 (2), 233-242.

[13]   Fountoulakis, M., Drillia, P., Pakou, C., Kampioti, A., Stamatelatou, K., & Lyberatos, G. (2005). Analysis of nonylphenol and nonylphenol ethoxylates in sewage sludge by high performance liquid chromatography following microwave-assisted extraction. Journal of Chromatography A, 1089 (1-2), 45-51.

[14]   Garcia, M. T., Campos, E., Sánchez-Leal, J., & Ribosa, I. (2006). Effect of linear alkylbenzene sulphonates (LAS) on the anaerobic digestion of sewage sludge. Water Research, 40 (15), 2958-2964.

[15]   Aparicio, I., Santos, J. L., & Alonso, E. (2007). Simultaneous sonication-assisted extraction, and determination by gas chromatography–mass spectrometry, of di-(2-ethylhexyl)phthalate, nonylphenol, nonylphenol ethoxylates and polychlorinated biphenyls in sludge from wastewater treatment plants. Analytica Chimica Acta, 584 (2), 455-461.

[16]   Santos, J. L., Aparicio, I., & Alonso, E. (2007). A new method for the routine analysis of LAS and PAH in sewage sludge by simultaneous sonication-assisted extraction prior to liquid chromatographic determination. Analytica Chimica Acta, 605 (1), 102-109.

[17]   Sułkowski, W., & Rosińska, A. (1999). Comparison of the efficiency of extraction methods for polychlorinated biphenyls from environmental wastes. Journal of Chromatography A, 845 (1-2), 349-355.

[18]   Pereira, M. S., & Kuch, B. (2005). Heavy metals, PCDD/F and PCB in sewage sludge samples from two wastewater treatment facilities in Rio de Janeiro State, Brazil. Chemosphere, 60 (7), 844-853.

[19]   Liu, H., Zhang, Q., Cai, Z., Li, A., Wang, Y., & Jiang, G. (2006). Separation of polybrominated diphenyl ethers, polychlorinated biphenyls, polychlorinated dibenzo-p-dioxins and dibenzo-furans in environmental samples using silica gel and florisil fractionation chromatography. Analytica Chimica Acta, 557 (1-2), 314-320.

[20]   Helaleh, M. I. H., Al-Omair, A., Nisar, A., & Gevao, B. (2005). Validation of various extraction techniques for the quantitative analysis of polycyclic aromatic hydrocarbons in sewage sludges using gas chromatography-ion trap mass spectrometry. Journal of Chromatography A, 1083 (1-2), 153-160.

[21]   Miège, C., Dugay, J., & Hennion, M. C. (2003). Optimization, validation and comparison of various extraction techniques for the trace determination of polycyclic aromatic hydrocarbons in sewage sludges by liquid chromatography coupled to diode-array and fluorescence detection. Journal of Chromatography A, 995 (1-2), 87-97.

[22]   Moreda, J. M., Arranz, A., De Betoño, S., Cid, A., & Arranz, J. F. (1998). Chromatographic determination of aliphatic hydrocarbons and polyaromatic hydrocarbons (PAHs) in a sewage sludge. The Science of the Total Environment, 220 (1), 33-43.

[23]   Alzola, R., Pons, B., Bravo, D., & Arranz, A. (2008). Determination of polynuclear aromatic hydrocarbons (PAHs) in sewage sludge by micellar electrokinetic capillary chromatography and HPLC-fluorescence detection: A comparative study. Environmental Technology, 29 (11), 1219-1228.

[24] Abad, E., Martínez, K., Planas, C., Palacios, O., Caixach, J., & Rivera, J. (2005). Priority organic pollutant assessment of sludges for agricultural purposes. Chemosphere, 61 (9), 1358-1369.

[25] Fernández-Sanjuan, M., Rigol, A., Sahuquillo, A., Rodríguez-Cruz, S., & Lacorte, S. (2009). Determination of alkylphenols and alkylphenol ethoxylates in sewage sludge: Effect of sample pre-treatment. Analytical and Bioanalytical Chemistry, 394 (6), 1525-1533.

[26] Gatidou, G., Thomaidis, N. S., Stasinakis, A. S., & Lekkas, T. D. (2007). Simultaneous determination of the endocrine disrupting compounds nonylphenol, nonylphenol ethoxylates, triclosan and bisphenol A in wastewater and sewage sludge by gas chromatography–mass spectrometry. Journal of Chromatography A, 1138 (1-2), 32-41.

[27] Riu, J., Martínez, E., & Barceló, D. (2001). LC-MS determination of linear alkylbenzene sulfonates and their carboxylic degradation products in influent and effluent water samples and sludges from sewage-treatment plants. Analytical and Bioanalytical Chemistry, 371 (4), 448-455.

[28] Reid, A. M., Brougham, C. A., Fogarty, A. M., & Roche, J. J. (2009). Accelerated solvent-based extraction and enrichment of selected plasticisers and 4-nonylphenol, and extraction of tin from organotin sources in sediments, sludges and leachate soils. Analytica Chimica Acta, 634 (2), 197-204.

[29] Eichhorn, P., López, O., & Barceló, D. (2005). Application of liquid chromatography–electrospray-tandem mass spectrometry for the identification and characterisation of linear alkylbenzene sulfonates and sulfophenyl carboxylates in sludge-amended soils. Journal of Chromatography A, 1067 (1-2), 171-179.

[30] Mansuy-Huault, L., Regier, A., & Faure, P. (2009). Analyzing hydrocarbons in sewer to help in PAH source apportionment in sewage sludges. Chemosphere, 75 (8), 995-1002.

[31] Villar, P., Callejón, M., Alonso, E., Jiménez, J. C., & Guiraum, A. (2004). Optimization and validation of a new method of analysis for polycyclic aromatic hydrocarbons in sewage sludge by liquid chromatography after microwave assisted extraction. Analytica Chimica Acta, 524 (1-2), 295-304.

[32] Núñez, L., Turiel, E., & Tadeo, J. L. (2007). Determination of nonylphenol and nonylphenol ethoxylates in environmental solid samples by ultrasonic-assisted extraction and high performance liquid chromatography-fluorescence detection. Journal of Chromatography A, 1146 (2), 157-163.

[33] Gfrerer, M., Gawlik, B. M., & Lankmayr, E. (2004). Validation of a fluidized-bed extraction method for solid materials for the determination of PAHs and PCBs using certified reference materials. Analytica Chimica Acta, 527 (1), 53-60.

[34] CEN/TC, TC WI: 2003 (E), EN 0000:2003. Determination of adsorbable organically bound halogens (AOX)-Solid materials. Agosto 2003.

[35] USEPA Method 8290. Polychlorinated dibenzodioxins (PCDDs) and polychlorinated dibenzofurans (PCDFs) by high resolution gas chromatography/high resolution mass spectrometry (HRGC/HRMS).

[36] Ahel, M., Giger, W., & Koch, M. (1994). Behavior of alkylphenol polyethoxylate surfactants in the aquatic environment - I. Occurrence and transformation in sewage treatment. Water Research, 28 (5), 1131-1142.

[37] Gomez-Rico, M. F., Font, R., Aracil, I., & Fullana, A. (2007). Analysis of organic pollutants in sewage sludges from the Valencian community (Spain). Archives of Environmental Contamination and Toxicology, 52 (3), 306-316.

[38] Jones, K. C., & Northcott, G. L. (2000). Organic contaminants in sewage sludges: a survey of UK samples and a consideration of their significance. WQD0001. Final report to the department of the Environment, Transport and the Regions. Water Quality Division.

[39] Denmark, 1996. Statutory Order no. 823 of September 16, 1996 on application of sludge, wastewater and compost for agricultural use: Bekendtgorelse om anvendelse af affaldsprodukter til jordbrugsformal (BEK no. 823 af 16/09/96).

[40] CSHPF 1998. Section des eaux; Risques sanitaires liés aux boues d'épuration des eaux usées urbaines; Editions Lavoisier Tec et Doc.

[41] Austria, 1994. Federal law Gazette 1007/1994; rev. NLG 240/1998; rev. NLG 277/1998, Fertilizer Ordenance.

[42] Germany, 1992. Sewage Sludge Ordinance of 15/4/92, BGBl I of 28/4/92 S. 912 (Klärschlammverordnung), modified by the Ordinance (Verordnung) of 6/03/97, BGBl I S.446.

In: Sewage Sludge Management
Editors: A. A. Zorpas and V. J. Inglezakis

ISBN: 978-1-61324-393-0
© 2012 Nova Science Publishers, Inc.

Chapter 17

PHOSPHORUS RECOVERY AS MAP-STRUVITE FROM DIGESTED SEWAGE SLUDGE

*Kenan Güney**

Universität Stuttgart Germany, Germany

ABSTRACT

Phosphorus is an essential nutrient for all forms of life. Rapid annihilation of economically extractable phosphate rocks forces researchers to investigate sustainable ways to use phosphorus. Since all removed phosphorus accumulates in digested sewage sludge in a domestic wastewater treatment plant, recovering phosphorus from digested sewage sludge can be a promising source to recycle phosphorus.

The mineral fertilizers account for approximately 80 % of phosphates used worldwide. Therefore it would be beneficiary to recover phosphorus from digested sewage sludge/sludge ashes as fertilizer. In this sense, one of the most popular phosphorus recovery applications is converting phosphorus into MAP-Struvite.

The importance of phosphorus sustainability, phosphorus potential of digested sewage sludge, MAP-Struvite, fertilizer value of struvite, and its production from digested sewage sludge will shortly be explained in this chapter.

"…life can multiply until all the phosphorus is gone, and then there is an inexorable halt which nothing can prevent…. We may be able to substitute nuclear power for coal, and plastics for wood, and yeast for meat, and friendliness for isolation - but for phosphorus there is neither substitute nor replacement."

Isaac Asimov [1]

---

* E-mail: kenan.guney@iswa.uni-stuttgart.de.

1. INTRODUCTION

Phosphorus (P) is a multivalent non-metal element which is a major constituent of DNA, RNA, and ATP. Hence it is essential for all forms of life. Mankind uses phosphorus extensively in different forms, in: foods, fertilizers, detergents, explosives, pesticides, etc. An average person utilizes 1 to 3 g phosphorus per day in the form of phosphate while the minimum phosphorus requirement is 0.8 g P per day for an adult person [26].

Since the life standards of a human life increased, society has moved from a phosphorus recycling loop where animal manure and human wastes were spread on farming land to recycle nutrients, to a use-dispose system where phosphates are extracted from mines (non-renewable phosphate rocks), used and disposed either to a domestic wastewater treatment plant, to a landfill, or to water bodies.

According to 2009 USGS report, the annual global production of phosphate is more than 40 million tons of P_2O_5, derived from 156 million tons of rock concentrate [27]. As the phosphate rocks are diminishing in amount, the world demand for phosphorus consumption is predicted to increase as 1.5 % per year [25] annihilating phosphate rocks in about 100 to 250 years [7]. Therefore, the reliance on phosphate rocks should be stopped and sustainable ways to use phosphorus should be researched.

A sustainable way for phosphorus usage can be recovering it from digested sewage sludge (will be referred as digested sludge). Phosphorus removal in a domestic wastewater treatment is made either biologically or chemically. Chemical phosphorus removal is done by adding chemicals -especially metal salts- which can react and precipitate with phosphorus while phosphorus is utilized by biomass and being removed from wastewater in biological phosphorus removal. In both methods, removed phosphorus is accumulated in sewage sludge. Berg and Schaum [2] stated that during the wastewater treatment 40-95 % of the incoming phosphorus load is transferred into the sewage sludge depending on the technology applied. Fricke and Bidlingmaier [9] showed that sewage sludge has the highest phosphorus potential in organic wastes, which can be more than 3 gr/l. Jaffer et al. [13] stated that digested sludge has the highest phosphorus (as phosphate) concentration in a domestic wastewater treatment plant. Pinnekamp et al. [16] stated that the recovery potential of phosphorus from the aqueous phase of biological waste water treatment plants reaches a maximum of about 40 % while the recovery potential of phosphorus from sewage sludge and sludge ashes reaches up to 90 %.

Mineral fertilizers account for approximately 80 % of phosphates used worldwide [15]. Although digested sludge can sometimes be applied directly as a fertilizer in few countries, -its usage is limited due to heavy metal concentrations in the sludge- it would be beneficiary to recover phosphorus from digested sludge as a fertilizer. In this sense, one of the most popular phosphorus recovery applications is converting phosphorus and ammonia from digested sludge into MAP-Struvite (will be referred as struvite).

The main driving forces behind the development of P recovery as struvite can be stated as: the diminishment of global high quality phosphate reserves; the reduction in the return of re-released phosphorus from anaerobic sludge digestion to the headworks of wastewater treatment plants; improved sludge management in advanced wastewater treatment plant; control of struvite encrustation in biological nutrient removal plants; and, marketability of struvite as a sustainable product [8].

2. MAP-STRUVITE

Magnesium ammonium phosphate (MAP), also named as struvite, is the result of a chemical reaction between magnesium, ammonia, and phosphorus, in equal stoichiometric proportions. It is a slow releasing [5, 6] high quality fertilizer with a chemical formula of $MgNH_4PO_4 .6H_2O$.

$$Mg^{+2} + NH_4^+ + PO_4^{-3} + 6 H_2O \rightarrow MgNH_4PO_4 \cdot 6 (H_2O)$$

Pure struvite is white in color, but yellowish white, brownish white or light gray struvites can be seen due to the impurities in it. Struvite is readily soluble in acid, while it is meagerly soluble in neutral and alkaline conditions allowing it to be used at high application rates without damaging plant roots.

Struvite precipitation is controlled by pH, super saturation, temperature and the presence of impurities. Many researchers stated pH value of minimum struvite solubility mainly between pH 8 and pH 10 [3, 4, 12, 14].

Struvite is used for agricultural and horticultural fertilizer applications and it is commercially available in market. It is especially used for magnesium poor soils. If struvite were to be recovered from wastewater treatment plants worldwide, 0.63 million tons of phosphorus (as P_2O_5) could be harvested annually, reducing phosphate rock mining by 1.6 % [24].

3. FERTILIZER VALUE OF STRUVITE

Plants can only acquire the phosphorus they need by roots taking it up in simple ionic forms ($H_2PO_4^-$ and HPO_4^{-2}) from the soil solution [19]. Thus the value of any soil amendment intended to supply phosphorus depends not only on its ability to release phosphorus in these ionic forms to the soil solution but also the duration of the phosphorus release. Struvite has small leaching rate, therefore prolonged release of nutrients occurs. This behaviour of struvite allows farmers to apply it not only in high amounts but also just once or twice in a growing season.

Bridger et al. [20] concludes in his fertilizer industry manual "New Fertilizer Materials 1968" that struvite is a highly efficient source of magnesium, nitrogen and phosphorus for plants. He states that the release of nutrients from struvite to plants can be controlled by use of different granule sizes, and it can be applied in massive doses without burning the plants.

Numerous research on struvite application have been done (suggesting that struvite is an effective agricultural, horticultural, and fischery fertilizer) not only about applicability of struvite in agriculture but also about comparison of struvite with other fertilizers, in the last fifty years. In these research [10, 11, 20, 21, 22, 23] struvite was stated as good as or better fertilizer than some commonly used commercial fertilizers.

As a result, as explained before, recovery of phosphorus from digested sludge as struvite is a promising route not only for sustainable usage of phosphorus but also to observe an economical benefit.

4. PHOSPHORUS RECOVERY AS STRUVITE

Processes investigated to recover phosphorus from digested sludge and sludge ashes are Krepro, Kemicond, Seaborne, Aqua Reci and BioCon, Sephos, Rotary Furnace respectively. [2]

The main steps of phosphorus recovery from digested sludge or sludge ashes are:

I.   Dissolution of biologically or chemically bound phosphorus, which is stored mainly in solid part of the sludge or directly in the ash, and obtaining phosphorus rich water phase.
II. Removal of impurities, which exist in phosphorus rich water phase and react with phosphorus and hence disturb the phosphorus removal process, to ease the recovery of phosphorus.
III. Recovery of phosphorus.

Dissolution of phosphorus: Dissolution of phosphorus is required to isolate and obtain phosphorus in water phase. Weidelener et al. [28] stated that under acidic conditions mainly chemically bound phosphorus is dissolved while under alkaline conditions augmented dissolution of phosphorus bound in biomass takes place in the process of phosphorus dissolution from digested sludge.

It is a common knowledge that more than 90 % P can be dissolved at pH below 2.

Removal of impurities: Al, Ca, and Fe ion concentrations are usually high in digested sludge, especially where phosphorus removal is made by addition of metal salts, which include either Al or Ca or Fe in high concentrations. These metal ions also got dissolved during phosphorus dissolution process. They react with phosphorus and re-precipitate as metal phosphates during pH adjustment for struvite production [13, 28]. Therefore these interfering ions have to be either removed or inactivated through complexation.

$$Me^{+3} + PO_4^{-3} \rightarrow MePO_4$$

Weidelener et al. [28] used complexing agents to remove Al, Ca, and Fe ions out of the phosphorus rich solution. They stated that complexing Al ions was effective but it disturbed struvite precipitation at the end, while complexing Fe and Ca ions were effective without a serious problem. As a result, they observed that complexing Al, Ca, Fe ions was not an effective way to eliminate impurities for struvite production.

Another way to remove Al, Ca, and Fe ions can be usage of an ion exchange membrane. Prakash et al. [15, 17, 18] stated that Nafion® 117 cation exchange membrane can be used to remove metallic ions, especially Al ion, effectively. Güney et al. [12] used Nafion® 117 membrane to remove metal ions prior to struvite production and observed 98 %, 97 %, 80 % Al, Ca, Fe removal respectively. After elimination of interfering ions, they produced high purity struvite in which total concentration of Al, Ca, and Fe ions was less than 0.1 % in weight.

Recovery of phosphorus: Struvite production dependent strongly on solution pH. Optimized struvite production is observed at pH between 8 and 10 however pH of 8.5-9 is recommended since calcium phosphate precipitation strongly occurs at pH between 9 and 10.5. During the struvite production:

I.   P:N:Mg molar ratio should be arranged at least to 1:1.1:1.1 to not to have phosphorus as a rate limiting element for struvite production,
II. PH tends to decrease with struvite formation so addition of alkaline is required to keep pH constant,
III. Stirring speed should not be high since it can affect the size and the form of struvite crystals.

Produced struvite can be filtered by using paper filtration or microfiltration depending on the size of the struvite crystals.

CONCLUSION

Phosphorus recovery as struvite from digested sewage sludge is a promising method which helps the management of a domestic wastewater treatment plant, supplies economical benefit, and maintains sustainability of phosphorus and nitrogen.

REFERENCES

[1]   Asimov, I. *Asimov on chemistry;* Doubleday: June 1974, Isbn 0385041004
[2]   Berg, U.; Schaum, C. Recovery of phosphorus from sewage sludge and sludge ashes-Applications in Germany and Northern Eurpe, 1. Ulusal Aritma Camurlari Sempozyumu ACS 2005, 23-25 March 2005, Izmir Turkey
[3]   Booker, N.A.; Priestley, A.J.; Fraser, I.H. Struvite formation in wastewater treatment plants: opportunities for nutrient recovery. *Environ. Technol.* 1999, 20, 777-782.
[4]   Buchanan, J.R.; Mote, C.R.; Robinson, R.B. Thermodynamics of struvite formation. Transactions of the American Society of Agricultural Engineers 1994, 37, 617-621.
[5]   Çelen, I.; Türker, M. *Recovery* of ammonia as struvite from anaerobic digester effluents. *Environ. Technol.* 2001, 22 (11), 1263-1272
[6]   Demirer, S.U.; Demirer, G.N.; Chen, S. Ammonia removal from anaerobically digested dairy manure by struvite precipitation. *Process Biochem.* 2005, 40, 3667–3674
[7]   European Fertilizer Manufacturers Association, *Phosphorus essential element for food production*, 9-10, 2000
[8]   Forrest, L.; Fattah, K.P.; Mavinic, D.S.; Koch, F.A. Optimizing Struvite Production for Phosphate Recovery in WWTP. *J. Environ. Eng.* 2008, 395-402
[9]   Fricke, K.; Bidlingmaier, W. Posphatpotenziale qualitativ hochwertiger organischer Siedlungsabfälle; In: Rückgewinnung von Phosphor in der Landwirtschaft und aus Abwasser, RWTH Aachen, Umweltbundesamt Berlin, S. 9-1 bis 9-15, 2003

[10] Ghosh, G.K.; Mohan, K.S.; Sarkar, A.K. Characterization of soil-fertilizer P reaction products and their evaluation as sources of P for gram (*Cicer arietinum* L.). *Nutr. Cycling Agroecosyst.* 1996, 46 (1)

[11] Goto, I. Application of phosphorus recovered from sewage plants. *Environ. Conserv. Eng.* 1998, 27 (6), 418-422

[12] Güney, K.; Weidelener, A.; Krampe, J. Phosphorus recovery from digested sewage sludge as MAP by the help of metal ion separation. *Water Res.* 2008, 42, 4692-4698

[13] Jaffer, Y.; Clark, T.A.; Pearce, P.; Parsons, S.A. Potential phosphorus recovery by struvite formation. *Water Res.* 2002, 36, 1834-1842

[14] Momberg, G.A.; Oellermann, R.A. The removal of phosphate by hydroxyapatite and struvite crystallisation in South Africa. *Water Sci. Techno.* 1992, 26, 987-996.

[15] "Phosphorus availability in the 21st century: Management of a non-renewable resource", http://www.nhm.ac.uk/research-curation/projects/phosphate-recovery/p&k217/steen.htm, (accessed 28.05.2007)

[16] Pinnekamp, J.; Baumann, P.; Buer, T.; Cornel, P.; Donnert, D.; Goettlicher-Schmidle, U.; Heinzmann, B.; Jardin, N.; Longdong, J.; Mueller, J.; von Sothen, F.; Temmink, H.; Berg, U.; Christian- Bickelhaupt, R.; Schaum C.; Weidelener, A.; Wichmann-Seidel, A. ATVDVWK-AK1.1 KA 50, 2003 (6) 805.

[17] Prakash, P.; Hoskins, D.; SenGupta, A.K. Application of homogenous and heterogenous cation-exchange membranes in coagulant recovery from water treatment plant residuals using Donnan membrane process. *J. Membr. Sci.* 2004, 237, 131-144

[18] Prakash, P.; SenGupta, A.K. Modeling Al^{+3} and H^+ ion transport in donnan membrane process for coagulant recovery. *AIChE J.* 2005, 51 (1)

[19] Richards, I.R.; Johnston, A.E. The effectiveness of different precipitated phosphates as sources of phosphorus for plants. 2001, Report on work undertaken for CEEP, EFMA, Anglian Water UK, Thames Water UK and Berlin Wasser Betriebe, (accessed 29.05.2007) http://www.nhm.ac.uk/researchcuration/projects/phosphate-recovery/Richardsjohnston.pdf

[20] Scope Newsletter 43, http://www.ceep-phosphates.org (accessed 30.05.2007)

[21] Scope Newsletter 50, http://www.ceep-phosphates.org (accessed 30.05.2007)

[22] Scope Newsletter 53, http://www.ceep-phosphates.org (accessed 30.05.2007)

[23] Scope Newsletter 60, http://www.ceep-phosphates.org (accessed 30.05.2007)

[24] Shu, L.; Schneider, P.; Jegatheesan, V.; Johnson, J. An economic evaluation of phosphorus recovery as struvite from digester supernatant. *Bioresour. Technol.* 2006, 97, 2211-2216

[25] Steen, I. Phosphorus availability in the 21st century management of a non-renewable resource. *Phosphorus and Potassium.* 1998, 217, 25–31

[26] "The Phosphate File", Centre Europeen d'Etudes des Polyphosphates. CEFIC publications, 1997

[27] U.S. Geological Survey - Mineral commodity summaries 2009,

[28] Weidelener, A.; Brechtel, K.; Maier, W.; Krampe, J.; Rott, U. Recovery of phosphorus from sewage sludge as MAP, *IWA-/WISA-Conference on the Management of Residues Emanating from Water and Wastewater Treatment*, Johannesburg, South Africa, Aug. 9-12, 2005

In: Sewage Sludge Management
Editors: A. A. Zorpas and V. J. Inglezakis

ISBN: 978-1-61324-393-0
© 2012 Nova Science Publishers, Inc.

Chapter 18

LIFE CYCLE SUSTAINABILITY ASSESSMENT OF SLUDGE MANAGEMENT

Lehmann Annekatri[2], Traverso Marzia[3] and Finkbeiner Matthias

Department of Environmental Technology, Technische Universität Berlin,
Berlin, Germany

ABSTRACT

Sludge management encompasses various processes from generation of the sewage sludge over sludge treatment to final recycling and disposal. The issue of sustainability is nowadays addressed in many regional, national and international programmes and policies and should be implemented and considered in the design of management systems. A sustainable sewage sludge management can be understood in such a way that environmental, economic and social aspects are jointly considered when decisions are made regarding alternative management options respectively alternative technologies. Various different methods for sustainability assessment exist. A commonly used and standardised tool for analysing environmental impacts is Life Cycle Assessement (LCA). By considering the whole life cycle of a product or a service, problem shifting in other parts of the system or other regions in the world can be avoided. The Life Cycle Approach is also reflected in economic respectively social assessment methods namely Life Cycle Costing (LCC) and Social Life Cycle Assessment (SLCA). Addressing the three dimensions of the sustainability, the integration of LCA, LCC and SLCA is also called Life Cycle Sustainability Assessment (LCSA). This chapter proposes LCSA as tool for sustainability assessment of sewage sludge management.

---

[2] E-mail: Annekatrin.Lehmann@campus.tu-berlin.de.
[3] E-mail: Marzia.traverso@tu-berlin.de.

1. INTRODUCTION

Sewage sludge is generated all over the world, whenever wastewater is collected and treated in waste water treatment plants. A sludge management encompasses the generation, treatment, application and disposal of the sludge and should ideally be organised in due consideration of the sustainability concept.

Sustainability is an overused term in society. The most common definition is probably the one from the Brundtland-report in which sustainable development is defined as a "…development that meets the needs of the present without compromising the ability of future generations to meet their own needs" (Brundtland 1987). The outcome document of the United Nations World Summit, 2005 additionally defines three components of sustainable development – "economic development, social development, and environmental protection (UN 2005).

Nowadays, sustainable development respectively a contribution on the way to more sustainability, is an objective stated on national as well as on international level worldwide and the principles e.g. expressed in the three pillars of sustainability should be considered in policies, management, product development, consumption; actually in all aspects of life. The question is how we know if we achieved our objectives for sustainability or if they are indeed feasible. Sustainability objectives or criteria, as quoted before, exist on regional, national or international level and they can be expressed by specific indicators. An example of sustainable indicators are those related to the millennium development goals (UNDP 2008), serving as "measuring instruments" by describing relevant sustainability aspects. Generally a millennium development goal (e.g. ensure environmental sustainability) is described by specific targets (e.g. reduce biodiversity loss; reverse loss of environmental resources) and measured with indicators (e.g. CO_2-emissions, total or proportion of total water resources used) (UNDP 2008). Further details on sustainability assessment methods and tools are explained in chapter 0.

Within the sustainability debate, the life cycle thinking gains – especially since the introduction of the action plan for "Sustainable Consumption and Production and Sustainable Industrial Policy (SCP/SIP) Action Plan" of European Commission in 2008 - more and more attention (Klöpffer and Heinrich 2009). The life cycle thinking approach intends to consider environmental, economic as well as social impacts arising over the whole life cycle, from the production phase (e.g. resource extraction), via use-phase to disposal phase. By this problem shifting in the product system to other phases of the life cycle or other regions in the world can be avoided (Finnveden, Hauschild et al. 2009). A worldwide accepted tool to assess environmental impacts related to products (processes and services) is the Life Cycle Assessment (LCA). A Sustainability assessment of products and technologies additionally should include the social and economic dimensions. Against this background and inspired by the method of (environmental) LCA, the "Guidelines for Social Life Cycle Assessment" (SLCA) were published recently (UNEP 2009). The method of Life Cycle Costing (LCC) is generally considered as method analysing economic aspects related to products. A complementary approach of these three methods, called Life Cycle Sustainability Assessment (LCSA), and its applicability with regard to sludge management systems is the main target of this chapter.

Management of resources and waste is characterized by the complexity of the system. Different field of activities and functions are included and have to be considered as have the interests of the various stakeholders. In sludge management systems, questions addressed are for example: What to do with the sludge after separating it from the waste water? Will it be applied e.g in agriculture as fertilizer (agricultural recycling)? Will it be disposed on landfills mixed with household waste for biogas generation or will it be – as final treatment – thermally treated and the mineral residues recycled or disposed on landfills? Which technologies respectively systems are accepted and preferred by the local community or institutions as usually main decision makers? And which system or which technologies in the system are recommended with regard to sustainability issues?

Since the application respectively disposal of sludge depends strongly on treatment-processes in the waste water treatment plant (WWTP), generation respectively treatment of sludge is regarded here as part of sludge management. Main objectives of development of new waste water and sludge treatment technologies are producing safer sludge, smaller amounts and a better quality in order to meet the general demands of sludge management namely safe disposal of sludge (e.g. to protect public health and environment) and the recovery of the scarce resource Phosphorus (Pinnekamp, Montag et al. 2007).

A sustainability assessment of sludge management systems encompassing treatment, recovery and reuse of sludge can contribute to a sustainable development by:

1)analysing environmental or social impacts as well as economic aspects connected with technologies and by this revealing improvement potential and
2)serving as supporting tool in decision processes when alternative technology are available and the most sustainable one should be implemented within a (sustainable) sludge management system.

With regard to the three dimensions of sustainability and life cycle thinking-approach, this book chapter intends to provide an overview about the methods of life cycle based sustainability assessment with special regard to sludge management. It starts with an introduction in the issue of sustainability (paragraph 0) before focussing on life cycle based assessment methods Life Cycle Assessment (LCA), Life Cycle Costing (LCC) and Social Life Cycle Assessment (0), already referring to sludge management. An insight into life cycle based studies is sludge management sector is given in paragraph 0. Conclusions and recommendations regarding sustainability assessment for certain sludge management system are drawn in (paragraph 0).

2. SUSTAINABILITY

The origin of the word sustainability are the Latin words "sustinere" (tenere: to hold; sus: up). As synonyms for sustainable, dictionaries offer e.g. bearable, endurable, prolong, keep up, support, maintain, preserve, conserve etc. The understanding of what is sustainability, which goals are pursued and how these should be achieved thus can be interpreted in various ways (e.g. (Kates, Parris et al. 2005; IISD 2009). Especially among stakeholders, the perception of what is sustainable can vary significantly, thus, talking about sustainability and sustainability assessment requires a common understanding. An overview about understandings of the term sustainability and how it is used in this chapter is presented in the following.

2.1. Sustainability – Definitions and Assessent Methods

As mentioned in the introduction, the most common definition of sustainability is probably the one from the report of the Worlds Commission of Environment and Development "Our common future", known as the "Brundtland-report": "Sustainable development is development that meets the needs of the present without compromising the ability of future generations to meet their own needs" (Brundtland 1987). The United Nations (UN) declared in Rio de Janeiro, 1992 (UN 1992) and in Johannesburg, 2002 (UN 2002) sustainability as general principle of the 21st century and as "...collective responsibility to advance and strengthen the interdependent and mutually reinforcing pillars of sustainable development - economic development, social development and environmental protection - at the local, national, regional and global level" (UN 2002). The mentioned three dimensions of sustainability have been presented in various illustrations, for example as overlapping circles, to demonstrate that the pillars are not mutually exclusive but can be mutually reinforcing, as stated above (Figure **8**). Another illustration shows the three dimensions as concentric circles, showing economy as a subsystem of human society, which is itself a subsystem of the environment (Figure 9).

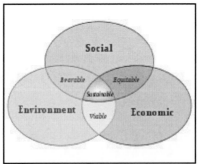

Figure 8. The three pillars of sustainable development (Adams 2006).

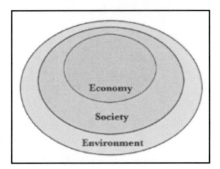

Figure 9. Sustainability - a presentation showing how both economy and society are constrained by environmental limits (Ott 2003).

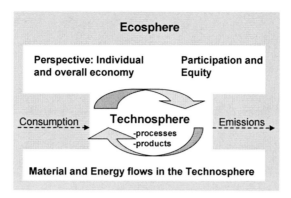

Figure 10. Conservation of nature as basis for sustainable development (Kloepffer 2008).

Figure 10 also embeds the economic dimension and social dimension – here combined in the technosphere - in the environment respectively the ecosphere. This 2-dimension model (technosphere and ecosphere) is similar to the 2-spheres - model of the Austrian Lebensministerium, which distinguishes between environment and human/society and allocates economy to the society (Lebensministerium 2006), emphasizing the integrated

Another approach addresses institutional conditions as fourth dimension and - according to its integrated perspective - projects three general sustainability objectives (protection of human existence, of social productivity potential and of development and action potential) on these dimensions (Kopfmüller, Brandl et al. 2001). Chapter 8 of the Agenda 21(an action plan addressing relevant policies and actions and passed at the Rio-Conference in 1992 Conference) addresses " Integrating Environment & Development in Decision-Making" and again states that economic, social and environmental aspects should be jointly considered at policy, planning and management level. Assessing the sustainability of sludge management and included technologies thus means analysing these aspects in order to 1) identify potential environmental impacts, costs and social impacts, 2) to reveal optimization potential, 3) to serve as decision support tools, in order to evaluate alternative technologies, in order to "find" the one with ideally least environmental damages, least costs and most contribution to society respectively least impairments and 4) to "monitor" if measures taken lead to an improvement of the situation.

For assessing sustainability different indicator-sets and indicator-systems have been proposed for companies and states to measure sustainable development. Generally, the current status is analysed and compared with regionally, national or international target values. Common indicators are for example the human development Index HDI, the Environmental Sustainability Index (ESI), the Environmental Performance Index EPI, the Genuine progress Index (GPI), indicators of the Global reporting Initiative (GRI 2002) and CSD-indicators of the Commission on Sustainable Development (CSD 2001; CSD 2008). Even if they are considered as indicators /indexes for assessing sustainability, most of them do not include the three dimensions of sustainability in their evaluation. According to Finkbeiner (2009) applicable, quantitative and product-related indicators and methods are still missing. As further challenges in sustainability assessment he mentions missing evaluation approaches for the social dimension, weighting of the single sustainability dimensions (Kopfmüller, Brandl et al. 2001; Ott 2001) and the presentation of the results.

In most approaches analysing sustainable aspects is somehow translated into assessing environmental impacts and it leads to a partial evaluation. With regard to sludge management sustainability refers amongst others to a safe disposal of sewage sludge (e.g. destruction of organic contaminants), recovery of nutrients (fertilizer production), energy production (e.g. incineration of sewage sludge) or protection of water resources (eutrophication). Typical criteria and indicators amongst others are (Lundin and Morrison 2002):

- production → indicator: waste water production per day
- recycling of nutrients → indicator: amount of Phosphorus and Nitrogen recycled
- quality of sludge → indicator: Cadmium content in sludge
- energy recovery → indicator: energy recovered, heating and power.

The growing interest to analyse consequences or impacts of products / technologies and by this provide a basis for decision making under consideration of certain sustainability aspects is also reflected by the increasing amount of studies, that has been carried out in recent times to determine environmental impacts of different recycling or disposal options for sewage sludge. Studies, focussing on economic aspects related to different recycling and disposal options have also been conducted, although to a lesser extent. Social aspects mentioned (and analysed) related to sludge management respectively technologies for sludge treatment and disposal are for example the contribution to regional disposal of waste, the compliance with valid safety regulations and hygienic aspects (e.g. with regard to odour) or preservation or creation of jobs.

3. LIFE CYCLE SUSTAINABILITY ASSESSMENT

The focus of this chapter is life cycle sustainability assessment a complementary approach of LCA, LCC and SLCA: This enables the analysis of environmental, economic and social impacts connected with technologies used in the sludge management and can also serve as decision support tool on the way to more sustainable sludge management strategies.

Sustainability assessment with regard to the Brundtlandt definition refers to three dimensions of sustainability and by this to different capitals or areas of protections (AoP), which should be maintained and protected.

Table 1. The three dimensions of sustainability and areas of protection (AoP) (according to (deHaes, Jolliet et al. 1999), (Jolliet, Müller-Wenk et al. 2004), (Weidema 2006), (Schmidt, Meurer et al. 2004), (Nazarkina and Le Bocq 2006), (Jørgensen, LeBocq et al. 2008)

Dimensions of sustainability	Description
Environmental sustainability	• Protection of the AoP nature and environment (biodiversity, climate protection,) and protection of the intrinsic value of nature (natural resources, biodiversity), of the economic value (e.g. protection of vital ecosystem functions) and of the social value (e.g. recreation).

Dimensions of sustainability	Description
Economic sustainability	▪ Protection of economic resource, economy should be organized in a way which ensures the ecosystem productivity (biodiversity, natural resources), human productivity (economic value of human health). With regard to technologies, economic sustainability can be understood, that they should be : ▪ Financially independent (amortisation of investment costs over lifetime) ▪ Noticable economic benefit
Social sustainability	▪ Protection and improvement of social, human and productive capital [1] ▪ Protection of the AoP to secure the intrinsic value (human health, well-being), the cultural heritage and the economic value With regard to technologies, economic sustainability can be understood in a way, that: ▪ the majority of stakeholders accepts the technologies ▪ institutional, political and organisational structures for implementation and usage of technologies exist or will be developed ▪ employment rate is increased ▪ no conflicts arise du to the implementation of the technologies

[1] Social capital: resources bound e.g. in networks, institutions by norms and relations.
Human capital: resources bound e.g. in productive abilities and talents of the people, in education and health.
Produktive capital: resources bound e.g. in machines, factories, buildings, infrastructure.

The methods Life Cycle Assessment (LCA), Life Cycle Costing (LCC) and Social Life Cycle Assessment (chapters 0-0) address these dimensions. This leads to a complementary approach of combining these methods to a Life Cycle Sustainability Assessment (LCSA) according to this equation (Kloepffer 2003, Kloepffer 2008, Finkbeiner 2008; Klöpffer and Renner 2007; Klöpffer and Grahl 2009):

$$(1) \quad LCSA = LCA + LCC + SLCA$$

Major assumption for applying this concept is similar system boundaries for the three balances. Ideally one inventory analysis could be used as basis for all three dimensions, but might not be feasible, since SLCA normally requires higher consideration of regional aspects than LCA or LCC (Klöpffer and Grahl 2009). The combination of the three methods, addressing the three sustainability dimensions is analysed and discussed in, e.g, the EU 6th Framework Co-ordination Action for innovation in Life-Cycle Analysis for Sustainability (CALCAS) (CALCAS 2009.) and the EU 7th Framework project Development and application of a standardized methodology for the PROspective SUstaInability assessment of TEchnologies (PROSUITE; (Patel 2009) (see (Finnveden, Hauschild et al. 2009). In the CALCAS project the term LCSA stands for Life Cycle Sustainable Analysis. The analysis aims at broadening the scope of indicators – including environmental, economic and social indicators, as formulated in equation (1). Additionally the object of analysis should be broadened from the product level to sector or even economic wide level. According to (Jørgensen, Hermann et al. 2010) sustainability can be understood as analysis to assess changes in these capitals. LCA considers changes of natural capital on global level; SLCA considers (at least partly) changes in the social, human and produces/physical capital. According to the authors LCC reflects more individual than (with regard to the AoP relevant global costs and thus delivers more information on sustainability of a company than on a sustainable development as understood in the Brundtland-definition.

Since economic aspects are essential to evaluate management systems (e.g. within decision support processes) and sustainable technologies have to be profitable and cost-efficient (Klöpffer and Grahl 2009), the method of LCC is included in the proposed sustainability assessment of sludge management system.

Generally LCSA contributes to answer the following questions:

- Which potential environmental impacts and costs occur within a sludge management system and related to specific technologies?
- Which social (negative or positive) effects go along with the technologies within the sludge management system?
- Which processes along the life cycle of sludge (generation, treatment, application/ disposal) have the highest impact?
- Which technologies respectively management systems are recommendable with regard to ecological, environmental and social aspects, respectively sustainability?

The life cycle approach of a (product related) sustainability assessment is crucial, since only by this "...problem shifting and (...) trade-offs can be identified and avoided", which is a precondition of a worldwide and intergenerative fairness, as required in the Brundtland-Commission (Klöpffer and Grahl 2009).

In the following the methodologies of LCA, LCC and SLCA are presented and related to sludge management. Furthermore, existing case studies (for LCA and LCC) are presented in the end of the respective paragraph. These sections do not claim to give a comprehensive review of studies related to life cycle thinking and sustainability assessment in sludge management but rather intend to provide an insight in current activities.

4. LIFE CYCLE ASSESSMENT (LCA)

Life Cycle Assessment is currently the only internationally standardised (ISO 14040 2006) method to analyse environmental aspects of products, processes and services. Main characteristics are the focus of the life cycle and the "relative approach, which is structured around a functional unit" ((ISO14044 2006). Life Cycle is understood as "consecutive and interlinked stages of a product system, from raw material acquisition or generation from natural resources to final disposal" („from cradle to grave"). The functional unit defines the subject of study and "all subsequent analyses are then relative to that functional unit, as all inputs and outputs in the LCI and consequently the LCIA profile are related to the functional unit." (ISO14044 2006). The method was defined by the Society of Environmental Toxicology and Chemistry (SETAC) (SETAC, 2003 (SETAC 2003), the US Environmental protection agency EPA (Vigon, Consoli et al. 1993) and the International Organization for Standardization (ISO14040 2006; ISO14044 2006). According to (ISO14044 2006). LCA can assist in:

- identifying opportunities to improve the environmental performance of products at various points in their life cycle,
- informing decision-makers in industry, government or non-government organizations (e.g. for the purpose of strategic planning, priority setting, product or process design or redesign),
- the selection of relevant indicators of environmental performance, including measurement techniques, and
- marketing (e.g. implementing an ecolabelling scheme, making an environmental claim, or producing an environmental product declaration).

Against this background and as already mentioned in paragraph 0, LCA is increasingly used decision support tool in wastewater and sludge management.

It is important to emphasize that an LCA addresses potential environmental impacts and "...does not predict absolute or precise environmental impacts due to the relative expression of potential environmental impacts to a reference unit, the integration of environmental data over space and time, the inherent uncertainty in modelling of environmental impacts, and the fact that some possible environmental impacts are clearly future impacts."(ISO14044 2006)

An LCA consist of four stages, presented in Figure 11. The first stage - *goal and scope* - defines amongst others the system boundaries for the product system and the functional unit (fu), which "...defines the quantification of the identified functions..." (e.g. function of drying x pair of hands),"...provides a reference to which the inputs and outputs are related." (ISO 14044:2006) and by this forms the basis for comparison of LCA-results (e.g. comparison of paper towel and an air-dryer system for hand drying). The reference flow – the "measure of the outputs from processes in a given product system required to fulfil the function expressed by the functional unit" can be determined for the systems to be compared (e.g. the average mass of paper or the average volume of hot air required for one pair of hand-dry, respectively) for the systems can be determined (ISO 14044:2006).

The functional unit in sludge management can be for example a safe disposal respectively recycling of a certain amount of sludge. As reference flow for example 1ton of sludge could be chosen. Possible system boundaries respectively an exemplarily product system are shown in Figure 4.

The second stage – the (life cycle) *inventory analysis* (LCI) – balances the input- and output flows of the product system related to the functional unit and includes data collection and calculation. Data can be gained for example directly from the plant (annual reports, interviews with plant operator, workers), from studies or from literature. An LCI is an iterative process, which means that new data and or limitations that might occur during the study may require a change in the initial goal and scope of the study. Results of the LCI are quantitative expressions of emissions, such as x kg CO_2/fu and resource consumption.

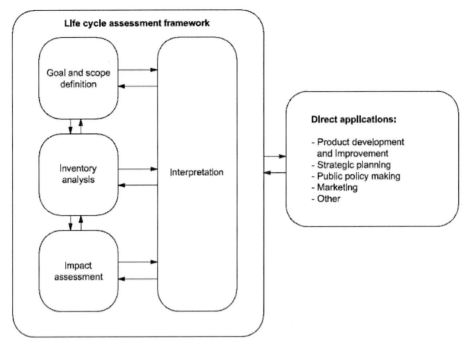

Figure 11. Stages of an LCA (ISO14040 2006).

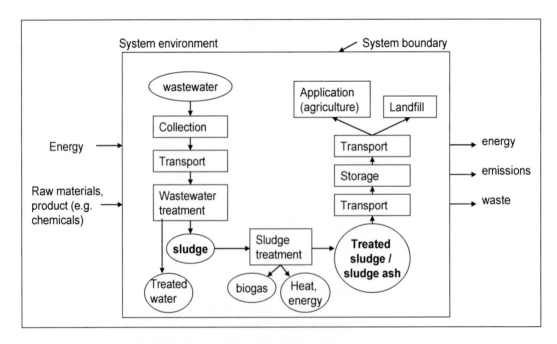

Figure 12. Example of a product system for LCA of sewage sludge.

The third stage, the life cycle *impact assessment*, aims at "evaluating the significance potential environmental impacts using the LCI results" (ISO14044 2006). The impact assessment consists of three steps:

1) Selection; impact categories to be considered in the study are chosen (e.g. climate change) within the goal and scope phase;

2) classification; the results from the LCI (emissions) are classified to the defined impact categories;

3) characterisation; the contribution of a certain emission to the impact category is quantified and expressed with the category indicator, which is calculated using (scientific) characterisation models. Step 2) and 3) are matters of research not of applied LCA. The elements of life cycle impact analysis are presented in Figure 13.

Different impact assessment methods exist for analysing environmental impacts, these methods are normally already integrated in common LCA-Softwares. Main differences lie in choice of impact indicators regarding whether they are related to midpoint or endpoint level and with regard to weightings used. Known impact assessment methods are: CML 2 baseline 2000 (CML 1992), Eco-Indicator 99, Eco-Indicator 95 EDIP 2003, ReCiPe endpoint, ReCiPe midpoint or IMPACT 2002+. The default list of impact categories (and respective unit) of the Centre of Environmental Science (CML) of Leiden University's methodology (CML 2001), which is often used in LCA studies are listed in Table 2.

Most of the existing impact assessment methods show according to (Klöpffer and Grahl 2009) an unbalance with regard to chemical, physical and biological emissions. Biological emissions such as microbiological contaminations (e.g. pathogens) or physical emissions such as heat (in water) and possible resulting impacts on human health or the ecosystem are not considered so far, which is mainly due to lack of appropriate characterisation models and which shows potential for further research.

It should be noted here again, that the choice of impact categories (selection) depends on goal and scope of the study. As mentioned in paragraph 0, based on the results from LCA studies specific indicators for describing environmental impacts within a certain product system can be deflected, such as amount of Phosphorus and Nitrogen recycled (Lundin and Morrison 2002).

Since LCA is only one of various methods to assess environmental impacts, it does not claim to cover all impacts related to the considered system. Generally, environmental impacts occur at a certain time at a certain place as a result of emissions deriving from various processes and an existing background load (contamination). An LCA typically does not

Table 2. Impact categories (CML 2001)

Impact category	Unit
depletion of abiotic resource	kg (antimony eq.)
climate change	kg (CO_2 eq.)
stratospheric ozone depletion	kg (CFC-11 eq.)
human toxicity	kg (1,4-DCB eq.)
ecotoxicity (freshwater aquatic, marine, terrestrial)	kg (1,4-DCB eq.)
photo-oxidant formation	kg (C_2H_4eq.)
acidification	kg (SO_2eq.)
eutrophication	kg (PO_4^{3-}eq.)
effects of land use	$m^3 \cdot yr$

provide information regarding time lapse or resulting concentration in the "receiving" environment or background loads and emissions of other processes outside the product system, localised impacts are thus typically not considered. Moreover LCA describes normally stationary situations/states, processes are considered as linear. Thus, the potential environmental impacts calculated with LCA represent the sum of impacts of past, present and future emissions in different eco systems in different parts of the world, which means that the LCA results show potential contributions to actual impacts or possible risks with regard to a defined reference situation (Finnveden, Hauschild et al. 2009).

For interpretation of the LCA results - in the fourth stage – the results of the LCI and LCIA are brought together with goal and scope of the study to gain conclusions and recommendations (from ecological point of view). Transparency, that means clear description of assumptions- is crucial in Life Cycle impact assessment since choice, modelling and evaluation of impact categories may introduce subjectivity in to the LCIA phase.

Also sensitivity analysis, uncertainty analysis and consistency analysis are mandatory in LCA-studies, they evaluate the influence of assumptions that have been made, determine the probable spread of results (considering the data quality) and analyse the compatibility of the data, assumptions and models from the overall view

A general overview about current LCA studies in the water sector is provided by (Lassaux, Renzoni et al. 2007). Most of the studies focus on technologies of waste water treatment and disposal e.g. (Tillman, Svingby et al. 1998), (Hospido, Moreirai et al. 2004), (Hospido, Moreira et al. 2008), (El-Sayed, Mahgouba et al. 2010). Studies focussing explicitly on

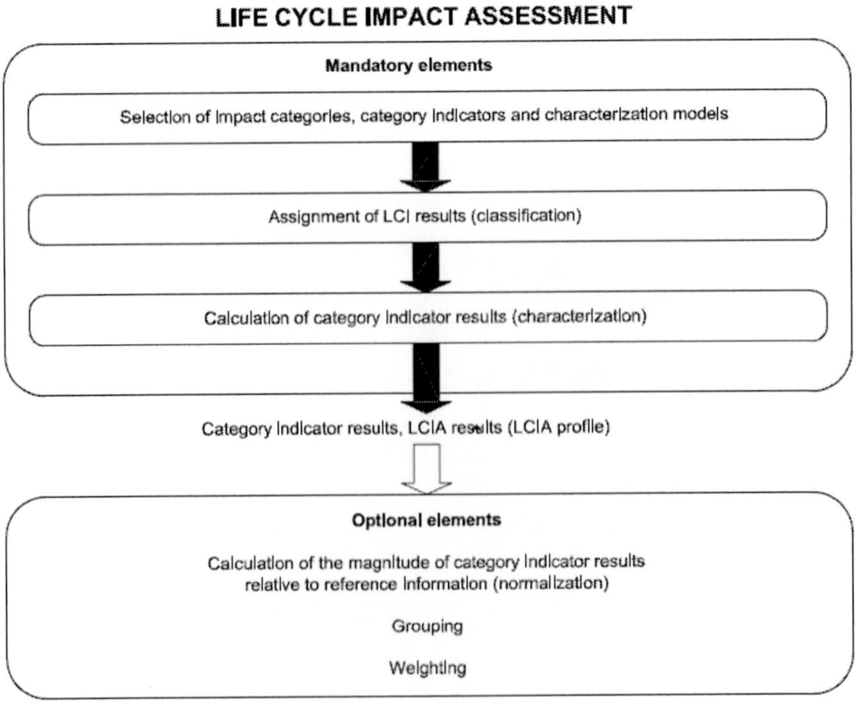

LIFE CYCLE IMPACT ASSESSMENT

Mandatory elements

Selection of impact categories, category indicators and characterization models

Assignment of LCI results (classification)

Calculation of category indicator results (characterization)

Category indicator results, LCIA results (LCIA profile)

Optional elements

Calculation of the magnitude of category indicator results relative to reference information (normalization)

Grouping

Weighting

Figure 13. Elements of the LCIA phase.

environmental impacts with regard to sludge disposal are presented below. The studies generally refer to actual scenarios in existing waste water treatment plants and compare different technologies such as biological and thermal processes, anaerobic digestion versus pyrolysis and incineration. Johanssona, Perzona et al (2008) analysed different handling options for sewage sludge (restoration of mining areas, composting with other biomaterials for use on golf courses, hygienisation through storage for agricultural use, and supercritical water oxidation with phosphorus recovery), from an environmental system's perspective using LCA. Hospido, Moreirai et al. (2004), (2005) and (2008) examined different municipal waste water treatment plants (WWTP) in Galicia, Spain for big centres of population, focussing on removal of organic matter in urban waste water including the option of anaerobic digestion of sludge, different types of units for sludge dewatering and deodoristaion towers for odour treatment.

Houillon and Jolleit (2005) compared six different scenario of sludge treatment with focus on energy and emissions regarding global warming potential. Suh and Rousseaux (2002) examined environmental impacts of five alternative treatment scenarios of sewage sludge in the French context, including processes of incineration, agricultural land application, or landfill, stabilization processes (lime stabilization, composting, or anaerobic digestion) and transports of sludge. Tillman, Svingby et al. (1998) analyse environmental consequences of changing an existing waste water system in Sweden including e.g. pre-treatment, anaerobic digestion or drying of solid fraction in sand filter beds. Lederer and Rechberger (2010) analyse conventional and alternative methods to treat and dispose of sewage sludges with regard to environmental impact, resource recovery (phosphorus) and material dissipation, based on material flow analysis, LCA impact categories and statistical entropy analysis.

5. LIFE CYCLE COSTING (LCC)

Economic aspects are one of the main drivers in decision support processes. A restructuring of sewage sludge management system for example by implementing new technologies for sludge treatment in order to improve nutrient recovery (as one target regarding sustainability of sludge management), thus necessarily has to consider the related economic aspects. Accordant to the life cycle thinking approach in LCA, the method of Life Cycle aims at determining the costs over the life cycle of the product, which is here the sludge.

Various approaches exist for analysing the life cycle costs due to different perspectives and interests which can be taken respectively followed (Finkbeiner 2009). Generally, in cost analysis a differentiation can be made between production costs (from company perspective) and life cycle costs (from customer perspective). Life cycle costs or total cost or total life costs describe costs, which occur over a defined life cycle. (Rebitzer and Hunkeler 2003) additionally differentiate between physical life cycle, as understood in LCA ((ISO14044 2006) and economic life cycle (market launch, growth phase and phase of saturation). A standardised method for LCC or does not exist yet. According to (Rebitzer and Hunkeler 2003) LCC can be defined as cost management method aiming at the assessment of costs related to the existence of a certain product. The standardisation of LCC is object of a SETAC

Europe Working group on Life Cycle Costing (since 2003), which proposes a method according to the LCA-method defined in (ISO14044 2006) and which includes the whole life cycle of a product (production, use and disposal) (Hunkeler, Rebitzer et al. 2008). The inventory of the LCA offers an "excellent basis for the identification and allocation of occurring costs" (Rebitzer and Hunkeler 2003). The LCC method examines cash flows and – as LCA – defines a functional unit. The results of an LCC are (positive/negative) costs in a certain currency which makes a separate impact assessment needless. According to (Klöpffer and Grahl 2009), LCC is a meaningful additional method within sustainability assessment, since, (more) sustainable products should be also cost-efficient. Relevant questions for conducting an LCC are shown in Table and related to sludge management system.

Combined analyses of LCA and LCC already exist for example as eco-efficiency-analysis (see (Saling, Kicherer et al. 2002), (Landsiedel and Saling 2002; Kicherer, Schaltegger et al. 2007).

Economic aspects of sludge management systems are addressed e.g. by Hong, Hong et al. (2009), Kalderis, Aivalioti et al. (2010) and Lim, Park et al. (2008). Hong, Hong et al. (2009) conducted LCA and LCC study analysing six alternative scenarios most often used in Japan: dewatering, composting, drying, incineration, incinerated ash melting and dewatered sludge melting, each with or without digestion as well as three end-of-life treatments landfilling, agricultural application and building material application. Kalderis, Aivalioti et al. (2010) analysed four essentially different and widely established methods for the treatment of sewage sludge with regard to applicability in the economical, geographical and environmental settings of the island of Crete in Greece. Lim, Park et al. (2008) studied the environmental and economic feasibility of a total wastewater treatment network system using LCA and LCC.

6. SOCIAL LIFE CYCLE ASSESSMENT

Social aspects or impacts are defined in the "Guidelines of Social LCA" (in the following named as UNEP-Guidelines) as „…consequences of positive or negative pressures on social endpoints (i.e. well-being of stakeholders)" respectively consequences of social relations

Table 3 . Relevant questions for conducting a LCC ((Huppes 2004; Rebitzer and Seuring 2004)

Questions	LCC within sludge management system
What is he subject of analysis?	Economic aspects (costs and benefits), which occur in sludge management system, e.g. with regard to treatment or disposal of a certain amount of sludge .
Which costs (cost type)	material- energy and transport costs, labor costs (construction, operatation and maintainance), research costs
Whose costs (cost benefector)	Institutions Population (fees) Subsidies
How should costs be measured (Estimation, discounting)	Estimation of costs based on current market prices for e.g. materials, energy, wages

(interactions) weaved in the context of an activity (production, consumption or disposal) and/or engendered by it and/or by preventive or reinforcing actions taken by stakeholders (ex. enforcing safety measures in a facility)."(UNEP 2009). As causes for social impacts the UNEP-Guidelines name:

- behaviors: social impacts are those caused by a specific behaviour or decision (e.g. forbidding employees to form unions, allowing illegal child labor).
- socio-economic processes: social impacts are the downstream effect of socio-economic decisions.
- capitals: (human, social, cultural): social impacts relate to the original context (attributes possessed by an individual, a group, a society e.g., education level).

Various authors address the question if and how social aspects, as third sustainability dimension, could be included in life cycle analysis (Klöpffer and Grahl 2009; Klöpffer and Heinrich 2009). A standardisation or Code of Practice for analysing the social dimension in life cycle thinking does not exist yet. The status quo is presented in the "Guidelines for Social Life Cycle Assessment of products" from SETAC/UNEO Life Cycle initiative (UNEP 2009), which also offer – in so called methodological sheets – methodological advises with regard to indicator selection, measuring and data collection. An actual review of SLCA-literature is given by (Jørgensen, LeBocq et al. 2008).

With regard to the variety of social impacts, one of the main questions for conducting Social LCA is which impact categories should be chosen and then finding indicators that are able to describe these impacts. One of the main challenges in SLCA is the description of relation between social indicators and the specific products/processes to be analysed.

Regarding the impacts it has to be differentiated whether, the impacts derive from the processes (product) itself (Schmidt, Meurer et al. 2004) or if they result from behaviour of the companies in which these processes occur (Dreyer, Hauschild et al. 2006), (Spillemaeckers, Vanhoutte et al. 2004). Studies to distinguish these types of social impacts does not exist so far, which is why a differentiation is done based on common sense and arguments ((Jørgensen, LeBocq et al. 2008). However, (Schmidt, Meurer et al. 2004) state, that an analysis according to the LCA methodology requires a focus on processes/product. Assuming that appropriate social indicators could be identified and measured (e.g. number of working hours) these indicator results, the task of the impact assessment is to relate them to the product system.

Main challenges of SLCA with regard to practicability lay in finding appropriate indicators to describe social aspects and in (often unsufficient) data availability and data quality.

An overview about impact categories included in current SLCA studies is presented in Table , also indicating which stakeholders might be affected.

SLCA aims at identifying social and socio-economic effects on stakeholders (e.g. workers or entire communities in which production takes place) deriving from certain choices been made.

Table 4. Overview of impact categories included in current SLCA
(excerpt from (Jørgensen, LeBocq et al. 2008) – addressed stakeholders

Impact categories	Stakeholder
Human rights	
Non-discrimination	worker
Freedom of association	society
Child labour	
Work conditions	
Wages	worker
Psychological and organi-satorical working conditions (e.g. maximum working time)	
Physical working conditions (e.g. accidents)	
Society	
Corruption (e.g. incidents, illegal price-fixing)	society, institutions
Development support towards society (e.g. job creation, support of local suppliers)	worker society, institutions
Acceptance in local community	society
	user
Product responsibility	
Integration of customers health an0d safety concerns in product	user

The UNEP-Guidelines choose the example of a cotton shirt retailer who intends to increase is social responsibility and consider also sections of the life cycle (production, use phase, disposal), which are typically not part of e.g. the label "Fair Trade", including shipping or disposal. Answers of the study might be for example how much of this labor is child labor free or if the practices comply standards for human rights and may by this influence the consumers choice.

With regard of the life cycle of sludge, the main processes - generation, treatment and disposal, as presented very simplified in Figure 14 usually take place in one region which allows to assume that standards like working conditions (e.g. human rights) are similar and most of the criteria/indicators listed in Table 4 might not be appropriate to assess the "socially friendly" products within a decision process. However, other aspects like number of created of jobs or acceptance of the local community can differ significantly and are worth to be assessed. Moreover SLCA can also be used to determine social effects resulting from pre-processes like chemical production.

Of major interest with regard to social impacts of sludge management systems is probably the acceptance of society respectively different stakeholders, which means any persons, groups or organisations that are positively or negatively affected by or may act positively or negatively on an activity. Stakeholders (within a sludge management system) include politicians, institutions, society, users, NGOs, scientists, companies, plant operators and workers. For a successful and sustainable (in terms of e.g. maintaining) implementation of technologies it is crucial to consider interests, priorities and values of the respective stakeholders. For example, it is very important to know, if farmers of local community are interested and willing in using sludge as fertilizer. Normally, a stakeholder analysis should be carried out before a project starts, accompanying it to the end and beyond to guarantee that stakeholders interest are considered and by this ensuring a "functioning" management. The analyses should be carried out together with economic and ecological analyses since these aspects are linked to each other and findings from one study (e.g. costs, potential environmental damages) influence the stakeholder's perception.

Currently, only a few SLCA studies have been conducted with social indicators mainly focussed on working conditions. With regard to sludge management no respective study was found. It can be argued that other methods such as social impact analysis or stakeholder analysis might be more appropriate tools to determine social aspects within decision processes regarding sludge management. However, since all methods address different but relevant topics, SLCA can deliver important additional information; an increased application can be expected in future.

7. LCSA IN SLUDGE MANAGEMENT

This paragraph intends to give an idea how and for which purposes a LCSA could look like for a sludge management system. Sludge management includes various processes, from sludge generation to sludge disposal, and involves many stakeholders and can thus be seen as complex system. As described in paragraph 0, complexity also quite adequately describes the subject of sustainability and sustainability assessment.

Details on sludge management and technologies for generation, treatment and disposal – which can be seen as life cycle of the sludge (and which are presented very simplified in Figure 8) - are discussed in the respective chapters in the book and are not subject of this chapter.

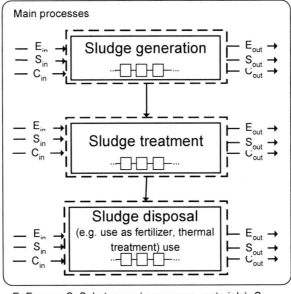

E: Energy; S: Substances (ressources, materials); C: Costs (positive and negative)

Figure 14. Life Cycle of sludge, simplified illustration.

Major intention of sludge management is safe disposal of sewage sludge as well as recovery of contained nutrients. To fulfil these objectives, various management options for sewage sludge exist, from sludge generation and treatment (e.g. process steps like thickening, conditioning, hygienization, dewatering and drying) to sludge application or disposal. The different options cause different environmental impacts and offer different possibilities with regard to resource and energy recovery. Generally, sludge can be used as a product and used for application e.g. in agriculture as fertilizer. The use of sewage sludge in agriculture is encouraged by the Sewage Sludge Directive 86/278/EEC. To avoid harmful effects on soil, vegetation, animals and man, sludge has to be treated (e.g. biologically, chemically, thermally) to reduce its fermentability and the health hazards resulting from its use. The Directive also sets requirements concerning composition and quality of sludge, such as limit values for concentrations of heavy metals in sewage sludge which is intended for agricultural use. The requirements vary between the European countries since some member states set their own stricter limit vales respectively added requirements for other contaminants. (http://ec.europa.eu/environment/waste/sludge/index.htm). Sludge can also be seen as waste, which has to be disposed. Disposal options for sludge that is not used as fertiliser is thermal treatment, including co-incineration in coal fired power plants, incineration in cement works (for energy recovery and substance recovery), Co-Incineration in waste incineration plants or mono-Incineration. Whether or not a thermal treatment is seen as recycling or disposal depends on calorific value of the sludge. Generally, a (high energy demanding) drying process is required to establish an adequate high calorific value.

With regard to this lots of various scenarios can be composed, examining how the environmental (economic and social) performance of a sludge management system could change if e.g. another sludge treatment, sludge application or disposal option or maybe different means of transportation are introduced.

Currently main activities are restricted to LCA studies analysing sludge management systems from an environmental point of view. LCC studies were barely found in literature, SLCA studies do not exist so far, according to the author´s knowledge. However, with increasing awareness of sustainability discussions in society further LCC and SLCA case studies analysing economic and social aspects of sludge management are to be expected. Generally, the studies vary significantly with regard to system boundaries, e.g. consideration of upstream activities (such as transport processes), consideration of infrastructure (e.g. sewers), consideration of background systems (e.g. electricity production, transports). Differences exist also with regard to the functional units chosen; Tilman et al, 1998, Hospido, Moreira et al. 2005) and (Hospido, Moreira et al. 2008) for example defined as functional unit the "amount of waste water from one person equivalent during one year", Suh&Rousseaux, 2001 the "quantity of inflow water in certain period of time". Also possible might be "a certain amount of (dried) sludge, which should be disposed of" to express the function of the system to "dispose sludge in an environmentally, economically and socially acceptable way ".

The studies normally use site specific data, collected directly from the plants, average data (annual data from plant) or data from databases and literature as well as estimation or simulation data. Since the scenarios considered refer mainly to waste water treatment plants, site specific data are necessary and results obtained cannot be easily transferred to similar plants in other regions in the world. A comprehensive assessment of environmental and economic aspects (social aspects anyway) in sludge management systems should thus be regular part when sludge management systems are planned and decisions are taken.

Decisions about the "final way" depend on various aspects, such as composition of sludge, location of the Waste water treatment plant, e.g. with regard to possible application fields in agriculture, economic aspects, political and societal preferences.

CONCLUSION

Nowadays, a contribution to sustainable development, understood as integration of environmental, economic and social aspects in decision processes is a target in many regional, national and international programmes and policies. However, decisions are still more based on practical feasibility and economic aspects (costs) than on environmental reasons (such as resource recovery) or social impacts. The increasing number of LCA studies on sewage sludge managements already indicates the growing interest of additionally consider environmental aspects in decision processes. The underlying life cycle approach also increasingly influences economic and social assessment tools and it can be expected that Life Cycle Costing and Social Life Cycle Assessment respectively the combination of the three methods described as Life Cycle Sustainability Assessment (LCSA) will be further developed to a usable tool for sustainability assessment.

Challenges remain with regard to selection and development of indicators able to appropriately describe sustainability aspects and the presentation of the results. It can not be assumed, for example that the results from environmental, economic and social assessment concordantly identify one sludge management system or more specifically a technological option as the "most sustainable" one. So, one technology might show the best environmental performance amongst the examined alternatives but at the same time be cost-intensive or inferior with regard to social impacts (e.g. protection of jobs). In complex management systems, various stakeholders (for example, institutions, politicians, plant operator, supplier, local community) are addressed whose targets and values with regarding environmental, economic or social aspects might differ significantly and who thus might prefer different management system. One task of sustainability assessment is to provide information by analysing and explaining possible positive and negative impacts of a decision (within the scope of the study) and their presentation with regard to sustainability aspects (e.g. potentially reduced CO_2-emissions, reduced costs or gained social benefits like creation of jobs).

To show, how prioritisation of technologies or alternative managements systems change depending on environmental, economic or social preferences the so called preference triangle can be used and which is presented in (Finkbeiner 2009). This allows a combination of the three sustainability dimension considering different weightings of stakeholders.

Sustainable development is a great task; fixed definitions, universal targets as well as universal methods and tools for assessing sustainability do not exist and probably cannot exist for a topic of such complexity. The methods and tools to be used also depend on the scope of the study or project. For sewage sludge management system respectively alternative technologies within the system the life cycle approach objectified in the life cycle sustainability assessment is proposed in this chapter. However, while LCA is already a commonly used tool within sustainability assessment and LCC considered partly in existing economic analyses, SLCA and the combined assessemt Life cycle sustainability assessment are still young disziplines and further research as well as practical application is of high importance on the way to a comprehensive sustainability assessment.

REFERENCES

Adams, W. M. (2006). The Future of Sustainability: Re-thinking Environment and Development in the Twenty-first Century. J. Report of the IUCN Renowned Thinkers Meeting.

Brundtland, G. H. (1987). Our Common Future, Chapter 2: Towards Sustainable Development. From A/42/427. Our Common Future (Brundtland-Report): Report of the World Commission on Environment and Development

CALCAS (2009.). "Co-ordination action for innovation in life-cycle analysis for sustainability".CML (2001). "CML Guide - Part2b: Operational Annex."

CSD (2001). "Commission of Sustainable Develoment: Indicators of Sustainable Development – Guidelines and Methodologies, New York ".

CSD (2008). Commission of Sustainable Develoment: Indicators of Sustainable Development. Internet (Stand 13.Mai2010).

deHaes, U., O. Jolliet, et al. (1999). "SETAC-Europe: Second Working Group on LCIA (WIA-2) -Best Available Practice Regarding Impact Categories. Category Indicators in Life Cycle Impact Assessment and Background Document for the Second Working Group on Life Cycle Impact Assessment " Int. J. LCA 4 (2) (1999).

Dreyer, L. C., M. Z. Hauschild, et al. (2006). "A Framework for Social Life Cycle Impact Assessment." Int J LCA 11 (2) 88 – 97 (2006).

El-Sayed, M., M. Mahgouba, et al. (2010). "Towards sustainability in urban water: a life cycle analysis of the urban water system of Alexandria City, Egypt." Journal of Cleaner Production. Volume 18, Issues 10-11, July 2010, Pages 1100-1106

Finkbeiner, M. (2007). "one of the authors, hat proposed the term SLCA (see Klöpffer&Renner, 2007) ".

Finkbeiner, M., K. Reimann, and R. Ackermann. (2008). Life Cycle Sustainability Assessment (LCSA) for products and processes. Paper presented at SETAC Europe 18th Annual Meeting, 25-29 May 2008, Warsaw, Poland

Finkbeiner, M. (2009). "Vom Umweltschutz zur Nachhaltigkeit - Die Nachhaltigkeitsnutzwertanalyse unterstützt die Verwirklichung eines zentralen Paradigmenwechsel in Politik und Wirtschaft." Müllmagazin 2/2009.

Finnveden, G., M. Z. Hauschild, et al. (2009). "Recent developments in Life Cycle Assessment." Journal of Environmental Management 91 (2009) 1–21.

GRI (2002). "Global Reporting Initiative: Sustainability reporting Guidelines. Boston, USA."

Hong, J., J. Hong, et al. (2009). "Environmental and economic life cycle assessment for sewage sludge treatment processes in Japan." Waste Management, Volume 29, Issue 2, February 2009, Pages 696-703.

Hospido, A., T. Moreira, et al. (2008). "A Comparison of Municipal Wastewater Treatment Plants for Big Centres of Population in Galicia (Spain). Int J LCA 13 (1) 57 – 64 (2008).

Hospido, A., T. Moreira, et al. (2005). "Environmental Evaluation of Different Treatment Processes for Sludge from Urban Wastewater Treatments: Anaerobic Digestion versus Thermal Processes. Int J LCA 10 (5) 336 – 345 (2005).

Hospido, A., T. Moreirai, et al. (2004). "Environmental Performance of a Municipal Wastewater Treatment Plant." Int J LCA 9 (4) 261 - 271 (2004).

Houillon, G. and O. Jolliet (2005). "Life Cycle Assessment of Processes for the Treatment of Wastewater Urban Sludge: Energy and Global Warming Analysis. ." Journal of Cleaner production 13: 287-299.

Hunkeler, D. (2006). "Societal LCA Methodology and Case Study " Int J LCA 11 (6) 371 – 382 (2006).

Hunkeler, D., G. Rebitzer, et al. (2008). Environmental Life Cycle Costing. , Taylor & Francis 2008.

Huppes, G. (2004). "Life Cycle Costing and Life Cycle Assessment in management accounting for sustainability". CML, Department Industrial Ecology." Presentation at EMAN-Conference Lüneburg, March 2004.

IISD (2009). What is Sustainable Development?International Institute for Sustainable Development, 2009.

ISO14040 (2006). "Environmental management – Life cycle assessment – Principles and framework (ISO 14040:2006)."

ISO14044 (2006). "Environmental management –Life cycle assessment – Requirements and guidelines (ISO 14044:2006)

Johanssona, K., M. Perzona, et al. (2008). "Sewage sludge handling with phosphorus utilization – life cycle assessment of four alternatives." Journal of Cleaner Production. Volume 16, Issue 1, January 2008, Pages 135-151.

Jolliet, O., R. Müller-Wenk, et al. (2004). "The LCIA Midpoint-damage Framework of the UNEP/SETAC Life Cycle Initiative." Int J Life cycle Assess 9 (6): 394-404.

Jørgensen, A., I. T. Hermann, et al. (2010). "Is LCC relevant in a sustainability assessment?" INT J LCA PublisHed online 27.April2010.

Jørgensen, A., A. LeBocq, et al. (2008). "Methodologies for Social Life Cycle Assessment." Int J LCA 13 (2) 96 – 103 (2008).

Kalderis, D., M. Aivalioti, et al. (2010). "Options for sustainable sewage sludge management in small wastewater treatment plants on islands: The case of Crete." Desalination, In Press, Corrected Proof, Available online 20 May 2010.

Kates, R., T. Parris, et al. (2005). "What is Sustainable Development? ." Environment 47(3): 8–21. .

Kicherer, A., S. Schaltegger, et al. (2007). "Eco-Efficiency - Combining Life Cycle Assessment and Life Cycle Costs via Normalization." Int J LCA 12 (7) 537 – 543 (2007).

Kloepffer, W., 2003: Life-Cycle Based Methods for Sustainable Product Development. Editorial for the Life Cycle Management (LCM) Section in International Journal of Life Cycle Assessment Int J LCA 8,157-159

Kloepffer, W. (2008). "Life Cycle Sustainability Assessment of Products (with Comments by Helias A. Udo de Haes)." Int J LCA 13 (2) 89–95.

Klöpffer, W. and B. Grahl (2009). Ökobilanz (LCA) - Ein Leitfaden für Ausbildung und Beruf, Wiley-VCH, Weinheim.

Klöpffer, W. and A. B. Heinrich (2009). "Our plans and expectations (for the 14th volume 2009 of Int J Life Cycle Assess)." Int J Life Cycle Assess (2009) 14:1–7.

Klöpffer, W. and I. Renner (2007). "Lebenszyklusbasierte Nachhaltigkeitsbewertung von Produkten. ." Technikfolgenabschätzung - Theorie und Praxis 16 (3).32-38.

Kopfmüller, J., V. Brandl, et al. (2001). Nachhaltige Entwicklung integrativ betrachtet – Konstitutive Elemente, Regeln, Indikatoren. Global zukunftsfähige entwicklung – Perspektiven für Deutschland 1. edition sigma, Berlin.

Landsiedel, R. and P. Saling (2002). "Assessment of Toxicological Risks for Life Cycle Assessment and Eco-efficiency Analysis." Int J LCA 7 (5) 2261-268.

Lassaux, S., R. Renzoni, et al. (2007). "Life Cycle Assessment of Water from the Pumping Station to the Wastewater Treatment Plant." Int J LCA 12 (2) 118 – 126 (2007).

Lebensministerium (2006). Monitoring nachhaltiger Entwicklung in Österreich. Indikatoren für nachhaltige Entwicklung. Wien, 07.07.06.

Lederer, J. and H. Rechberger (2010). "Comparative goal-oriented assessment of conventional and alternative sewage sludge treatment options " Waste Management Volume 30, Issue 6, June 2010, Pages 1043-1056.

Lim, S.-R., D. Park, et al. (2008). "Environmental and economic feasibility study of a total wastewater treatment network system." Journal of Environmental Management, Volume 88, Issue 3, August 2008, Pages 564-575.

Lundin, M. and G. M. Morrison (2002). "A life cycle assessment based procedure for development of environmental sustainability indicators for urban water systems." Urban Water 4 (2002) 145–152.

Nazarkina, L. and A. Le Bocq (2006). "Social Aspects of Sustainability Assessment: Feasibility of Social Life Cycle Assessment (S-LCA)." EDF 2006, Moretsur-Loing.

Ott, K. (2001). Eine Theorie "starker" Nachhaltigkeit. Ethik und Nachhaltigkeit, Grundsatzfragen und Handlungsperspektiven im universitären Agendaprozess. Frankfurt am Main, 30-63 G. Altner and G. Michelsen.

Ott, K. (2003). The Case for Strong Sustainability. Greifswald's Environmental Ethics. K. Ott and P. T. (eds.). Greifswald: Steinbecker Verlag Ulrich Rose.

Parent, J., C. Cucuzzella, et al. (2010). "mpact assessment in SLCA: sorting the sLCIA methods according to their outcomes." Int J Life Cycle Assess (2010) 15:164–171.

Patel, M. (2009). "Sustainability Assessment of Technologies – mid term and long term Research to improve the Sustainability Evaluation of Technologies." CALCAS Workshop-Life Cycle Analysis for Sustainability Governance Brussels, 17-18 February 2009.

Pinnekamp, J., D. Montag, et al. (2007). Rückgewinnung eines schadstofffreien, mineralischen Kombinationsdüngers "Magnesiumammoniumphosphat - MAP" aus Abwasser und Klärschlamm

UBA Texte Texte 25/07. Forschungsbericht 202 33 308, Umweltforschungsplan des Bundesministerium für Umwelt, Naturschutz und Reaktorsicherheit. .

Rebitzer, G. and D. Hunkeler (2003). "Life Cycle Costing in LCM: Ambitions, Opportunities, and Limitations Discussing a Framework." Int J LCA 8 (5) 2003.

Rebitzer, G. and S. Seuring (2004). The SETAC Working Group on Life Cycle Costing. L. Präsentation, 04.03.2004.

Saling, P., A. Kicherer, et al. (2002). "Eco-efficiency Analysis by BASF: The method " Int J Life Cycle Assess 7:203-218.

Schmidt, I., M. Meurer, et al. (2004). "SEEbalance - Managing Sustainability of Products and Processes with the Socio-eco-efficiency Analysis by BASF." Green Manage In 45:79–94.

SETAC (2003). Life-Cycle Assessment and SETAC: 1991–1999, Pensacola (Florida). CD-ROM.

Spillemaeckers, S., G. Vanhoutte, et al. (2004). "Integrated Product Assessment – The Development of the Label "Sustainable Development" for Products Ecological, Social and Economical Aspects of Integrated Product Policy. ." Belgian Science Policy, Belgium.

Suh, Y.-J. and P. Rousseaux (2002). "An LCA of Alternative Wastewater Sewage Sludge Treatment Scenarios." Resources, Conservation and Recycling 35 (3):191-200.

Tillman, A.-M., M. Svingby, et al. (1998). "Life Cycle Assessment of Municipal Waste Water Systems." Int. J. LCA 3 (3) 145 - 157 (1998).

UN (1992). United Nations Conference on Environment and Development (Earth Summit), June 1992, Rio de Janeiro.

UN. (2002). "World Summit on Sustainable Development, 26 August–4 September 2002, Johannesburg, South Africa." from http://www.earthsummit2002.org/

http://www.un.org/jsummit/html/documents/summit_docs/0409_l6rev2_pol_decl.pdf.

UN (2005). 2005 World Summit Outcome. Integrated and coordinated implementation of and follow-up to the outcomes of the major United Nations conferences and summits in the economic, social and related fields. Follow-up to the outcome of the Millennium Summit. U. N. G. A. A/60/L.1.

UNDP (2008). "Millenium development goals (MDG) - official list of MDG indicators."

UNEP (2009). "Guidelines for social life cycle assessment of products - Social and socio-economic LCA guidelines complementing environmental LCA and Life Cycle Costing, contributing to the full assessment of goods and services within the context of sustainable development." United Nations Environment Programme, Paris, 2009

Vigon, B., F. Consoli, et al. (1993). Guidelines for Life Cycle Assessment. A Code of Practice

Weidema, B. P. (2006). "The Integration of Economic and Social Aspects in

Life Cycle Impact Assessment." Int J LCA 11 • Special Issue 1 (2006) • 89 – 96.

In: Sewage Sludge Management
Editors: A. A. Zorpas and V. J. Inglezakis

ISBN: 978-1-61324-393-0
© 2012 Nova Science Publishers, Inc.

Chapter 19

THE POTENTIAL OF ANAEROBIC DIGESTION FOR ENERGY RECOVERY FROM SLUDGE

Lise Appels [*] *and Raf Dewil* [†]

Chemical and Biochemical Process Technology and Control Section,
Department of Chemical Engineering, Katholieke Universiteit Leuven,
W. De Croylaan, Heverlee, Belgium

1. INTRODUCTION

Anaerobic digestion plays an important role in the processing of sludge for its abilities to convert the present organic matter into biogas and at the same time reduce the amount of sludge solids for final disposal. Moreover, the process destroys most of the pathogens present in the sludge and limits possible odour problems associated with the residual matter.

Anaerobic sludge digestion can optimise WWTP costs and is considered to be a major and essential part of a modern WWTP. In the process, a large fraction of the organic matter is degraded into biogas, mainly consisting of methane (CH_4) and carbon dioxide (CO_2), and small quantities of H_2, N_2, H_2S and H_2O. The high concentration of methane (60-75%) gives the produced gas a high calorific value. The potential of using the produced biogas as energy source is widely recognised. Biogas is currently produced mostly by digestion of sewage sludge, with minor contributions from fermentation or gasification of solid waste or of lignocellulosic material (processes currently being further developed). It is considered an important future contributor to the energy supply of Europe, although upgrading of the biogas is mostly needed. The annual potential of biogas production in Europe is estimated in excess of 200 billion m³.

---

[*] E-mail: lise.appels@cit.kuleuven.be.
[†] E-mail: raf.dewil@cit.kuleuven.be.

Despite the advantages described above, some limitations are inherently attached to the process, e.g. (i) only a partial decomposition of the organic fraction; (ii) the rather slow reaction rate and associated large volumes and high costs of the digesters; (iii) the vulnerability of the process to various inhibitors; (iv) the rather poor supernatant quality produced; (v) the presence of other biogas constituents such as carbon dioxide, hydrogen sulphide and excess moisture; (vi) the possible presence of volatile siloxanes in the biogas that can cause serious damage in the energy users (generator, boiler) due to the formation of microcrystalline silica; and (vii) the increased concentration of heavy metals and various industrial "organics" in the residual sludge due to the significant reduction of the organic fraction during digestion, leaving the mineral and non degradable fraction untouched. Currently, most research on anaerobic digestion is focussed on overcoming these problems, with specific attention to improving the overall degradation efficiency and the digestion rate.

The current chapter will provide an overview on the current status of anaerobic digestion of sewage sludge. First, the basic mechanisms will be discussed shortly, after which the technical aspects of an anaerobic digester will be reviewed. The second part is dedicated to the energetic use of the produced biogas and the possibilities for enriching the biogas. Finally, sludge pre-treatment methods that have the possibility to enhance anaerobic digestion will be overviewed.

2. TECHNOLOGY OF ANAEROBIC DIGESTION OF SEWAGE SLUDGE

2.1 Basic Mechanism

The anaerobic digestion of organic material consists mainly of the following four steps: hydrolysis, acidogenesis, acetogenesis and methanogenesis as shown in Figure 1 [1].

Anaerobic digestion is a complex process that requires strict anaerobic conditions and depends on the coordinated activity of a complex microbial community to convert organic matter into mostly carbon dioxide (CO_2) and methane (CH_4). The first phase, hydrolysis, is considered to be rate-limiting for the overall process, since in this stage, complex refractory (organic) matter such as lipids, polysaccharides, proteins and nucleic acids, has to be degraded into simple, readily available monomers (e.g. amino acids and fatty acids). These components formed during hydrolysis are further split during acidogenesis, the second step. Volatile fatty acids (VFA) are produced by acidogenic (or fermentative) bacteria along with ammonia (NH_3), carbon dioxide (CO_2), hydrogen sulphide (H_2S) and other by-products.

The third stage in anaerobic digestion is acetogenesis, where the higher organic acids and alcohols produced by acidogenesis are further digested by acetogens to produce mainly acetic acid as well as CO_2 and H_2. This conversion is controlled to a large extent by the partial pressure of H_2 in the gas mixture.

The final stage of methanogenesis produces methane by two groups of methanogenic bacteria: a first group splits acetate into methane and carbon dioxide and the second group uses hydrogen as electron donor and carbon dioxide as acceptor to produce methane.

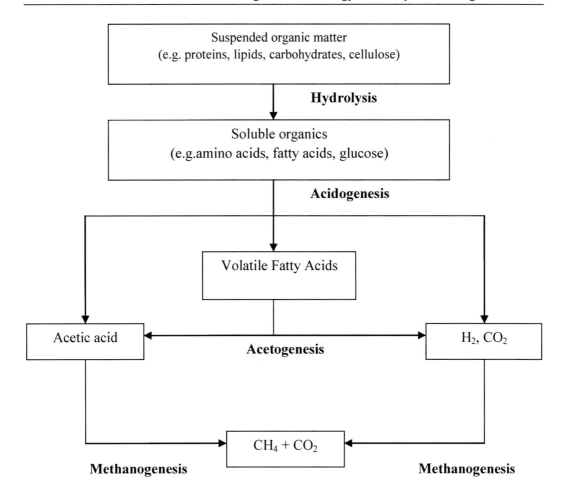

Figure 1: Subsequent steps in the anaerobic digestion process.

2.2 Technical Aspects

Types of Anaerobic Digesters

Different types of anaerobic digesters are being used in practice. Standard-rate (cold) digestion is the simplest type. The sludge content is usually neither heated nor mixed and requires a long digestion period of 30 to 60 days since only psychrophylic micro-organisms are active at these low temperatures. Although the generated biogas provides some form of mixing, stratification occurs in four zones: (i) a scumlayer, (ii) a liquid layer (or supernatant), (iii) a layer of digesting solids and (iv) a layer of digested solids. The supernatant is withdrawn and recycled to the wastewater treatment plant. The accumulated digested solids are periodically extracted at the bottom of the digester. This type of digester is seldom built because of its low efficiency and extremely long retention times. Most of them are encountered at smaller WWTPs and in countries having a warm climate.

The so called high-rate digester is a major improvement of the standard-rate digestion. The sludge is heated to a temperature that is optimal for the anaerobic organisms (either mesophilic or thermophilic – see later) and completely mixed. This type of digester has become the standard technology for anaerobic digestion and is therefore frequently encountered in practice. This digester is fed with thickened raw sludge with a solid concentration of 3-6%. The sludge mixture in the digester is mixed by gas recirculation, pumping or draft-tube mixers. The reaction mixture is heated to a pre-set constant temperature mostly by external heat exchangers. Other ways of heating include internal heat exchangers and steam injection. Part of the energy recovered from the biogas valorisation can be used for this purpose. Uniform feeding is very important: the sludge should be fed continuously or at regular intervals to help maintain steady-state conditions in the digester and reduce shock loadings. This steady draw and fill mode is also important to improve pathogen kill [1].

A high-rate digester is sometimes coupled with a second tank, called a secondary digester although the latter is merely used to store the digested solids and decant the supernatant: it is neither heated nor mixed. The tanks may have fixed roofs or floating covers. If the secondary tank is of the floating cover type, it can also be used to store digester gas. Very little solids reduction and gas production take place in the second tank. Sometimes, primary and secondary tanks are of equal design, each with heating and mixing capacity to serve as a standby digester. Due to the mixing and natural breakdown of the solids, fine particles or flocs are produced which do not settle easily. The application of a second, cold digester was very popular in the past, but is rarely used in newer plants.

Mesophilic and Thermophilic Digestion

Most high-rate digesters are operated in the mesophilic range, with a temperature between 30 and 38°C. Anaerobic digestion can also take place at higher temperatures, in the thermophilic region, where digestion occurs at temperatures between 50 and 57°C suitable for thermophilic bacteria. Thermophilic digestion is faster than mesophilic digestion. Other advantages are an increased solids reduction, improved dewatering, and increased destruction of pathogenic organisms. The use of thermophilic temperatures, however, requires a higher energy input, produces a lower quality supernatant with large quantities of dissolved solids and has a higher odour potential. The process stability is much poorer due to the fact that thermophilic bacteria are far more sensitive to fluctuations in substrate composition and digestion temperature. Because of the before mentioned reasons, the use of thermophilic digestion is still marginal compared to mesophilic digestion.

Tank Design

Anaerobic digestions tanks are mostly cylindrical or egg-shaped [2]. A cylindrical tank has a typical diameter between 6 to 40 m, a conical floor with a slope of about 15 %, and a withdrawal of the sludge in the centre of the tank. The reaction mixture requires a minimum depth of 7,5 m to allow proper mixing but can be as high as 15 m. Some digesters are equipped with a so called "waffle" bottom to minimize grit accumulation and reduce digester cleaning [1].

Digesters are covered to maintain the operating temperature and anaerobic conditions and of course to collect the digester gas. The tanks can have either a fixed or a floating roof. Fixed covers are dome-shaped or flat and are built of reinforced concrete, steel or fibreglass-reinforced polyester. Floating covers are normally used for single-stage digester and for the second stage of two-stage digesters. A variation of the floating cover is the floating gas holder, consisting of a floating cover with an extended skirt, so that gas can be stored during periods when the supply of digester gas exceeds the demand. A recent development in gas-holder covers is the membrane cover. It consists of supported, flexible gas and air membranes. When the gas storage volume decreases or increases in the space between the liquid surface and the membranes, the space between the membranes is pressurized or depressurized using an air-blower bleed-valve system [3]. Floating covers directly float on the liquid and generally have a maximal vertical ravel of $2 - 3$ m [2,4]. The gas pressure under a digester cover is typically in the range of $0 - 3.7$ kN/m² [2]. In egg-shaped digesters, there is only limited storage available for gas and the provision of external gas storage is needed [3].

Digester Mixing

Mixing provides intimate contact between the substrate (feed sludge) and anaerobic biomass, yielding uniformity of temperature, of substrate concentration, of other chemical, physical and biological aspects throughout the digester, and preventing both the formation of surface scum layers and the deposition of sludge on the bottom of the tank. The natural mixing in the digestion tank created due to the rise of gas bubbles and thermal convection currents is generally insufficient for an optimal performance. Therefore, auxiliary mixing is needed, mostly generated through external pumped recirculation, internal mechanical mixing and internal gas mixing.

By external pumped recirculation, digested sludge is withdrawn from the centre of the digester and pumped through external heat exchangers where the digested sludge is blended with raw sludge and heated. This mixture is then pumped back in the digestion tank through nozzles at the base of the digester or at the top to break the possibly present scum layer. The sole use of this method for mixing the tank is limited, since the flow rate in the recirculation should be very high to guarantee a complete mixing of the digester. The minimum required power is $0.005 - 0.008$ kW/m³ of digester volume . Other disadvantages of external pumped recirculation are plugging of the pumps by rags, impeller wear from grit and bearing failures. The mechanical pumping action is provided by centrifugal pumps, generally set up in an internal or external shaft tube to support vertical mixing. Mixing is supported by the circulation of the sludge. These systems are most suited for digesters with fixed covers [2,3,4].

Internal mechanical stirring systems generally use low-speed flat-blade turbines. Here, the sludge is transported by rotating impeller(s), thus mixing the content of the digestion tank. The use of high speed propeller mixers is not recommended because they generate high shear forces that will very likely damage the anaerobic micro-organisms.

Internal gas mixing uses the produced biogas for mixing the digester contents. It is a successful method of mixing the digester content and avoid the build-up of scum. Gas mixing systems can be confined and unconfined. In the latter, the gas is collected at the top of the digestion tank, compressed and then released through a pattern of diffusers or a series of radially placed lances suspended from the digester cover. The digester content is mixed, releasing gas bubbles that rise and push the sludge to the surface. Scum has to be specifically controlled as it causes roof fracture, gas surging etc. The unit gas flow requirement for unconfined systems is 0.0045 to 0.005 m^3 gas/m^3 digester volume.min [3].

In confined systems the gas is collected at the top, compressed and discharged through confined tubes. The gas lifter system is composed of flooded gas pipes placed in an eductor tube or gas lifter. The compressed gas is released from these pipes and gas bubbles rise, creating an air-lift effect. The gas piston system releases gas bubbles intermittently at the bottom of the piston, hereby creating piston pumping action of the bubbles and pushing the sludge to the surface. These confined systems generally have a low power requirement and a gas flow rate of 0.005 to 0.007 m^3 gas/m^3 digester volume.min [3,4].

3. BIOGAS ENRICHMENT, COMPRESSION AND STORAGE

3.1 Generalities

Biogas that is produced from anaerobic digestion of sludge is a clean and environmentally friendly fuel. It contains about 55 – 65% of CH_4, 30 – 40 % of CO_2, fractions of water vapour, traces of H_2S and H_2, and possibly other contaminants (e.g. siloxanes). The heating value of biogas is determined by its CH_4 content, with the higher heating value being the energy released when 1 Nm^3 of biogas is combusted and the water vapour formed within combustion is condensed. The lower heating value omits the vapour condensation. Methane gas itself has a higher heating value of 55.50 MJ/kg.

For energetic valorisation at the place of production it can in most cases be used without further treatment. To make it economically and energetically sensible to transport the biogas over larger distances, there is a great need to increase its energy content. Ultimately, the compression and use of gas cylinders or introduction into the gas network are future targets. For this enrichment and enhanced potential of use, the removal of CO_2 and contaminants is a prerequisite. Important here is the methane number that describes the gas resistance to knocking when used in a combustion engine. Methane has per definition a methane number of 100 and H_2 a methane number of 0. CO_2 increases the methane number because it is a non-combustible gas with a high knocking resistance. Upgraded biogas therefore has a methane number in excess of 100.

Biogas contains various sulphur compounds, mostly sulphides, but also traces of disulphides and thiols are detected. Especially oxidized sulphur (sulphate and sulphite) is corrosive with the presence of H_2O. H_2S itself is reactive with most metals and the reactivity is enhanced by concentration and pressure, by the presence of H_2O and at elevated temperature. Halogenated compounds are only rarely encountered in biogas from digestion of sewage sludge since if present, they would have killed off the digester in the first place. High ammonia concentrations form a problem for gas engines (generally 100 mg/Nm^3 NH_3 is acceptable). The combustion however leads to NOx formation. Finally siloxanes (volatile compounds of silicon bound by organic radicals) are present and can cause problems through coating of the gas beneficiation equipment with a layer of abrasive silica. Especially gas engines are susceptible for damage.

All biogas plants are equipped with some kind of filter to avoid the presence of fine particles in the gas. These filters, with a 2 – 5 μm mesh size, not only remove particulates, but also droplets of water or foam.

3.2. Biogas Utilisation

Biogas can in general be used in all applications that were originally developed for natural gas. There are four basic ways of biogas utilisation, including the production of heat and steam; the generation or co-generation of electricity; the use as transport fuel; and possibly the production of chemicals. The final choice of the type of valorisation is governed by national frameworks including the tax system, subsidies, green energy certificates and increased feed-in tariffs for electricity, availability of heat or gas grids.

Conventional gas burners can easily be adjusted to biogas by changing the air-to-gas ratio. Burning biogas is an established and reliable technology, with low demands on biogas quality. Pressure usually has to be between 8 and 25 mbar. It is recommended to reduce the level of H_2S below 1000 ppm to maintain the dew point at approximately 150°C.

Biogas is mainly used as fuel for combined heat and power (CHP) applications. Although gas turbines could be used (micro-turbines, 25 to 100 kW; large turbines, > 100 kW) with low emissions, efficiencies comparable to spark-ignition engines and low maintenance, the latter are less preferable because rather high investment costs. CHP applications depend mostly on the use of either spark-ignition or dual fuel engines. Dual fuel engines, with e.g. injection of diesel ($\geq 10\%$) are, although much less economic, very popular in smaller scales, with good power efficiency (up to 40%). They have high emissions, unless a treatment of combustion gas is used, but allow easy start-up by using diesel only (when the biogas production is started) and are less susceptible for fuel impurities like siloxanes. Spark-ignition engines can be stoichiometric or lean-burn engines, the latter common for larger sizes and having a higher efficiency.

Fuel cells have the potential to reach very high efficiencies (> 60%) and low emissions and are considered to become the small scale power plant of the future. Especially hot fuel cells (> 800°C) are interesting for biogas utilisation since CO_2 does not inhibit the electrochemical process at these temperatures, but rather serves as a heat carrier. Both the solid oxide fuel cell (for small applications of a few kW) and the molten carbonate fuel cells (up to 250 kW and more) can be envisaged.

Various E.U. countries have embarked on programmes to use a growing portion of the biogas in the transport sector, especially attractive in view of the steady increase of the cost of fossil fuels. Biogas can serve as a fuel for vehicles that use natural gas (NGVs) [5], provided it is upgraded to natural gas quality. At the end of 2005 there were more than 5 million NGVs in the world. The number of public transport vehicles driven on gas such as buses and waste trucks is increasing considerably. Most of the gas driven personal cars are converted vehicles that have been retro-fitted with a gas tank in the luggage compartment and a gas supply system in addition to the normal petrol fuel system. Dedicated gas vehicles run at a better efficiency and also allow for more convenient placement of the gas cylinders without losing luggage space.

Gas is stored at 200 to 250 bars in pressure vessels made from steel or aluminium composite materials. Gas vehicles have substantial advantages over vehicles equipped with diesel or petrol engines, since CO_2 emissions are reduced by more than 95%, together with emissions of particles and soot. Heavy duty vehicles are normally converted to run on methane gas only, but in some cases dual fuel engines can also be used. The dual fuel engine still has the original diesel injection system and gas is ignited by injection of a small amount of diesel oil. Dual fuel engines normally require less engine development and maintain the same driveability as a diesel vehicle. However, emission values are not as good as for the corresponding dedicated gas vehicle and the engine technology remains a compromise between spark ignition and diesel engine. Beside the close to 100% CO_2 reduction, pure gas engines with catalytic converters demonstrate far better emission values than the most modern diesel engines (EURO 4 and 5) tested according to the European Transient Cycle (ETC) or the Enhanced Environmental friendly Vehicle (EEV) standard at the EMPA, Switzerland. Stoichiometric gas engines with an air-to-fuel ratio of 1 demonstrate a better emission pattern than lean engines. However, both are far better than dual fuel engines although at a reduced efficiency.

The number of biogas and natural gas filling stations is still limited in Europe and the rest of the world, although the situation is improving significantly with the number of pumping stations multiplied over the last few years: at the end of 2005 there were 1600 pumping stations in Europe.

Biogas injection in the gas grid is possible, and various countries of the EU have proposed standards for injecting upgraded biogas into the grid to avoid contamination of the grid. These standards of e.g. Sweden, Switzerland, Germany and France, set limits for e.g. sulphur, oxygen, particles and dew point. Upgrading methods must allow treated biogas to meet these stringent quality standards. This upgrading and associated cost outweigh the rising costs of fossil fuels [1].

3.3. Biogas Upgrading Technologies

Various advantages are associated with biogas upgrading: (i) various applications require the removal of trace elements, (ii) the heating value of the biogas is increased, and (iii) the biogas quality is standardised/uniformised what is an important factor for its industrial use. The following subsection will discuss the most important separation processes that are used for biogas upgrading.

Carbon Dioxide Removal

The removal of CO2 increases the heating value and leads to a gas quality similar to natural gas. It is important to limit methane losses when using removal techniques for economical (fuel losses) and environmental reasons (CH4 is a greenhouse gas 21 times stronger than CO2).

Various removal methods can be applied, ranging from absorption and adsorption (most commonly performed), to cryogenic separation (albeit expensive) and membrane separation [6].

Absorption processes remove CO_2 and H_2S simultaneously due to the difference in binding forces of the polar CO_2 and H_2S and the non-polar CH_4. Water is the most common solvent for counter-current scrubbing of pre-compressed biogas (4 – 7 bar). The design of a water scrubbing system depends on the solubility of CO_2, as solubility is governed by pressure, temperature and pH: as the pressure increases, the solubility of CO_2 in water increases; but decreases as the temperature increases.

After pressure scrubbing, CO_2 and H_2S are desorbed in a flash tank, where the pressure is reduced and the temperature possibly increased. H_2S, which is released to the air can create an emission problem. Some of the sulphur accumulates in the water and can cause problems of fouling or corrosion of piping. It is hence recommended to separate H_2S beforehand. Air or vacuum stripping are seldom used since they introduce O_2 in the system. Results show that 5 – 10% of CO_2 remains in the biogas.

The absorption efficiency can significantly be improved (to a nearly complete CO_2 removal) if $Ca(OH)_2$ solutions are used to remove both CO_2 and H_2S, resulting in the formation of insoluble $CaCO_3$ and CaS. Also organic solvents, such as polyethyleneglycol (Selexol®, Genosorb®) and alkanol amines (mono-ethanol-amine, or di-ethanol-amine) can be used as absorbents. The solubility of CO_2 and H_2S in these solvents is significantly higher than in water, what makes a low pressure operation possible. The solubility of CH_4 is very limited, hence only small amounts of CH4 are removed. The desorption (regeneration) takes place in a second column through increasing the temperature. Reductions of CO_2 to 0.5 – 1 vol% in biogas are possible. The organic solvent removal units are however more expensive than those using water as a solvent, and suffer from the need to periodically partly discharge, dispose and replace its solvents.

The removal of CO_2 by Pressure Swing Adsorption on solids such as activated carbon or molecular sieves is possible. The selectivity is achieved with different mesh sizes. Various processes are described in literature [7]. Adsorption is generally accomplished at high temperature and pressure. It is simple in design and easy to operate, but is a costly process with high pressure drops and high heat requirements. Desorption is performed by depressurisation or even by using a slight vacuum. The process needs dry biogas, hence the need to remove the water vapour as pre-treatment step.

Cryogenic distillation can be used since CH_4 has a boiling point of -160°C at 1 atm, whereas CO_2 has a boiling point of -78°C. CO_2 can be removed as liquid by cooling the biogas mixture at elevated pressure. Until now, this expensive method has only been tested in pilot plants in Europe and in the USA. More than 97% pure CH4 is produced. Investment and operational costs are high and limit its current application [8].

Membrane separation is a rather new technology, but gains interest rapidly [9,10,11]. The method is based on the difference in retention for the various biogas components. The transport of each component is driven by the difference in partial pressure over the membrane and is highly dependent on the permeability of the component in the membrane material [12]. For high methane purity, the difference in permeability must be high. Solid membranes constructed from acetate-cellulose polymer have permeabilities for CO_2 and H_2S up to 20 and 60 times the value for CH4. However, high pressures (up to 25 bar) are required for the process. Although the gas flux across the membrane increases proportionally with the pressure difference, thus reducing the size of the membrane, there is a maximum pressure that the membrane can withstand. Since some CH_4 passes through the membrane to the permeate stream, methane losses occur. If the permeate can be used in a CHP (combined with raw gas), these CH_4 losses can be recovered.

Additional techniques are under investigation such as the chemical conversion by e.g. catalytically reacting CO_2 and H_2 to CH_4 [13]. This process is extremely expensive and the need of H_2 makes the process generally unsuitable. In-situ CH_4 enrichment is another technique that is under development. Sludge from the digestion chamber is contacted with air in counter-current. Carbon dioxide that is dissolved in the sludge is desorbed. The CO2-lean sludge is led back to the digestion chamber where more carbon dioxide can now dissolve into the sludge, resulting in CH4 enriched gas in the chamber. The results from lab scale test in Sweden indicate that it is technically possible to construct a system that increases the methane content of the gas to 95% and still keeps the methane losses below 2%.

Removal of Water

The drying of biogas is generally needed or recommended since it is saturated with water vapour when it leaves the digester. Refrigeration or sensible pipework design is a common method to condense the water. In order to reach higher dew points, the gas can be compressed before cooling.

Adsorption on silicagel or Al_2O_3 is applied when very low dew points need to be achieved. An alternative method of drying biogas can be the absorption in glycol or hygroscopic salts, which can be recovered at elevated temperatures.

Removal of H_2S

It is important to realise that the release of H_2S to the biogas can be limited by an appropriate conditioning of the sludge [14]. The addition of Fe^{3+}-salts to the sludge can indeed produce insoluble sulphides and reduce the amount of H_2S released. Concentrations below 150 ppm are achievable (depending on the amount of Fe3+ added). An excess of Fe3+ salts added can however inhibit the anaerobic micro-organisms, so an optimal dosing is required. For removing H_2S from the biogas, absorption processes are frequently used (as discussed before).

H_2S can also be adsorbed on activated carbon [15]. Activated carbon acts as a catalyst to convert H_2S into elemental S. Impregnation with KI is needed. Impregnated activated carbon is a common method of removal of H_2S before upgrading with Pressure Sweep Adsorption.

Micro-organisms, belonging to the Thiobacillus family, can be used to reduce the level of sulphides in biogas, by oxidising it mainly to elementary sulphur and some sulphates. These bacteria are commonly present in the digestion material and thus do not have to be inoculated. Furthermore, most of them are autotrophic, which means that they use carbon dioxide from the biogas as carbon source. Oxygen needs to be added to the biogas for biological desulphurisation and the level needed depends on the concentration of hydrogen sulphide, usually around 2 – 6 vol% air in biogas. The simplest method for desulphurisation is to add oxygen or air directly into the digestion chamber. With this method, H_2S level can be reduced by up to 95% to levels lower than 50 ppm, however function of temperature, place and amount of air added and reaction time. When adding air into the biogas, safety measures need to be taken into consideration to avoid overdosing of air in case of a pump failure. Methane is explosive in the range 5 – 15% in air. Biological desulphurisation can also take place in a separate bio-filter filled with plastic bodies on which desulphurising micro-organisms are attached. In the unit up-flowing biogas meets a counter flow of liquid consisting of gas condensate and liquid from effluent slurry separation or a solution of minerals. Before the biogas enters the unit, $5 - 10$ vol% air is added. The H_2S level can be reduced from $3000 - 5000$ ppm to $50 - 100$ ppm. Ammonia is separated at the same time.

H2S can also be reduced by NaOH scrubbing to form Na2S of NaHS, both unsoluble salts.

Removal of Siloxanes

It was already mentioned that siloxanes can be present in the biogas. The reduction of their concentration and/or abatement processes were described in detail by Dewil et al. [16].

The presence of siloxanes in biogas gives rise to some problems regarding its thermal valorisation. These silicon containing compounds are widely used in various industrial processes (e.g. for replacing organic solvents) and are frequently added to consumer products (e.g. detergents, personal care products, etc.). Moreover they are released as a residue in the production of silicon-containing chemicals. The consumption of siloxanes is growing steadily, e.g. wet wipes, disposable nappies, etc. A significant amount of siloxanes reaches the wastewater and are not decomposed in a conventional activated sludge wastewater treatment plant. Although a large part is volatilised to the atmosphere during the treatment, a significant amount is adsorbed to the sludge flocs.

During the anaerobic digestion of the sludge, siloxanes are released from the sludge and volatilise due to the breakdown of the organic material and the elevated temperature in the digester. Therefore the biogas is enriched with siloxanes. The siloxane concentrations typically found in biogas are between 30 and 50 mg/m3 with peaks up to 400 mg/m^3 in some WWTPs [16]. Only volatile siloxanes are detected in the biogas.

During the combustion of the biogas, these siloxanes are converted into a hard and abrasive microcrystalline silica which gradually coats the gas beneficiation equipment. The coating leads to serious motor damage by abrasion of gas motor surfaces, the overheating of sensitive motor parts (thermal insulator) and by depressing the function of spark plugs. Moreover the catalytic gas exhaust treatment is affected. The crystalline sand moreover accumulates in lubricant oil and coats turbine surfaces. For the biogas of Trecatti (UK), the presence of up to 400 mg/m3 of volatile siloxanes led to a major engine failure within 200 hours of operation. The problem is moreover enhanced by the use of efficient spark ignition engines which are fast running, operate at high temperature and use biogas only. Previous dual-fuel engines (slow, low temperature, with fuel oil to aid ignition) were less prone to silica deposits.

Although it is difficult to measure the concentration of siloxanes in sludge and biogas, a recent testing of analytical methods revealed that extraction followed by GC-MS was an adequate method [17].

All the currently used treatment techniques are end-of-pipe and remove siloxanes from the biogas. The most frequently used method for removing siloxanes is the adsorption on activated carbon. Since biogas contains a broad range of compounds (H_2S, siloxanes, organics) with concentrations covering several orders of magnitude, a competitive adsorption of siloxanes and a variety of trace compounds must be considered leading to large adsorption capacities needed for the target silicon compounds when using adsorbent materials in biogas pre-treatment [18]. Active sites of the adsorbent will retain water vapour and other pollutants, thus decreasing adsorbent-life. Moreover the adsorbent beds have to be replaced regularly because siloxanes are difficult to desorb from the material,. At Trecatti (UK) for example, a weekly change of activated carbon is necessary, with a one-day downtime at the adsorber and a cost of nearly € 2000 per change.

Other possible adsorbents are molecular sieves and polymer pellets. Polymer beads, silica gel and activated carbon can be used, all exhibiting large adsorption capacities for the siloxane D5. Especially silica gel seems to be promising and a highly cost-effective candidate, since this can be used simultaneously for biogas drying.

Absorption in non-volatile organic solvents has also been reported in both spray and packed columns (e.g. with Raschig rings). A major drawback of this gas pre-treatment method is the fact that complete siloxane elimination is difficult to obtain since the highly volatile siloxanes are easily stripped from the solvent at elevated gas flow rates. This problem does not arise if the siloxanes are chemically absorbed, i.e. they are converted to compounds of low volatility [18].

The cryogenic condensation of the siloxanes from the gas is a feasible, but an expensive alternative. When the temperature of the biogas is decreased, a condensate is formed which contains part of the siloxanes that are present.

A final reported method for removing siloxanes from biogas is chemical abatement. The caustic- or acidic-catalysed hydrolysis of the silicon-oxygen bond seems to be useful. The high stability of these compounds however requires high or low pH-values and/or high temperatures. Sulphuric, nitric and phosphoric acid were reported as well as sodium hydroxide. The removal efficiencies of these methods are however rather low.

4. Enhancing Anaerobic Digestion through Pre-Treatment Methods

Although anaerobic digestion is a promising technology for the processing of excess sludge, several disadvantages are attached to the process, as mentioned above. The limiting factors are generally associated with the hydrolysis stage [19]. During hydrolysis, cell walls are ruptured and extracellular polymeric substances (EPS) are degraded resulting in the release of readily available organic material for the acidogenic micro-organisms. This mechanism is particularly important in the digestion of sludge, since the major constituent of its organic fraction are cells, being a relatively unfavourable substrate for microbial degradation [20]. The cell envelope of micro-organisms is a semi-rigid structure which provides sufficient intrinsic strength to protect the cell from osmotic lysis. Microbial cell walls contain glycan strands crosslinked by peptide chains, causing resistance to biodegradation.

Various sludge disintegration methods have been studied as a pre-treatment: they disrupt cell walls which results in a lysis or disintegration of sludge cells. Slowly degradable, particulate organic material is converted to low molecular weight, readily biodegradable compounds, thus bypassing the rate limiting hydrolysis stage. Possible pre-treatments include mechanical, thermal, chemical and biological action, as reviewed in the present section with their working mechanism and potential.

4.1. Thermal Pre-Treatment

The heat treatment of waste activated sludge (WAS) was shown as early as 1970 to be an effective pre-treatment method for anaerobic digestion. The sludge is generally subjected to a temperature in the range of 150-200°C, although lower temperatures have also been applied. The pressures adjoining these temperatures are in the range of 600-2500 kPa. Heat applied during thermal treatment disrupts the chemical bonds of the cell wall and membrane, thus solubilising the cell components.

The optimum conditions and magnitude of the improvement vary considerably since temperature and duration of the optimum pre-treatment depend on the nature of the sludge: the greater the proportion of difficulty in hydrolysing biological sludge substances, the higher intensity of pre-treatment needed. In general, thermal pre-treatment of WAS can considerably increase methane production for mesophilic anaerobic digestion and to a lesser extent for thermophilic anaerobic digestion, showing that the impact of preconditioning is more significant in a low-rate system such as in a mesophilic digestion. Thermophilic digestion is already more efficient at VSS reduction and methane production as compared with mesophilic digestion, and reduced benefits of pre-treatment can be expected.

Evidently, the thermal pre-treatment requires the input of a considerable amount of heat, since the sludge feedstock needs to be preheated to the operating temperature (~ 700 kJ/m3) at the expense of using some of the biogas produced.

Some commercial processes were developed based on thermal pre-treatments. The Norwegian company Cambi developed a system based on thermal hydrolysis (180 °C, 30 min). A solids solubilisation and associated significant increase of biogas production could be achieved. A similar thermal treatment is sold as BioThelys® by Krüger Inc., a subsidiary of Veolia Water, yielding an increased biogas production and improved dewaterability.

4.2. Mechanical Pre-Treatment

Mechanical sludge disintegration methods are generally based on the disruption of microbial cell walls by shear stresses. Stirred ball mills, high-pressure homogenizers and mechanical jet smash techniques have been used for mechanical treatment application [21]. This methods only transfers mechanical energy to the sludge flocs and current applications are only limited, compared to other methods.

A number of technologies use high pressure gradients to rupture cell walls. Most methods disrupt the sludge by jetting it to collide with a collision plate or impact ring. Sludge is pumped under high pressure (typically between 30 and 70 bar) into a homogenizer. As the sludge reaches the impact plate or ring, the velocity increases rapidly and the pressure decreases. The resulting intense energy release results in severe turbulence and localized pressure differences, which also lead to the formation of cavitation bubbles. When applied on sewage sludge, all these forces tear apart cell walls, releasing cellular cytoplasm [22].

The combination of high-pressure homogenization and alkaline hydrolysis prior to disintegration has been patented under the trade name MicroSludge™ [23]. The process has been implemented in some full scale WWTPs (e.g. in Chilliwack WWTP, British Columbia, Canada). The MicroSludge™ pretreatment has been reported to improve VS reduction in the mesophilic digester by 18 percentage points to 78%. It was estimated that the HRT in the mesophilic digesters needed to be only 13 days as compared with the existing 18 days as a result of preconditioning.

Another technology using pressure gradients to solubilize the sludge is the Biogest Crown Disintegration System [24] The process has been reported to improve solids reduction by 20% after digestion with 30% increase in gas production.

Through milling, sludge is subjected to pure mechanical forces by squeezing the flocs between two hard surfaces or subjecting it to high shear forces. Wet milling or stirred ball milling uses small beads (mostly made of steel) for the rupturing of cell walls, with the size of the beads used as an important factor for sludge disintegration.

Although less results are available than for the other pre-treatment methods, it is seen that their efficiency of improving anaerobic digestion of sewage sludge is rather low, compared to the other methods. Application for anaerobic digestion has not been reported, partly due to the fact that this technology is heavily power-consuming, however it does not require the addition of chemicals or heat.

4.3 Chemical Pre-Treatment

Chemical pre-treatment hydrolyses the cell wall and membrane and thus increases the solubility of the organic matter in the cells. Various chemical methods have been developed, based on different operating principles.

In (thermo)chemical hydrolysis methods, an acid or base is added to solubilise the sludge. The addition of acid or base avoids the necessity of high temperatures and these methods are thus mostly carried out at ambient or moderate temperatures. The methods are shown to be an effective albeit cumbersome method for sludge solubilisation since required pH levels are extreme, and sludge needs subsequently to be re-neutralised. Their use as a pre-treatment for anaerobic digestion is therefore rather limited.

Oxidative waste sludge destruction was first practised in the aerobic Zimpro process originally designed as a wet oxidation method in the USA (1954). This process uses oxygen or air at high temperatures (260°C) and pressures (10 MPa) [25]. An effective solubilisation of a large part of the sludge was achieved. Problems with odour, corrosion and high energy cost however restrict the practical applications of this process. A modern method using wet oxidation is the Vertech process, achieving 20% solubilisation and 75% complete oxidation.

The most frequently studies oxidative methods are ozonation and peroxidation, belonging to the advanced oxidation processes and based on the generation of hydroxyl (OH*) radicals which are extremely powerful oxidants (oxidation potential 2.8 V). Due to the oxidative power, hazardous by-products were not detected [26].

These radicals are frequently generated using hydrogen peroxide H_2O_2 in combination with transition metal salts. Generally, Fe^{2+}-ions are used in combination with H_2O_2. This reaction is referred to as the Fenton peroxidation. A major drawback of this method is the necessity of bringing the sludge to a very low pH (optimum at 3). More recent research uses alternative peroxidants such as peroxymonosulphate POMS and dimethyldioxirane (DMDO) which do not require stringent reaction conditions and significantly increase the biogas production during the anaerobic treatment of raw secondary sludge. Additional tests are currently carried out using thickened sludge. Although oxidative treatments are considered promising, additional research is needed to avoid extreme reaction conditions in terms of pressures and temperatures, or pH (Fenton).

Ozone (O_3) is a powerful oxidant which is commonly used for the disinfection of drinking water and the destruction of pathogens. The treatment can also be applied to the destruction of cellular material in WAS. During ozonation, organic contaminants are oxidized in two ways. Ozone itself can directly react with dissolved chemicals, at varying rates, and is a highly selective oxidant. In addition to direct oxidation, ozone decomposes via a chain reaction mechanism to form OH radicals, which in turn can oxidize the pollutant. The two pathways can lead to different products and display different transformation kinetics.

4.4. Ultrasound

Sonication is no doubt the most powerful method to disrupt sludge cells. Although cell disintegrations of 100% can be obtained at high power levels, power consumption then becomes a serious drawback [20]. The principle of ultrasonic treatment relies on the induced cavitation process. Through subsequent compression and expansion of the fluid under the effect of the ultrasonic waves, implosions are generated which give rise to local extreme conditions (temperatures of several thousand degrees centigrade and pressures of up to 500 bar). Ultrasound treatment units are commercially available in a wide range of capacities (between 1 and 20 kW) and modular lay-out. The use of ultrasound enhancement has been tested in several WWTPs, where significant improvements in biogas production, volatile solids reduction and dewatering were obtained.

Since the degradation rate is accelerated, the dimensions of the digesters can moreover be reduced for a given load, thus reducing the impact of high capital requirements.

4.5. Enzyme Hydrolysis

Biological hydrolysis with or without enzyme addition relies on the enzymatic lysis to crack the cell-wall compounds by an enzyme catalysed reaction. Analytic processes can be used at ambient temperatures or external enzyme can be added.

CONCLUSION

It is clear from the previous discussion that biogas produced from anaerobic digestion of sewage sludge has a high potential for energy recovery. However, there is still need for additional research, especially for increasing the digestibility of the sludge and for upgrading the produced biogas in a economical viable way.

REFERENCES

[1]   Appels, L.; Baeyens, J.; Degrève, J.; Dewil, R. Prog. Energy Combust. Sci. 2008, 34, 755-781.

[2]   Qasim, S.R. Wastewater Treatment Plants: Planning, Design and Operation, 2nd ed. Boca Raton: CRC Press; 1999.

[3]   Turovskiy, I.S; Mathai, PK. Wastewater Sludge Processing. New York: John Wiley & Sons, 2006.

[4]   Metcalf & Eddy. Wastewater Engineering: Treatment and Reuse, 4th ed. New York: McGraw-Hill; 2003

[5]   SenterNovem. Biogas as vehicle fuel, a trend setter report; at http://www.novem.nl (assessed on 6th January 2010)

[6]   Degrève, J.; Everaert, K.; Baeyens, J. Filtr. Sep. 2001, 38, 49-54.

[7]   Pande, D.R.; Fabian, C. Gas Sep. Purif. 1989, 3, 143-147.

[8]   Deublein, D., Steinhauser, A. Biogas from waste and renewable resources, 1st ed. Weinheim: Wiley-VCH; 2008.

[9]   Lastella, G.; Testa, C.; Cornacchia, G.; Notornicola, M.; Voltasio, F.; Sharma, V.K. Energy Conv. Manag. 2002, 43, 63-75.

[10]   Stern, S.A.; Krishnakumar, B.; Charati, S.G.; Amato, W.S.; Friedmann, A.A.; Fuess, D.J. J. Membrane Sci. 1998, 151, 63-74.

[11]   Li, K.; Teo, W.K. J. Membrane Sci. 1993, 78, 181-190.

[12]   Toshima, N. Polymers for gas separations. VCH Publishers, NY, 1992.

[13]   Glub, J.C.; Diaz, L.F. Biogas purification processes. Biogas and alcohol fuels production, vol.II, JP Press Inc., 1991.

[14]   Dewil, R.; Baeyens, J.; Roels, J.; Van De Steene, B. Env. Eng. Sci. 2008, 25, 879-889.

[15]   Hagen, M.; Polman, E. Adding gas from biogas to the gas grid. Final Report Danish Gas Agency, 2001, 26-47.

[16]   Dewil, R.; Appels, L.; Baeyens, J. Energy Conv. Manag. 2006, 47, 1711-1722.

[17]   Dewil, R; Appels, L.; Baeyens, J.; Buczynska, A.; Van Vaeck, L. Talanta 2007, 74, 14-19.

[18]   Schweigkofler, M.; Niessner, R. J. Hazard. Mater. 2001, 83, 183-196.

[19]   Tiehm, A.; Nickel, K.; Zellhorn, M.; Neis, U. Water Res. 2001, 35, 2003-2009.

[20]   Weemaes, M.P.J.; Verstraete, W. J Chem. Technol. Biotechnol. 1998, 73, 83-92.

[21]   Climent, M.; Ferrer, I.; Baeza, M.D. Chem. Eng. J. 2007, 133, 335-342

[22]   Elliott, A.; Mahmooda, T. Water Res. 2007, 41, 4273-4286.

[23]   Stephenson, R.; Rabinowitz, B.; Laliberte, S.; Elson, P. 2005. Teaching an old digester new tricks: full-scale demonstration of the MicroSludge process to liquefy municipal waste activated sludge. In: WEF Proceedings of the Residuals and Biosolids Management Conference, Covington, KY.

[24]   Froud, C.; Weber, R.; Schmitt, W. 2005. Collective experience of the crown sludge disintegration system for carbon release for improved biological treatment-final results. In: Proceedings of the 10th European Biosolids and Biowaste Conference, Wakefield, UK.

[25]   Zimpro Environmental, Inc. Rotschild, Wis., Wet air oxidation cleans up black wastewater. Chem Eng. 1993; 175-176.

[26]   Weemaes, M., Grootaerd, H., Simoens, F., Verstraete, W. Water Res. 2000, 34, 2330-2336.

In: Sewage Sludge Management   ISBN: 978-1-61324-393-0
Editors: A. A. Zorpas and V. J. Inglezakis   © 2012 Nova Science Publishers, Inc.

Chapter 20

SEWAGE SLUDGE TREATMENT: COST BENEFITS ANALYSIS AND ECONOMIC ASPECTS

Costas J. Athanasiou[*]

Department of Environmental Engineering Democritus University of Thrace,
Xanthi, Greece

ABSTRACT

Today most of the sludge is landfilled, used in agriculture or incinerated, after treatment involving stabilization, mechanical dewatering, composting or thermal drying. The present chapter attempts to address the economic issues related to both the established and the novel options for sludge treatment and form a context for their cost-benefit evaluation. The cost estimations of the dominant sludge treatment integrated schemes exhibit considerable variations, even for nominally the same options, due to the variation of sludge properties and the process capacity and specific duty, the geographical variation of technology and labor costs, as well the corresponding fees for sludge handling. Nevertheless, anaerobic digestion presents improved economics and moreover it can be combined to other treatment process for both performance and costs optimization. Land application, in case it becomes both environmentally and socially accepted, it can maximize the economic feasibility of sludge treatment schemes, while incineration or other thermal options (gasification or pyrolysis) gain an increasing share, mostly because of the minimized solid effluents and despite its high costs.

---

[*] E-mail: costath@cperi.certh.gr / kathanas@env.duth.gr.

1. INTRODUCTION

Sewage sludge is in a liquid or semi-liquid form (0.25 to 12 %w solids), and contains the organic and inorganic content of the raw wastewater as well as the organic content generated during the wastewater treatment. The treatment of the sludge itself is a complex integrated process, the structure of which depends onto the variable composition and properties of the sludge, as well as the variable characteristics (controlled by local legislations) of the process outcomes (pollutants and/or potential products). These integrated processes involve some initial steps of thickening, conditioning, dewatering and drying to reduce sludge moisture at acceptable levels, for disposal or for the steps of treatment to follow. The latest, like composting, digestion, incineration and possibly pyrolysis or gasification, are used primarily to stabilize the organic fraction of the solid or semi-solid outcomes and potentially to transform these outcomes to products and energy. Sewage sludge management accounts for about 50 % of the total operating costs of wastewater treatment facilities, and refers to a wide range of integrated process schemes. Figure 1 presents the main paths for sludge handling and disposal and the various steps that can be involved in these paths [1,2].

There is a growing body of literature that examines the costs and benefits of the ways to take advantage of the energy and the materials content of the sludge [1-10]. Several of these ways manage to hand out both the energy and the materials content at high capital and/or operational costs, like sludge dewatering and incineration followed by ash utilization in constructing materials, or anaerobic digestion followed by drying and land application. Other options, like composting, only take advantage of the nutrients resourced in the sludge, at moderate costs, whereas a third category of options, like dewatering and landfilling, take no advantage of the sludge in an effort to overcome environmental constrains and possible social reactions at minimal economic expenditure. Due to:

- the numerous sewage sludge treatment options and process combinations,
- the high variability of sludge composition and
- the variations of the cost elements due to local market characteristics and, regarding the technology costs, the specialized equipment/processes specifications

The evaluation and selection of the most appropriate process compilation, for each specific application, should begin with a technical and economical feasibility study for the specific application. Among the various issues related to these pathways, this chapter focuses to their economic aspects, and attempts to economically evaluate the most common among the established process compilations and end-use options.

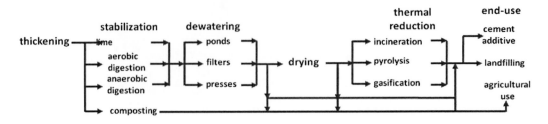

Figure 1. Sewage sludge treatment and end-use options.

Agriculture application, incineration and landfilling are the main end-steps for sludge management. The amount of incinerated sludge increases significantly when agricultural application is prohibited by local regulations. On the other hand, landfilling is increasingly restricted to the disposal of ash from sludge incineration, while ash incorporation into building materials is a rather marginal option.

2. SEWAGE SLUDGE TREATMENT OPTIONS

Almost all sewage sludge disposal routes end at landfilling, land application or incineration (due to the minimal amounts of ash effluent, incineration can also be considered as an end-use option). However, these end-use options have certain limitations. Landfilling is still the most frequent alternative but not always the economically optimum one, due to strengthening restrictions on sludge volumes and characteristics and the increasing landfilling fees. Land application results in toxics accumulation in the soil and it is also subjected to strict legislation. Finally incineration minimizes solid volumes and when combined with cogeneration recovers the energy content of the sludge, but its gaseous emissions are highly regulated.

Thus, in order to achieve sludge suitability for the aforementioned end-use options, almost all of the integrated process compilations start with sludge thickening (usually by gravity or centrifugation), a process which removes a portion of the liquid content of the sludge and increase the solids concentration. This step is crucial because a modest increase in solids concentration can significantly decrease the total sludge volume and reduce the size requirements of the downstream units. After thickening, the organic phase of the sludge ought to be stabilized for the reduction of the pathogens and the offensive odors, through a variety of potential technologies such as:

- composting for both stabilization and final disposal (through composting, organic material undergoes biological degradation, resulting in a 20 to 30 per cent reduction of volatile solids, and the end product is stable and suitable for soil conditioning)
- lime addition, in order to raise the pH and inhibit the survival of micro-organisms, which are responsible for sludge digestion, putrefaction and odor creation (lime can be added prior or after dewatering)
- aerobic digestion in open sinks for an extended period of time (from several days to months) for the total oxidation of the biodegradable matter and the microbial content, and the
- the anaerobic digestion, which reduces organic solids by about 50 % and gives the opportunity to exploit the a part of the heat content of the sludge through the utilization of released biogas in CHP engines

Solid and liquid outcomes of both aerobic and anaerobic digestion, as well as lime stabilized sludge should be further treated for final disposal or agricultural application. Following stabilization, a step of sludge dewatering is used in most cases. Dewatering involves both chemical (flocculants addition) and physical (draining in drying lagoons, centrifugation, vacuum filtration or belt pressing) phases, and further reduces the water content of the sludge to 65 – 85 % (just below the highly viscous "plastic" form of the semi-dried sludge).

After dewatering, the sludge can be disposed to a landfill or used as a soil conditioner, in case it fulfils certain (but locally specified) legislation. Or it can be led to a thermal drying stage, in which it remains for 15 – 40 minutes at 120 – 250 °C (possibly under high pressure), to be sanitized and to further reduce its moisture content (in most cases well below 10 %), according to the requirements for disposal, land application, or for the potential steps of further treatment. Thermal drying is a process of intensive heat consumption, and this heat can be provided by auxiliary fuels or by upstream (anaerobic digestion) or downstream (incineration) steps, which release the thermal content of the sludge.

These subsequent steps for the thermal reduction of the sludge's organic content usually involve the incineration of the carbonaceous solid phases, and provide the ability to gain the calorific content of the sludge in the form of high quality heat. This heat is usually enough to sustain the heat consuming stages of the integrated process and for the cogeneration of heat and electricity, through steam turbines and generators. Incineration releases elevated NOx, other oxidic and organic pollutants (SOx, dioxins, fourans and others) and particulates which contain considerable amounts of heavy metals. Thus, it requires sophisticated and expensive emissions control systems, to fulfill environmental regulations. Besides incineration, which is a wide spread and upcoming option, two other thermal processes, i.e. sewage sludge gasification and pyrolysis, are at technology demonstration level, and appear to be promising.

Gasification, at 800 °C or higher temperatures and sometimes with added steam, transforms the calorific content of the sludge into a gaseous energy carrier of H_2, CO and light hydrocarbons, which can fuel gas-turbines of higher efficiency and flexibility than steam turbines. Gasification can proceed to completion, and its ash residues can be essentially the same with those of incineration, though in practice it retains some carbonaceous phases and it is characterized as char. Both Incineration and gasification more or less liberate the same amounts of energy, but the electricity production from gasification can be considerably higher than from incineration.

Pyrolysis on the other hand, at 500 °C and in absence of any oxidizing agent (oxygen or steam), converts sewage sludge to a phase of gasses and hydrocarbon vapors, which by rapid cooling produce a mixture of char, oil (termed bio-oil or bio-cude) and permanent gases, all considered as biofuels [11-15]. Pyrolysis is of interest because it potentially possesses the ability to maximize both environmental and economic benefit of sludge treatment, in near future applications. It minimizes NOx and SOx emissions and avoids the formation of toxic organics, with low capital and operating costs, compared to incineration and gasification [12]. Despite that interest, full scale implementation has been limited [16], and in fact, the opinion that it has proved too problematic to hold a promise for the future, is quite common [4]. Pyrolysis char still contains the inorganic content of the sludge, together with non-volatile and potentially harmful organic compounds, the final combustion of which might lead to environmental impacts comparable to incineration. The acceptance of the technology is also limited by the low economic value of the produced bio-crude, as well as the complexity of the process. The economic viability of pyrolysis may be improved if bio-crude yield and quality are stable, and if value-added products could be produced from both oil and char [4,12].

Concerning these novel options for sludge handling, both gasification and pyrolysis have not been developed at any significant scale, and cannot yet be considered as technologically and economically mature [4].

3. ECONOMICS OF SEWAGE SLUDGE TREATMENT

In this section, a cost estimation of the main sludge treatment integrated schemes is attempted, in general terms, and with preliminary accuracy. The main cost factors considered included:

• construction (capital cost – CC) and operating and maintenance (O&M) costs
• sludge transportation costs and
• costs of disposal of the final by-products

The total cost for sludge treatment primarily depends on the sludge flow-rate and characteristics. The flow-rate, including the solid and organic loading, determines the equipment/process sizing of the various steps of the sludge treatment plant, and consequently the associated capital and operation costs. The six-tenths rule of thumb:

$$C_A = C_B \times (Q_B/Q_A)^{0.6} \tag{1}$$

Is applied to describe cost variation (C_A and C_B) of the same process at different capacities (Q_A and Q_B, usually expressed in tones of dry solids input) [65G++], within the preliminary accuracy standards of this chapter, in case of absence of more specified correlations.

An indicator of the economic costs of sludge handling can be expressed as the total specific cost for handling 1 ton of dry solids, although specific cost values of essentially the same process or process scheme may vary a lot. This is due to the considerable effect of sludge characteristics (for example the concentration of hazardous materials that require special treatment), the significant variations of the involved technologies, duties and specifications, as well as the variation of regulations for sludge treatment and disposal.

3.1. Investment and Operation Costs of Integrated Processes

A 2003 study of the United Nations [1] estimates the capital and O&M costs of integrated solutions for sludge dewatering (which is quite common for sludge elementary treatment and landfill disposal), anaerobic stabilization and sludge dewatering (for land application), and sludge dewatering and incineration. Based on those data, Equation 1 can lead to the capital and operation cost profiles, shown in Figure 2. These profiles denote that Incineration requires excessively higher capital and O&M costs, than anaerobic digestion, due not only to the higher capital costs of the incinerator (with sophisticated flue gas cleaning), but also due to the intensive dewatering requirements for incineration. The capital costs of these two processes can reach 60 – 70 M€ and 25 – 30 M€, respectively, for a plant capacity of 90 ton

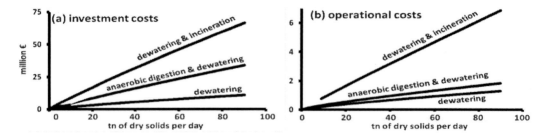

Figure 2. Indicative capital (a) and operational (b) cost estimations for three sludge treatment integrated processes (based on Ref. 1).

of dry solids per day (rough equivalent of 1 million population equivalent). The annual operation costs lay one order of magnitude lower, than capital costs, and still they are considerably higher in case of incineration.

Murray et al. [3], based on real sludge treatment plants and the corresponded literature, presented an organized set of data for the capital, operation and transportation costs related to a number of integrated processes for sewage sludge treatment. Based on those data and using Equation 1, the cost profiles of Figures 3 – 5 were estimated. According to these estimations, mechanical dewatering and sludge stabilization processes (belt pressing, lime stabilization, aerobic digestion, thermal drying and composting – Figure 3a) exhibit considerably lower capital costs than processes, which aim to take advantage of the thermal content of the sludge (anaerobic digestion and incineration – Figure 3b). In agreement to Figure 2, incineration options exhibit almost double capital costs, compared to the anaerobic digestion ones. The capital costs of the latest are closed to aerobic digestion, despite the fact that aerobic digestion does not involve equipment for energy recovery and utilization or electricity generation. In general, anaerobic digestion exhibits lower capital and treatment costs than the aerobic digestion, because aeration is a process of high installation and energy costs. Moreover, since the aerobic digestion is a slow process, it requires considerably larger sinks to store raw sludge, and this mean higher construction and land costs. Among the integrated schemes of Figure 3a, composting appears to be the least expensive way to gain a marketable product from the sludge (although heat drying might also – in some cases – lead to soil conditioners or cement additives), since other options can only lead to dewatered and stabilized sludge landfilling and land application (end use options of no income generation). According to Figure 3b, energy oriented processes exhibit a considerable difference, regarding their investment costs, which renders incineration prohibited for small scale applications.

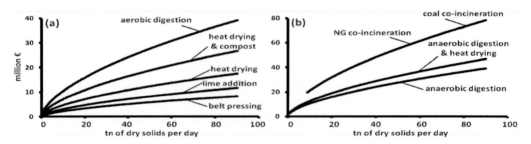

Figure 3. Indicative capital costs for several sewage sludge treatment integrated options: a) schemes which do not gain the energy content of the sludge and b) energy gaining schemes (based on Ref. 3).

Figure 4 presents the estimated annual O&M costs, for the same integrated alternatives, and shows that composting and coal co-incineration have remarkably higher operation expenses, compared to the dewatering/stabilization options (Figure 4a) and the energy utilization ones (Figure 4b). Nonetheless, with the exception of these two, energy utilization processes involve considerably lower operation costs than simple dewatering and stabilization, due to the heat recycle within the process and the potential income generation from heat and electricity selling (incomes were subtracted from the operation costs of Figure 4). Especially in the case of anaerobic digestion and further stabilization without heat drying, the annual revenues from energy and soil conditioner overcome the annual O&M expenses, leading to net annual incomes. The biogas production in both anaerobic digestion schemes (with or without heat drying) offsets fuel demand for raw sludge heating to mesophilic temperatures, and drying heat in the corresponding option, and the remainder was used to produce electricity. At least in the specific cases, on which the presented data are based [3], the utilization of generated heat for drying within the process, does not lead to an improvement in annual operational costs (or earnings), despite the fact that the heat-dried effluents might lead to marketable products (fertilizers or cement additives) with higher price than digested sludge conditioner. Moreover, heat dried digested sludge has half the volume of mechanically dewatered digested sludge, leading to considerably lower transportation costs to carry the end product to the place of its use or to dispose it (Figure 5). Concerning the low capital cost solutions of Figure 4b, heat drying exhibit higher both investment and operation costs, compared to lime addition and belt pressing dewatering, the latest consisting the most economical solution.

The operation costs of Figure 4 do not include the costs for transportation of the final solid and/or liquid products/effluents. These costs depend highly on the quantities, and they are solely presented in Figure 5, for an indicative distance of 25 km [3]. The transportation of sludge ashes is of course minimal, while the anaerobic digestion and heat drying option involves almost two thirds of the corresponding costs without drying. At the same low order of magnitude are the transportation costs of the dewatering/stabilization process schemes (Figure 6a), with the exception of belt pressing, for which annual transportation costs correspond to almost 10 % of the total O&M costs. Nevertheless, annual transportation costs are low compared to annual operation costs, and in most cases they are not determinative for the overall annual expenses (they represent 0.1 – 2.5 % of annual O&M costs).

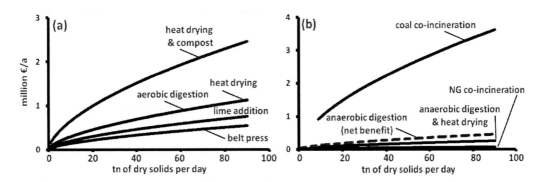

Figure 4. Indicative operation and maintenance costs for several sewage sludge treatment integrated options (based on Ref. 3).

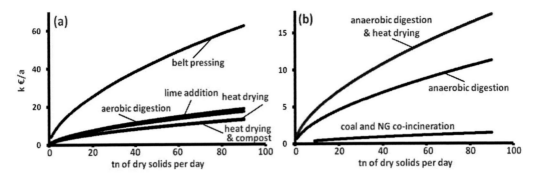

Figure 5. Indicative transportation costs of treated effluents of several sewage sludge treatment integrated options (based on Ref. 3).

In agreement to the aforementioned, Kalderis et al [18], reported that, from an economic point of view, lime stabilization is neither the most expensive nor the most economically viable, as shown in Figure 6, because the constant need for CaO may rise the total cost above that of thermal drying. Concerning composting, the installation is more expensive compared to the installations of other options, however it has quite low operating costs (it must noted that, in contrary to Figure 4, the operational costs of Figure 6b do not include any income from the potential marketability of the final effluents).

On the other hand, it is generally accepted that both investment and operational costs of thermal drying are higher than lime stabilization and composting. Nevertheless, thermal drying is an efficient method that produces a biologically stable materials of low volumes, but it requires significant fuel consumption, which rises annual costs (this rise is more profound in Figure 6b – Ref.O – than in Figure 4a – Ref. B). Compared to lime stabilization and composting, the installation costs of thermal drying are of the same order of magnitude (in rough agreement to Ref B – Figure 3). The more recent method of solar drying can reduce the moisture content of the sludge below 10 % and obtain stabilization, although its final product does not yet have any commercial value. Considering economics, the use of solar energy, where available, to increase the solids content of intermediate or the final product is advantageous [18].

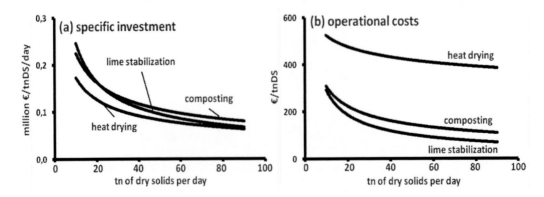

Figure 6. Investment and annual operation costs estimation, for heat drying, composting and lime stabilization (based on Ref. 18).

Along with solar drying, fry-drying options, i.e. the direct contact of sludge with heated used oils, exhibit a promising potential for a major decrease in both drying capital and operational costs. Unlike thermal drying, the energy cost of fry-drying is not the most important contributor to the overall capital and operation cost. Oil penetration and adhesion to the fried sludge contributes to its energy value. On the other hand, this consumes waste oil in the process and adds a considerable annual cost element, which represents about 50 % of the direct O&M costs [19].

Comparing Figures 4 and 6a, the first reports higher investment costs (by a factor of almost 2) for thermal drying than the second one, while for lime stabilization the corresponding values are more consisted. The same stands, more or less, for the annual operating costs for drying and lime addition between Figures 5 and 6b (i.e. Figure 5 reports higher values for drying and the values for lime addition are comparative in the two Figures). These inconsistencies between cost values of nominally the same processes are quite common in literature and they represent that:

- swage sludge specific characteristics
- the variation of technologies and process duties and
- the variation of effluent specifications

Result in remarkable variations in investment and operational costs. As for all other sludge treatment processes, the capital and operating costs for thermal drying are difficult to be compared, because of the highly variable drying duties of each drying installation. In this context, Figure 7 represents these variations for two of the most commonly used thermal drying technologies, i.e. drum and disk drying. Despite the variability of the presented capital costs, Figure7 can provide some elements to economically asses the competing systems. First of all, disk drying is, on average, 10 % more expensive than drum drying, while the required specific investment (investment over drying capacity, the latest expressed as tons of dry solids treated per day), decrease with the capacity of the installation. Nevertheless, process duty (crucially affected by raw and treated sludge moisture contents) is much more determinative for the capital investment, than the plant's capacity or technology. Other assessments, for drying capacities of the order of 3 – 10 tons of dry solids per day, report specific capital investment costs between 0.45 – 0.70 M€/tn DS/d, in agreement with Figure 7 [19].

In the same context, the capital and annual operating costs for composting may vary a lot, depending primarily on the plant capacity, the type of the installation and the technology used and the moisture content of the sludge. In this context, Figures 8 and 9 shows the variation of investment and annual operation costs, on the composting plant capacity (according to Eq. 1), the technology used and the moisture content [20]. The overall capital and operation costs, presented in these figures, are in relative consistency to the corresponding costs of Figures 3, 4 and 6.

According to Figure 8a, the capital cost for a composting plant, for the same capacity, can vary by almost 100 %, while the corresponding variation of the annual operating costs is less than 10 % (Figure 8b). Moreover, capital costs can be increased by almost 50 % if the moisture content of the feeding sludge increases from 70 to 80 % (Figure 9a). The effect of the moisture content on the annual composting costs is even stronger, and they can be

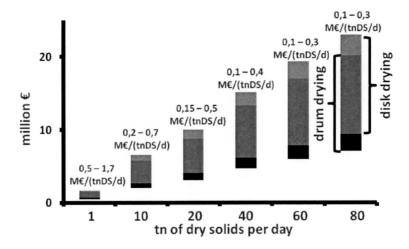

Figure 7. Capital investment variations for two of the most common thermal drying technologies (based on Ref. 19).

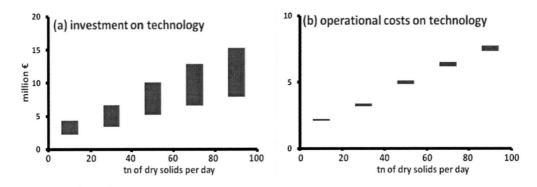

Figure 8. Composting investment cost (a) and annual operational cost (b) dependence on capacity and the technology used (based on Ref. 20).

increased by even 75 % for the aforementioned increase of moisture (Figure 9b). Regarding capital costs, the most volatile and moisture-depending components are the required land area and equipment, the cost of which can be doubled for handling sludge of 80 % moisture, compared to sludge of 70% moisture, while the dominant components of annual operating costs are the bulking agents and labor [20].

The variation of cost data for sewage sludge handling options affects all the involved technologies. Thus, Figure 10 presents the investment and annual operating cost estimations for sewage sludge landfilling and co-incineration with municipal solid wastes [6], which deviate from the corresponding estimations of Figures 3, 4 and 5 [3]. Landfilling involves sludge spreading over municipal wastes, and takes advantage of the capacity of the latest to absorb the moisture content of the sludge. This process is functional when the moisture content of the sludge is between 65 and 80 % and the mixing ratio is of the order of 1 ton of sludge per 5 tons of waste. These parameters depend upon the depth, the surface stability and

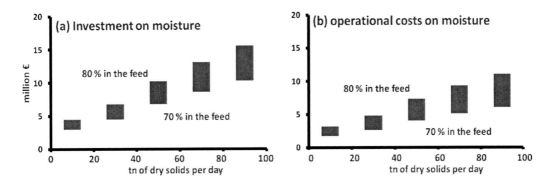

Figure 9. Composting investment cost (a) and annual operational cost (b) dependence on plant capacity and moisture content of the feed (based on Ref. 20).

the drainage behavior of the landfill, as well as the absorbent characteristics of wastes and the climatic conditions, since wet weather may reverse on-site dewatering. In any case, sludge landfilling employs mechanical sludge dewatering of variable duties and variable transportation costs. Investment costs, for landfilling include the dewatering infrastructure and equipment, and the annual operating costs usually refer to power consumption for mechanical dewatering, chemicals, labor and maintenance. Comparing dewatering and belt-pressing capital costs of Figures 2a and 3a with the investment costs for landfilling of Figure 10a, no major differences are profound and the same stands for the annual operational costs of Figures 2b and 3b, when compared to Figure 10b [6].

On the other hand, incineration investment costs of Figure 10a are about 50 % higher than the corresponding costs of Figures 2a and 3b. This difference reflects the fact that Figure 10 refers to 50 % oversized incineration plants (oversizing incineration plants is a common practice) as well as the variable nature and calorific value and the of the municipal waste, along with the requirement of more sophisticated effluent gas cleaning systems. It might also reflect differences regarding the acceptable level of sludge moisture and the corresponding dewatering requirements for sludge co-incineration with wastes, and the sludge-to-waste or carbon mixing ratios. In general, the moisture content of the sludge should be below 70 %, prior mixing with wastes and sludge/waste ratio lies in the region of 20 %. When mixed with municipal solid waste, the inert ash proportion in the feed may exceed 10% weight, increasing ash management costs. The overall sludge incineration costs include the incineration plant investment, labor and maintenance [6,17].

The gain and utilization of the calorific content of the sewage sludge for heating or cogeneration may proceed through various process combinations. Among these combinations, anaerobic digestion of raw sludge along with incineration of the digested sludge, which obtains cogeneration through biogas and digested sludge combustion leads to both the maximization of energy output and the optimization of the involved economics. In general, electricity generation from biogas is considerably higher than incineration, but it is maximized by their combination and the same stands for heat generation. Among those options, sludge incineration is the least economically viable one, while the combined process of anaerobic digestion and incineration of the digested sludge competes effectively to anaerobic digestion [7].

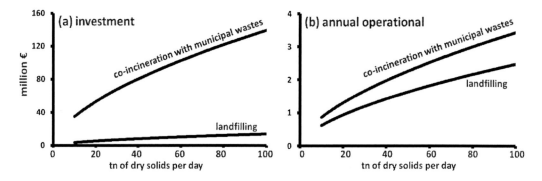

Figure 10. Investment (a) and annual operational (b) costs for sludge landfilling and co-incineration with municipal solid wastes (based on Ref. 6).

According to the above, the most expensive option for sludge treatment is its co-incineration with waste, which to a large extent is due to the costs of expanding the waste incineration capacity. This cost is dominated by the capital cost for the extra incineration capacity (estimated at 65% of the total cost), and the infrastructure for receiving and mixing sludge. Both these costs are sensitive to the life-time of the installation. On the other hand, the revenues for energy recovery are relatively small (about 10 % of the annual operating and the annualized investment costs). Potential energy recovery through incineration is very sensitive to the water content and the LHV of the sludge, which can vary considerably, and depends upon the wastewater source and treatment processes. Regarding the novel and more sophisticated options of sludge gasification and pyrolysis, they generally exhibit 30 – 40 % higher capital costs than incineration (both with co-generation), especially for small scale (few MW) applications [8,17].

3.2. Specific Life Cycle Costs

Along with the investment capital costs and the annual operation costs, probably a more thorough and sufficient way to describe the economics of sewage sludge treatment options is to divide the total internal life cycle costs (i.e. the sum of investment costs, annual operation costs, transportation costs and any other economic cost elements throughout the expected lifetime of the installation) with the total dry sludge solids, expected to be treated in the lifetime of the installation.

In this context, the previously mentioned UN study (Figure 2) reports life cycle costs (LCC), per tn of treated dry solids (DS), for several options of sludge handling, denoting that these costs may vary a lot, due to the special conditions and capacities of each specific application. According to this study, incineration specific (per ton of treated dry solids) LC costs appear to be the higher ones (of the order of 200 – 350 €/ton DS) and compare only to those of thermal sludge drying (120 – 300 €/ton DS). Specific costs for composting and landfilling appear to be of the same order of magnitude (120 – 250 €/ton), while the raw or mildly (lime pasteurized and/or mechanically dewaterd) treated sludge application in agriculture seem to be the most economic one (10 – 150 €/ton), in terms of internal LCC (i.e. not taking into account any environmental costs) [1].

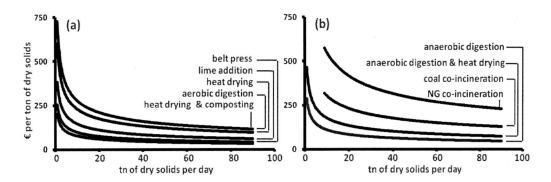

Figure 11. Indicative life cycle costs (considered installation lifetime: 20 years) per ton of treated dry solids, for several sewage sludge treatment options (based on Ref. 3).

Murray et al. [3] report 20 years life cycle internal cost estimations, for the treatment schemes of Figures 3 – 5, and the corresponding specific (per ton of dry solids treated) LCC values are shown in Figure 11. It must be reminded here, that the variation of these costs with plant capacity is based on Equation 1. In this holistic approach of the life cycle internal costs (i.e. the direct economic costs for plant construction, operation, maintenance and effluent transportation costs), coal/sludge co-incineration becomes competitive (but still more expensive) to the simple dewatering/stabilization options and the energy gaining anaerobic digestion ones, as the capacity of the plant increases. On the other hand natural gas co-incineration preserves its considerable economic feasibility objection, regardless plant capacity. Anaerobic digestion, either with or without digested sludge drying, appears more competitive than thermal drying, composting and aerobic digestion, while belt pressing and lime addition remain the most economic solutions (as in Figures 3 – 5), even after combining investment and operation costs on a life cycle basis.

According to the 2010 EU report on the environmental, economic and social impacts of the land application of sludge [4], the use of treated, usually digested, sludge in agriculture exhibits life cycle internal costs of the order of 150 – 210 €/tnDS, while LCCs for co-incineration with municipal wastes or for dedicated incineration lie in the region of 250 – 320 €/tnDS (Figure 12). These values remain in agreement with References 1 and 3, provided that the land application end-use option involves any possible treatment scheme that does not involve incineration (and in most cases not even thermal drying – Figure 12) . Moreover, and in accordance to the sewage sludge handling economic literature (Table 1), Figure 12 shows that lime stabilization is cheaper than sludge thermal drying. This is generally attributed to the fact that drying is an energy intensive process and any route that involves it can be almost as expensive as even co-incineration. Despite its high LC costs (primarily due to high annual operational costs for heat), drying is used quite frequently since it offers great flexibility in terms of storage and final disposal of the treated sludge [4].

Figure 12 also distinguishes between annualized capital costs and annual operating costs, for each treatment scheme. Thus, landfilling of mildly treated sludge (i.e. mechanically dewatered and/or lime stabilized) is shown to exhibit very high operating costs which render it at the same cost level with waste co-incineration. Apart the high operational costs, for transportation and fees, landfilling generates emissions into the air (mainly greenhouse gases like methane and carbon dioxide) and into the soil and water (ions, heavy metals, organic compounds and micro-organisms), along with other impacts such as odors, disturbance of

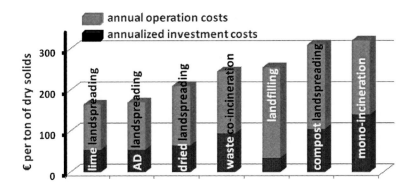

Figure 12. Specific internal life cycle costs of the main sewage sludge treatment and end-use options (based on Ref. 4).

land use and near-by vegetation. It must also be mentioned that all routes, except landfilling, differences between the highest (or lowest) LC costs and the average are of the order of 25%. For landfilling, this difference can reach 80%, due to the highly variable fees for landfilling [10]. The operating costs of landfilling include the transportation and handling of large volumes of sludge as well as the possible fees which has to be paid for landfilling. Over the entire life cycle, landfilling costs are important for both dewatering and drying, and electricity consumption is important for dewatering, composting and drying. The internal LC specific costs of the routes leading to agricultural application include extended storage periods (of up to 9 months) for treated sludge, and in case that storage requirements are restricted, costs could be further reduce by another 50 €/tnDS. In the same context, the major cost factor for Incineration is the extensive maintenance, along with the fact that the process requires extra standby capacities, which can increase LC costs by even 50%. It must be noted here that the LC specific internal costs of Figure 12 include potential benefits from energy recovery but not the potential value of the solid or semi-solid products, as soil-conditioners and/or fertilizers [4,5,9,10].

Compared to incineration, the agricultural application of sludge is a cost-effective solution, which can be beneficial to the soil provided that the levels of contaminants in the sludge are below certain and locally specified limits and that the migration and accumulation of contaminants into the environment and the food chain are addressed by adequate regulatory measures and reliable practices. Application of 10 tnDS/ha onto soil can lead to observe detectable effects of organic matter on soil physical properties. The overall LC costs of agricultural recycling are dominated by the transport requirements to the farmland (almost 50%), followed by the capital costs for the pasteurization processes (about 20%), and the payment to the farmers for receiving and using the sludge (20%). Costs of land-spreading routes are also very sensitive to the type and duration of storage, when the agricultural application is directly not possible (especially in winter). If storage was not necessary LC internal costs could significantly decreased by more than 30%, while on the contrary, if storage has to be more sophisticated (cover, odor treatment), these costs could be increased increase by 30% [4,5,8,10,21].

**Table 1. Literature estimations for internal specific LC costs
for sludge handling end-use options [3,4,6,8,9,18]**

	internal specific LCC, €/tnDS
agricultural application	75 – 225
incineration	180 – 350
landfilling	100 – 140

It could be argued that this last cost could be changed into revenue by letting the farmers pay for the nutrient content of the sludge. Treated sludge contains compounds of agricultural value and it can partially replace conventional fertilizers, and if sludge could be accepted for fertilizing arable land, there could be a potential for lowering LC costs, by creating a market demand. However, land-spreading applies to the soil the pollutants that the sludge contains, which can undergo several transfer paths, including leaching to groundwater, towards the food chain. The prevention of soil pollution implies the prevention of pollution of the waste water at source, improved sludge treatment as well as ensured monitoring of sludge quality, technical solutions which increase investment costs. Considering current regulations for sludge application onto soil, the benefit for fertilizer savings does not occur. It may also be stated that the water that sludge contains can also be considered as valuable source, especially for dry climates [5,8,10].

Provided that the quality of the sludge is confident, the value of the displaced chemical fertilizers, for a range of sludge products (including both lime-stabilized and composted sludge), can be even of the order of 75 €/tnDS. Thus, internal benefits from agricultural recycling can be of the same order. Composted sludge can have considerably improved conditioning/fertilizing properties, especially in case of the addition of vegetal co-substrates, which, nevertheless, raise all capital and annual operating costs. Nevertheless, compost selling price can very well exceed 50 € per ton and it can save a much greater portion of fertilizers, while the cost of producing sewage sludge compost can be as low as 40 €/tnDS, and it can be partially or totally offset [4,5,21].

In case the available land is limited and the quality of sludge questionable (for example for large cities) incineration might be a better alternative than land recycling. Regarding incineration, different techniques are currently in practice, divided into mono-incineration, when the sludge is incinerated in dedicated plants, or co-incineration with other wastes. The total net cost for these options are dominated by the capital cost for investing for extra capacity in waste incineration (65% of the total net cost), and are sensitive to the life-time of the investment. Moreover, incinerators should be designed with an extra standby capacity and costs are very sensitive to this parameter, too. For 100 % extra capacity, it is estimated that the LC costs of incineration could be increase by 50% (140 €/tnDS on average). For incineration options, energy generation is generally counterbalanced by the energy consumption for sludge drying and revenues for energy recovery are relatively small (they might reach about 10 % of the total costs). Dried sludge can be used as a substitute co-fuel in power stations and in cement industry, but in general, there are restrictions for burning waste materials, which makes this route unattractive [4,5,8,10,17].

Incineration also involves serious environmental concerns since it produces significant amounts of air emissions (particles with toxic metals, acid and greenhouse gases, heavy metals, dioxins, furans, etc.) and soil pollutants (ashes and flue gas treatment residues, atmospheric deposition of particulate emissions, etc), the nature and the quantities of which depend on the process and on the sludge characteristics [10].

Regarding the developing processes of pyrolysis and gasification, the first is probably the most promising, and might be proved of lower cost than incineration. However, the solid char residues of pyrolysis are not that easy to be disposed. They can be incinerated, but this solution might remove much of the advantage of pyrolysis. Besides pyrolysis, there are several sludge gasification options. Nevertheless, there a number of issues (feedstock preparation and treatment, emissions and ash management, plant capacity and the options for co-gasification, etc), which affect both performance and economic viability of sewage sludge gasification or co-gasification. The high ash content of the sludge leads to operational problems such as bed sintering, ash fouling and high concentrations of particulates in the product gas. Moreover, due to the high process temperature and the oxygen lean environment considerable amounts of trace metals are released as vapors. Thus the required gas cleaning systems should be highly sophisticated and expensive. In order for the process to be technologically and economically effective, the feedstock should be thermally dried or mixed with wood or coal to improve its calorific value. Co-gasification options and especially those of coal, allow larger and more efficient plants. Since these technologies are not yet applied at an extend that could provide economic elements for comparison to co- or mono-incineration, a rough approximation for capital and LC costs for sewage sludge pyrolysis and gasification could be based on relevant applications for other biomass resources. In this context, the capital costs for cogeneration from biomass pyrolysis and gasification are reported to exceed the corresponding costs for biomass combustion by about 5 % for pyrolysis and 35 % for gasification. Sewage sludge anaerobic digestion can reduce total LC costs for all the thermal reduction options (incineration, pyrolysis and gasification), due to the reduction of the sludge volumes and consequent decrease the required capacities and costs of the downstream processes. Thus, these options, after raw sludge digestion, can be of improved economic feasibility [4,9,22].

The so far analysis of LC costs involved only the so called internal economic costs (investment and operation costs), and have not been concerned with the external costs (i.e. the quantifiable or not environmental costs, over the life cycle of the application). Quantifiable environmental impacts, for example greenhouse gaseous emissions, losses of arable land nearby or land depreciation, are generally estimated to represent less than 15% of total LC costs (both internal and external). However, environmental impacts such as impacts on soil biology, ecosystems and some long-term effects on human health could not be quantified, and the importance of environmental costs and benefits are in fact underestimated. Taking into account both internal and external costs and benefits, landspreading routes rank best than landfilling and incineration. Landspreading of composted sludge record intermediate total LC costs, and landfilling, mono-incineration and co-incineration still entail the highest ones [5,10].

4. OUTLOOK

The economics of the potential sludge handling schemes depend crucially on whether or not the end use option of land application is allowed and under what regulative restrictions. These restrictions usually refer on the heavy metals and pathogens, as well as the types of soil and crops on which sludge may be used. Combining these restrictions with the fact that landspreading of sludge is seasonal, results in serious constraints regarding the economics of this option. Along with landspreading, legal limitations are increasingly set on the amounts and characteristics of sludge that can be disposed to landfills, and they play a crucial role in determining the economic sustainability of the sewage sludge management options. Nevertheless, in case the aforementioned constrains are fulfilled, and the nutrients content of the sludge could gain some fertilizing and economic value, land application could optimize both environmental and economic performance of a range of sludge treatment schemes. In this context, certain policies, including source control of treated wastewaters, tax incentives to sludge final acceptors and users, or even international carbon markets could form an economic background to encourage the economic viability of sludge handling solutions with minimal environment costs.

If land application is somehow feasible, anaerobic digestion is generally recognized as the economically optimum treatment scheme. Compared to heat drying and compost, anaerobic digestion is advantageous due to the significant energy savings and the considerable sludge volume reduction. However, heat drying and composting may be environmentally beneficial over anaerobic digestion, regarding the pathogen standards, the moisture content and the odors of the final product, which makes it easier to store and more socially acceptable. These factors could elevate the economics of heat drying and compost, in case the end products a capable to attract and sustain a market demand from farmers or fertilizer industries.

If land application is not feasible, the optimal sludge handling scheme is still anaerobic digestion but it requires a heat drying step of the digested sludge, in order to minimize transportation and landfilling costs. This combination leads to the elimination of fuel consumption and still preserves the flexibility to manage with the lower volumes of stabilized solids. The optimal degree of heat drying results from a trade-off between fuel and transportation costs. Compared to anaerobic digestion with heat drying, sludge incineration is more than two times as expensive, although it seems capable to achieve the largest electricity offsets. Despite large capital and operation costs, incineration is rapidly spreading because the minimized volumes of the end product (ashes are only about 30% of the initial solids content of sewage sludge). On the other hand, landfilling of dried and semi-dried sludge requires the minimum capital costs, although sludge storage and transporting might worsen economics and it attracts considerable external costs.

REFERENCES

[1] *Waste-water treatment technologies:a general review*; UN, New York, 2003.
[2] Kroiss, H. *Water Sci. Technol.* 2004, *49 (10)*, 1-10.
[3] Murray, A.; Horvath, A.; Nelson, K. *Env. Sci. Technol.* 2008, *42*, 3163-3169.

[4]   Gendebien, A *Environmental, economic and social impacts of the use of sewage sludge on land: Final Report;* Milieu Ltd (for the European Commission), Belgium, 2010.

[5]   Aubain, P.; Gazzo, A.; Le Moux, J.; Mugnier E.; Brunet, H.; Landrea, B. *Disposal and Recycling Routes for Sewage Sludge: Synthesis Report;* European Commission, 2002.

[6]   Hau, K. C.; Sculli, D. *Eng. Costs and Prod. Econ.* 1991, *21*, 133-141.

[7]   Houdkova, L.; Boran, J.; Ucekaj, V.; Elsaber, T.; Stehlik, P. *Applied Thermal Engineering* 2008 *28*, 2083-2088.

[8]   Lundin, M.; Olofsson, M.; Pettersson, G. J.; Zetterlund, H. *Resources, Conservation and Recycling* 2004 *41*, 255-278.

[9]   Hong, J.; Hong, J.; Otaki, M.; Jolliet, O. *Waste Manag.* 2009 *29*, 696-703.

[10] *Disposal and Recycling Routes for Sewage Sludge: Economic sub-component report* EU, 2002.

[11] Park, H. J.; Heo, H. S.; Park, Y. K.; Yim, J. H.; Jeon, J. K; Park, J. H.; Ryu, C. K.; Kim, S. S. *Bioresource Technology* 2010*101*, S83-S85.

[12] Kim, Y.; Parker, W. *Bioresource Technology* 2008 *99*, 1409–1416.

[13] Gasco, G.; Blanco, C. G.; Guerrero, F.; Lazaro, A. M. *J. Anal. Appl. Pyrolysis* 2005 *74*, 413-420.

[14] Mundez, A.; Gasco, G.; Freitas, M.M.A.; Siebielec, G.; Stuczynski, T.; Figueiredo, J.L. *Chem. Eng. J.* 2005*108*, 169-177.

[15] Bandosz, T.J.; Block, K.; *Appl. Catal. B* 2006 *67*, 77-85.

[16] Bridle, T.; Skrypski-Mantele, S.; *Water Sci. Technol.* 2004 *49 (10)*, 217.

[17] Werther, J.; Ogada, T. *Prog. Energ.Combust. Sci.* 1999 *25* , 55–116.

[18] Kalderis, D.; Aivalioti, M.; Gidarakos, E. *Desalination* 2010 *260*, 211-217.

[19] Peregrina, C.; Rudolph, V.; Lecomte, D.; Arlabosse, P. *Journal of Environmental Management* 2008 *8* , 246-261.

[20] Wei, Y. S.; Fan, Y. B.; Wang, M. J. *Resources, Conservation and Recycling* 2001 *33*, 203–216

[21] Song, U.; Lee, E. J. *Res. Conserv. and Recycl.* 2010 *54*, 1109-1116.

[22] Bridgwater, A. V. *Chemical Engineering Journal* 2003 *91*, 87-102.

In: Sewage Sludge Management
Editors: A. A. Zorpas and V. J. Inglezakis

ISBN: 978-1-61324-393-0
© 2012 Nova Science Publishers, Inc.

Chapter 21

GASEOUS EMISSIONS AND POLLUTION PREVENTION FROM THE INCINERATION OF SEWAGE SLUDGE

Stavros G. Poulopoulos[*]

Environmental Specialist at Hellenic Ministry of Infrastructure,
Transport and Networks, Greece

ABSTRACT

Sewage sludge is inevitably produced during the treatment of sewage in wastewater treatment plants. The huge amounts produced worldwide with increasing trend over the years makes the case of the sustainable management of sewage sludge a real challenge. For example, 7.58×10^6 tonnes are produced annually in EU-15 Member States. Among the various options regarding the management of sewage sludge, incineration exhibits a number of advantages, such as considerable reduction of sewage sludge volume, complete destruction of pathogen and organic pollutants and energy recovery. However, incineration of sewage sludge is also accompanied by the formation of hazardous air pollutants like nitrogen oxides, heavy metals, particulate matter and volatile organic compounds. The composition of sewage sludge plays a significant role on the pollutants emitted in terms of pollutant type and concentration in exhaust gases. For example, the emissions of nitrogen oxides, released to the atmosphere as a result of sewage sludge incineration, are significant due to the high content of nitrogen into sewage sludge. In the present article, the major air pollutants emitted during sewage sludge incineration in terms of amounts produced and health risk are presented as well as the relevant air pollution control techniques available.

---

[*] E-mail: s.poulopoulos@yme.gov.gr - stpoulgr@yahoo.gr.

1. INTRODUCTION

The necessity to preserve aquatic life and the quality of water in rivers, seas and surface waters in general as well as groundwater reserves imposed the treatment of wastewaters before their discharge to the aquatic environment. Sewage is the wastewater that comes from homes, factories, offices and shops and is collected and carried through the sewerage system to a wastewater treatment plant. It is about 99.9% water as it contains also rainfall run-off from roads and other areas of concrete or tarmac as well as domestic water and factory waste.

The treatment of sewage in a wastewater treatment plant results in the unavoidable production of sewage sludge on a continuous, 24 hours a day basis. Failure to regularly remove the sludge from the treatment units is a threat to the successful operation of the whole plant.

Sewage sludge (SS) is a nutrient-rich organic material, about 60% of which originates from the sewers and up to 40% of which comes from the micro-organisms, or biomass, that grows during the treatment process (EC, 2008). In general, three kinds of SS can be distinguished according to the pretreatment steps: wet (mechanically dewatered), dry, and semi-dried sludge. In *Table 1*, the proximate and elementary analysis of each type of sludge (wet, semi-dried, and semi-dry) is shown. It has to be noted that wet sludge has higher N-content than the dried one because part of the ammonia is released during the drying process (Svoboda et al., 2006). Moreover, these numbers concerning sewage sludge analysis can vary considerably locally and from installation to installation.

Table 1. Proximate and Elementary Analyses of the Wet, Semi-Dry, and Dry Sewage Sludge (Svoboda et al., 2006)

Analysis (%weight)	Wet SS	Semi-dried SS	Dry SS
			Proximate
Water	76.0	68	13
Ash (wf)	51.8	31	45.0
Volatiles (waf)	92.4	90	92.7
NH_3-N (waf)	3.2	0.9	1.19
			Elementary (waf)
C	51.9	47.8	53
H	7.8	7.68	7.8
O	29.8	38.4	31.13
N	8.8	4.6	6.5
S	1.7	0.77	1.4

* wf = on water-free basis, waf = on water- and ash-free basis.

Table 2. Amounts of sewage sludge (SS) in eleven EU Member States (EC, 2006)

Country	Total estimated SS production (in 10^6 tonnes as dry solids)	Year of data source
Austria	0.39	1999
Belgium	0.85	1997
Denmark	0.15	1997
Finland	0.14	1997
France	0.82	1997
Germany	2.48	1998
Ireland	0.39	1997
Portugal	0.24	2000
Sweden	0.23	1997
Netherlands	0.69	1999
United Kingdom	1.20	1999

Huge amounts of sewage sludge are produced annually across Europe, with a total 7.58×10^6 tonnes per year in EU-15 (at least), which makes the case of the sustainable management of sewage sludge a real challenge. Sludge production in other countries is similarly excessive. In the United States, roughly 6.2 million dry tons per year of domestic sewage are produced (Marmo, 2001). According to Dai et al. (2007), more than four million tons (dry weight) of municipal sewage sludge are produced annually in China, whereas Deng et al. (2009) raises this number to 9.18 million tons. In Taiwan, the annual volume of sludge originating from domestic and industrial wastewater treatment plants was estimated at 2.9×10^7 and 7.3×10^7 m^3, respectively (Lo and Chen, 1990). The estimated volume of sewage sludge produced in each Member State in EU-15 is presented in *Table 2*.

The available disposal methods for such large quantities of sludge include land application, landfilling, incineration, ocean dumping, and lagooning (Babel and del Mundo Dacera, 2005; Park et al., 2009). As environmental legislation became more stringent, especially in Europe, imposing increasing restrictions in relation with landfilling or utilization of SS in agriculture and forbidding disposal by sea, whereas the SS amounts produced globally have been increased dramatically, incineration attracted more attention for SS management (Suh and Rousseaux, 2002; EC, 2008).

2. INCINERATION OF SEWAGE SLUDGE

There are various options concerning the management of sewage sludge (Suh and Rousseaux, 2002):

- Incineration
- Anaerobic digestion
- Composting
- Agricultural land application
- Landfill

In *Table 3*, the most common disposal methods applied across Europe are presented. It is obvious that incineration is the first choice in some countries, such as Austria and Germany.

Table 3. Disposal methods for sewage sludge in EU Member States as percentage (EC, 2008)

Country	Agriculture	Landfill	Incineration	Other	Year of data
Austria	18	1	47	34	2005
Belgium	32	2	66		2002
Denmark	55	2	43		2002
Finland	12	6		80	2000
France	62	16	20	3	2002
Germany	30	3	38	29	2003
Greece		>90			
Ireland	63	35		3	2003
Italy	32	37	8	22	
Luxembourg	47		20	33	2004
Netherlands			60	40	2006
Sweden	10-15		2	85-90	
UK	64	1	19.5	15.5	2004
Bulgaria	40	60			2006
Czech Republic	45	28		26	2004
Hungary	26	74			2006
Poland	14	87		7	2000
Slovenia	1	50		49	2006
Slovakia		17		83	2006

Another benefit of the incineration of dry sewage sludge is the co-production of energy. Sewage sludge belongs to the category of waste and thus falls under the scope of Directive 2000/76/EC on the incineration of waste. This Directive sets several standards and technical requirements (air emissions, water discharges contamination, plant designs) that have to be respected by the operators of the plants, where incineration of dry sewage sludge takes place (EC, 2008).

In some locations sewage sludge is incinerated separately from other wastes in dedicated installations, whereas in other ones SS is incinerated along with other wastes such as municipal solid waste. Therefore, two main technologies exist for incineration of sewage sludge (Svoboda et al., 2006; EC, 2008):

- Mono-incineration, and
- Co-incineration (with coal, municipal solid waste or in other processes).

The furnace systems commonly used for SS incineration are (Svoboda et al., 2006; EC, 2008):

- Fluidised Bed Furnace,
- Multiple Hearth Furnace,
- Multiple Hearth Fluidised Bed Furnace, and
- Cycloid Furnace.

In the case of mono-incineration, fluidised beds are considered to be the best available technique because of the higher combustion efficiency and lower flue-gas volumes that generally result from such systems (EC, 2008). Consequently, the system mainly used for the incineration of sewage sludge is fluidized-bed combustion and most plants are stationary fluidized-bed furnaces, usually operated at combustion temperatures in the range of 850-900°C. The waste gas volume from this process is generally assumed to be about 8,000 m^3 (dry) per Mg of sewage sludge (dry matter). The sewage sludge delivered to the incineration plants in de-watered and/or partially dried condition usually has water content 50-70%. The content of mineral and inorganic components in sludge can be as high as 30%. The carbon content of sludge is generally about 30% (Johnke, 2000). In *Table 4*, the typical operation parameters for a fluidised bed dedicated to SS incineration are presented.

Table 4. Typical operational parameters for a fluidised bed SS incinerator (Johnke, 2000; EC, 2008)

Parameter	Value
Steam Load	300-600 kg/m^2h
Feed air amount	1000-1600 Nm^3/m^2h
Heat turnover	3-5 GJ/m^3h
Final incineration temperature	850-950°C
Residence time, open space and afterburner zone	minimum 2 sec
Preheating of atmospheric oxygen	400-600°C

In Europe, most of the incinerators that are dedicated to sewage sludge incineration are found in Germany (23), with Switzerland and UK following with 14 and 11 incinerators, respectively.

The major advantages of using incineration as a method for sewage sludge management are (Shao et al., 2008; Deng et al., 2009; Poulsen and Hansen, 2009):

- Reduction of its volume,
- Complete destruction of pathogen and organic pollutants, and
- Energy recovery.

These benefits make incineration a very important disposal method in Japan, U.S., and some EU countries. It is estimated that 38% of the sludge produced by the Member States of the European Union in 2005 had been incinerated (Sänger et al., 2001).

The main disadvantage of sewage sludge incineration is the formation of hazardous air pollutants. Specifically, the most important compounds concerning environmental and health concerns, which are emitted in the atmosphere during sewage sludge incineration, are:

- Polychlorinated dibenzo-Pdioxins dibenzofurans (PCDD/Fs),
- Polycyclic aromatic hydrocarbons (PAHs),
- Heavy metals like mercury, and
- Nitrogen oxides (NO_x).

According to the U.S. Environmental Protection Agency (*www.epa.gov*), the major air pollutants emitted by sewage sludge incinerators are:

- Particulate matter,
- Metals,
- Carbon monoxide (CO),
- Nitrogen oxides (NO_x),
- Sulphur dioxide (SO_2), and
- Unburned hydrocarbons.

In the present article, the most important gas emissions in terms of amounts emitted and environmental and health risk are classified as:

- Nitrogen oxides and nitrous oxide,
- Heavy metals, and
- Hazardous organic air pollutants.

3. AIR POLLUTANTS EMISSIONS

Nitrogen oxides and nitrous oxide. The emissions of nitrogen oxides (NO_x) and nitrous oxide (N_2O) that accompany the combustion of sewage sludge are high due to the high content of nitrogen into sewage sludge (typically 5-8 mass %) (Svoboda et al., 2006). These emissions may significantly vary from installation to installation as a result of the different compositions of the sewage sludge incinerated (Sänger et al., 2001). Nitrous oxide is involved in global warming as well as in ozone layer depletion, whereas nitrogen oxides contribute also to the greenhouse effect and may lead to the problem of acidification (Inglezakis and Poulopoulos, 2006).

Werther and Ogada (1999) reported that during the combustion of pre-dried sludge in a fluidised bed operating at $850^{\circ}C$ and oxygen concentration higher than 6%, the emissions of N_2O were 250-350 mg m^{-3}, whereas NO_x emissions were as high as 1000-1200 ppmv. The respective emissions in the case of wet sludge combustion were considerably lower. Specifically, NO_x emissions were less than 200 mg m^{-3}, whereas those of N_2O were in the range of 400-700 mg m^{-3}. NO_x emissions decreased with oxygen increasing concentrations whereas N_2O levels remained practically unaffected. The combustion of semi-dried sludge led to NO_x levels close to those of wet sludge, reaching 400 mg m^{-3}. On the other hand, N_2O emissions were lower than in the case of wet sludge with typical concentrations lower than 250 mg m^{-3} (Sänger et al., 2001; Gutierrez et al., 2005; Svoboda et al., 2006).

Pohorely et al. (2005) investigated the air pollutants releases from the combustion of dried sewage sludge in a bench-scale fluidized-bed combustor, operated under steady-state conditions. NO_x concentration measured was approximately between 600 and 1100 ppm, whereas that of N_2O was 150-400 ppm. They observed interdependence between CO and NO_x concentration in flue gas; the lower the CO concentration was in flue gas, the higher the concentration of NO_x was.

Heavy metals. The removal of heavy metals from wastewater prior to its discharge to aquatic environment is necessary if water deposits are to be protected. It is also very important to immobilise the metal removed. However, where sewage solids are incinerated, a heavy metal like mercury can be released to the atmosphere and ultimately be deposited in the environment posing a long-term threat (Balogh and Liang, 1995). For example, mercury discharged to the atmosphere can reside there for months before being deposited at local, regional or global scale, where it poses a risk to life (Lindqvist and Rodhe, 1985). Moreover, mercury deposited to lakes led to fish consumption advisories across North America and northern Europe (Lindqvist et al., 1991; Swain et al., 1992), which allows the further transport of mercury across food chain (Clarkson, 1990).

The emissions of heavy metals during sewage sludge combustion constitute an issue of great environmental importance, as the volatility of heavy metals is enhanced at high temperatures (Lopes et al., 2003). The same problems related to heavy metals emitted in the atmosphere during combustion of coal and municipal solid wastes apply to SS combustion. The most volatile metals, such as Hg, Cd, and Pb, among others, may be emitted as vapours or be enriched in small size particles during SS combustion and may even escape the air pollution control devices (Gulyurtlu, 2006). So, during combustion of SS, heavy metals may be emitted in gaseous forms or with fine ash particles (Gulyurtlu, 2006).

According to Balogh and Liang, dewatered sewage sludge from Metropolitan Wastewater Treatment (Metro) Plant in St. Paul, Minnesota in USA, which was diverted to the incineration installation, contained 399 g Hg per day, all of which had been transferred to the gas emissions during combustion. The conclusion of the authors was that almost all of the Hg entering the large municipal wastewater treatment plant was emitted to the atmosphere via sewage sludge incineration exhaust.

Shao et al. (2008) used a bench-scale fluidized bed combustor to investigate the partitioning of nine heavy metals (Cd, Co, Cr, Cu, Mn, Ni, Pb, V, and Zn) in the gas and solid phases of emissions from combustion of seven sewage sludge samples. They reported that the concentration of zinc was the highest in bottom and fly ashes, ranging from 1600-1900 and 2200-2900 mg/kg, respectively. The average concentrations of the metals analysed in both bottom ash and fly ash followed the order of Zn > Cu > Mn > Cr > Ni > Pb > V > Cd > Co. Co and Cu were mostly found in bottom ashes, whereas Pb and Zn were mostly found in fly ashes; other heavy metals were almost equally distributed between bottom and fly ashes.

Generally, the behaviour of heavy metals during combustion depends on various factors (Gulyurtlu, 2006):

- Cl content, which seems to dictate the degree of volatilization, due to the formation of volatile chlorinated species, which in turn influences the partitioning of heavy metals and the content of heavy metals in the fine fly ashes.
- temperature of combustion, since volatilization is significantly promoted for temperatures above 750-850°C for certain metals.
- atmosphere of combustion, i.e., the oxygen content and its concentration profile inside the reactor,
- nature of the fuels, as the composition of gases inside a reactor depends not solely on the air staging, but also on the characteristics of the fuels and on their behaviour during combustion.

It is apparent that the content of heavy metals in SS greatly affects the emissions of these elements. In *Table 5*, a typical composition of SS regarding the heavy metals content is shown along with the respective emissions during combustion.

The authors reported that in the case of mercury the cyclone installed was not sufficient to reduce emissions below the limits set in the relevant legislation (Gulyurtlu, 2006).

Table 5. Typical composition of SS concerning heavy metals and the respective emissions (Gulyurtlu, 2006)

Metal	Composition in mg/kg (on dry basis)	EMISSIONS IN FLUE GASES in mg/kg (on dry basis)
Mn	394	13
Co	5.3	n.d.
Ni	43	n.d.
Cr	210	n.d.
Pb	365	8.6
Cu	377	29
Cd	10.1	0.9
Zn	1132	111
Hg	2.77	193
As	<0.3	n.d.

* n.d.: non-detected.

Table 6. Heavy metals concentrations in cyclone ash samples (Marani et al., 2003)

Metal	ppm
Cd	7-21
Cr	218-650
Mn	68-121
Ni	34-69
Pb	27-120
Ti	856-1984
Zn	1894-4461

The study conducted by Wang et al. (2001) on heavy metal emissions from incinerators indicated that the partitioning of heavy metals among the incinerator discharges might vary during the combustion process depending on the composition of the wastes, the operational conditions, the physical and chemical properties of the heavy metals, and the compounds formed during combustion (Barton et al., 1990). In particular, the volatility of heavy metals and the presence of waste-derived chlorine significantly affect heavy metal emissions by the formation of volatile metallic chlorides (Fernandz et al., 1992).

Marani et al. (2003) studied the incineration of sewage sludge and reported the concentrations of seven metals in the cyclone ash after a circulating fluidised bed furnace, as shown in the *Table 6*.

The authors pointed out that their results were in close agreement with previous studies conducted by Gerstle and Albrink (1982), which suggested that among the metals considered, only Cd and Pb underwent significant volatilisation in the furnace and could be potentially emitted from the stack.

Hazardous Organic Air Pollutants. Polychlorinated dibenzo-p-dioxins and polychlorinated dibenzofurans (PCDF) occur as unwanted trace contaminants in many industrial and thermal processes. In addition to these primary sources dioxins can be released from secondary sources, such as landfills and contaminated areas or re-enter the environment via the application of sewage sludge, compost or liquid manure (Fiedler, 1996). Dioxins may undergo long-range transport to even remote areas after their release to the atmosphere. Soils and sediments are the ultimate sinks for these dioxins. Vegetation is known to filter-off lipophilic compounds from the atmosphere and may incorporate particle-bound organic pollutants. Thus, besides the primary sources, the so-called sinks for dioxins may represent secondary sources of PCDD/PCDF (Fiedler and Hutzinger, 1992).

Incineration of SS is accompanied by the release of polychlorinated dibenzo-p-dioxins dibenzofurans (PCDD/Fs) and polycyclic aromatic hydrocarbons (PAHs) to the atmosphere. These compounds pose a serious threat to human health and environment in general (Deng et al., 2009).

Mininni et al. (2004) examined the SS incineration in a pilot circulating fluidized bed furnace and a rotary kiln furnace. The results showed that organic chlorine had an apparent effect on PCDD/Fs emissions. Fullana et al. (2004) studied the formation and destruction of chlorinated pollutants in detail during sewage sludge incineration in a dual chamber reactor. Their results showed that the ash in the post-combustion zone induced the yields of PCDD/Fs, whereas the presence of HCl or $CuCl_2$ enhanced the formation of PCDD/Fs. Samaras et al. (2000) investigated the emissions of PCDD/Fs during SS combustion using a quartz-tube furnace. The emissions of PCDD/Fs were in the range of 0.5-300 ng I-TEQ/kg fuel.

Table 7. PAH concentration in the sewage sludge in five installations in Korea (mg/kg) (Park et al., 2009)

PAHs	Concentration variation (mg/kg)
Naphthalene	0.114-0.238
Acenaphthylene	0.000-0.004
Acenaphthene	0.006-0.085
Fluorene	0.030-0.851
Phenanthrene	0.078-1.110
Anthracene	0.004-0.975
Fluoranthene	0.029-0.602
Pyrene	0.209-1.458
Benzo(a)anthracene	0.009-0.926
Chrysene	0.049-1.859
Benzo(b)fluoranthene	0.061-2.315
Benzo(k)fluoranthene	0.051-3.915
Benzo(a)pyrene	0.002-0.864
Dibenz(a,h)anthracene	0.000-0.360
Indeno(1,2,3-cd)pyrene	0.004-0.900
Benzo(g,h,i)perylene	0.061-1.306

Table 8. PAH concentration in the inlet and outlet flue gas at three installations ($\mu g\ m^{-3}$) (Park et al., 2009)

Installation	A		C		E	
PAHs	Inlet	Outlet	Inlet	Outlet	Inlet	Outlet
Naphthalene	66.192	0.490	496.615	123.872	904.238	183.855
Acenaphthylene	n.d.	n.d.	0.370	n.d.	n.d.	0.061
Acenaphthene	0.182	0.096	7.525	0.084	0.265	0.060
Fluorene	n.d.	0.037	8.769	0.292	1.815	0.243
Phenanthrene	1.561	0.186	5.256	1.748	9.073	1.565
Anthracene	0.109	0.161	0.190	0.062	1.096	0.394
Fluoranthene	1.555	0.022	0.088	0.236	0.529	0.139
Pyrene	1.955	0.030	0.860	0.994	1.399	0.523
Benzo(a)anthracene	0.876	n.d.	n.d.	n.d.	n.d.	n.d.
Chrysene	1.348	n.d.	n.d.	n.d.	n.d.	n.d.
Benzo(b)fluoranthene	2.498	n.d.	0.807	0.123	1.852	0.381
Benzo(k)fluoranthene	1.072	n.d.	1.154	0.241	3.780	0.783
Benzo(a)pyrene	1.157	0.053	2.543	0.269	1.701	0.942
Dibenz(a,h)anthracene	n.d.	0.056	n.d.	n.d.	n.d.	n.d.
Indeno(1,2,3-cd)pyrene	n.d.	n.d.	n.d.	n.d.	n.d.	0.458
Benzo(g,h,i)perylene	2.460	0.022	n.d.	n.d.	n.d.	0.045
Total	*80.966*	*1.153*	*524.176*	*127.821*	*925.748*	*189.449*

n.d.: non-detected.

The content of PAHs varies and depends on the type of sewage, the size of the basin and the type of the sewage system applied as well. Obviously, the PAH presence in stack flue gas during incineration is closely related to PAHs content in SS. PAHs are produced by high-temperature reactions, such as incomplete combustion or the pyrolysis of organic materials. In general, PAHs are emitted primarily as gases during the combustion process (Park et al., 2009).

Park et al. (2009) determined the concentrations of 16 PAHs in the sewage sludge fed in five incineration facilities in Korea as shown in *Table 7*.

The corresponding concentrations measured in the inlet and outlet of the air control devices installed at the sewage sludge incineration facilities are shown in *Table 8*. Concentration of PAHs emissions at the inlet ranged from 3.926 to 925.748 µg m^{-3}.

Deng et al. (2009) employed an electrically heated lab-scale fluidised bed incinerator to study the emissions of PCDDs, PCDFs and PAHs during the combustion of pre-dried sewage sludge in five operating conditions. Their results are presented in *Tables 9-10*. According to their findings, 3-ring PAHs were dominant in the flue gas from sewage sludge mono-combustion with a total PAHs concentration of 106.7 µg m^{-3}, much higher than the values obtained from coal mono-combustion and sludge and coal co-combustion (conditions A, C, D and E). They attributed this observation to the properties of the sewage sludge, which was characterized by high volatile and low fixed carbon content. It was found that the total concentration of 2,3,7,8-substituted PCDD/Fs in flue gases released from the mono-combustion of the sewage sludge was 8996 pg m^{-3}.

Table 9. PAHs concentration in flue gases and fly ashes (Deng et al., 2009)

Combustion condition	PAHs concentration in flue gases (µg/m^3)					PAHs concentration in fly ashes (µg/g)				
	A	B	C	D	E	A	B	C	D	E
Naphthalene	0.04	2.58	0.72	0.10	n.d.	n.d.	n.d.	n.d.	n.d.	n.d.
Acenaphthylene	0.48	1.84	1.11	0.02	0.38	n.d.	n.d.	n.d.	n.d.	n.d.
Acenaphthene	0.66	1.70	4.51	0.01	2.36	0.01	n.d.	0.09	n.d.	0.01
Fluorene	1.23	24.14	5.12	0.20	1.53	0.03	0.16	0.02	n.d.	n.d.
Phenanthrene	4.81	34.55	4.25	3.65	8.87	2.44	0.53	0.57	0.09	1.35
Anthracene	0.80	12.73	0.69	3.52	0.01	0.06	0.08	n.d.	n.d.	0.01
Fluoranthene	5.16	4.10	2.53	0.91	5.08	1.11	0.45	0.98	0.11	0.59
Pyrene	3.45	2.49	3.57	0.50	4.75	0.65	0.13	0.74	0.11	1.12
Banzo[a]anthracene	2.31	6.26	1.72	0.95	0.88	0.18	0.03	0.17	0.07	0.24
Chrysene	0.92	6.73	0.99	0.07	0.54	0.42	0.30	1.16	0.18	0.56
Benzo[b]fluoranthene	0.02	2.04	1.74	0.07	0.11	0.41	1.64	1.71	0.51	0.25
Benzo[k]fluoranthene	0.01	2.40	1.20	0.02	0.32	n.d.	0.62	n.d.	0.03	0.01
Benzo[e]pyrene	1.52	1.76	2.28	0.06	0.30	0.07	0.01	0.12	0.84	0.12
Benzo[a]pyrene	0.05	1.40	0.09	n.d.	0.05	0.85	0.01	n.d.	0.17	0.65
Indeno[1,2,3-cd]pyrene	0.98	0.23	0.81	0.81	n.d.	n.d.	38.94	3.37	n.d.	n.d.
Dibenzo[a,h]anthracene	n.d.	0.33	n.d.	0.59	0.19	n.d.	0.13	n.d.	n.d.	n.d.

n.d.: non-detected.

**Table 10. 2,3,7,8-Substituted PCDD/Fs concentrations in flue gases
and fly ashes (Deng et al., 2009)**

Combustion condition	PCDD/Fs concentrations in flue gases (pg/m³)			PCDD/Fs concentration in fly ashes (pg/g)		
	A	B	C	A	B	C
2,3,7,8-TCDDg	n.d.	n.d.	0.02	n.d.	0.02	0.02
1,2,3,7,8-PCDD	n.d.	n.d.	0.25	0.56	0.01	0.22
1,2,3,4,7,8-HexCDD	1.64	3.68	1.01	n.d.	1.99	0.01
1,2,3,6,7,8-HexCDD	2.30	18.31	2.18	5.46	1.48	6.89
1,2,3,7,8,9-HexCDD	n.d.	n.d.	5.69	9.45	1.63	15.66
1,2,3,4,6,7,8-HepCDD	77.87	178.96	32.43	17.97	24.32	14.84
OCDD	1352.39	7265.29	781.67	280.56	289.15	210.23
2,3,7,8-TCDF	n.d.	n.d.	0.12	0.01	0.01	0.03
1,2,3,7,8-PCDF	6.34	62.36	0.03	2.21	0.23	0.22
2,3,4,7,8-PCDF	n.d.	59.93	0.89	n.d.	2.56	0.12
1,2,3,4,7,8-HexCDF	6.33	37.61	5.18	2.45	3.36	0.02
1,2,3,6,7,8-HexCDF	4.64	146.03	10.23	1.36	2.12	1.23
1,2,3,7,8,9-HexCDF	n.d.	15.19	n.d.	5.28	3.25	8.88
2,3,4,6,7,8-HexCDF	n.d.	n.d.	n.d.	n.d.	9.45	0.18
1,2,3,4,6,7,8-HepCDF	17.72	416.45	2.58	3.78	7.89	1.19
1,2,3,4,7,8,9-HepCDF	n.d.	n.d.	n.d.	n.d.	7.56	2.26
OCDF	65.23	752.11	13.95	25.58	56.66	42.62
PCDD/Fs	1534.46	8955.93	856.23	354.66	411.71	304.61
I-TEQconcentration	4.18	69.14	4.18	3.32	4.39	3.93

* n.d.: non-detected

Shao et al. (2008) reported the emissions of hazardous air pollutants including PAHs, besides heavy metals, from the combustion of seven sewage sludge samples in a bench-scale fluidized bed combustor. The concentrations of four types of hazardous air pollutantss (classified as aliphatics, monoaromatic compounds, cyclic hydrocarbons, and PAHs) varied in flue gases and fly ashes with sewage sludge samples studied.

4. AIR POLLUTANTS REDUCTION TECHNIQUES

There are two approaches for reducing the air pollutants releases in the atmosphere (Inglezakis and Poulopoulos, 2006):

- Pollution prevention measures, and
- Waste treatment (end-of-pipe techniques).

The first approach involves measures that lead to cleaner processes and less waste to be discharged. That means that the operating conditions of the incineration process, such as temperature and oxygen concentration, are controlled so that air pollution is prevented or minimized at least. The second approach is an end-of-pipe treatment of the flue gases for the destruction of air pollutants inevitably produced during the process. It is obvious that both approaches have their own usefulness and have to be employed for achieving less and harmless emissions in the atmosphere.

Both waste minimisation and waste treatment techniques depend on the type of pollutant. As an example of the first approach, nitrous oxide emissions can be reduced by applying staged combustion in the case of pre-dried sludge. In contrast, this technique is not effective for wet and semi dried sludge. An increase in temperature of the fluidised bed can reduce substantially N_2O emissions during the combustion of sewage sludge. It is reported that N_2O concentrations decreased from 160 mg m^{-3} to 20 mg m^{-3} as the freeboard temperature was increased from 900∘C to 934∘C (Svoboda et al., 2006).

On the other hand, suitable gas treatment equipment like cyclones is required for the abatement of mercury emissions. Mercury is considered to be one of the most harmful substances for the environment and human health. Therefore, legislation is setting increasingly strict flue gas emission limits. It is expected that older incineration installations cannot meet new standards regarding mercury emissions. For example, in Germany the average daily emission limit of 30 µg mercury/m^3 flue-gas is set by the 17th regulation for the Federal Emission Law for waste incineration. That limit cannot be achieved without special abatement facilities (Malerius and Werther, 2003).

According to Gutierrez et al. (2005), two factors affect considerably the nitrous oxide emissions from the incinerations of wastes with high nitrogen content as sewage sludge:

- Incineration temperature, and
- Avoiding the selective non-catalytic reduction de-NO_x method based on urea or ammonia treatments.

Utilisation of combustion temperatures above 900°C and installation of modern selective catalytic reduction de-NO_x systems may lead to negligible emissions of N_2O.

The selection of the incineration equipment as well as the pretreatment of the sewage sludge to be combusted may be also considered important regarding the exhaust gases emitted. According to the reference document on the best available techniques (BAT) for waste incineration released in August 2006 by the European Commission, the best available techniques for sewage sludge incineration are considered to be the following:

- Fluidised bed technology is considered as BAT for installation dedicated to the incineration of sewage sludge because of the higher combustion efficiency and lower flue-gas volumes that generally result from such systems, and
- The drying of the sewage sludge to the extent that additional combustion support fuels are not generally required for the normal operation of the installation.

As far as particulate matter is concerned, which may be enriched with volatile trace metals such as cadmium, lead, and zinc, wet scrubbers are the most common option for minimising their emissions from sewage sludge incinerators. Low pressure drop spray towers, wet cyclones and venturi/impingement tray scrubber combinations are among the available solutions. In the case of co-combustion of sewage sludge with municipal solid waste, baghouses and electrostatic precipitators are the normally selected equipment. For multiple hearth incinerators, the impingement tray scrubber is the most common option, whereas most electric incinerators and many fluidised bed incinerators utilise venture scrubbers only.

According to Turovskiy and Mathai (2006), control technologies for particulate matter from flue gases include:

- Mechanical collectors,
- Wet scrubbers,
- Fabric filters, and
- Electrostatic precipitators.

The exact nature of the particulate matter, the emission limits imposed as well as the conditions of the flue gases play the most significant role on the selection of the specific control technology.

Mechanical collectors, such as settling chambers, impingement separators and cyclones, exert inertia forces for particle separation, whereas water is utilised in wet scrubbers for removing dusts from gas streams. Wet scrubbers include spray towers, cyclone scrubbers, ejector venture scrubbers as well as venture scrubbers and their use can be also advantageous in removing water soluble pollutants like hydrogen chloride, sulphur dioxide and ammonia. Fabric filters (baghouses) collect particulates in the gas stream by passing the gas through a filter medium or fabric, whereas electrostatic precipitators, including dry electrostatic precipitators and wet electrostatic precipitators, electrostatic forces are utilised for collecting particulates from the exhaust gases (Turovskiy and Mathai, 2006).

In most of the cases, it is not possible for waste incinerations to meet the air emission limit values for PCDD/Fs only by means of pollution prevention measures (i.e. combustion related measures). Abatement measures are, therefore, necessary (EC, 2006). Selective catalytic reduction systems, which are primarily used for NO_x reduction, can also be used to destroy gas phase PCDD/Fs through catalytic oxidation. Typically, 2-3 SCR catalyst layers are required to provide combined NO_x and PCDD/Fs reduction. It has to be pointed out that in waste incineration the majority of airborne PCDD/Fs may be adhered to particulates, which means that techniques applied to remove dust will also remove PCDD/Fs, whereas SCR and other catalytic methods will destroy the PCDD/Fs proportion in the gas phase. Consequently, suitable combination of particulate matter removal techniques with catalytic destruction will generally give the lowest overall emissions of PCDD/Fs to air (EC, 2006).

In the reference document on BATs for waste incineration (EC, 2006), an exhaustive reading on pollution prevention and pollution control measures for waste incineration is provided.

CONCLUSION

Incineration is an advantageous method for treating the increasing amounts of sewage sludge produced worldwide from wastewater treatment plants. However, the incineration process is accompanied by the production of undesired air pollutants, whose reduction can be accomplished by employing both pollution prevention and pollution control techniques. The most important issues regarding sewage sludge incineration and the respective air pollution issues are the following:

i.   The emissions of nitrogen oxides and nitrous oxide that accompany the combustion of sewage sludge are high due to the high content of nitrogen into sewage sludge (typically 5-8 mass %). NO_x emissions can be as high as 1000-1200 ppmv and these of N_2O as 250-350 mg m^{-3} during the combustion of pre-dried sludge in a fluidised bed operating at 850°C (Werther and Ogada, 1999). In contrast, wet sludge combustion leads to considerably lower NO_x and N_2O emissions.

ii.   The emissions of heavy metals during sewage sludge combustion constitute an issue of great environmental importance, as the volatility of heavy metals is enhanced at high temperatures (Lopes et al., 2003). The most volatile metals, such as Hg, Cd, and Pb, may be emitted as vapours or be enriched in small size particles during sewage sludge combustion and may even escape the air pollution control devices. Heavy metals may be emitted in gaseous forms or with fine ash particles (Gulyurtlu, 2006). The content of heavy metals in sewage sludge greatly affects the emissions of these elements. Depending on the metal, these emissions can vary typically in the range of 1-200 mg/kg on dry basis (Gulyurtlu, 2006).

iii. Polychlorinated dibenzo-p-dioxins dibenzofurans (PCDD/Fs) and polycyclic aromatic hydrocarbons (PAHs) can also be released to the atmosphere during sewage sludge incineration. These compounds belong to hazardous organic air pollutants and pose a serious threat to human health and environment in general (Deng et al., 2009). Typical concentrations of PAHs in exhaust gases range in 3-1000 μg m^{-3} depending on the composition of sewage sludge.

iv.   The employment of fluidised bed technology along with the drying of sewage sludge prior to its combustion can be considered as the best available techniques for air pollution prevention, whereas mechanical collectors, wet scrubbers, fabric filters and electrostatic precipitators are the most commonly used equipment for air pollution control.

REFERENCES

Babel S. and del Mundo Dacera, D., Heavy metal removal from contaminated sludge for land application: a review, Waste Management, vol. 26, 2005, 988-1004.

Balogh S. and Liang L., Mercury pathways in municipal wastewater treatment plants, Water Air and Soil Pollution, vol. 80, 1995, 1181-1190.

Clarkson T. W., Human health risks from methylmercury in fish, Environmental Toxicology Chemistry, vol. 9, 1990, 957-961.

Dai J., Xu M., Chen J., Yang X. and Ke Z., PCDD/F, PAH and heavy metals in the sewage sludge from six wastewater treatment plants in Beijing, China, Chemosphere, vol. 66, 2007, 353-361.

Deng W., Yan J., Li X., Wang F. Chi Y. and Lu S., Emission characteristics of dioxins, furans and polycyclic aromatic hydrocarbons during fluidized-bed combustion of sewage sludge, Journal of Environmental Sciences, vol. 21, 2009, 1747-1752.

European Commission (EC), Integrated Pollution Prevention and Control, Reference Document on the Best Available Techniques for Waste Incineration, August 2006.

European Commission (EC), Environmental, economic and social impacts of the use of sewage sludge on land, Final Report, 2008.

Fernandz M., Martinez L., Segavra M., Garcia J. and Espiell F., Behavior of heavy metals in the combustion gases of urban waste incinerators, Environmental Science & Technology, vol. 26, 1992, 1040-1047.

Fiedler H. and Hutzinger O., Sources and sinks of dioxins: Germany, Chemosphere, vol. 25(7-10), 1992, 1487-1491.

Fiedler H., Sources of PCDD/PCDF and Impact on the Environment, Chemosphere, vol. 32(1), 1996, 55-64.

Fullana A., Conesa J.A., Font R. and Sidhu S., Formation and destruction of chlorinated pollutants during sewage sludge incineration, Environmental Science and Technology, vol. 38(10), 2004, 2953-2958.

Gerstle R.W. and Albrinck D.N., Atmospheric emissions of metals from sewage sludge incineration. Journal of the Air Pollution Control Association, vol. 32, 1982, 1119-1123.

Gulyurtlu I., Lopes M., Abelha P., Cabrita I., Oliveira J., The Study of Partitioning of Heavy Metals During Fluidized Bed Combustion of Sewage Sludge and Coal, Journal of Energy Resources Technology, vol. 128, 2006, 104-110.

Gutierrez M.J.F., Baxter D., Hunter C. and Svoboda K., Nitrous oxide (N_2O) emissions from waste and biomass to energy plants, Waste Management Research, vol. 23, 2005, 133-147.

Hall J.E. and Dalimier F., Waste management-sewage sludge: survey of sludge production treatment, quality and disposal in the EC. EC reference no. B4-3040/014156/92, 1994.

Inglezakis V.J. and Poulopoulos S.G., Adsorption, Ion Exchange and Catalysis: Design of Operations and Environmental Applications, Elsevier, Amsterdam 2006.

Johnke B., Good Practice Guidance and Uncertainty Management in National Greenhouse Gas Inventories, Waste Sector, Germany, 2000.

Lindqvist O. and Rodhe H., Atmospheric mercury – a review, Tellus, vol. 37B, 1985, 136-159.

Lindqvist O., Johansson K., Aastmp M. and Andersson A., Mercury in the Swedish environment – Recent research on causes, consequences and corrective methods, Water Air and Soil Pollution, vol. 55, 1991, 1-261.

Lo K.S.L. and Chen Y.H., Extracting heavy metals from municipal and industrial sludges, Science of Total Environment, vol. 90, 1990, 99-116.

Lopes M., Abelha P., Lapa N., Oliveira J., Cabrita I. and Gulyurtlua I., The behaviour of ashes and heavy metals during the co-combustion of sewage sludges in a fluidised bed, Waste Management, vol. 23, 2003, 859-870.

Malerius O. and Werther J., Modeling the adsorption of mercury in the flue gas of sewage sludge incineration, Chemical Engineering Journal, vol. 96, 2003, 197-205.

Marani D., Braguglia C., Mininni G. and Maccioni F., Behaviour of Cd, Cr, Mn, Ni, Pb, and Zn in sewage sludge incineration by fluidised bed furnace, Waste Management, vol. 23, 2003, 117-124.

Marmo L., Towards a revision of the sewage sludge directive 86/278/EEC. In: Proceedings of the Conference on Researching the Sludge Directive, October 30- 31, 2001, Brussels.

Mininni G., Sbrilli A., Guerriero E. and Rotatori M., Dioxins and furans formation in pilot incineration tests of sewage sludge, spiked with organic chlorine. Chemosphere, vol. 54(9), 2004, 1337-1350.

Park J., Lee S., Kim J., Kim M., Kwon O. and Jung D., Behavior of PAHs from sewage sludge incinerators in Korea, Waste Management, vol. 29, 2009, 690-695.

Pohorely M., Svoboda K., Trnka O., Baxter D. and Hartman M., Gaseous Emissions from the Fluidized-Bed Incineration of Sewage Sludge, Chemical Papers, vol. 59(6b), 2005, 458-463.

Poulsen T. and Hansen J., Assessing the impacts of changes in treatment technology on energy and greenhouse gas balances for organic waste and wastewater treatment using historical data, Waste Management & Research, vol. 27, 2009, 861-870.

Samaras P., Blumenstock M., Schramm K.W. and Kettrup A., Emissions of chlorinated aromatics during sludge combustion, Water Science and Technology, vol. 42(9), 2000, 251-258.

Sänger M., Werther J. and Ogada T., NO_x and N_2O emission characteristics from fluidised bed combustion of semi-dried municipal sewage sludge, Fuel, vol. 80, 2001, 167-177.

Shao J., Yan R., Chen H., Yang H., Lee D. and Liang D., Emission Characteristics of Heavy Metals and Organic Pollutants from the Combustion of Sewage Sludge in a Fluidized Bed Combustor, Energy & Fuels, vol. 22, 2008, 2278–2283.

Svoboda K., Baxter D. and Martinec J., Nitrous Oxide Emissions from Waste Incineration, Chemical Papers, vol. 60(1), 2006, 78-90.

Suh Y. and Rousseaux P., An LCA of alternative wastewater sludge treatment scenarios, Resources, Conservation and Recycling, vol. 35(3), 2002, 191-200.

Swain E. B., Engstrom D. R., Brigham M. E., Henning T. A. and Brezonik P. L., Increasing Rates of Atmospheric Mercury Deposition in Midcontinental North America, Science, vol. 257, 1992, 784-787.

Turovskiy I. S. and Mathai P. K., Wastewater Sludge Processing, John Wiley & Sons, New Jersey 2006.

Wang K., Chiang K., Tsai C., Sun C., Tsai C. and Lin K., The effects of $FeCl_3$ on the distribution of the heavy metals Cd, Cu, Cr, and Zn in a simulated multimetal incineration system, Environment International, vol. 26, 2001, 257-263.

Werther J. and Ogada T., Sewage sludge combustion, Progress in Energy and Combustion Science, vol. 25(1), 1999, 55-116.

In: Sewage Sludge Management
Editors: A. A. Zorpas and V. J. Inglezakis

ISBN: 978-1-61324-393-0
© 2012 Nova Science Publishers, Inc.

Chapter 22

GLOBAL CHANGE AND SOIL CARBON

Jose Navarro Pedreño, José Martin Soriano Disla and Ignacio Gómez Lucas

Departamento de Agroquímica y Medio Ambiente,
Universidad Miguel Hernández de Elche, Alicante, Spain

ABSTRACT

Global change includes a number of processes where climate change may be the most important. Climate change is associated to the rising levels of CO_2 in the atmosphere which are causing important environmental problems. So, one of the solutions to this problem may be the storage of atmospheric carbon in the most important terrestrial ecosystem sink, the soil. Soils have lost important amounts of organic matter for decades and sewage sludge production has been increased due to the necessary treatment of waste water. Thus, the use of organic residues as soil amendments (i.e. sewage sludge) is a key factor to recover the soil organic matter and adequate agricultural management can contribute positively to combat the presence of CO_2 in the atmosphere.

1. INTRODUCTION

Global change is the name given to a group of processes affecting the Earth environment: abrupt land-use changes due to poor management techniques including overexploitation, development and international market pressures, natural habitats degradation and loss, and pollution. (Regato, 2008) including global warming (caused by greenhouse effect which is the mechanism whereby incoming solar radiation reflected by the earth surface is trapped by the presence of carbon dioxide and other greenhouse gases) and its consequence, the climate change (the solar heat gain could disrupt climate patterns). Most of these processes are interacting between them but climate change may be the most important and the key to solve problems derived from other processes.

Climate change causes environmental problems like the rapid increment of temperature, changes in atmospheric and meteorology phenomena and the increase of geological risks associated to climatic events (i.e. floods, landslides and others). The main characteristic of the group of processes associated to global change is that man is the main vector causing the change. The atmospheric system is finely tuned and there are limits to what it, or any other natural system, will tolerate. The current global warming trend shows that humankind cannot really control atmospheric processes and should exercise much more care (Porteous, 2008).

One of the human impacts related to global change is the displacement of carbon from land to atmosphere. Approximately, half of all soil carbon in managed ecosystems has been lost to the atmosphere for the past two centuries due to cultivation; a loss which represents an opportunity for carbon sequestration. Carbon sequestration refers to the storage of carbon in stable solid form (McCarl *et al.* 2007).

The climate change is attributed to the increase and persistency of greenhouse gases in the atmosphere and the consequences derived from the rapid change of atmosphere composition. The sources of these gases are mainly human activities.

The increment in greenhouse gas levels for the last centuries is due principally to CO_2 arising from the use of fossil fuels since the industrial revolution (18th century) and losses from soils (mainly from green revolution in mid 20th century after World War II and the use of industrial fertilizers), and other minor sources. However, the increase in atmospheric CO_2 levels has proved to be less than was previously in ancient Earth times but nowadays, the problem may be the quick increment of the concentration of carbon dioxide (CO_2) in a small geological period of time.

The atmospheric concentration of CO_2 has increased from 280 ppmv in 1750 to 367 ppmv in 1999 and is currently increasing at the rate of 1.5 ppmv year^{-1}; the average global surface temperature has increased since late 19th in 0.6 °C, with the current warming rate of 0.17 °C decade^{-1} (IPCC, 2001). These changes due to anthropogenic activities are the subject of almost unanimous consensus in the international scientific community.

Carbon cycles globally among three distinct pools: the atmosphere, the ocean, and terrestrial ecosystems. The atmosphere currently contains 720-760 Pg (Pg = 1 Gton) of carbon with nearly all of it as CO_2. The oceans store about 38,000-40,000 Pg and terrestrial systems contain about 2,000-2500 Pg, most of which resides in the soil (1,500 Pg) and about 560-650 Gt in vegetation (Lal, 2004; McCarl *et al.*, 2007; Paul *et al.*, 2008; Read, 2009). Globally, the soil C pool is almost two-fold bigger than the atmospheric pool.

2. SOIL CARBON AND SOIL ORGANIC MATTER IN SOIL

Soil organic carbon (SOC) in the top 1 meter comprises about ¾ of the earth's terrestrial carbon (Soil Science Society of America, 2001). Carbon is the main element that constitutes the organic matter of the soils and both are closely related.

The organics forms presented in the soil constitute the soil organic matter which is subjected to changes due to natural conditions and the anthropogenic management of the lands, determining the carbon cycle. The incorrect management and mismanagement of soil promote unsustainable land use and important losses of carbon from the soil system.

The carbon cycle in terrestrial ecosystems is determined by the photosynthesis, which is the main natural pathway for the uptake of atmospheric CO_2 in these ecosystems and transform it into organic compounds. Direct plant carbon sequestration occurs as plants photosynthesize atmospheric CO_2 into plant biomass. Basic equation of the photosynthesis is represented as:

$$6CO_2 + 6H_2O + Energy\ (sun) \rightarrow C_6H_{12}O_6 + 6O_2$$

The organic matter (biomass synthesized) is deposited in oceans and soils. This organic matter in the soil is always ultimately degraded at one level or another of the trophic networks by respiration (with the release of CO_2) or, under anaerobic conditions, by fermentation (with the release of CH_4). This organic matter can also be destroyed by combustion (i.e. forest fires), which also gives rise to CO_2. Average annual exchanges between the continental biosphere and the atmosphere reach 120 GtC/year (Arrouays *et al.*, 2002).

The term "soil C sequestration" implies removal of atmospheric CO_2 by plants and storage of fixed C as soil organic matter (Lal, 2004). Carbon sequestration promotes the long term storage of carbon or other forms of carbon and the mitigation of global warming. The amount of carbon sequestered at a site reflects the long-term balance between carbon uptake and release mechanisms (McCarl *et al.*, 2007). To determine a positive carbon sequestration, the balance between uptake and release has to be favourable to the uptake considering a long time period.

Carbon dioxide is usually captured from the atmosphere through biological processes but also by chemical or physical processes. It occurs through direct and indirect fixation of atmospheric CO_2. Direct soil carbon sequestration occurs by inorganic chemical reactions that convert CO_2 into soil inorganic carbon compounds such as calcium and magnesium carbonates.

Carbon can be accumulated in the soil from biomass, mainly in an organic form. This organic matter arises from plant parts (leaves, roots, etc.), dead organisms, animal waste and rhizodeposition (deposition of organic compounds excreted by roots in the soil), and also include microbial biomass. Nowadays, another source of organic carbon in soil is the application of biosolids as sewage sludge. These residues are subjected to several processes that determine the carbon content and the fraction which is stabilized. However, when biosolids are added, SOM dynamic is the main pathway to produce stable forms of SOC.

The factors influencing the SOM may be divided into two groups: natural factors and human-induced factors (Jones et al., 2004):

The most important *natural factors* are:

1. Climate: temperate or Mediterranean for example;
2. Soil parent material: acid or alkaline (or even saline);
3. Land cover and/or vegetation type;
4. Topography: slope and aspect.

The *Human-induced factors* can be summarised as follows:

1. Land use and nature of farming systems;
2. Land management;
3. Degradation of soil and land.

Thus soil carbon storage potential depends on external factors as climate (temperature and precipitation), the nature of parent geological materials (i.e., texture and mineralogy), and vegetation (crop) type and land management practices. Net sequestration occurs with management systems that increase plant material or the organic wastes from several origins that are returned to the soil, and reduce carbon loss or both. However, soils have a limited possibility to ware organic carbon. Long term experiments developed in IACR-Rothamsted (Harpenden, UK) have demonstrated that soils have a finite capacity to increase in SOC, but most of the soils cultivated are enough far away from their limit to sequester carbon.

It is a key factor for carbon sequestration in the soil to know that it is affected by the inner processes affecting organic matter. The three main mechanisms of SOM stabilization proposed are: chemical stabilization, physical protection and biochemical stabilization (Stevenson, 1994; Six *et al.*, 2002; Mikutta *et al.*, 2006).

Soil organic matter can be stabilized through the chemical stabilization refers to intermolecular interactions between organic and inorganic substances that decreases the availability of the organic substrate due to complexation of functional groups and charges in conformation (Guggenberger and Kaiser, 2003); organic matter can be protected by intimate association with silt and clay particles forming physical barriers between microbes and enzymes) and physically stabilized through aggregate formation (John *et al.*, 2005); and the formation of recalcitrant SOM compounds (recalcitrant connotes the preservation of SOM caused by structures inherently stable against biochemical decay such condensed and lignin-derived aromatic carbons or aliphatic compounds);

Thus the main factors affecting the stabilization of SOM are mainly associated to the following processes:

1. Aggregate turnover which favours the physically protection.
2. Adsorption and desorption in soil mineral surfaces, especially in small particles (silt and clay).
3. Condensation and complexation forming recalcitrant compounds of large molecular weight.

The stabilization of SOM is the important step for enhancing SOC and C sequestration. It is remarkable to consider that many of the mechanisms involving stabilization and decomposition are acting at the same time, which makes difficult to study the dynamic of the organic matter in soils. For instance, "dead" organic matter undergoes a series of biotransformations, including decomposition and finally mineralization by micro-organisms, with the release of CO_2. The rate of these phenomena (stabilization *vs.* decomposition) depends on the composition of the organic matter and local physicochemical conditions (humidity, temperature, oxygen, etc.). If the organic matter is stabilized, for instance when organic matter is associated with mineral particles (particularly clay, forming humus-clay complex) which provide physical protection, this organic matter resists against the activity of microorganisms.

The stabilization of SOM is the most important step for enhancing SOC and C sequestration. It is necessary to consider that many of the mechanisms involving stabilization and decomposition are acting at the same time, which makes difficult to study the dynamic of the organic matter in soils. For instance, "dead" organic matter undergoes a series of biotransformations, including decomposition and finally mineralization by micro-organisms, with the release of CO_2. The rate of these phenomena (stabilization vs. decomposition) depends on the composition of the organic matter and local physicochemical conditions (humidity, temperature, oxygen, etc.). If the organic matter is stabilized, for instance when organic matter is associated with mineral particles (particularly clay, forming humus-clay complex) which provide physical protection, this organic matter resists against the activity of microorganisms.

Most of the stable organic carbon (about 73%) was associated with soil mineral phase rather than being recalcitrant, but recalcitrant fraction had on average longer mean residence times (Mikutta et al., 2006).

There is no definitive storage of organic carbon in the soil, because in the long term, any organic matter is mineralized. However, the residence times for organic carbon in the soil, which average some decades, can range from a few hours to several millennia.

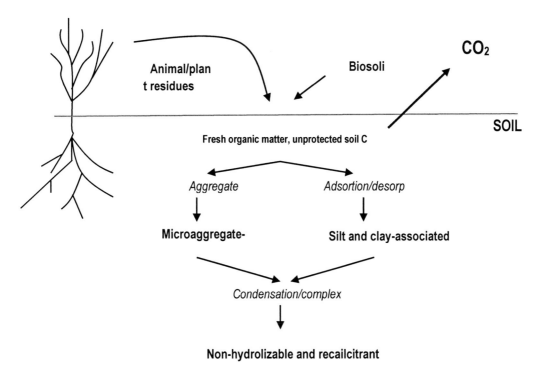

Figure 1. Soil organic matter dynamics for the stabilization of carbon.

Figure 2. Accumulation of carbon found in a paleontological site in Crevillente (SE Spain) (about 6-7 million of years) with an important amount of organic matter presented in an ancient horizon (authors, 2008).

Thus the evolution of carbon stocks is determined by the balance between the input of organic matter and the output of CO_2 and controlled by the stabilization of SOM. So, as it was stated in the introduction, soils have lost half of the soil organic carbon and this fact represents an opportunity for carbon sequestration

There are several strategies for enhancing SOM pool (Smith *et al.*, 1997; Houghton and Goodale, 2004; Lal, 2009) and as a consequence can increment SOC and carbon sequestration.

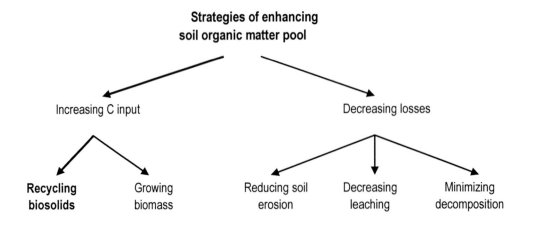

Figure 3. Several strategies of enhancing and preserving SOM based in several authors.

3. AGRICULTURE, SOIL CARBON AND GLOBAL CHANGE

The reduction of the CO_2 content in the atmosphere is an international challenge. For instance, the European Union greenhouse gas reduction target is for a 20% by 2020 (actually CO_2 reduction) and 60% reduction is promised by 2050.

Soils and agriculture can offer several ways to solve, in part, the problem. The Intergovernmental Panel on Climate Change (IPCC) identifies three main carbon mitigation options for agriculture (IPCC, 1996):

1. The reduction of agriculturally related emissions.
2. The use of biofuel to replace fossil fuels.
3. The sequestration of carbon in soils.

There is an enormous potential to reduce the carbon content in the atmosphere due to the great losses of carbon from terrestrial ecosystems in the past centuries, especially from agricultural soils. Any agro-ecosystem of sustainable productivity focuses on goals to maintain and improve soil organic matter (SOM) content (Gregorich *et al.*, 1994). An increase in the organic matter content of soil through a single organic amendment can reinitiate nutrient cycling and may also improve water retention capacity (Fierro *et al.*, 1999).

Several studies have demonstrated that soil is a potential carbon sequester, slowing down the current rise in the greenhouse effect and being environmentally acceptable due to the recovery of soil carbon levels with a minimum impact in the environment.

The loss of carbon for the last three centuries now represents an opportunity for carbon storage in the main pool of the terrestrial ecosystems, the soil. This incites the importance of enhancing the soil organic carbon to combat climate change.

If we are able to increment the soil organic matter in agricultural soils meanwhile the use of organic residues, like sewage sludge, it could be expected a reduction of the carbon presence in other environmental compartments but especially in the atmosphere. Thus an important contribution of soil carbon to combat global change could be expected.

4. MANAGEMENT PRACTICES AND SOIL CARBON STOCKS

Managing soils to increase their carbon storage capacity has been proposed as a means to reduce the rise of CO_2 concentrations in the atmosphere (Mikutta *et al.*, 2006). The implementation of effective land/agriculture management practices, especially through stewardship activities such as the conservation of soils, increasing wetland reserves, favouring forestry incentive and conservation tillage, leads to increased soil organic carbon (SOC). This increment of SOC can favour ground carbon sequestration.

The type of land use is an important factor controlling organic matter storage in soils since it affects the amount and quality of litter input, the litter decomposition rates and the processes of organic matter stabilization in soils (John *et al.*, 2005). Adequate agricultural practices are a recognized method of carbon sequestration as soil can act as an effective carbon sink. The strategies to increase the amount of SOC are basically associated to both the increment of the inputs of exogenous organic matter and the reduction of the mineralization levels:

1. The increase of carbon pool in soils by the input of organic matter coming from crop residues and animal wastes or spreading organic wastes from industrial or urban origin is one of the methods that can be used. Since the early 1970s, scientist, engineers, regulators, and interested parties in the waste management field have met to access the body of knowledge on land application of municipal wastewaters and sludges (O'Connor et al., 2005) in order to determine the best use of these wastes.
2. Agricultural management practices may help to reduce the rate of mineralization facilitating the physicochemical conditions for the protection of the soil organic matter (physically, chemically and biochemically protected).

Increased organic matter input can be achieved in a number of ways including selection of high-residue-producing crops, residue retention via lessened tillage intensity, use of crop rotations, improved management of nutrients and water, and adding adequate doses of organic wastes as sewage sludge. Moreover, some of the plant biomass is indirectly sequestered as soil organic carbon (SOC) during decomposition processes. Many agronomic, forestry, and conservation practices including best management practices lead to a beneficial net gain in carbon fixation in soil, favouring the incorporation of dead plants.

In general, there is a favourable interplay between carbon sequestration and various recommended land management practices related to soil fertility (i.e. adding mineral fertilizers, manures, sludge/biosolids), tillage, grazing, and forestry. Recommended agronomic, grazing land and forestry practices also enhance land sustainability, wildlife habitat and water quality. In most locations, especially environmentally sensitive settings, these practices also result in decreased water and wind erosion that degrade soil carbon stocks. So, environmental positive effects not only control climate change but also recovering of soil properties can be achieved.

Decreasing soil carbon loss is accomplished by minimizing soil disturbance and erosion through reduced tillage intensity, especially no-till practices. While well management is understood on a technical level, enhancing soil carbon sequestration can be achieving (McCarl et al. 2007).

Soils gaining SOC are also generally gaining in other attributes that enhance plant productivity and environmental quality (SSSA, 2001). Increases in SOC generally improve soil structure, increase soil porosity and water holding capacity, as well as improve biological health for life forms in soil (Navarro-Pedreño et al., 1996; Mataix-Solera et al., 2001; García-Orenes et al., 2005; Bronick and Lal, 2005;)

Arrouays et al. (2002) studied the land uses in French soils and determined three groups depending on their capacity to sequester carbon, which can be applied for other countries. Average stocks of C according to land use ranged from 30 to 90 tC hectare^{-1}, and can be divided into these groups:

1- Land under annual crops and perennial crops with bare soil, where stocks are lower than 45 tC/ha. Vineyards and orchards and crops with a very low organic return have the lowest stocks: approximately 32 tC/ha. Arable land is also characterised by relatively low stocks: 43 tC/ha on average.

2- Land under permanent grassland and forests (excluding litter) exhibits average stocks of nearly 70 tC/ha.

3- Land under high-altitude pastures and in wetlands, where stocks are higher than 90 tC/ha (due to low temperatures and the effect of anoxia on carbon mineralisation, respectively).

According to these groups, the most marked effect in terms of carbon accumulation was obtained by a change from the first (arable) to the second group. Grasslands and forests exhibit very similar soil carbon stock potentials. Soil type conditions the carbon stock. In general, average carbon stocks by soil type range from 40 tC/ha (sandy or skeletal soils) to 100 tC/ha (clay or hydromorphic soils). Texture strongly influenced soil C storage in soils, indeed increasing silt+clay contents were associated with higher soil C storage (Paul *et al.*, 2008). A high clay content is the principal factor correlated with high soil carbon levels. High contents of calcareous (rendzinas) or aluminium (podzols) also make it possible to attain medium stock levels in low-clay soils. In European soils, 35% of total carbon is held within high organic matter soils (≥ 8 % of SOM) that cover only 13 % of the area of the European Union (Smith *et al.*, 1997).

It is not easy to determine the capacity of a soil to storage carbon in the long-term. Few studies have been performed because, obviously, long time experiments are needed. According to the results of Arrouays *et al.* (2002), soil type and land use appear to be the principal factors controlling the C stock levels, but the dispersion of these values remains considerable. This marked residual variability is indicative of the importance of other parameters which are not taken into account, and also on the fact that carbon stocks measured at time under a given land use does not often correspond to "steady state" stocks and reflects in part the previous uses or other events of the site.

The C stock is always based on point measurements of soil carbon contents, which are then converted into a pool (change from content with reference to a soil mass to a pool with reference to a volume using bulk density); mean stock values are then extrapolated to surface areas considered as homogeneous. However, it is important to consider the temporal and geographical variations in soil carbon stocks, as well as a marked but variable vertical gradient (higher carbon levels on the surface, which decline with depth). Factors which may affect carbon stocks are numerous and their interactions complex.

In Western Europe, more carbon is sequestered by conversion of arable land into grassland, while in Eastern Europe more carbon is sequestered by the annual application of farmyard manure per hectare;. the carbon saving effect of reduced tillage and leaving behind cereal straw is considerably lower than the other two measures (Vleeshouwers and Verhagen, 2002).

The data available are usually too few in number (insufficient sampling in the context of the existing variations), not always reliable and complete (no measurement of bulk density and other parameters) and not often comparable between each other (differences in analytical techniques, soil depth taken into account, etc.). The risks of errors and inappropriate extrapolations are therefore high. However, several studies and data elaborated like the

European map of organic carbon content (%) in the surface horizon of soils in Europe, can be acceptable in order to have an overview of the problem of losses of SOM (Maps can be consulted in: S.P.I.04.72, Jones *et al.* 2004). In contrast, it is difficult to demonstrate variations in stocks. The relative amplitude of the changes is generally low and the evolution slow, often masked by spatial and seasonal variations. Best estimates are obtained from long-term experiments that are difficult to maintain along the time for practical problems and financial support.

CONCLUSION

As a summary of the management practices, the increased long term (20-50 years) sequestration of carbon in soils, plants and plant products will benefit the environment and agriculture. Crop, grazing, and forestlands can be managed for both economic productivity and carbon sequestration. In many settings, this dual management approach can be achieved by applying currently recognized best management practices such as conservation tillage, efficient nutrient management, erosion control, use of cover crops and restoration of degraded soils. One of the best strategies is the conversion of marginal arable land to forest or grassland can rapidly increase soil carbon sequestration. However, and without any doubt, agricultural soils are needed to keep the production of food and strategies to increment SOM are crucial.

More research is needed to better quantifies carbon sequestration obtained by these practices; this research should culminate in a scientifically defensible soil carbon sequestration accounting system that also would be suitable to the business sector, soil carbon become a marketable commodity. Many countries also can save and improve their accounts of CO_2 emissions based in Kyoto and Post-Kyoto agreements. Implementation of these practices will integrate a wide range of disciplines in the basic, agricultural, silvicultural, and environmental sciences as well as in the social, economic and political sciences. The Soil Science Society of America advocates for a global increase in soil organic matter as a timely benefit way of reducing the levels of atmospheric CO_2 and increasing the productivity of soil, particularly in many areas with degraded soils.

REFERENCES

Arrouays, D.; Balesdent, J.; Germon, J.C.; Jayet, P.A.; Soussana, J.F.; Stengel, P. (eds) (2002). Mitigation of the greenhouse effect. Increasing carbon stocks in French agricultural soils? An assessment report compiled by the French Institute for Agricultural Research (INRA) on the request of the French Ministry for Ecology and Sustainable Development. Ed. Institut National de la Recherche Agronomique (IRNA), Paris, 32pp.

Bronick , C.J. and Lal, R. (2005). Soil structure and management: a review. Geoderma 124: 3-22.

Fierro, A.; Angers, D.A.; Beauchamp, C.J. (1999). Paper mill sludge as a revegetation tool in an abandoned sandpit : project outline and preliminary results. In "Remediation and management of degraded lands", Wong M.H., Wong J.W.C. and Baker A.J.M. (eds.), p. 89-95. Ed. Lewis Publishers, Boca Ratón p.364.

García-Orenes, F.; Guerrero, C.; Mataix-Solera, J.; Navarro-Pedreño, J.; Gómez , I.; Mataix, J. (2005). Factors controlling the aggregate stability and bulk density in two different degraded soils amended with biosolids. Soil and Tillage Research 82 (1): 65-76.

Gregorich, E.G.; Carter, M.R.; Angers, D.A.; Monreal, C.M.; Ellert, B.H. 1994. Towards a minimum data set to assess soil organic matter quality in agricultural soils. Canadian Journal of Soil Science 74: 367-385.

Guggenberger, G.; Kaiser, K. (2003). Dissolved organic matter in soils. Challenging the paradigm of sorptive preservation. Geoderma 113: 293-310.

Houghton, R.A.; Goodale, C.L. (2004). Effects of land-use change on the carbon balance of terrestrial ecosystems. In "Ecosystems and land use change" Geophysical Monograph Series 153, 85-98. American Geophysical Union 10.1029/153GM08.

Intergovernmental Pannel on Climate Change (IPCC) 1996. Climate Change 1995. Impacts, adaptations and mitigation of climate change. Scientific-Technical Analyses, Cambridge University Press, Cambridge (UK), 878 pp.

Intergovernmental Pannel on Climate Change (IPCC) 2001. Climate Change: The scientific basis. Cambridge University Press, Cambridge (UK).

John, B.; Yamashita, T.; Ludwig, B.; Flessa, H. (2005). Storage of organic carbon in aggregate and density fractions of silty soils under different types of land use. Geoderma 128: 63-79.

Jones, R.J.A., Hiederer, R., Rusco, E., Loveland, P.J. and Montanarella, L. (2004). The map of organic carbon in topsoils in Europe, Version 1.2, September 2003: Explanation of Special Publication Ispra 2004 No.72 (S.P.I.04.72). European Soil Bureau Research Report No.17, EUR 21209 EN, 26pp. and 1 map in ISO B1 format. Office for Official Publications of the European Communities, Luxembourg.

Lal, R (2009). Challenges and opportunities in soil organic matter research. European Journal of Soil Science 60: 158-169.

Lal, R. (2004). Soil carbon sequestration to mitigate climate change. Geoderma 123: 1-22.

Mataix-Solera, J.; Navarro-Pedreño, J.; Guerrero, C.; García, E.; Jordán, M.M.; Gómez I. (2001). Application of different organic wastes to three soils of degraded areas: effect of some physical, chemicals and biological soil properties. Advances in Ecological Sciences 10: 321-330.

McCarl, B.A.; Metting F.B.; Rice, C. 2007. Soil carbon sequestration. Climatic Change 80: 1-3.

Mikutta, R.; Kleber, M.; Torn, M.S.; Jahn, R. (2006). Stabilization of soil organic matter: association with minerals or chemical recalcitrance? Biogeochemistry 77:25-56.

Navarro-Pedreño, J.; Gómez, I.; Moral, R.; Mataix, J. (1996). Improving the agricultural value of a semi-arid soil by addition of sewage sludge and almond-residue. Agriculture, Ecosystems and Environment 58(2-3): 115-119.

O'Connor, G.A.O.; Elliott, H.A.; Basta, N.T.; Bastian, R.K.; Pierzynski, G.M.; Sims, R.C.; Smith, J.E. (2005). Sustainable land application: an overview. Journal Environmental Quality 34: 7-17.

Paul, S.; Flessa, H.; Veldkamp, E.; López-Ulloa, M. (2008). Stabilization of recent soil carbon in the humed tropics following land use changes: evidence from aggregate fractionation and stable isotope analyses. Biogeochemistry 87: 247-263.

Porteous, A. (2008). Dictionary of Environmental Science and Technology (4th edition). Ed. John Wilye & Sons, Chichester, 797 p.

Regato, P. (2008). Adapting to Global Change. Mediterranean Forest. Ed. International Union for Conservation of Nature and Natural Resources, Gland, p. 256.

Read, P. 2009. Reducing CO_2 levels – so many ways, so few being taken. Climate Change 97: 449-458.

Rothamsted Research (2006). Guide to the Classical and other Long-term Experiments, Datasets and Sample Archive. Ed. Lawes Agricultural Trust Co. Ltd.

Six, J.; Connat, R.T.; Paul, E.A.; Paustian, K. (2002). Stabilization mechanisms of soil organic matter: implications for C-saturation of soils. Plant and Soil 241: 155-176.

Soil Science Society of America (SSSA) (2001). Carbon sequestration: Position of the Soil Science Society of America. SSSA Ad Hoc Committee S893.

Stevenson, F.J. (1994). Humus Chemistry: genesis, composition, reactions. Ed. John Wilye & Sons, New York, 496 p.

Vleeshouwers, L.M. and Verhagen, A. (2002). Carbon emission and sequestration by agricultural land use: a model study for Europe. Global Change Biology 8: 519-530.

INDEX

B

E

N

O

Q

U